Life and Death of Coral Reefs

A service of ITP®

Life and Death of Coral Reefs

Charles Birkeland
Editor

University of Guam

I(T)P® International Thomson Publishing

New York • Albany • Bonn • Boston • Cincinnati • Detroit • London • Madrid • Melbourne
Mexico City • Pacific Grove • Paris • San Francisco • Singapore • Tokyo • Toronto • Washington

Cover design: Sabina Kahn, emDASH inc.
Art Direction: Andrea Meyer
Photo credit: "Before" photo, Great Barrier Reef Marine Park Authority (G.B.R.M.P.A.)
"After" photo, Rob van Woesik, Department of Marine Science, *University of the Ryukyus*, Okinawa

Printed in the United States of America

Chapman & Hall
115 Fifth Avenue
New York, NY 10003

Chapman & Hall
2-6 Boundary Row
London SE1 8HN
England

Thomas Nelson Australia
102 Dodds Street
South Melbourne, 3205
Victoria, Australia

Chapman & Hall GmbH
Postfach 100 263
D-69442 Weinheim
Germany

International Thomson Editores
Campos Eliseos 385, Piso 7
Col. Polanco
11560 Mexico D.F
Mexico

International Thomson Publishing–Japan
Hirakawacho-cho Kyowa Building, 3F
1-2-1 Hirakawacho-cho
Chiyoda-ku, 102 Tokyo
Japan

International Thomson Publishing Asia
221 Henderson Road #05-10
Henderson Building
Singapore 0315

1 2 3 4 5 6 7 8 9 10 XXX 01 00 99 98 97

Library of Congress Cataloging-in-Publication Data

Birkeland, Charles.
Life and death of coral reefs / Charles Birkeland.
p. cm.
Includes bibliographical references and index.
ISBN 0-412-03541-3 (alk. paper)
1. Coral reef ecology. 2. Coral reef biology. I. Title.
QH541.5.C7B57 1996
574.5'26367—dc20 95-19293
CIP

British Library Cataloguing in Publication Data available

To order this or any other Chapman & Hall book, please contact **International Thomson Publishing, 7625 Empire Drive, Florence, KY 41042.** Phone: (606) 525-6600 or 1-800-842-3636. Fax: (606) 525-7778. e-mail: order@chaphall.com.

For a complete listing of Chapman & Hall titles, send your request to **Chapman & Hall, Dept. BC, 115 Fifth Avenue, New York, NY 10003.**

For Charles J. and Wilma F. Birkeland

Contents

Acknowledgments xv

Contributors xvii

Chapter 1. Introduction (by Charles Birkeland) 1

1.1 The Value of Reefs 2
1.2 Present Conditions of Coral Reefs 6
1.3 Shift in Controlling Factors 10

Chapter 2. Reefs and Reef Limestones in Earth History (by Pamela Hallock) 13

2.1 Biogenic Sediments and Bioherms 14
2.2 Basic Carbonate Chemistry 16
2.3 Limestones and Earth History 17
2.4 The Atmosphere and the Evolution of Life 19
2.5 CO_2 and Time Scales 22
2.6 Atmospheric CO_2 as an Evolutionary Driving Force 24
2.7 What It Takes to Accumulate $CaCO_3$ 26
2.8 History of Biogenic Reefs Through Time 29
2.8.1 Paleozoic Era—Time of Ancient Animal Life 31
2.8.2 Mesozoic Era—Time of Intermediate Animal Life 33
2.8.3 Cretaceous Boundary Extinctions 34
2.8.4 Cenozoic Era—Time of Recent Animal Life 36
2.9 Modern Reefs 38
2.10 Conclusion 42

Chapter 3. Reefs as Dynamic Systems (by Dennis K. Hubbard) 43

3.1 Process: The Ultimate Control of Reef Development 44
3.1.1 Macroscale Controls 44
Tectonics 44

Sea Level 46
Reef Accretion Under the Influence of Relative Sea-Level Change 48
3.1.2 Mesoscale Controls 50
Temperature 50
Salinity 52
Wave Energy 52
3.1.3 Microscale Controls 56
Light 56
Nutrients 56
Sediment 57
Antecedent Topography 58
3.2 Reef Accretion: How It Works 60
3.3 Reefs and Global Warming 64
3.3.1 Reefs in Flux 64
3.3.2 Reefs as Mediators of Global Climate 65
3.3.3 The Reef Response 66
3.3.4 Evolving Methodologies 66
3.4 Where Do We Go from Here? 67

Chapter 4. Bioerosion and Coral-Reef Growth: A Dynamic Balance (by Peter W. Glynn) 68

4.1 Bioeroder Diversity 70
4.1.1 Bacteria 71
4.1.2 Fungi 73
4.1.3 Algae 73
4.1.4 Sponges 75
4.1.5 Polychaete Worms 77
4.1.6 Crustacea 77
4.1.7 Sipuncula 79
4.1.8 Mollusca 79
4.1.9 Echinoidea 84
4.1.10 Fishes 84
4.2 Conditions Favoring Bioerosion 85
4.3 Variety of Effects 89
4.4 Case Studies 91
4.4.1 El Niño–Southern Oscillation 91
4.4.2 Crown-of-Thorns Seastar *(Acanthaster)* 91
4.4.3 Runoff (Eutrophication, Sedimentation, Freshwater, and Pollutants) 92
4.4.4 Overfishing 92
4.5 Conclusion 94

Chapter 5. Interactions Between Corals and Their Symbiotic Algae (by Gisèle Muller-Parker and Christopher F. D'Elia) 96

5.1 Description of the Symbiosis 96

5.1.1 Coral Anatomy and Location of Zooxanthellae 96
5.1.2 Zooxanthellae 98
5.1.3 Acquisition of zooxanthellae by corals 99
5.2 Nutrition and Adaptations to Environmental Factors 100
5.2.1 Coral Nutrition 100
5.2.2 Productivity of Corals and Role of Zooxanthellae in Calcification 102
5.2.3 Effect of Light and Temperature on Productivity of Zooxanthellae 102
5.2.4 Effect of Nutrient Supply on Zooxanthellae in Corals 104
5.3 Stability of the Symbiosis 105
5.4 Cost-Benefit Analysis of the Symbiosis 106
5.5 Environmental Effects on the Symbiosis 110
5.5.1 Local and Regional Stresses to Symbiotic Corals 110
5.5.2 Global Stresses on Symbiotic Corals 111
5.6 Conclusion 112

Chapter 6. Diseases of Coral-Reef Organisms (by Esther C. Peters) 114

6.1 Diseases of Reef Plants 116
6.2 Diseases of Reef Invertebrates 118
6.2.1 Corals 118
6.2.2 Sponges 122
6.2.3 Molluscs 123
6.2.4 Crustaceans 126
6.2.5 Echinoderms 127
6.3 Diseases of Reef Vertebrates 128
6.3.1 Fishes 129
6.3.2 Sea Snakes 131
6.3.3 Sea Turtles 131
6.4 Unanswered Questions 132
6.5 Ecological Implications 136

Chapter 7. Organic Production and Decomposition (by Bruce G. Hatcher) 140

7.1 Definitions and Methodologies of Production 144
7.1.1 Currencies of Production 145
7.1.2 Pools of Organic Materials in Coral Reefs and Control Volumes 146
7.1.3 The Concepts of Turnover in Coral-Reef Ecology 147
7.1.4 Scaling Up 151
7.2 Producers and Consumers 153
7.2.1 Microbial Communities Rule 153
7.2.2 Algae: The Engines of the Reef 155
7.2.3 Macroconsumers: Pretty Fishes and Others 157
7.3 Rates of Coral-Reef Metabolism and Their Variation in Space and Time 158

7.3.1 Habitat-Scale Variation in Community Metabolism and Export 158
7.3.2 Reef-Scale Variation in Community Metabolism and Export 162
7.4 Controls on Coral-Reef Production 165
7.4.1 Limitation of Gross Production 166
7.4.2 Nutrient Limitation of Excess Production 167
7.4.3 Organic Inputs Support Excess Production 169
7.4.4 Secondary Production and Decomposition 171
7.5 Interactions Between Humans and Coral-Reef Production 173
7.5.1 Anthropogenic Effects on the Trophic Status of Reefs 173
7.5.2 Harvestable Yields from Reefs 174

Chapter 8. Reproduction and Recruitment in Corals: Critical Links in the Persistence of Reefs (by Robert H. Richmond) 175

8.1 Coral Reproduction 176
8.1.1 Asexual Reproduction 176
8.1.2 Sexual Reproduction 177
8.1.3 Gonochorism Versus Hermaphroditism 177
8.1.4 Brooding Versus Spawning 178
8.1.5 Self-Fertilization Versus Outcrossing 187
8.1.6 Hybridization 189
8.1.7 Larval Development 189
8.2 Larval Recruitment 190
8.2.1 Settlement 190
8.2.2 Metamorphosis 191
8.3 Reproductive and Recruitment Failure of Corals 192
8.3.1 Terrestrial Runoff and Water Clarity 193
8.3.2 Water Pollution 195
8.3.3 Population Depletion 195
8.3.4 Prevention and Mitigation 196

Chapter 9. Invertebrate Predators and Grazers (by Robert C. Carpenter) 198

9.1 Invertebrates Having Minor Effects 200
9.1.1 Corallivores 200
9.1.2 Herbivores 204
9.2 Invertebrates Having Major Effects 207
9.2.1 Corallivores 207
Eucidaris thouarsii 207
Acanthaster planci 210
9.2.2 Herbivores 220
Echinometra mathaei 220
Diadema antillarum 221
9.3 Disproportionate Effects of Some Invertebrate Species 226

Chapter 10. Effects of Reef Fishes on Corals and Algae (by Mark A. Hixon) 230

10.1 The Players: Corallivorous and Herbivorous Reef Fishes 231
10.2 Fish Effects on Algae 233
10.2.1 Schooling Herbivores 233
10.2.2 Territorial Damselfishes 236
10.2.3 Conclusion 239
10.3 Fish Effects on Corals 241
10.3.1 Direct Consumption 241
10.3.2 Indirect effects 242
10.3.3 Conclusion 244
10.4 Fish Effects on Invertebrate Corallivores and Herbivores 245
10.5 Ramifications for Reef Management 246

Chapter 11. Indirect Interactions on Coral Reefs (by Steven C. Pennings) 249

11.1 Definition, Variety, and Examples of Indirect Effects 250
11.1.1 Trophic-Linkage Indirect Effects 251
Competition Between Fish and Urchins for Algae and Its Mediation by Predators 254
Herbivore Mediation of Algal-Coral Competition 256
Cascading Effects of Predation 257
Cleaning Symbioses 258
11.1.2 Behavioral Indirect Effects 260
Associational Defenses 260
Mixed-Species Foraging Schools 261
Coral Zonation Patterns Indirectly Created by Damselfish 262
Mediation of *Acanthaster* Predation on Corals by Coral Symbionts and Damselfishes 262
Presence of Predators Affects Grazing Behavior of Herbivores 263
11.1.3 Chemical-Response Indirect Effects 264
Sequestering of Dietary Secondary Metabolites 264
Use of Chemically Defended Plants as Domiciles 265
Induced Defenses Affecting Multiple Grazers 265
11.2 Relative Importance on Coral Reefs 266
11.3 Some Practical Considerations 268
11.4 Future Directions 271
11.5 Conclusion 272

Chapter 12. Geographic Differences in Ecological Processes on Coral Reefs (by Charles Birkeland) 273

12.1 Physical Processes That Bring About Geographic Differences 276
12.1.1 Longitudinal Differences 276
12.1.2 Latitudinal Differences 279

12.2 Geographic Differences in Characteristics of Dominant Species 280
12.3 Regional Differences in Ecological Processes 282
12.3.1 Grazing and Predation Pressure 282
12.3.2 Food Webs 284
12.3.3 Diversity and Ecosystem Function 285
12.4 Management Considerations 286

Chapter 13. Ecosystem Interactions in the Tropical Coastal Seascape (by John C. Ogden) 288

13.1 Biological Interactions in the Seascape 288
13.1.1 Edge Effects 288
13.1.2 Nurseries 289
13.2 Fluxes of Nutrients and Organic Material in the Seascape 290
13.3 Physical Interactions in the Seascape 292
13.3.1 Lagoons 292
13.3.2 Buffers and Sinks 293
13.4 Reef Management, Global Change, and Comparative Research 295

Chapter 14. Diversity and Distribution of Reef Organisms (by Gustav Paulay) 298

14.1 Definitions and Scales 299
14.1.1 What Is Diversity? 299
14.1.2 Taxonomic Diversity: Spatial Scales 299
14.1.3 Taxonomic Diversity: Phylogenetic Scales 300
14.2 Biota of Reefs 301
14.2.1 Species Richness on Reefs 301
14.2.2 Ecological Restriction to Reefs 304
14.2.3 Why Are Coral-Reef Environments so Diverse? 306
14.3 Role of Diversity in Reef Ecosystems 308
14.3.1 Diversity in Photosymbiotic Relationships 308
14.3.2 Diversity Among Reef Constructors and Eroders 309
14.3.3 Diversity and Ecosystem Function 312
14.4 Practical Biodiversity: Identification of Reef Organisms 314
14.5 Distribution of Diversity: Large-Scale Patterns 316
14.5.1 Variation in Diversity with Latitude 316
14.5.2 Variation in Diversity with Longitude 320
14.6 Distribution of Diversity: History of Regionalization 326
14.6.1 Tectonic Provincialization 327
14.6.2 Dynamics of Species Biogeography During the Late Cenozoic 327
Western Atlantic 328
Eastern Pacific 328
Eastern Atlantic 331
Indo-West Pacific 332

14.7 Distribution of Diversity: Evolutionary Diversification 332
14.7.1 Interregional Speciation 333
14.7.2 Speciation within the Indo-West Pacific 334
14.7.3 Speciation within Other Regions 340
14.8 Effects of Human Activities 341
14.8.1 Extinction 341
14.8.2 Introduced Species 344
14.9 Conclusion 345

Chapter 15. Disturbances to Reefs in Recent Times (by Barbara E. Brown) 354

15.1 Conceptual Framework 355
15.2 Tolerances of Reef Corals 357
15.2.1 Geographic Variations in Environmental Tolerances 358
15.2.2 Within-Site Variation in Environmental Tolerances 359
15.2.3 Between Species Variations in Environmental Tolerances 359
15.2.4 Within-Species Variations in Environmental Tolerances 360
15.2.5 Within-Colony Variations in Environmental Tolerances 361
15.3 Natural Disturbances on Coral Reefs 362
15.3.1 Effects of Storms, Cyclones, and Hurricanes 363
15.3.2 Coral Bleaching 365
15.3.3 Outbreaks of the Crown-of-Thorns Starfish *Acanthaster planci* 366
15.3.4 Mass Mortality of the Sea Urchin *Diadema antillarum* in the Caribbean and Consequences on Coral Communities 368
15.4 Human Disturbances to Coral Reefs 370
15.4.1 Eutrophication 370
15.4.2 Sedimentation 372
15.4.3 Oil Pollution 373
15.4.4 Coral Mining 374
15.5 Human Versus Natural Influences on Coral Reefs 375
15.6 Prediction of Responses of Reefs to Climate Change 377
15.6.1 Sea-Level Rise 377
15.6.2 Effects of Increased Seawater Temperature Increase 378
15.6.3 Carbonate Mineral Saturation State 378
15.6.4 Ultraviolet Radiation 378

Chapter 16. Traditional Coral-Reef Fisheries Management (by Robert E. Johannes) 380

16.1 Value of Traditional Knowledge 381
16.2 Putting Management Back at the Village Level 383
16.3 Applying Traditional Knowledge in Modern Settings 383

Chapter 17. Resource Use: Conflicts and Management Solutions (by Gregor Hodgson) 386

17.1 Sustainable Use 390
17.2 Resource-Use Conflicts 390
17.3 Case Study: Logging Versus Fisheries and Tourism in the Philippines 392
17.3.1 Logging and Siltation 392
17.3.2 An Integrated Assessment 394
17.3.3 Coral-Reef Damage 395
17.3.4 Effects of Sediment on Fish 397
17.3.5 Assumptions in Estimation of Effects 398
17.3.6 Economics of Management Alternatives 399
17.3.7 Intergenerational Equity and Sustainable Development 401
17.3.8 El Nido Today 403
17.4 Marine Parks 403
17.5 Implementation of Sustainable Development 406
17.6 Conclusion 409

Chapter 18. Implications for Resource Management (by Charles Birkeland) 411

18.1 Export Fisheries Not Sustained 412
18.2 High Productivity, Low Yield 413
18.2.1 Recycling 415
18.2.2 Constraints of Complexity 416
18.2.3 Life-History Traits 417
18.3 Reserves for Breeding Stock 418
18.4 Economics of Harvesting Diverse Resources 422
18.4.1 High Values 423
18.4.2 Subsidies 424
18.4.3 "Ratchet Effect" from Population Growth 425
18.5 Village-Based Control 426
18.6 Exemplary Cases: Bermuda and Palau 428
18.7 Nonextractive Commercial Uses of Coral-Reef Resources 429
18.8 Integrative Coastal-Area Management 432
18.9 Conclusion 433

References 437

Index 527

Acknowledgments

The authors thank the many colleagues and friends who generously helped with this book in many ways. Kelly K. Gathers, C. Messing, Daniel Norris, and John Starmer skillfully produced some of the figures. Other illustrations were contributed by Greta S. Aeby, Lucy Bunkley-Williams, T. J. Done, J. W. Focke, C. Louise Goggin, John C. Halas, M. Harmelin-Vivien, R. C. Highsmith, Harold H. Hudson, T. P. Hughes, P. L. Jokiel, H. A. Lowenstam, I. G. Macintyre, M. J. Risk, Klaus Rützler, B. Salvat, T. P. Scoffin, C. R. C. Sheppard, James W. Warren, C. R. Wilkinson, and J. D. Woodley. We thank Mei Tsu Birkeland, H. Carpenter, S. Longmore, Martha M. Martin, and C. Zilberberg for their assistance with the references and other editorial matters.

S. S. Amesbury, Marlin J. Atkinson, Robert Buddemeier, M. H. Carr, M. W. Colgan, L. Collins, J. Connell, C. B. Cook, Paul Copper, D. G. Fautin, R. N. Ginsburg, John S. Glazebrook, R. W. Grigg, J. A. Harding, Rosalind Hunter-Anderson, P. Hutchings, David Hopley, T. P. Hughes, Noah Idechong, A. T. Kaltenberg, W. E. Kiene, Mark M. Littler, P. S. Lobel, I. G. Macintyre, Harley Manner, Harry B. McCarty, D. W. Meadows, S. G. Nelson, C. Osenberg, R. T. Paine, Valerie J. Paul, Rob Rowan, Garry R. Russ, P. W. Sammarco, Michael C. Schmale, T. P. Scoffin, H. Galt Siegrist, Jr., G. D. Stanley, Jr., Ernest H. Williams, Jr., and the students in the University of Guam class on Advances in Ecology provided constructive reviews of drafts of certain chapters. T. Schoener allowed us to use his unpublished manuscript.

Cindy Ahearn, Frederick Bayer, Steve Cairns, Lu Eldredge, Bert Hoeksema, Arthur Humes, Alan Kabat, Roy Kropp, Rich Mooi, Rob Myers, Jack Randall, and Klaus Rützler helped with references and provided information on a variety of taxa.

Nancy Budd, Bern Holthuis, Alan Kohn, Brian Rosen, Geerat Vermeij, and Charlie Veron contributed valuable comments and discussion. The authors of the chapters in this book also provided constructive critiques, advice, and illustrations for chapters other than their own. We thank Rob van Woesik and G.B.R.M.P.A. for the photographs on the cover of this book.

Contributors

Dr. Charles Birkeland
Marine Laboratory
University of Guam
Mangilao, Guam 96923

Dr. Barbara E. Brown
Director, Centre for Tropical Coastal Management Studies
Department of Marine Sciences and Coastal Management
Ridley Building
The University
Newcastle-upon-Tyne NE1 7RU
England, U.K.

Dr. Robert C. Carpenter
Department of Biology
California State University
Northridge, CA 91330

Dr. Christopher F. D'Elia
Sea Grant College Program
0112 Skinner Hall UMCP
University of Maryland
College Park, Maryland 20742

Dr. Peter W. Glynn
Division of Marine Biology and Fisheries
Rosenstiel School of Marine and Atmospheric Science
University of Miami
4600 Rickenbacker Causeway
Miami, Florida 33149-1098

Dr. Pamela Hallock
Department of Marine Sciences
140 Seventh Avenue South
University of South Florida at St. Petersburg
St. Petersburg, Florida 33701-5016

Dr. Bruce G. Hatcher
Department of Oceanography
Dalhousie University
Halifax, Nova Scotia
B3H 4J1 Canada

Dr. Mark A. Hixon
Department of Zoology
Cordley Hall 3029
Oregon State University
Corvallis, Oregon 97331-2914

Dr. Gregor Hodgson
Coastal Systems Research Ltd
G.P.O. Box 3534
Hong Kong

Dr. Dennis K. Hubbard
V. I. Marine Advisors, Inc.
5046 Cotton Valley - 66
St. Croix, USVI 00820-4519

Dr. Robert E. Johannes
8 Tyndall Court
Bonnet Hill
Tasmania 7053
Australia

Dr. Gisèle Muller-Parker
Shannon Point Marine Center
Western Washington University
1900 Shannon Point Road
Anacortes, Washington 98221-4042

Dr. John C. Ogden
Director, Florida Institute of Oceanography
830 First Street South
St. Petersburg, Florida 33701

Dr. Gustav Paulay
Marine Laboratory
University of Guam
Mangilao, Guam 96923

Dr. Steven C. Pennings
Marine Institute
University of Georgia
Sapelo Island, GA 31327

Dr. Esther C. Peters
Tetra Tech, Inc.
10306 Eaton Place, Suite 340
Fairfax, VA 22030

Dr. Robert H. Richmond
Marine Laboratory
University of Guam
Mangilao, Guam 96923

Life and Death of Coral Reefs

1

Introduction

Charles Birkeland

Living coral is a thin veneer, measured in millimeters. Yet this thin film of living tissue has shaped the face of the Earth by creating limestone structures sometimes over 1,300 m thick from the surface down to its base on volcanic rock (Enewetak Atoll), or over 2,000 km long (Great Barrier Reef). About half the world's coastlines are in the tropics and about a third of the tropical coastlines are made of coral reef. Archipelagoes of hundreds of atolls such as the Marshalls, the Maldives, the Tuamotus, and most of the Carolines and Kiribati have been formed by coral. In addition to enlarging high islands (such as the entire northern end of Guam) and extending and protecting coastlines, ancient biogenic reefs have formed even larger areas on the present continents. Shallow living coral reefs are estimated to presently cover over 600,000 km^2 (Smith, 1978).

Coral reefs are dynamic systems, producing limestone at the rate of 400–2,000 tons per hectare per year (Chave et al., 1972). The Great Barrier Reef dominates 230,000 km^2 and has grown to this size in a geologically brief period of a few million years. Coral reefs influence the chemical balance of the world's oceans. Roughly half the calcium that enters the sea each year around the world, from the north to south poles, is taken up and temporarily bound into coral reefs (Smith, 1978). With each atom of calcium, a molecule of CO_2 is also deposited, with gross CO_2 fixation estimated on the order of 700 billion kg carbon per year.

Coral reefs and many of the present genera of corals have been around since long before the prairies or other ecosystems of grasses existed. The genera *Leptoria* and *Montastrea* were around before the class Angiospermae (flowering plants) dominated the terrestrial realm. Yet coral reefs and other systems based on plant-animal symbioses are especially sensitive and vanish about a million years before other groups of organisms each time there is a global mass extinction (Copper, 1994). Reefs of plant-animal symbioses have occupied over 10 times their present area at some times during the Paleozoic, yet at times they have vanished completely or have occupied only a hundredth of their present area.

Although coral reefs and other plant-animal reef systems are fragile, they are resilient. Coral reefs vanished from the record during a period of about 10 million years following the late Cretaceous, but many of the same families and even some of the same genera of scleractinian corals eventually returned.

Coral reefs are among the most biologically productive ecosystems in the world (see Fig. 18–1). The global potential for coral-reef fisheries has been estimated at 9 million tons per year, which is impressive in view of the relatively small area of coral reefs compared to the world ocean, and to the total marine fisheries of the world being about 75–100 million tons per year (Smith, 1978; Munro, 1984). Coral reefs, on the other hand, are vulnerable to overexploitation if harvested repeatedly (section 18.1).

So here is the paradox. Although coral reefs are the most productive communities in the sea, the fisheries of coral reefs are among the most vulnerable to overexploitation. Despite having the power to create the most massive structures in the world made by living creatures (including humans), the thin film of living tissue of coral reef seems particularly vulnerable to natural disturbances and the effects of human activities. Coral reefs and other animal-plant reefs have been the first to go during periods of climate change, but they have always come back (although not always with the same components). This combination of attributes—creative power and fragility, resilience and sensitivity, productivity and vulnerability to overexploitation—makes management of coral-reef systems a special challenge to science.

1.1. The Value of Reefs

Coral reefs are found in over 100 countries, mostly in the less economically developed tropical regions. Despite the vulnerability of reefs to overharvesting for export, reef fisheries have served for hundreds and, in some locations, thousands of years as major sources of food. The economy of atolls is dependent on the coral reefs, whether the economy is based on fisheries, aquaculture, or tourism. People depending on coral reefs for part of their livelihood and for obtaining part of the protein in their diet are estimated to number in the tens of millions (Salvat, 1992a).

The standing stocks of coral-reef fishes are about 30–40 times greater than standing stocks on demersal fishing grounds in Southeast Asia, the Mediterranean, or other temperate regions (Russ, 1984a). The standing stocks of reef fishes have been estimated to be as high as 160 metric tons km^{-2} in the Atlantic (Randall, 1963) and as high as 93–239 metric tons km^{-2} in the Pacific (Goldman and Talbot, 1976; Williams and Hatcher, 1983). About 60,000 small-scale fishery operations harvest coral reefs in the Caribbean and 14,000 people fish the reefs of the Gulf of Lingayen in the Philippines (Wells and Hanna, 1992). In the Caribbean region, about 180 species of reef-associated fishes are marketed, with catch figures for coralline areas (and excluding oceanic pelagic fishes) often ranging from 15 to 19 kg per ha per year (Munro, 1983). The potential harvest of Jamaican fishermen by traps alone should exceed 95,000 tons per year (1.4 kg km^{-2} yr^{-1}). Of about 900 species of fishes around Guam, fully a third of the species are taken for consumption (Myers, 1993). Of 286 species harvested at one local fishery at Bolinao, Philippines, 209 are reef associated (McManus et al., 1992). About 250 reef-fish species are taken in the Tigak Islands in Papua New Guinea (Wells and Hanna, 1992). The mean harvest on reefs of a couple of small islands in the central Philippines over a 5-year period was found to be 11.4 and 16.5 metric tons km^{-2} yr^{-1} (Alcala and Luchavez, 1982). The shoreline fishery of American Samoa has yielded up to 26.6 metric tons km^{-2} yr^{-1} (Wass, 1982).

Coral reefs also benefit fisheries by supplying food for more wide-ranging pelagic or inshore pelagic fishes. Sudekum et al. (1991) calculated that just two of the species of jacks, *Caranx ignobilis* and *C. melampygus,* together eat 30,600 metric tons per year of benthic fauna off French Frigate Shoals, an atoll in the northwestern Hawaiian Islands.

Many reef invertebrates are harvested mainly for subsistence and so there are few data available for estimates of the magnitude of the resource. About 5,500 tons of mother-of-pearl for the curio trade were collected from the coral-reef gastropod *Trochus niloticus* in 1978 alone, and this figure does not include the weight of meat for this common source of food in the tropical Pacific (Craik et al., 1990). Other reef invertebrates that provide substantial amounts of food for humans include other gastropods (e.g., the queen conch *Strombus gigas* in the western Atlantic and the green snail *Turbo marmoratus* in the western Pacific), bivalves (e.g., several species of giant clams, rock oysters, pearl oysters, *Spondylus*), octopus, squid, cuttlefish, lobsters, prawns, and sea cucumbers.

A variety of seaweeds are also collected from coral reefs and used for food, folk medicine, and fertilizer, and sold for agar and carrageenan. In 1989 alone, the Philippines produced 65,600 metric tons of algae of the genus *Eucheuma,* worth tens of millions of dollars (South, 1993).

Corals deposit tremendous quantities of limestone. Large amounts of the coral limestone also contribute to coral rubble and sand. Blocks of living or dead coral are used for building materials, breakwaters, and cement. In Sri Lanka in the 1980s, over 2,000 metric tons of live coral skeletons, 7,000 metric tons of coral rubble, and 34,000 m^3 of sand were removed from the coast each year (Wells and Hanna, 1992). However, the economic value in many of the resources is far less when extracted than when left in place (Chapter 17).

Coral reefs can also produce a substantial revenue from jewelry and curios. The pearl culture of the coral-reef oyster *Pinctada margaritifera* is bringing millions of dollars annually to the Tuamotu Archipelago. Curios and souvenirs made from black corals, gorgonaceans (especially sea fans), seashells, giant clams, dried fishes, and echinoderms support a multimillion-dollar international business. During the mid-1980s, 4,500 metric tons of shellcraft per year (not including mother-of-pearl) were being exported from the Philippines (Wells and Hanna, 1992). In 1988, 1,456 metric tons of ornamental corals were imported to the United States (Wells and Hanna, 1992).

The marine aquarium trade is a rapidly growing business, with 871 live coral imported to the United States in 1984, 40,000 in 1988, and about 250,000 in 1991 (Derr, 1992; Wells and Hanna, 1992). About 1,500 people make a living collecting live aquarium fishes in the Philippines. About 50,000 persons are employed in the aquarium trade in Sri Lanka (Craik et al., 1990). The export of aquarium fishes from the Philippines brought in nearly $2.5 million in 1978 and $2.75 million in 1979 (Salm, 1984). The coral-reef aquarium trade operates in the tens of millions of dollars annually, and is growing very rapidly.

Scuba diving on coral reefs forms the main base of the economies of a number of tropical developing countries. Tourism of coral reefs brings in about $13 million a year to the national economy of Palau, a Pacific country with a population of about 14,000. Scuba-related tourism brings in about $21 million annually to Bonaire, about half its gross domestic product. In developed countries like the United States and Australia, the economic value of coral reefs for tourism is also large. Reef tourism is estimated to be worth about $1.6 billion a year for Florida's economy, with over 2 million tourists visiting John Pennecamp Coral Reef State Park and Key Largo National Marine Sanctuary alone (Wells and Hanna, 1992).

The potential for pharmaceuticals from natural products from coral reefs would seem to be greater than from other systems because biodiversity and ecosystem complexity of coral reefs is on a higher scale than in other ecosystems (Chapter 14), both from the perspective of evolutionary potential and natural products chemistry. Rain forests are considered to have a greater biodiversity at the species level because of insects and flowering plants. However, coral-reef communities have the greater diversity in terms of prevalent phyla and kingdoms, a greater diversity of basic body plans, and chemistry. Of 33 animal phyla, 16 are exclusively marine and only one, the relatively obscure Onycophora, is exclusively terrestrial (although it, like all phyla, originated in the ocean). Of the phyla found outside the marine environment, the majority are still most prevalent in the oceans (e.g., sponges, coelenterates, bryozoans).

In terms of phyla and classes, coral reefs have by far the greatest diversity per hectare of any ecosystem in the oceans, and it has recently been suggested that the actual species diversity on coral reefs may be even three to five times greater than previously recognized (Knowlton and Jackson, 1994).

Intracellular symbiotic relations among kingdoms facilitate evolutionary advances above the species level (Margulis and Sagan, 1986; Margulis and Fester, 1991), which suggests that coral-reef communities might have special potential for evolutionary innovation. The potential of coral reefs as a source of new chemicals for pharmaceuticals has compelled the National Institutes of Health (including the National Cancer Institute and the National Institute of General Medical Sciences) to fund the establishment of a laboratory in Micronesia. The marine laboratory on Chuuk was contracted to provide specimens of at least 5,000 different species of coral-reef organisms as material for chemical exploration.

Some natural-product chemicals have undergone clinical analysis. A number of prostaglandins were discovered in large quantities in the common gorgonacean *Plexaura homomalla* (Bayer, 1974). Prostaglandin is a potent pharmaceutical that affects a wide range of clinical applications in humans including assisting the process of childbirth, terminating pregnancies, and treatment of cardiovascular disease, asthma, and gastric ulcers. A chemical from the red alga *Portieria hornemannii* has shown antitumor activity for a variety of human tumors (Fuller et al., 1992). Didemnin B from the coral-reef ascidian *Trididemnum solidum* has demonstrated activity against leukemia, a variety of human tumors, viruses, and carcinomas (including melanoma) in clinical trials. The purity of $CaCO_3$ produced by corals makes it valuable for use in bone-marrow transplants.

Reefs serve as protection against wave action. During typhoons, the damage from wave action to coastal communities is much less where there are reefs. On Guam, the damage from wave action in areas protected by extensive reef flats was minor, but in areas around the villages of Inarajan and Merizo, where the fringing reefs are narrow, wave action damaged homes, removed buried caskets from cemeteries, moved automobiles, carried a refrigerator away from inside a home, and caused damage as far as a kilometer inland. In addition, coral reefs protect mangroves and seagrass beds in some localities, and thus they provide protection for nurseries of commercially important fishes. Coral reefs are self-repairing, and the cost of building and maintaining equivalent breakwaters is nearly always omitted in the consideration of the commercial value of coral reefs.

As the value of reefs to the economic and social well-being of human communities in coastal regions becomes apparent, the cost assessments of damages to coral reefs become large. The government of Egypt claimed $30 million for damage to 340 m^2 of coral reef in the Strait of Tiran, although the final settlement was out of court for $600,000, or $1,765 m^{-2} (Spurgeon, 1992). This was considerably less than the cost calculated for reef damage in Florida. Assuming the minimum nonmarket value for live coral, assuming the reef would recover naturally rather quickly, and assuming a financial rate of return for lost revenue from tourism of only 3% the calculations presented in a legal journal for the minimum damage to the local economy caused by the grounding of the M/V WELLWOOD on Molasses Reef off the Florida Keys was $2,833 m^{-2} (Mattson and DeFoor, 1985). By these calculations, Molasses Reef, just one of many reefs off the Florida Keys, is worth about $400 million.

Perhaps the most important role of coral reefs in the lives of local people is usually not recognized by outsiders. This is the stabilizing effect of reefs on social structure. Fishing is often a cooperative activity in which each of the family members has a clearly recognized role. It has been discerned from interviews of fishermen in Palau that fishing activities help solidify the roles and importance of members of the family. It was stated that reefs may be more important in providing the opportunity for fishing activities than in providing the catch. Fishing and reef-gleaning are often perceived as fun and wholesome. In cases where large developments such as resorts or military bases obstruct access of local people to traditional fishing or reef-gleaning areas, the effects cannot be overcome simply with jobs providing wages by which foods can be purchased. As social structure deteriorates, the numbers of suicides and criminal acts increase. The economic costs of such societal maladies are rarely taken into account in the evaluation of coral reefs, but nevertheless, these costs of the deterioration of coral reefs are ultimately paid by all of us.

1.2. Present Conditions of Coral Reefs

Human activities have been affecting coral reefs and tropical coastal ecosystems far more than has been generally recognized. Sediments and accompanying nutrients are usually considered the greatest threat to coral reefs (Johannes, 1975; Hatcher et al., 1989; Rogers, 1983, 1990; Hallock et al., 1993). Approximately 75–80% of the sediment entering the world's oceans (from the Arctic to the Antarctic) comes off land in the tropical western Pacific, with half the global sediment discharge coming off continental high islands (e.g., Papua New Guinea, the Philippines, and Indonesia) and the other 25–30% from Southeast Asia (Milliman, 1992). These are the areas of greatest diversity of coral-reef communities. The mean soil erosion rates in Asia are 30–40 metric tons per hectare per year because of agricultural practices (Pimentel et al., 1995). In contrast, soil erosion rates in undisturbed forests range from 0.004 to 0.05 metric tons per hectare per year (Pimentel et al., 1995). Coral reefs do better along coasts occupied by vegetation (Chapter 13; see Fig. 13–3).

The soil erosion and transport of sediment to the coastal marine ecosystems has increased dramatically in recent times. Depending on the type of human activity, rates of erosion can increase by as much as 100-fold (Doolette and Magrath, 1990). Milliman et al. (1987) calculated that the sediment load of the Yellow River has increased by an order of magnitude because of farming practices in northern China. The fluvial discharge rates from the coast of Australia have been calculated to be four times greater than they were before humans began altering the drainage basins through deforestation and farming. Approximately four times as much sediment, nitrogen, and phosphorus enter the marine environment off the Queensland coast than before western agriculture began (Brodie, 1995).

On the basis of using percent cover of living coral as an index of reef health, about 60–70% of the reefs in Indonesia and the Philippines have been concluded to be seriously degraded, with only about 5% still in excellent condition (Yap and Gomez, 1985; Sukarno et al., 1986). However, we should be cautious about relying on percent cover of living coral as an indicator of the "health" of the coral-reef community. A coral community with a relatively low proportion of living corals could still be a "healthy" community, depending on the particular circumstances (e.g., vertical wall). Change through time of living coral cover and age distribution of corals in the community are more reliable indicators of the state of the reef.

Increased sedimentation and nutrient input into the coastal marine ecosystems may have been causing broadscale changes in the biotic communities of coastal regions. Paralytic shellfish poisoning resulting from blooms of toxic dinoflagellates was unrecorded in the tropical western Pacific before 1972, but the number of deaths and hospital cases from dinoflagellate toxicity has been increasing rapidly in the continental high islands of the western Pacific and along the coasts of Southeast Asia (Maclean, 1984) where so much of the world's sediment input into the world's oceans is occurring.

There are also indications that outbreaks of animals with planktotrophic larvae have become more frequent in recent years (see Fig. 15–1; Birkeland and Lucas, 1990). This is possibly because increased nutrient input fertilizes the waters and the resulting increased density of phytoplankton, bacteria, and organic matter provides food for the larvae of marine animals. However, overfishing of the predators of the key invertebrate grazers may also free the invertebrate populations to grow beyond their food resources.

Large-scale population outbreaks, mass mortalities, and community disturbances have been occurring on a regional scale in coral-reef ecosystems during recent decades (Chapter 15). Changes in abundances of key species, such as mass mortality of the urchin *Diadema antillarum* in the Caribbean (section 15.3.4) and outbreaks of the starfish *Acanthaster planci* in the Pacific (section 15.3.3), have shifted reef communities from coral to algal dominance over large areas and over long periods of time (Chapter 9). There has been no indication of a return to the previous state of dominance by corals in some sites in the Caribbean for over a decade (Hughes, 1994). The sea urchin *Echinometra mathaei* has been eroding reefs in Okinawa, Kuwait, and Kenya (Downing and El-Zahr, 1987; McClanahan and Muthiga, 1988, 1989). The reefs in the eastern Pacific, along the coast of Panama and in the Galápagos, are being eroded away through grazing by another sea urchin, *Eucidaris thouarsii* (Chapter 4). The bioerosion is occurring to the extent that some of the eastern Pacific coral reefs are being eroded away faster than they are being deposited, and so these reef frameworks are presently being reduced to unconsolidated sediment on a large scale (section 4.5.1; Glynn, 1988a; Eakin, 1993).

Although coral reefs have encountered natural disturbances and recovered in the past (Chapter 2), the evidence at hand indicates that the changes in community structure and shifts in the balance of coral-reef processes have increased in scale and frequency in recent decades, recovery is delayed more often, and situations that used to be acute are now often chronic (Chapter 15, see Fig. 15–1; Glynn, 1993; Hughes, 1994). Knowlton et al. (1990) documented continued mortality of corals 10 years after Hurricane Allen; the mortality from secondary factors was an order of magnitude greater than the direct effects of the hurricane. Coral bleaching (loss of zooxanthellae and/or pigment) has been increasingly widespread and frequent (Chapter 15, see Fig. 15–1; Glynn 1990a, 1993; Williams and Bunkley-Williams, 1990). Bythell and Sheppard (1993) surveyed more than 2,000 sites in the British Virgin Islands and found over 95% of the *Acropora* were dead. The primary reef-building corals in the Caribbean, *Acropora palmata, Acropora cervicornis,* and *Montastrea annularis,* had become scarce over large areas. Perhaps a third of the 400 species of corals in Japanese waters are in danger of local extinction unless effective coastal management practices are established (Veron, 1992a).

The tropical western regions of the oceans are the centers of diversity of coral-reef organisms (Chapter 14). About 40% of the coral reefs of the world are in the continental-shelf regions of the western Pacific and western Atlantic (Smith, 1978). As a result of geographic processes interacting with climatic patterns (Chapter 12), the coasts of western tropical oceans are also the regions where most of the terrestrial runoff occurs. While most of the terrestrial input of sediment is discharged into the western Pacific, about half of the remaining 20–25% comes off northern South America from the Amazon, Orinoco, and Magdalena rivers. Many reefs near human population concentrations in the western Atlantic have become degraded by sedimentation, for example, the Caribbean (Rogers, 1985), the Florida Keys (Marszalek, 1981; Jaap, 1984), Bermuda (Dodge and Vaisnys, 1977), and the Virgin Islands (Hubbard et al., 1987).

Deterioration of reefs in the western Atlantic is widespread and has been found as far as tens to hundreds of kilometers from concentrations of humans (Hallock et al., 1993). The damages to reefs far from human activities show signs of having been caused by increases in nutrification caused by long-range effects of anthropogenic nutrient input into marine ecosystems (Hallock et al., 1993). Dodge and Lang (1983) found a negative correlation between coral growth rates on Flower Gardens Bank and the discharge of the Atchafalaya River 280 km to the northeast.

The nutrient input from rivers to coastal regions has been increasing to an even greater extent than has sedimentation. Following deforestation in the watershed in the Amazon Basin, the volume of runoff has increased by 30%, but the concentration of nitrogen doubled and the concentration of phosphate increased nearly eight-fold (Williams, 1991, in Hallock et al., 1993). The Mississippi River now carries 10 times the concentrations of nitrates and phosphates that it did in the late 1960s (Redalje et al., 1991, in Hallock et al., 1993). The influence of nutrification extends farther downstream than sedimentation. This has been suggested as a primary cause of widespread coral-reef degradation far from concentrations of humans (Hallock et al., 1993).

Nutrient input contributes to coral-reef degradation by favoring the growth of algae and suspension-feeding animals over coral recruits in competition for space

(Chapters 8, 9, and 12), by increasing the rates of bioerosion by boring bivalves and sponges (Chapter 4), and through bacterial infection of corals (Chapter 6).

The coral reefs in the eastern tropical Pacific have been the most severely degraded by climatic events. There was an extensive mortality of hermatypic corals (>95% in the Galápagos) following the El Niño of 1982–1983 (Glynn, 1990a). Many of the reefs in the eastern Pacific have continued to deteriorate since the El Niño because recruitment by corals has been sparse and sea urchins continue to erode away the reef framework (section 4.5.1 in Chapter 4; Glynn, 1990a; Eakin, 1993).

While the effects of increased sedimentation and nutrient input extend over broad areas in the tropical western Atlantic and the continental tropical western Pacific, the effects of poor land management on the coral reefs of Oceania are concentrated mainly near urban areas. **Oceania** is comprised of island nations in the Pacific away from continents (Polynesia [excluding New Zealand], Melanesia [excluding Papua New Guinea], and Micronesia). The human population growth rates in some regions of Oceania are among the greatest in the world. But of more immediate concern for coral-reef management is that the rates of urbanization are increasing even faster (sections 18.4.3 and 18.5). Because the rates of urbanization are exceeding both population and economic growth, a large number of families concentrated in urban areas turn to fishing and gleaning of reefs as openings for salaried jobs become relatively scarce. Urbanization of the regional population and immigration of foreign workers both undermine the authority and effectiveness of traditional management practices (section 18.5).

Overexploitation may be having as great an effect on coral-reef communities as sedimentation and excess nutrient input because coral-reef ecosystems are driven more by predation than by upwelling or other changes in the oceanographic climate (Sherman, 1994). Although the diversity, standing stock, and yield of coral-reef fishery resources are spectacular, the impressions of a cornucopia have been misleading (Chapters 16 and 18). Most coral-reef fisheries with high yields have not been sustained when exploited commercially (section 18.1). The fishes and invertebrates have major influences on the functioning of coral-reef ecosystems (Chapters 9, 10, 11, and 15). **Ecosystem overfishing** occurs when the removal of resource species causes substantial changes in the species composition of the community and major changes in ecosystem processes (Hughes, 1994).

Human activities have been affecting coral reefs for hundreds of years, much longer than scientists have been actively observing the conditions of coral reefs. Scientists may be too late to observe the nature of "pristine" reefs, except possibly on some isolated atolls or islets in Oceania. It may be that scientists have been trained in substantially altered coral-reef ecosystems, so the altered state is the baseline against which reefs are judged to be "normal." For example, when the sea urchin *Diadema antillarum* suffered extensive mortality in the tropical western Atlantic in 1983, the community of herbivorous fishes was unable to control the algae. Although over 13 years have passed since the mass mortality of *D. antillarum,* the reef communities are still dominated by algae rather than corals in large areas scattered throughout the tropical western Atlantic (Hughes, 1994; Chapter 9). This suggests an "unnatural" situation in which the herbivorous fishes and perhaps the predators of urchins were overexploited. It may be that the dominance by *Diadema antillarum* in the western Atlantic coral-reef communities was not the "normal" condition that scientists had been assuming, but the urchin may have became established decades or centuries ago as a result of overfishing (Hay, 1984b).

Likewise, the large-scale increase in rates of sedimentation in the western tropical Pacific occurred with the advent of agriculture and deforestation, before scientists began monitoring the conditions of coral reefs. The additional 6–11 billion tons of sediment being discharged annually into the coastal marine communities of the tropical western Pacific (Milliman, 1992) could have changed the nature of coral-reef communities in ways about which we can only speculate for example, perhaps there has been a greater preponderance of sponges, bivalves, and ascidians now than there was before agriculture.

The importance of coral reefs to the livelihoods of millions of people and the extent of deterioration of the coral reefs of the world are not generally recognized by the public. The scale of change in conditions of coral reefs resulting from human activities over the past centuries is too large in both time and space for the matter to be easily perceived, especially when the major changes have occurred prior to scientific investigations of coral reef.

1.3. Shift in Controlling Factors

Unlike the terrestrial world for which a concept of a "balance of nature" was developed, coral-reef systems are rarely in "balance." They are often growing, adding structure to a reef over hundreds of years, and ultimately over millions of years, until the massive geological structures upon which nations are built have been formed. In contrast, there are times in which conditions change and the boring and eroding organisms gain the upper hand; massive geological structures can be reduced to rubble, sand, and silt. The purpose of this book is to present what is known about the factors that "shift the balance" back and forth between accretion and erosion, recruitment and mortality, prevalence of stony corals versus filamentous algae, recovery and degradation—the life and death of coral reefs.

This shift in the balance can occur at many levels. In Chapter 2, we look at coral reefs and other hermatypic (reef-building) communities from the perspective of geological and evolutionary time, reviewing episodes of extinctions and radiations, and considering algal symbiosis as an evolutionary driving force. Chapter 3 is set in the perspective of geological structure, and is concerned with factors affecting the relative importance of accretion and erosion at different locations in the reef structure and at different scales of reference. Chapter 4 is involved with the natural controls of the potentially delicate balance of reef growth and bioerosion, and how human activities influence them.

Both the massive geological structure and the high level of biological productivity of coral reefs are created by symbiotic interactions among animals and plants. Symbioses between representatives of kingdoms and phyla dominate the functioning of the system, and the factors regulating the interactions among partners in the symbioses are fundamental in the operation of the coral-reef ecosystem. In Chapter 5, we examine factors that shift the balance from benefit to cost in symbiotic relationships, and how human activities that influence corals at the cellular level can affect the community at the ecosystem level. Diseases of marine organisms, and how stress and pollution facilitate disease, are covered in Chapter 6.

The types and fates of production on coral reefs and how human activities influence the balance of production and consumption are presented in Chapter 7. Coral reefs can become degraded without obvious coral mortality when corals gradually die natural deaths but are not replaced by recruits. Chapter 8 explains reproduction and recruitment, critical links in the persistence of reefs.

In recent years we have seen that the mass mortality or population outbreak of a single species of invertebrate predator or grazer can cause drastic changes in coral-reef communities over large geographic regions. Chapter 9 covers the roles of invertebrate predators and grazers and how certain species can be extraordinarily influential in coral-reef ecosystems.

Coral reefs have the greatest species diversity of vertebrates per square meter of any community on Earth because of the spectacular array of hundreds of species of fishes. These fishes are also the main target of humans who harvest them. Chapter 10 tells how the changes in the fish communities from overexploitation by humans can shift the balance of the key processes in the coral-reef ecosystem. In the complex reef ecosystem, you can never change just one thing. Chapter 11 organizes our concepts of the indirect interactions in the reef community.

Chapter 12 presents evidence that reef processes vary so drastically over space that we must use entirely different management strategies in different situations. A particular factor to take into account when attempting to manage coral reefs is the interaction among neighboring habitats (Chapter 13). This is because some species recruit to neighboring habitats, others forage in neighboring habitats, and under some circumstances, the entire coral-reef community is protected by neighboring habitats.

The natural and man-made disturbances that can tip the balance among coral-reef ecological processes are reviewed in Chapter 15. Chapters 16, 17, and 18 present conflict management in resource use of coral reefs by integrating traditional ownership and multiuse strategies.

The diversity of symbiotic interactions among kingdoms and the basis of symbiotic interactions in the processes leading to the development of the geological structures and biological communities of coral reefs make the terms used in studies of other communities problematic when dealing with coral reefs. The "individuals" in population ecology of coral reefs can often be a combination of organisms of different kingdoms. Conversely, the "individual" genetic code may be embodied in numerous scattered colonies, many of which live long after the original colony has died. The concepts of "species," "trophic levels," and "communities" become all the more diffuse when trying to understand the factors that tip the balance in processes of coral-reef systems.

2

Reefs and Reef Limestones in Earth History

Pamela Hallock

To seafarers, a **reef** is a submerged hazard to navigation, usually a ridge of rocks or sand at or near the surface of the water. This is why the Exxon *Valdez* could hit a "reef" in Prince William Sound, Alaska, thousands of miles from the nearest coral reef. Historically, tropical waters were particularly treacherous for mariners because reefs constructed by coral communities may lurk just below the surface in otherwise open seas. In both calm seas and storms, often the first indication that a reef was nearby was when a ship ran aground. The earliest European "settlers" in the Florida Keys made their living salvaging shipwrecks. Today the most visible indication of many Pacific atolls is a rusting freighter.

The characteristic that distinguishes coral, oyster, and other **biogenic reefs** from sand and rock reefs is that biogenic reefs are produced by biological as well as geological processes. Another term for a limestone structure or buildup produced by biological activity is **bioherm**. Ideally, a biogenic reef is a significant, rigid skeletal framework that influences deposition of sediments in its vicinity and that is topographically higher than surrounding sediments. For example, a **coral reef** is a rigid skeletal structure in which stony corals are major framework constituents. Less rigid accumulations of biologically produced sediments are sometimes called reef mounds (James, 1983).

The history of biogenic limestones is a topic of considerable economic importance because many ancient reef provinces are major oil and gas reservoirs today. As a result, the literature on ancient reefs and lesser carbonate buildups is vast. A few of the multitude of useful compilations include Milliman (1974), Wilson (1975), Bathurst (1976), Toomey (1981), Scholle et al. (1983), Crevello et al. (1989), and Riding (1991). In particular, Fagerstrom (1987) details *The Evolution of Reef Communities* by examining modern and ancient reefs in the context of the changing organisms through time and the roles those organisms played in reef communities.

This summary cannot provide the details available in such larger compilations.

Instead it summarizes the significance of biogenic reefs and limestones within the context of Earth history. Limestones are predominantly calcium carbonate ($CaCO_3$), and are therefore intimately related to carbon dioxide (CO_2) concentrations in the atmosphere and oceans. Because many limestones are made up of the shells and skeletons of calcareous plants and animals, a substantial part of the marine fossil record is found in limestones. The relationships between atmospheric CO_2 and reef building are not clear-cut; the mechanisms involved are complex and full of poorly understood feedback mechanisms. Nevertheless, the history of biogenic reefs provides a fascinating glimpse at the major events in Earth history, including why life can exist at all.

2.1. Biogenic Sediments and Bioherms

Calcareous shells and skeletons of a wide variety of protists, plants, and animals become biogenic sediments upon the death of those organisms. The metabolic activities of certain bacteria and microalgae also contribute to the biogeochemical precipitation of calcareous (lime) muds in seawater overlying shallow banks and shelves. Biogenic sediments are most prevalent in marine environments that are separated by distance or physical barrier from the influx of sediments from land. Nearly 50% of the modern ocean floor is covered by foraminiferal ooze, the empty shells of protists that live as plankton in the surface waters of the open ocean. Shells and skeletons of benthic organisms are also important sediment constituents, especially on continental shelves, in some coastal areas, and on oceanic banks and shoals.

Whether these biogenic constituents make up most of the bottom sediments or whether they are only minor contributors depends on several factors. One factor is the rate at which sediments from land are entering the marine environment via runoff from rivers and streams. Another factor is the rate at which shells and skeletons are being produced by the biotic communities living in the marine environment. A third factor is the rate at which sediments, both terrigenous and biogenic, are removed from that environment by transport or dissolution. The biotic community not only produces sediments but also affects rates of dissolution of sediments, as well as rates of physical breakdown and transport of sediments. Thus, composition of the benthic community strongly influences rates of sediment accumulation.

Lees (1975) recognized three classes of shallow-water carbonate sediments, based on their major constituents. He called the simplest group foramol sediments after two of the most important constituents: benthic foraminifera and molluscs, especially fragments of snail and bivalve shells. Lees noted that foramol sediments are characteristic of temperate shelves, but sometimes dominate in tropical areas where reefs do not occur. Other important constituents of foramol sediments are fragments of coralline red algae, sea urchin spines and plates, bryozoa, barnacles, and worm tubes. Lees's second sediment type is called chloralgal, for its dominant constituent, the remains of calcareous green algae such as *Halimeda.* Chloralgal sediments have foramol constituents as secondary components. Chloralgal sediments are prevalent in expansive shallows like Florida Bay and the Bahama Banks, and in deep-euphotic settings including the lagoon behind the Ribbon Reefs of the northern Australian Great Barrier Reef. Lees's final sediment category is chlorozoan, which is the typical sediment around coral reefs. Coral and calcareous algal remains are the dominant constituents. Coralline algae, foraminifera, mollusc, and urchin fragments are secondary components. Bryozoa, barnacle, and worm shell debris are typically scarce in chlorozoan sediments because these organisms thrive best in waters with richer food supplies than do corals.

Foramol, chloralgal, and chlorozoan sediments are produced on shelves, oceanic banks and atolls, and nearshore environments. Whether these sediments accumulate in place or whether they are transported away depends on the strengths of waves and currents and the ability of the benthic community to hold sediments in place. Sediments are accumulated and bound by the presence and growth of organisms. Those that project upward from the sediment, slowing water motion and providing quieter places for sediments to settle, can be termed bafflers; those that live in or directly on the sediment, holding or encrusting it in place, can be referred to as binders (Fagerstrom, 1987).

Binders such as microalgae and bacteria grow and develop mats directly on sediments accumulating where wave and current motion is limited or intermittent. Bacterial filaments provide strength to these mats, which can resist as much as 10 times more wave or current energy than is required to move similar unbound sediments (Grant and Gust, 1987). Stromatolites, which are biogenic reefs constructed by this process, are layered accumulations of sediment and algal-bacterial mats. Ancient stromatolites were the first bioherms in the fossil record. Modern stromatolites are found in Shark's Bay, Australia (Logan et al., 1974), and at several localities on the Bahama Banks (Dill et al., 1986; Reid and Browne, 1991).

In some current-swept environments, specialized sponges live in and on the surface layers of sediment, binding it in place. Sediments may consist of coarse accumulations of *Halimeda* segments. Coralline red algae may colonize the surface of sediment-filled sponges, forming solid substrata upon which other organisms settle and grow. These communities produce sponge-algal mounds along the margins of some western Caribbean banks (Hallock et al., 1988; Hine et al., 1988), which show similarities to fossil sponge-algal reef mounds.

A variety of elongate, upward-projecting plants and animals baffle water motion and trap sediments. On modern shallow shelves, seagrass beds effectively stabilize sediment over vast areas. Seagrass blades slow water flow, allowing suspended sediments to settle out. Sediments are then held in place by extensive seagrass root and rhizome systems, as well as by the holdfasts of algae living within the seagrass bed. Sediment-dwelling macroalgae are also effective bafflers and binders, as are sponges, sea whips, and sea fans. In fossil reefs, a variety of less familiar organisms performed similar roles.

The ultimate bafflers are the biogenic framework constructors (Fagerstrom, 1987), which in modern tropical shallow-water environments are the stony corals. These organisms grow upward or outward in branching, massive, or platy morphologies, secreting substantial quantities of calcium carbonate, while trapping even greater quantities of sediment within the lee of the reef framework. Encrusting coralline algae bind the reef framework and enclose sediments into the massive, wave-resistant structures we recognize as coral reefs.

The three-dimensional topography of the reef provides abundant habitats for the diverse array of species that dwell within the reef structure. All contribute to the reef community in some way—many to the reef structure itself, all to energy flow within the community. Some of these species are encrusters, some are sediment producers, and some are wholly soft bodied and have little direct influence on the reef structure. Many species even contribute to the breakdown of the reef structure by boring into it or scraping away at it as they graze. Such organisms are known collectively as bioeroders (Neumann, 1966) or destroyers (Fagerstrom, 1987).

Bioeroding organisms are a diverse and important component of the reef community (Hutchings, 1986). Organisms that bore or etch their way into the reef include bacteria, fungi, several varieties of sponges, worms, clams, and urchins. Organisms that scrape away limestone as they graze algae include urchins, chitons, and some snails. Many reef fish feed by breaking or scraping off bits of coral or coralline algae. In a healthy, actively accreting reef, bioeroders contribute to the diversity of habitats within the massive reef structure. However, if reef growth slows in response to natural or anthropogenic environmental stresses, the rates of destruction can exceed rates of accretion and the reef may cease to exist (Glynn, 1988; Hallock, 1988).

2.2. Basic Carbonate Chemistry

The major chemical constituent of calcareous sediments and limestones is $CaCO_3$. Organisms secrete $CaCO_3$ either as calcite or aragonite. The obvious difference between these minerals is their crystal structure. Calcite forms rhombohedral crystals whereas aragonite forms orthorhombic crystals. Aragonite is structurally stronger than calcite. But the most important difference is in the chemical stability of the minerals at temperatures and pressures found on land and in the oceans. Aragonite more readily precipitates in warm seawaters that are supersaturated with $CaCO_3$, but it is less stable in cooler seawaters and in freshwater. Through time, most aragonite either dissolves or recrystallizes, so calcite predominates in ancient limestones.

The solubility of $CaCO_3$ is easily misunderstood if this chemical is expected to behave like other familiar solids. For example, table salt and sugar both dissolve faster in hot water than in cold. $CaCO_3$ is more soluble in cold water. The key to this intuitive discrepancy is in the reaction of carbon dioxide with water:

$$CO_2 + H_2O \Leftrightarrow H_2CO_3 \Leftrightarrow H^+ + HCO_3^- \Leftrightarrow 2H^+ + CO_3^{2-} \qquad (2\text{–}1)$$

That is, carbon dioxide and water combine to form carbonic acid, which can then dissociate to hydrogen ions and bicarbonate (HCO_3^-) or carbonate (CO_3^{2-}) ions. The two-headed arrows indicate that the reaction can go in either direction depending on environmental conditions. The dissolved inorganic carbon concentration of a sample of seawater is the sum of the carbon in these four states. In surface seawaters, CO_2 is only a small fraction of the total, often less than 1% (Riebesell et al., 1993).

How much CO_2 can be dissolved in water and which state predominates depends primarily on temperature, pressure, and concentrations of other dissolved materials. Cold water can hold far more CO_2 in solution than warm water; an example is a cold carbonated beverage, which loses CO_2 as it warms. Similarly, water under pressure can hold more CO_2 in solution. Addition of dissolved salts (i.e., increased salinity) decreases the ability of water to dissolve CO_2. Surface waters of the ocean can hold less CO_2 in solution than deeper waters, and tropical waters hold less than temperate or polar waters.

Carbon dioxide and water react with calcium carbonate in the following way:

$$CO_2 + H_2O + CaCO_3 \Leftrightarrow Ca^{2+} + 2HCO_3^- \qquad (2\text{–}2)$$

The more CO_2 that is dissolved in the water, the more readily the water can dissolve $CaCO_3$. Conversely, any process that removes CO_2 from solution promotes the precipitation of $CaCO_3$. Since calcium ion (Ca^{2+}) and bicarbonate ion (HCO_3^-) are both abundant in seawater, modern tropical ocean-surface waters are most conducive to the precipitation of $CaCO_3$, whereas deeper and colder ocean waters are more apt to dissolve $CaCO_3$.

2.3. Limestones and Earth History

Limestones are one of the major reasons that life can exist on Earth. Since limestones are $CaCO_3$, they provide a mechanism for storage of great quantities of CO_2 in the Earth's crustal rocks. Without limestones, the concentration of CO_2 in the Earth's atmosphere would be about 100 times higher, similar to that of Venus, and the surface of the Earth would be nearly as hot as that of Venus (425°C) (Jastrow and Thompson, 1972; Condie 1989).

Carbon dioxide is referred to as a "greenhouse gas" because it absorbs heat energy. Most of the energy reaching a planet's surface from the Sun is visible light. The planet's surface absorbs that radiation, is warmed by it, and reradiates heat (infrared radiation) back into space. Carbon dioxide, when present in the atmosphere, acts as an insulating blanket, trapping part of the infrared radiation the planet would otherwise lose into space.

Venus and Earth are quite similar planets. A major factor in their histories, which culminated in life on Earth but not on Venus, is distance from the Sun. This determines the intensity of solar radiation falling on the planet's surface. Based on distance from the Sun and without considering atmospheric effects, the average surface temperature of primordial Venus is estimated as 60°C, while the Earth as −30°C. But these planets probably had early atmospheres. Volcanic eruptions released water vapor, hydrogen sulfide, methane, ammonia, carbon dioxide, and other gases from these planets' interiors. Approximately 10% of gaseous volcanic emissions was CO_2. So both planets had insulating atmospheres and likely were somewhat warmer than estimates based only on distance.

Carbon dioxide is removed from the atmosphere during weathering of rocks (Berner et al., 1983). Rainwater, in which carbon dioxide is dissolved, falls on rock (represented here by basalt—$CaSiO_3$), slowly weathering it away to dissolved silica, calcium ions, and bicarbonate ions:

$$3H_2O + 2CO_2 + CaSiO_3 \rightarrow H_4SiO_4 + Ca^{2+} + 2HCO_3^- \qquad (2\text{–}3)$$

In lakes or oceans, dissolved silica precipitates to form opal

$$H_4SiO_4 \Leftrightarrow SiO_2 + 2H_2O \qquad (2\text{–}4)$$

and calcium ions and bicarbonate ions react to form calcium carbonate (equation 2–2 read right to left). Through geologic time, the opal crystallizes to quartz and the calcium carbonate accumulations become limestones and marbles.

The solubility of CO_2 is strongly temperature dependent and so is the weathering reaction (equation 2–3), which further explains why Earth supports life while Venus does not. Because Venus is closer to the Sun, the average surface temperature of Venus has been in excess of 60°C since the formation of the planets in the Solar System. At such high temperatures, there was little dissolution of carbon dioxide in liquid water, so weathering of crustal rocks and formation of limestones was minimal. Carbon dioxide rapidly accumulated in the atmosphere of Venus, resulting in runaway greenhouse effect that precluded the development of life. Because the Earth is further from the Sun, less solar radiation reaches the Earth, so average surface temperatures are lower. Even with the pre-Archean atmosphere, daytime temperatures were probably in the 0–25°C range, which is ideal for CO_2 dissolution in liquid water, for weathering of rocks, and for the formation of limestones. As a result, throughout Earth history, CO_2 accumulated in the crustal rocks of Earth, rather than in the atmosphere. At present, only about 0.03% of the atmosphere is CO_2 and the average temperature of the Earth's surface of 14°C (Axelrod, 1992) readily supports life.

2.4. The Atmosphere and the Evolution of Life

During the first 2 billion years of Earth's history, most precipitation of calcium carbonate likely occurred when concentrations of calcium and carbonate ions supersaturated the water. The evolution of life on Earth profoundly altered this relationship (Fig. 2–1), particularly the activities of photosynthetic microorganisms from 2.5 billion years ago to the present (Lovelock, 1988). During the process of photosynthesis, carbon dioxide is directly removed from the air or water to produce organic matter (abbreviated as CH_2O) and oxygen:

$$H_2O + CO_2 \Leftrightarrow CH_2O + O_2 \tag{2–5}$$

If the volume of seawater is limited and the rate of photosynthesis is high, rapid uptake of CO_2 promotes calcium carbonate precipitation. For example, on a warm, shallow, subtidal flat where cyanobacterial (blue-green "algal") mats cover the bottom, $CaCO_3$ crystals may form in the water, or within or on the mats (Pentecost, 1991). The result can be the formation of stromatolites. By this mechanism, tremendous volumes of limestones were deposited in shallow-shelf seas from about 2.5 billion to 600 million years ago (Grotzinger, 1989). And during this time, photosynthesis forever changed the Earth's atmosphere to the oxygen-bearing mixture necessary for the evolution and survival of multicellular life forms.

During the 4.7-billion-year lifetime of the Sun, its luminosity has increased roughly 40% as a part of the natural aging process of a star (Gilliland, 1989). If the process of carbon dioxide removal was purely geochemical, the rate of removal would have declined as solar radiation intensified, and the Earth should have become much warmer. But since the evolution of photosynthesis, life forms influenced the concentrations of carbon dioxide and oxygen in the atmosphere. As solar radiation has intensified, photosynthesis rates have likely increased, since photosynthesis is driven by solar radiation. This may account for the overall decline in carbon dioxide concentrations through geologic history of the Earth, to the very low levels of today.

However, the decline in CO_2 concentrations in the atmosphere has been neither uniform nor continuous. Global volcanic activity, which has not been constant through geologic time, adds CO_2 to the atmosphere. Over Earth history (4.5 billion years; see Fig. 2–1), volcanic rates have generally declined as the Earth's interior has progressively cooled. But over the time scales of large-scale plate tectonic processes (10s–100s of million years), volcanic rates have varied, slowing as large continental masses collide, and speeding up as they rift apart. For example, during the Cretaceous period (the Age of the Dinosaurs), 150 to 65 million years ago, the rates of oceanic rifting and subduction were substantially faster than modern rates. Atmospheric CO_2 concentrations as much as 5–10 times higher than present produced "greenhouse world" conditions in which high

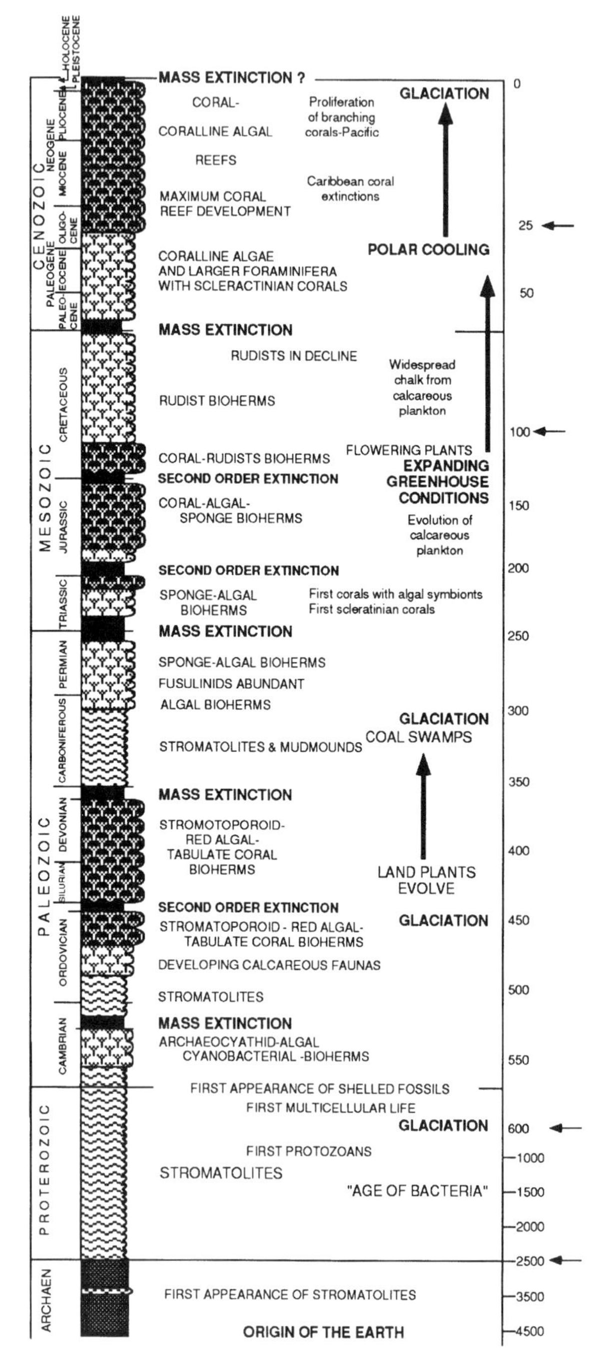

Figure 2–1. The geological time scale illustrating major reef-related events. The arrows along the right side of the figure note scale changes, with the scale greatly expanded over the past 100 million years, especially the past 25 million years.

latitudes enjoyed temperate climates and polar regions were ice-free (Worsley et al., 1986). Thus, during intervals such as the Cretaceous, the general long-term decline in atmospheric CO_2 concentrations temporarily reversed.

A second factor that may have influenced rates of CO_2 decline has been the successive evolution of groups of photosynthetic protists and plants with new and more efficient pigments and enzyme systems. Proliferation of these new organisms may have contributed to declines in atmospheric CO_2 concentrations that triggered global cooling and culminated in "icehouse world" conditions characterized by major episodes of high-latitude glaciation. Prior to the late Proterozoic, which began roughly 1 billion years ago (Fig. 2–1), bacteria were the major photosynthesizers. The evolution of the nucleated cell from the symbiosis of three or four different kinds of bacteria provided the evolutionary breakthrough that later made multicellular life possible (Margulis, 1993). The evolution of green and red algae (Schopf and Oehler, 1976), and of primitive phytoplankton known as acritarchs, may have increased the efficiency of photosynthesis and therefore rates of CO_2 extraction. A major glacial event occurred in the late Proterozoic, ending approximately 600 million years ago. Global cooling associated with this early glacial event would have slowed the rates of extraction of atmospheric CO_2 by the biotic community.

The appearance of multicellular life following that glacial event (Cloud and Glaessner, 1982) has been attributed to rising atmospheric oxygen concentrations. Shelled organisms appeared 570 million years ago (Conway Morris, 1993). Rising atmospheric oxygen concentrations probably also supported primitive land photosynthesizers, including lichens and cyanobacteria (Fischer, 1965). The organic acids and primitive soils produced by these early land plants likely increased rates of rock weathering. Major diversifications and proliferations of marine life were occurring by the Ordovician period, which culminated in a major glacial episode 458 million to 428 million years ago (Frakes and Francis, 1988).

The evolution and diversification of land plants more than 350 million years ago, which led to the accumulation of vast quantities of Carboniferous coals worldwide, removed tremendous quantities of CO_2 from the atmosphere and may have contributed to late Carboniferous glaciation. Another factor was the increasing intensity of rock weathering as soils became better developed.

In the Mesozoic, many new taxa evolved with the potential to alter the distribution of CO_2 and HCO_3^- in the atmosphere, oceans, and sediments. The middle Triassic to early Jurassic saw the evolution and diversifications of several new groups of $CaCO_3$ producers that profoundly changed oceanic sedimentation. Coccolithophorids, which represented a new lineage of phytoplankton, and planktonic foraminifera began producing calcite plates and shells in the surface waters of the open ocean, so that today calcareous sediments cover half the area of the ocean floor. On the shallow shelves, scleractinian corals and a variety of larger foraminifera developed symbioses with microalgae, greatly increasing their potential for carbonate production. CO_2 emission by the rapid volcanism of the Cretaceous more than compensated for carbonate production by these new groups of organisms. However, along with the new calcareous organisms, the proliferation of flowering plants on land and the extremely fast-growing diatoms in aquatic environments during the late Cretaceous and early Paleogene may have played a role in the global cooling that has occurred over the past 50 million years, culminating in the Ice Ages of the past few million years (Volk, 1989).

Atmospheric CO_2 concentrations have varied on much shorter time scales. Concentrations were about 200 ppm during glacial advances and about 280 ppm during interglacials (Delmas, 1992). These differences, their causes, and their significance is at the heart of modern studies of global climate change. Vegetation has such a strong influence on atmospheric CO_2 concentrations that differences between winter and summer can be detected in the Northern Hemisphere (Heinmann et al., 1989).

2.5. CO_2 and Time Scales

The weathering of rocks and the accumulation of limestones have different effects on atmospheric CO_2 concentrations, depending on the time scale being considered (Kinsey and Hopley, 1991). On the scale of Earth history, limestones are clearly important reservoirs of CO_2. This process can be summarized by the simplified expression:

$$CO_2 + CaSiO_3 \Rightarrow SiO_2 + CaCO_3 \quad (2\text{–}6)$$

On time scales relevant to humans, Berger (1982), Opdyke and Walker (1992), and others have argued that production of limestones is a net producer of CO_2, based on the relationship shown in equation 2–2 that for each $CaCO_3$ secreted, one HCO_3^- is converted to CO_2.

The key to understanding this apparent paradox is recognizing that carbon exists in a variety of forms (Fig. 2–2). The ultimate source of carbon is from within the Earth. Volcanic activity delivers carbon to the atmosphere, where it mostly occurs as CO_2, or to the ocean, where it mostly occurs as HCO_3^-. Organic carbon, produced mostly by photosynthesis (equation 2–5), occurs in short-term forms including living organisms and wood, and in water and sediments, and in long-term forms of carbon-rich rocks like coal and oil shales. Carbon stored in $CaCO_3$ can be recycled quickly, such as when aragonitic lime muds from the Bahama Banks are carried by currents into the deep ocean where they dissolve at depths greater than about 3,000 m. $CaCO_3$ in limestones can be recycled during glacial advances when reefs are exposed to the air and eroded; or $CaCO_3$ in limestones can be stored for millions to thousands of millions of years until plate tectonic activity uplifts them from the sea and exposes them to erosion, or until they are melted or metamorphosed by volcanic activity.

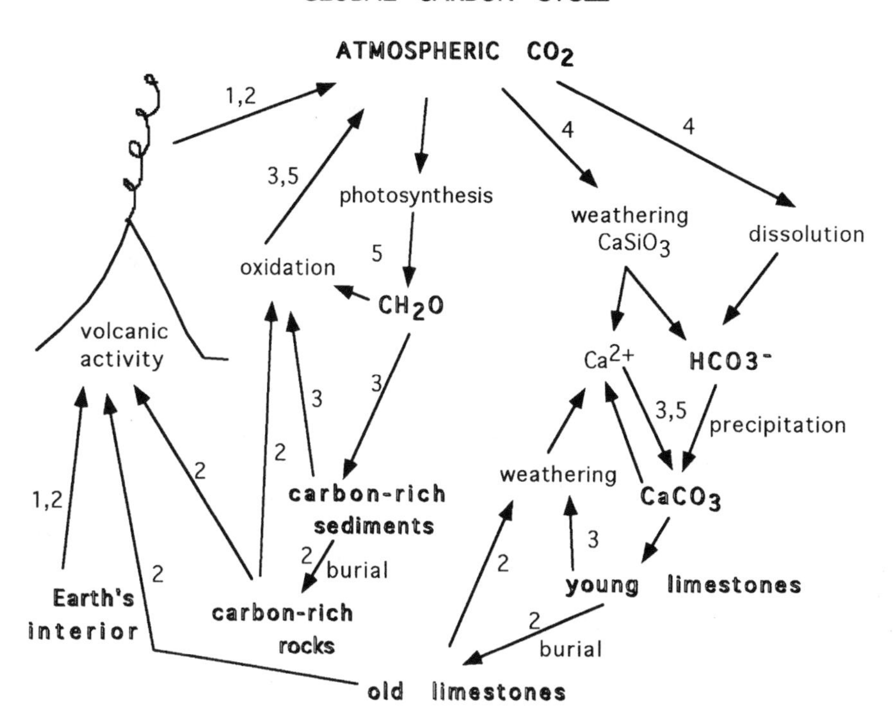

1. Rates have slowed through Earth history.
2. Rates controlled by plate tectonics.
3. Rates have changed with sea-level fluctuations.
4. All of these processes influence rates.
5. Biological scale processes.

Figure 2–2. Simplified representation of the global carbon cycle.

Figure 2–2 is a simplified representation of the CO_2 cycle, which is characterized by processes operating on four major time scales:

1. On the scale of Earth history and the evolution of life, CO_2 concentrations have been declining in the atmosphere to compensate for increasing solar output. Carbon has been stored in the Earth's crust as limestones and carbon-rich materials such as coal, oil shale, oil, and gas.
2. On the scale of 10s to 100s of millions of years, CO_2 has varied in response to plate movements and changes in rates of volcanic activity. Limestones and other carbon-rich rocks can be melted or metamor-

phosed by volcanic activity, recycling stored CO_2 back to the atmosphere-ocean pool.

3. On the scale of 10s to 100s of thousands of years, CO_2 in carbon-rich sediments is recycled as shallow-marine sediments are alternately deposited and eroded in response to sea-level rise and fall. During interglacial times when reefs are actively accreting, the global HCO_3^- pool is more rapidly converted to CO_2 and $CaCO_3$, so atmospheric CO_2 concentrations rise. During glacial events, when sea level is low and reef limestones are more actively eroded by atmospheric CO_2 and water, atmospheric CO_2 is decreased and oceanic HCO_3^- is increased.
4. Organisms typically act on biological time scales ranging from hours for bacteria to thousands of years for some long-lived trees.

Humans, however, have recently begun to perturb long-term cycles by extracting fossil fuels, burning them, and releasing huge quantities of CO_2 into the atmosphere. This process shifts carbon from long-term storage in the Earth's crustal rocks to CO_2 in the atmosphere. At the same time, we are cutting and burning forests. This shifts carbon from short-term storage in vegetation to CO_2 in the atmosphere. This process also reduces the rate of CO_2 removal from the atmosphere to short-term storage. By damaging coral reefs, humans have also influenced rates of $CaCO_3$ production. By perturbing both long- and short-term cycles, human activities over the past 100 years have increased CO_2 concentrations in the atmosphere more than reef growth increased CO_2 in the past 15,000 years.

2.6. Atmospheric CO_2 as an Evolutionary Driving Force

A geochemical paradox of declining concentrations of atmospheric CO_2 through time is that $CaCO_3$ has become increasingly easier for organisms to precipitate. The Proterozoic oceans, responding to a more CO_2-rich atmosphere, were more acidic and therefore able to hold in solution relatively high concentrations of Ca^{2+} and HCO_3^-, even in shallow water. $CaCO_3$ precipitation was primarily a "daytime" activity in restricted shallow-shelf areas in response to rapid CO_2 removal for photosynthesis by dense mats of cyanobacteria. At night and possibly seasonally during the day, seawater became undersaturated when photosynthetic rates slowed or stopped.

Then, sometime between 600 and 570 million years ago, a critical point was reached in the atmospheric-oceanic chemical system; shells evolved in several different groups of organisms. This indicates that atmospheric CO_2 concentrations must have dropped sufficiently so that low-latitude, shallow-water systems were consistently saturated with $CaCO_3$. Possibly the protective or supportive advantages of a mineralized shell finally exceeded the energetic costs of shell precipitation and maintenance. Biomineralized organisms have since flourished in the seas, secreting predominantly calcium carbonate shells or skeletons.

The succession of carbonate minerals through the geologic record also indicates a geochemical influence (Sandberg, 1983; MacKenzie and Morse, 1992). Though some of the earliest animals secreted calcium phosphate, calcite rapidly became the dominant shell material. Aragonite, which is stronger but less stable than calcite, also appeared as shell material in the early Paleozoic. But aragonitic algae did not produce significant bioherms until the Carboniferous. Aragonitic scleractinian corals did not evolve and construct reefs until the Triassic, only about 230 million years ago. This trend reversed as aragonite producers lost ground to calcite producers during the warm Cretaceous period, when atmospheric CO_2 concentrations rose in response to high rates of volcanic activity. Aragonite-producing corals did not fully regain a dominant position until 40 million years ago, with the global cooling that led to present glacial conditions.

A second implication of the reduction of CO_2 in the atmosphere, and therefore the partial pressure of CO_2 in surface waters of the ocean, is the possibility that shortage of CO_2 can limit photosynthesis (Riebesell et al., 1993). In shallow, warm, brightly illuminated waters, HCO_3^- is abundant and dissolved CO_2 is scarce. By using energy from photosynthesis to actively uptake Ca^{2+} ions from seawater, these organisms can use the calcification process to convert bicarbonate ions (HCO_3^-) to CO_2 needed for photosynthesis (McConnaughey, 1989). Calcareous algae, which may have been the first organisms to utilize calcification this way, first appeared in the Cambrian. The Mesozoic diversifications of more modern calcifying algae, including the coccolithophorids in the plankton, and melobesian green and coralline red algae in the benthos, along with the proliferation of algal symbiosis in foraminifera, corals, and some bivalves, may be further evidence of biotic response to declining CO_2 concentrations in the atmosphere.

There appear to be at least three major mechanisms of calcification in global oceans (Fig. 2–3). The first mechanism, geochemical precipitation of $CaCO_3$ in response to CO_2 uptake by photosynthesis, was particularly effective under relatively high atmospheric pressures of CO_2. The second mechanism, biomineralization by protozoan and animal cells, probably appeared when CO_2 concentrations declined sufficiently that the expenditure of energy for shell construction and maintenance became energetically feasible. The third mechanism, use of calcification to provide CO_2 for photosynthesis, was also likely related to declining atmospheric CO_2 concentrations, when reduced atmospheric concentrations began to limit aquatic photosynthesis in warm, shallow seas.

This third mechanism apparently arose independently in at least three groups of algae (reds, greens, and coccolithophorids) and several additional times when algal symbioses developed in calcified animals and protists. This mechanism has profound and almost paradoxical implications for atmospheric CO_2 concentrations. The very process that makes CO_2 instantaneously available for photosynthesis is responsible for the long-term removal and burial of CO_2 as $CaCO_3$.

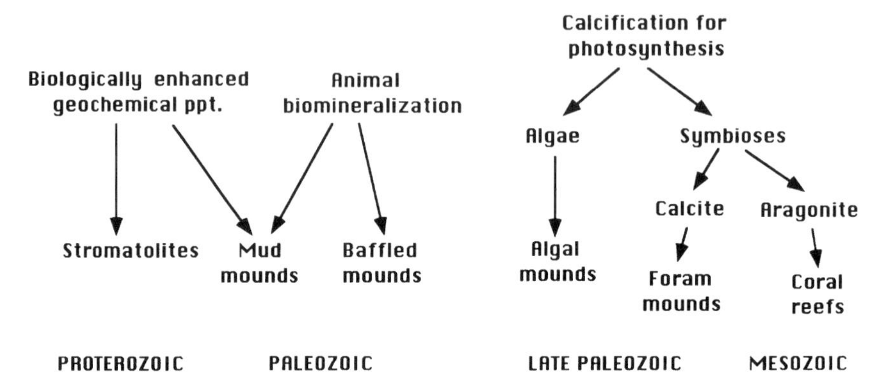

Figure 2–3. Calcification mechanisms and when they became important in the construction of biogenic reefs.

One mechanism that reduces the rate of CO_2 burial by calcification is global cooling. If organisms draw too much CO_2 out of the atmosphere, onset of glaciation lowers sea levels and reduces the habitats of warm-water calcifying organisms, as well as that of terrestrial plants. Lowered sea levels also prevent burial of some limestones, instead exposing them to erosion and returning the CO_2 to the ocean-atmosphere system. Interestingly, as atmospheric CO_2 concentrations have declined, aragonite is easier to precipitate in warm waters, and is more rapidly recycled by freshwater weathering.

2.7. What It Takes to Accumulate $CaCO_3$

Although calcium carbonate sediments have been precipitating and limestones have been forming for at least 3.5 billion years of Earth history, limestones in the rock record represent deposition during relatively small proportions of geologic time. This is because preservation of $CaCO_3$ is as important as production for accumulation and persistence in the rock record. Several factors influence both production and preservation potential. One very important condition is the persistence or repeated submergence of substantial areas of shallow shelf at mid to low latitudes over sufficient time so that limestones can accumulate. A second condition is relatively low input of terrigenous sediments and inorganic nutrients, especially nitrogen and phosphorus. A third condition is the presence of biota that can precipitate or enhance the precipitation of $CaCO_3$.

The persistence or repeated submergence of substantial areas of shallow shelf at mid to low latitudes is controlled by tectonics and climate, which together control sea level. Rates of seafloor spreading influence the average depth of the ocean (Fig. 2–4), pushing water higher onto the continental shelves when spreading rates are high (Kennett, 1982). The relative proportions of continents that are colliding or overlying rifting centers also influence global and local sea level. For example, all of the major continents came together in the late Paleozoic to form the supercontinent of Pangaea (Fig. 2–5). This can be compared to India colliding with Asia to form the Himalayas, only on a much larger scale. As the continents pushed together, sea level relative to the continents was low, so there were few shallow-shelf areas, and most that occurred were being buried in sediments being eroded from the highlands. Through the latest Paleozoic and into the early Mesozoic, Pangaea remained as one supercontinent. Heat from the Earth's interior was building up underneath Pangaea, pushing it up and beginning to break it apart. Sea level remained low and shallow-shelf area was limited. Pangaea at this stage may have been analogous to East Africa today, where eroding highlands plunge into deep rift valleys. As the continents rifted apart and began to move away from each other during the Cretaceous (Fig. 2–5), they slowly subsided and became generally similar to modern Australia. Most of the mountainous areas were island arcs analogous to the Marianas Islands or Japan.

Seafloor spreading rates and climate interact through a variety of feedback mechanisms to influence sea level (Worsley et al., 1986). As the continents

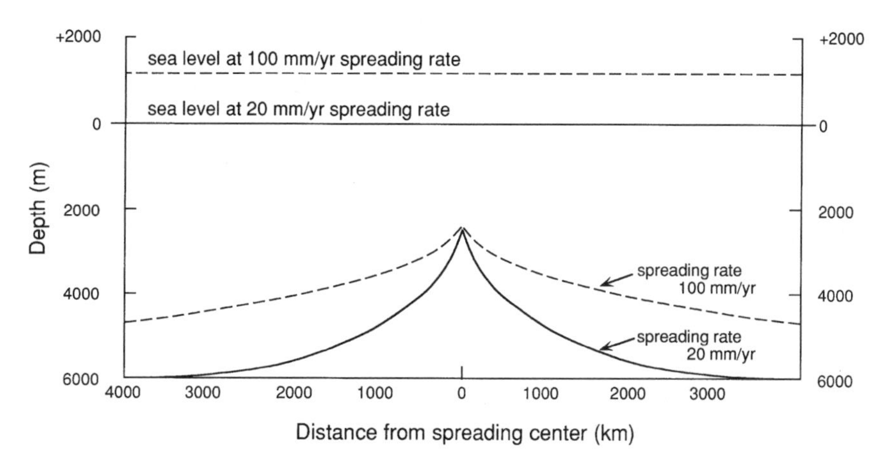

Figure 2–4. The effect of seafloor spreading rate on sea level. Oceanic crust that is newly formed at the mid-ocean ridge crest is hotter and less dense than older crust; oceanic crust cools and subsides with age (Kennett, 1982). The solid profile illustrates oceanic floor formed at 20 mm yr^{-1}, which is a common rate for the present mid-Atlantic ridge. If the spreading rate increased to 100 mm yr^{-1}, within 40 million years the seafloor profile would change to that shown by the dashed line. If the surface area of the ocean basin remained constant, sea level would rise more than 650 m, to the dashed sea-level line. In reality, when sea level rose in Earth history, lower-lying areas of the continents were flooded, spreading ocean waters over much larger areas. Thus, sea-level rises were probably never as much as 650 m; sea level during the Cretaceous was about 300 m higher than today.

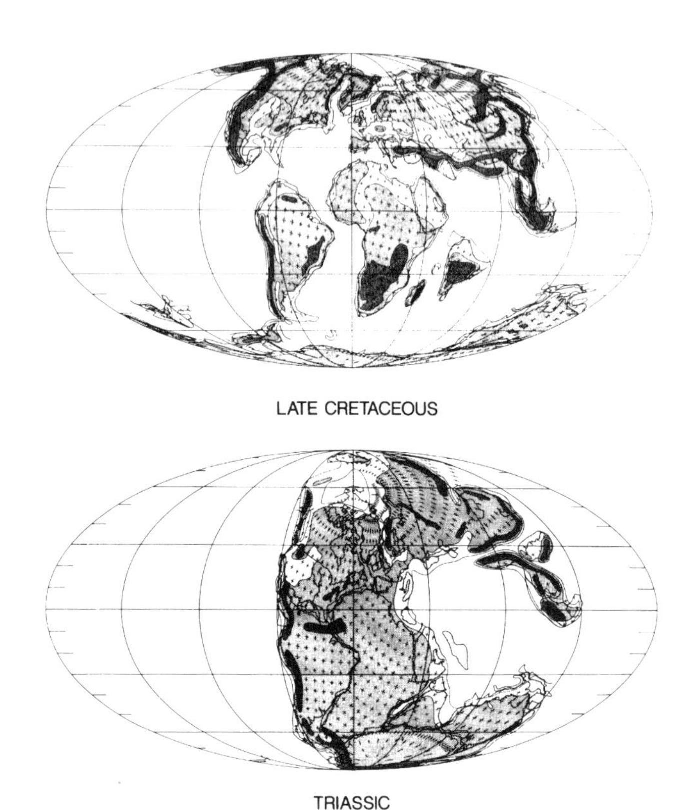

Figure 2–5. Plate reconstructions for the Triassic (lower), illustrating the supercontinent Pangaea that formed during the late Paleozoic; and for the Late Cretaceous (upper), after the breakup of Pangaea (from Ziegler et al., 1982). Oceans are shown in white and land areas are shaded; highlands are densely shaded.

ground together to form Pangaea, seafloor spreading and overall volcanic rates slowed, further lowering sea level. Slower volcanic rates resulted in slower emissions of volcanic gases, including CO_2. As a result, Earth's climate was cooler. Conversely, as continents moved apart, seafloor spreading rates increased, volcanic rates increased, and CO_2 emissions increased, triggering global warming in the Jurassic and Cretaceous. Relative sea level influences global climate because it alters the reflectivity (albedo) of the Earth; land is more reflective than water and ice is most reflective. The higher the sea level, the more solar energy the Earth can capture and thus the warmer the climate. Warm climates promote even higher sea level, because all polar ice is melted.

The differential heating of the equator relative to the poles drives atmospheric and oceanic circulation. Plate tectonics controls the positions of landmasses relative to the oceans and so further influences ocean circulation and climate by providing passageways and barriers. For example, the opening of the Drake Passage between Antarctica and South America, which occurred from 40 to 20 million years ago (Fig. 2–6), allowed the development of the circumpolar current in the southern oceans (Kennett, 1982). This led to the climatic isolation of Antarctica and was a major driving force for polar cooling. Closure of the Isthmus of Panama over the past few million years stopped flow of the Caribbean current into the Pacific, diverting it northward to accelerate the Gulf Stream. Delivery of warmer water to high northern latitudes increased snowfall, which may have triggered glaciation (Kennett, 1982).

Global climate is strongly influenced by CO_2 concentrations in the atmosphere, and therein lies another paradox of limestones and coral reefs. Times of global warming, such as the Cretaceous and early Paleogene (Fig. 2–1), are times of widespread limestone deposition, but not of coral-reef development. Geochemical evidence for ancient temperatures indicate that the Earth was so warm during these times that seawaters in the polar regions were 12–16°C; they are less than 1°C today. What was happening in the tropics is more controversial. Kauffman and Johnson (1988), Adams et al. (1990), and many others contend that distributions of fossil organisms indicate warm tropics, in some cases even warmer than today. However, some geochemical measurements have indicated that tropical seas may have been cooler, on the order of 18°C instead of 25–30°C common today (Shackleton, 1984). The explanation is that warming the polar regions required more effective heat transport, and therefore cooling of the tropics.

2.8. History of Biogenic Reefs Through Time

Although cyanobacteria have been available to build stromatolites for nearly 3.5 billion years, and there have been shell-forming animals for 570 million years (Fig. 2–1), biogenic reefs are sporadically scattered through the rock record (James, 1983; Copper, 1988). Factors that have controlled the formation of biogenic reefs through time include climate, ocean circulation, availability of habitat, and existence of reef-forming biotas. The history of reef development and reef-building biotas has not been continuous. Instead, it has been characterized by long periods of persistence of simple reef mound–producing communities, followed by longer episodes of development of complex baffler and reef-building communities, and termination of complex community development, followed by major extinction events (Stanley, 1992; Copper, 1994).

The fossil record is characterized by four kinds of reeflike structures: stromatolites, simple reef mounds, biotically complex reef mounds constructed by baffler/binder communities, and true framework reefs. These categories are more convenient than mutually exclusive. Within reef structures, there may be a colonizing stage by a binder community, a diversification stage by a baffling and binding community, and a climax stage of framework builders, binders, dwellers, and destroyers (Fagerstrom, 1987).

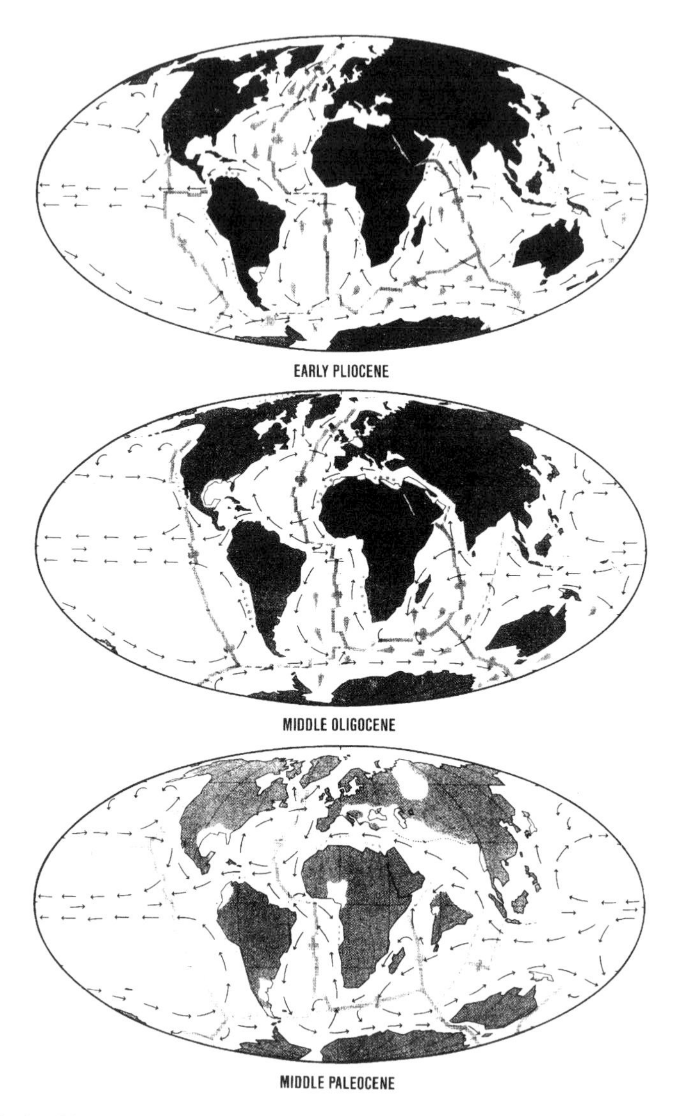

Figure 2–6. Plate reconstructions and inferred circulation patterns of oceanic surface waters (arrows) from Haq and Van Eysinga (1987): middle Paleocene (60 million years ago)—note circumtropical circulation and the lack of south circumpolar circulation; middle Oligocene (30 million years ago)—note that circumtropical circulation diminished as India and the Middle East were approaching Asia, while south circumpolar circulation was complete; and early Pliocene (5 million years ago)—note that the Isthmus of Panama is substantially restricted but not yet fully closed. Oceans are shown in white and land areas are shaded. Oceanic ridge crests are indicated by lighter shading.

From 3.5 to 6 billion years ago, whether stromatolitic reefs developed or not was likely controlled primarily by suitability of the physical and chemical environment. Limestones accumulated where seawater was warm enough and shallow enough for photosynthesis by cyanobacteria to promote $CaCO_3$ precipitation over a substantial area and over sufficient time to be preserved (Grotzinger, 1989). Following the evolution of multicellular life such as worms and trilobites that bulldozed through the algal mats eating and disrupting them, the chances for stromatolite preservation declined (Conway Morris, 1993). However, stromatolites continued to be an important type of bioherm through the early Paleozoic, and some still occur today (Dill et al., 1986).

2.8.1. Paleozoic Era—Time of Ancient Animal Life

The first animal communities to act as significant bafflers appeared in the Cambrian period, nearly 560 million years ago. The archaeocyathids or "ancient cups" are thought to have been calcified sponges (Wood et al., 1992); their erect skeletons trapped calcareous muds, which may have been chemically precipitated in response to photosynthesis in cyanobacterial mats and in the plankton. Calcified cyanobacterial/algal mats and encrusters formed a major component of the archaeocyathid reef mounds, which persisted almost 30 million years. Trilobites and sponges were inhabitants of these simple mounds. "Reef tracts" produced by these communities were extensive about 530 million years ago in Siberia, southern Australia, and Antarctica, which lay in low latitudes at that time (Copper, 1994). By the middle Cambrian these primitive reef communities suffered extinctions, so cyanobacterial mats and encrusters once again formed what reef mounds were preserved.

The early Ordovician began the first major diversification of calcified algae and animals capable of producing, trapping, and binding large volumes of calcareous sediments. Sponges and the earliest corals played the major baffler roles, cyanobacteria persisted as binders, while trilobites and primitive snails grazed the algae or ate the algal/bacterial-rich muds. By the middle Ordovician, a variety of calcified animals had evolved. These organisms were more prolific sediment producers, more effective sediment bafflers, and some were even framework constructors. Important contributors included bryozoa; stromatoporoids, which were a kind of sponge; and a calcareous red alga called *Solenopora*. These early reef mounds provided a multitude of niches for trilobites, brachiopods, snails, cephalopods (ancestors of the chambered *Nautilus*), other animals that fossilized, and many soft-bodied organisms that left no fossil record.

By the late Ordovician, tabulate and rugose (horn) corals were becoming important components of the reef community. A typical succession consisted of a colonizer community of stalked crinoids and bryozoa that trapped muds and began to accumulate a structure. Soon this habitat was invaded by sponges, solitary horn corals, small tabulate corals, and massive stromatoporoids, which together constructed larger skeletal structures and trapped calcareous muds and sands. Algae, smaller corals, bryozoa, brachiopods, clams, snails, trilobites, and other arthropods found shelter or food within the reef. Encrusting stromatoporoids bound the skeletal elements together, and typically overgrew and eventually dominated the whole structure. In this climax phase, which may have been very wave resistant like a modern coralline algal ridge, the diversity of species was low (e.g., James, 1983; Copper, 1988).

Interesting size comparisons can be made between the middle Paleozoic (Ordovician to Devonian) bioherms and modern coral reefs. Individual structures were often similar in size to small patch reefs, connected by expanses of calcareous sands or muds, so the total accumulation of limestones was comparable to a modern reef tract (Fagerstrom, 1987). However, there is no strong evidence for algal symbiosis in the calcifying animals, and calcareous algae are of limited importance. Thus, calcification mechanisms responsible for precipitation of most of the early to middle Paleozoic limestones were probably biologically enhanced geochemical precipitation and biomineralization by animals. Extensive Devonian reef complexes along the northern margin of the Canning Basin, Western Australia, as well as others found in Alberta, Canada; Belgium; and Germany, are comparable in size to, or even larger than, modern western Atlantic reef systems like the Florida Keys reef tract (Stanley, 1992).

In the late Devonian, an extinction event occurred that eliminated most of the reef-building species, setting back reef-building communities to simple stromatolites. Latest Devonian limestones in the Canning Basin are cyanobacterial/algal in origin, with a few sponges providing the little diversity. Early Carboniferous buildups scattered around the world are mud mounds, containing mostly calcareous muds of unknown origin, possibly trapped into buildups by crinoids and bryozoa, the only major skeletal components of the mounds. Rugose corals, chaetitids (an unusual group of calcified sponges), and brachiopods were among the first taxa to reappear to colonize the crinoid-bryozoan mud mounds as carbonate-producing communities evolutionarily began to recover.

Late Carboniferous biotas contain many unusual and taxonomically problematic groups that mostly represent binder and baffler communities. A major change seen in the Carboniferous is the prevalence of calcareous algal species that produced aragonite. The evolution and diversification of the first large, structurally complex foraminifera, the fusulinids, represents the earliest strong evidence for algal symbiosis in an important calcareous sediment producer. These two events may represent the first major proliferation of organisms calcifying to enhance photosynthesis. On a global scale, the continents were coming together to form Pangaea. Glaciation was occurring, indicating reduced atmospheric CO_2 concentrations, which was conducive to aragonite precipitation (Sandberg, 1983) and calcification to enhance photosynthesis.

One of the best known Paleozoic reef complexes in the world is that in western Texas and southeastern New Mexico (e.g., Wilson, 1975), the Permian-age bioherms of the Guadalupe Mountains and Carlsbad Caverns National Parks. These limestones have been intensively studied by petroleum geologists because subsurface limestones in this region have produced tremendous quantities of oil and natural gas. These are well-developed reef mounds in which a variety of sponges and sometimes algae, bryozoa, crinoids, and brachiopods acted as bafflers, trapping muds and coarser sediments produced by the community. Algae and sponges served to bind the trapped sediments. Diverse communities of foraminifera, snails, clams, small corals, brachiopods, cephalopods, and arthropods thrived in these buildups. In the Glass Mountains of western Texas, an unusual group of large, spiny brachiopods clustered to form a kind of reef framework, further developing the sponge-algal baffled structure. Late Permian (Guadalupian) communities are among the first to show a major contribution by encrusting red algae.

2.8.2. Mesozoic Era—Time of Intermediate Animal Life

The latest Permian and earliest Triassic witnessed the most extensive extinction event in Earth history and a prolonged episode in which limestones are missing from the rock record. The prolonged absence of limestones indicates widespread unfavorable paleoenvironmental conditions, possibly including geochemical conditions that suppressed carbonate precipitation or inhibited its preservation. When carbonate buildups reappeared in the middle Triassic, they were baffled and bound sediments similar to those of the late Permian, except simpler and with fewer species. Sponge-algal communities gradually resumed production of reef mounds, and included among their subsidiary fauna the first scleractinian (modern stony) corals, with their aragonite skeletons.

By the late Triassic, the scleractinian corals had diversified, and new forms restricted to shallow, brightly illuminated waters appeared that are believed to have hosted zooxanthellae (Stanley, 1981). Corals, sponges, and stromatoporoids produced framework; other sponges and stromatoporoids, as well as bryozoa, serpulid worms, and sponges served as bafflers and binders. Associated fossilized taxa were a diverse group of ammonites, brachiopods, bivalves, snails, echinoderms, foraminifera, and worms.

Following widespread extinctions in the latest Triassic, Jurassic reef-building communities reestablished to a diversity of types comparable to that of modern oceans. Besides coral-stromatoporoid reefs, there were sponge-dominated, sponge-algal, sponge-bivalve, and stromatolitic reef mounds, as well as deep-water coral-sponge mounds. In the late Jurassic, two related trends began that continued and fully developed in the Cretaceous period. Despite a continuing increase in the variety of corals, there was limited coral-reef construction, while large bivalve species proliferated as carbonate producers and sediment trappers. As a result, corals became subsidiary to the bivalves in the Cretaceous.

The major group of bivalves in Cretaceous buildups are the rudistids. Kauffman and Johnson (1988) interpret the rudists as having had algal symbionts to enhance calcification, and suggest that they displaced corals as reef builders during the Cretaceous. European reef researchers contend that there is little evidence for assuming that rudists had algal symbionts, and the prevalence of muddy sediments in most rudist buildups indicates that water transparency may have been poor. For example, Skelton et al. (1992) interpret rudists as having been superbafflers, trapping huge quantities of muds, and actually growing upward supported in muds. Characteristically, thickets and buildups have only one or a few rudist species and few associated organisms. Banktop habitats occupied by many rudists may have been similar to those of modern seagrass beds; the rudist thickets trapped sediments carried in by the currents as well as sediments produced in situ by the breakdown and bioerosion of the rudists themselves. Because of the exceptional high-latitude warming that occurred in the Cretaceous, Kauffman and Johnson (1988) suggested that low-latitude banktop habitats were consistently warm and probably somewhat hypersaline. The more biologically complex bivalves may have been better able to survive these extremes than corals, whose symbiotic relationship is particularly sensitive to temperatures above 30°C (Glynn and D'Croz, 1990).

Furthermore, geochemical factors may have favored rudists over corals. The rudists secreted an outer shell of calcite, with an inner layer and muscle insertion sites of aragonite (Skelton, 1976). Proportions of aragonite to calcite varied in different lineages of rudists. Some secreted shells were 60% aragonite; many were less than 30% (Kauffman and Johnson, 1988). Whatever the proportions, the predominance of rudists over wholly aragonitic corals represents a decline in aragonite production. This may reflect the higher concentrations of CO_2 in the Cretaceous atmosphere that made aragonite more soluble and therefore less energetically advantageous to produce.

2.8.3. Cretaceous Boundary Extinctions

A major biotic crisis occurred at the end of the Cretaceous Period. This extinction event is best known for the demise of the dinosaurs. Rudists also became extinct, as did nearly all shallow-water, tropical, carbonate-producing protists and animals. Coccolithophorids and planktonic foraminifera were so abundant in shallow-water plankton during the Cretaceous, and produced such tremendous accumulations of chalks throughout the world, that the Cretaceous actually means "time of chalk terrains." Only a few species of each survived the extinction event.

Fortunately for the scleractinian corals, during the Cretaceous many had retreated to deeper shelf-slope environments, perhaps to escape warm, saline waters or direct competition with rudists (Kauffman and Johnson, 1988). As a result, a few species survived the extinction.

The terminal Cretaceous event was similar to those of the early Cambrian, late Ordovician, late Devonian, and late Permian in that reefs generally vanished a million years or more before the final extinction events (Copper, 1994). Thus, periods of environmental perturbation or climatic deterioration that caused collapse of reef ecosystems must have preceded these major extinction events that geologists use to define significant boundaries in geologic time.

The two major hypotheses for this mass extinction event are a bolide (meteor or comet) impact (Alvarez et al., 1980) or an extensive volcanic episode (Officer and Drake, 1985). Because the bolide impact would have been an instantaneous event, while extensive volcanism would have lasted from 100,000 to a million years or more, scientists with evidence for a more gradual event favor volcanism. While this controversy has been raging in geology since the mid-1980s, strong evidence now indicates a massive impact site on the Yucatán Peninsula in Mexico (Alvarez et al., 1992). However, one can argue that a bolide impact, which occurred during a period of climatic deterioration resulting from volcanism, might have had compounding effects that pushed already stressed ecosystems into collapse.

As for shallow-water, photosynthetic, carbonate-producing, reef organisms, either event would have been equally devastating for some of the same reasons. Either event would have profoundly influenced ocean chemistry in ways detrimental to $CaCO_3$ production. The energy of a meteor passing through the atmosphere would have oxidized nitrogen in the atmosphere, resulting in global acid rain that would have acidified shallow oceanic waters, dissolving $CaCO_3$. Volcanism emits large quantities of hydrogen sulfide (H_2S), which would have oxidized to sulfur dioxide (SO_2) and hydrated to sulfuric acid, resulting in global acid rain. A meteor would have damaged the Earth's protective ozone layer; volcanic emissions are also corrosive to stratospheric ozone. Furthermore, the earthquake caused by a 10-km-diameter bolide hitting the Earth could have eliminated all nutrient-depleted shallow-water environments worldwide, just by stirring the oceans. Deeper oceanic waters, where there is insufficient light for photosynthesis, are substantially richer in inorganic nutrients (dissolved nitrogenous and phosphatic compounds needed by plants to photosynthesize) than photic surface waters. These nutrient-rich waters represent approximately 98% of the total ocean volume. Such a tremendous shock would have generated tsunamis and internal waves, rapidly mixing deeper waters into the shallow waters and stimulating red tidelike blooms of a few, fast-growing species of phytoplankton (disaster species, see Fischer and Arthur, 1977).

Whatever occurred, among the lessons to be learned from the latest Cretaceous and other major extinction events are that the most prolific carbonate-producing organisms are particularly sensitive to environmental perturbations because they thrive within relatively narrow environmental limits. Waters in which they proliferate are warm, but not too warm; clear and well illuminated but with not too much biologically damaging ultraviolet radiation; nutrient depleted; and highly supersaturated with respect to Ca^{2+} and HCO_3^-. Most reef-building organisms live at low latitudes on shelves and banks in the shallowest waters 100 m (330 ft), and thrive at depths less than 20 m (66 ft), in an ocean that averages 3800 m (12,500 ft) in depth. Suitable waters make up less than 1% of the ocean volume under optimum conditions; suitable benthic habitat far less than that. Thus, regional or global events that eliminate habitat for reef-building and reef-dwelling plants and animals can eliminate huge numbers of species and are recorded in the rock record as mass extinction events. Because reef species are often highly specialized to their environment, many specializations are lost in these events, while less specialized, more opportunistic species typically survive.

2.8.4. Cenozoic—Time of Recent Animal Life

Earliest Paleocene shallow-water limestones are dominated by bryozoans, coralline algae, and miliolid foraminifera (Hallock et al., 1991; Copper, 1994). The latter group live in warm, shallow-water environments, especially thriving on fleshy algae and seagrass.

Despite the mass extinction event, the high atmospheric CO_2 levels and greenhouse climate, which had developed during the Cretaceous, continued for another 20 million years. Recovery of diverse coral assemblages occurred much faster than reef-building potential. All the major circumtropical frame-building genera of scleractinian corals had evolved by the latest Eocene (Frost, 1977). Thus, the Eocene, like the Cretaceous, was a time of diverse coral assemblages and widespread occurrences of "reef-associated" biotas, but limited coral-reef production. Frost suggests that the capacity to construct massive wave-resistant reef structures developed long before such structures were widely produced.

Consistent with higher atmospheric CO_2 concentrations, prolific calcification by calcite-secreting larger foraminifera (e.g., Plaziat and Perrin, 1992) and coralline red algae appears to have recovered faster than comparable aragonite production by corals and calcareous green algae. The best known larger foraminiferal limestones are the Eocene nummulitic limestones of Egypt, from which the Pyramids were built. Larger foraminifera-rich, shallow-water limestones are widespread in mid to low latitudes. Fossils of larger foraminifera of this age can be found in Oregon, southern England, and other localities as high as 51°N latitude (Adams et al., 1990). Their complex shells, by analogy with modern larger foraminifera, are believed to have been highly adapted to house algal symbionts. Shallow-water limestones dominated by coralline algal nodules were also common (Bryan, 1991).

High-latitude cooling began in the middle Eocene, culminating with the Eocene–Oligocene boundary events, which appears to have been one of the coolest times of the past 200 million years (Shackleton, 1984). This boundary is characterized by extinctions of larger and planktic foraminifera (Hallock et al., 1991) and an interval of shallow-water coralline algae and miliolid limestones (Adams et al., 1986).

Following this setback, coral communities flourished circumtropically in the middle and late Oligocene. The sudden expansion in reef-building capacity may have resulted from falling atmospheric CO_2 concentrations and rising tropical sea-surface temperatures that accompanied high-latitude cooling (Shackleton, 1984). In the Caribbean, coral reefs reached their acme of development in the late Oligocene (Frost, 1977). By the early Miocene, reefs and reef-associated biotas worldwide extended their distributions more than 10° north and south into higher latitudes (Adams et al., 1990).

At least one factor accounting for the widespread distribution of tropical biotas in the Eocene and Oligocene was circumtropical oceanic circulation (Fig. 2–6). While western Pacific reef biotas have consistently been more diverse, most late Eocene and Oligocene coral in the Caribbean region were cosmopolitan taxa. With the breakup of circumtropical circulation, Atlantic biotas became increasingly isolated. Atlantic reef-associated biotas have lost so many taxa that many groups now have only a fraction of the species found in the Pacific. Roughly half of the coral genera were lost from Caribbean faunas at the end of the Oligocene, and many more became extinct during the Miocene (Frost, 1977; Edinger and Risk, 1994). There were also extinctions of Indo-Pacific corals, but far fewer. Larger foraminifera suffered similar losses in the Caribbean.

Progressive blockage of circumtropical circulation (Fig. 2–6) may account for some of these extinctions (Edinger and Risk, 1994). The trade-wind-driven, east-west–flowing circumtropical current passed through two oceanic gateways (Berggren, 1982): the Eastern Tethys through what is now the Middle East and the Central American seaway. The Eastern Tethys closed around the Oligocene-Miocene boundary with the development of the Qatar arch (Berggren, 1982). Although the central American landmass was developing and reducing exchange of Caribbean and Pacific waters, the isthmus of Panama did not close until the middle Pliocene (Keigwin, 1982a).

Because the Atlantic has higher evaporation rates compared with rainfall rates than the Pacific, sea level in the Pacific is actually slightly higher than the Atlantic. As a consequence, when the Central America seaway was open, water had to flow from the eastern Pacific into the Caribbean (Luyendyk et al., 1972). However, the easterly trade winds forced surface waters to flow westward from the Caribbean to the eastern Pacific. As long as the Central American passageway was wide and deep, two-way flow could be accommodated by westward surface flow and eastward subsurface flow, probably at depths in excess of 50–100 m or more (Maier-Reimer et al., 1990). However, as the passageway constricted, flow was constricted and surface flow probably reversed during calm weather. Nutrient-laden eastern tropical Pacific waters were more frequently and consistently introduced into Caribbean surface waters.

Edinger and Risk (1994) observed that shelf-edge and slope-dwelling coral genera were nearly eliminated in the Atlantic in the early Miocene. The corals that survived were mostly banktop species capable of tolerating higher nutrients and higher sedimentation rates that would have accompanied higher bioerosion rates (e.g., Hallock, 1988a). Larger foraminifera show the same trends. Slope-dwelling taxa were nearly eliminated, while shallower-dwelling taxa of banktops and restricted environments thrived and are actually more diverse in the Caribbean than in the Indo-Pacific today (Hallock and Peebles, 1993). Other banktop biota, including octocorals and calcareous algae, became more important components of Caribbean biotas than on comparable Pacific reefs.

Since the middle Miocene, there has been progressive high-latitude cooling, compression of tropical habitats, and increasing temperature gradients between high and low latitudes (Shackleton, 1984). Northern Hemisphere climatic deterioration began approximately 2.7 million years ago and culminated in the Pleistocene Ice Ages during the past million years (Kennett, 1982). The overall cooling and fluctuating climate has strongly influenced tropical reef-associated biotas in a variety of ways. Some taxa have become extinct. Others, like the fast-growing *Acropora* and *Montipora,* have diversified and now account for 25% of the Indo-Pacific coral species (Veron and Kelley, 1988). Regional isolation of populations in sufficiently different environmental conditions can promote speciation (Mayr, 1971). Reconnection of regions then can mix similar but reproductively distinct species. Successive isolation and mixing, which accompanies fluctuations in sea level, has long been recognized as a potential mechanism driving evolutionary diversifications (Hallam, 1985).

2.9. Modern Reefs

Another paradox of limestones and climate is that the Neogene represents a time of active reef building and coral evolution, despite high-latitude climatic deterioration that produced as many as 21 glacial advances and retreats (Delmas, 1992). Sea-level fluctuations of up to 140 m (450 ft), caused by repeated accumulation and melting of Northern Hemisphere continental glaciers, accompanied cyclic reduction and expansion in the areas of warm, tropical seas. The solution to this paradox may be that tropical climatic, geochemical, and topographic conditions during interglacials are close to optimum for reef building. Atmospheric CO_2 concentrations are relatively low (though not as low as during glacial advances) and subtropical/tropical sea-surface temperatures are optimal. Both factors promote aragonite calcification by corals and calcareous green algae, which in turn enhances their rates of photosynthesis and biological productivity.

Periods of rising sea level are also particularly favorable for reef building because there is space for accumulation of substantial thicknesses of reef limestone, which have the best chance for preservation in the rock record. Except for active tectonic areas, most shallow shelves are slowly subsiding at rates of a few centimeters per 1,000 years (Schlager, 1981). The most actively accreting portion of a reef system can grow upward at rates of at least 4 m/1000 years, and under exceptional conditions perhaps as much as 20 m/1,000 years. But they can only grow at those fast rates when growing in water depths of less than 20 m, where there is plenty of sunlight for photosynthesis. The only way they can grow upward at high rates for more than about 5,000 years, and produce more than 20–30 m thicknesses of limestone, is if the shelf subsides or sea level rises. Thus, during interglacial sea-level rise episodes, the margins of continental shelves, limestone banks, and volcanic islands provided ideal locations for thick sequences of reef-associated limestones to accumulate.

Fluctuating sea level strongly influences the morphologies of reefs. If the rate of sea-level rise at any time is too rapid, upward growth of coral reefs simply cannot keep pace, and one of three things happens (Neumann and Macintyre, 1985). That is, during the primary glacial melting event 10,000–12,000 years ago, sea level rose at least 20 m. Very few reefs could keep pace with that rise, so most backstepped (Fig. 2–7), if there was a higher substratum to which to backstep. Others did not fully keep pace during rapid melting events, but caught up when rates of sea-level rise slowed, while some simply failed to catch up and became submerged banks.

On most bank- or shelf-margin reefs, the most rapidly accreting part of the reef system is the zone of living coral that faces toward the prevailing wind and current. Sands and muds in the backreef and lagoon may accumulate much more slowly, providing the profile of the barrier reef or atoll with a deep lagoon (Fig. 2–7C). In such a situation, the reef margin may keep up or catch up, but the rest of the bank will continue to fill, possibly until sea level begins to fall with the next glacial advance. In cases such as Little Bahama Bank, the whole bank is near sea level, and the main direction available for accretion is lateral (Fig. 2–7B). When sea level falls, the most recently deposited limestones are exposed to erosion and the reef-building organisms are forced to relocate downslope (Fig. 2–7A).

Where a reef begins to grow is often dependent on the topography of the seabed (Longman, 1981). Slight topographic highs, particularly if the substratum is rock or coarse shell debris, favor recruitment and growth of colonizing reef builders. The length of time the reef has been growing under stable or slowly rising sea level is reflected by how much reef growth has modified the underlying topography (Fig. 2–8). Even in cases where reef development seems independent of underlying topography (Fig. 2–8C), original colonization probably occurred on minor topographic highs.

Modern reefs and reef shorelines are characterized by two major events: the last interglacial episode 120,000 years ago and the initiation of growth of modern reef structures less than 9,000 years ago (Davies, 1988; Shinn, 1988). During the last interglacial, sea level was at least 6 m higher than today and the limestone was formed that makes up many of the islands associated with reef tracts today. During the last glacial episode, which lasted more than 80,000 years, sea level fell to 130 m below the present level. Continental shelves were dry land; reef growth was limited to steep island or continental slopes. Why present reef growth began almost worldwide about 6,000–9,000 years ago is a fascinating mystery.

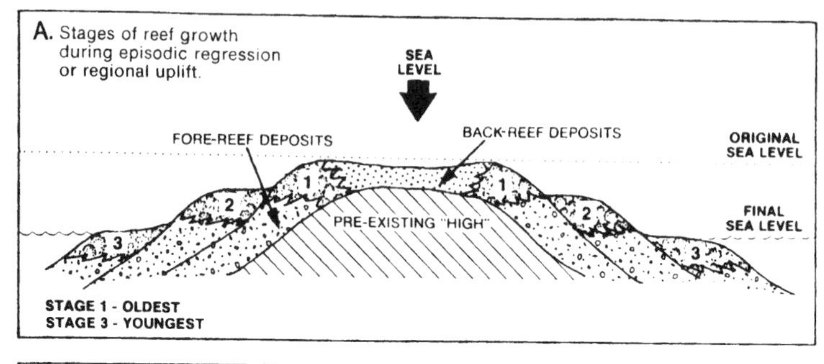

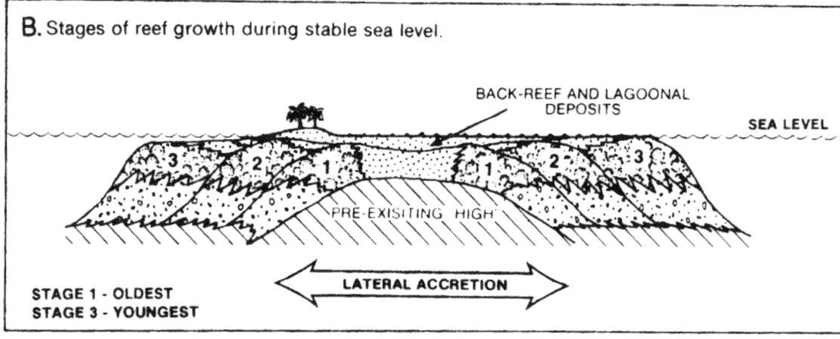

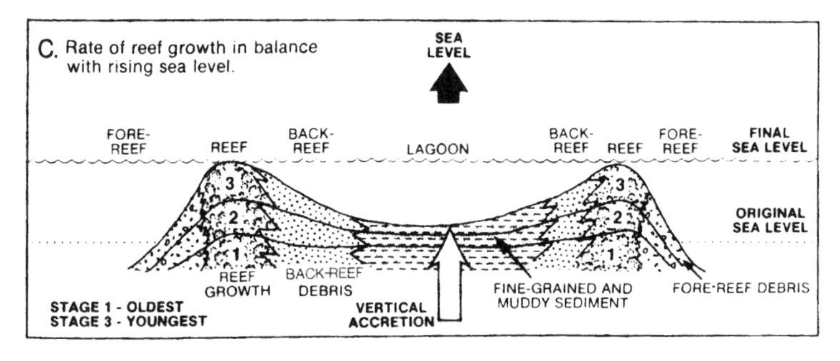

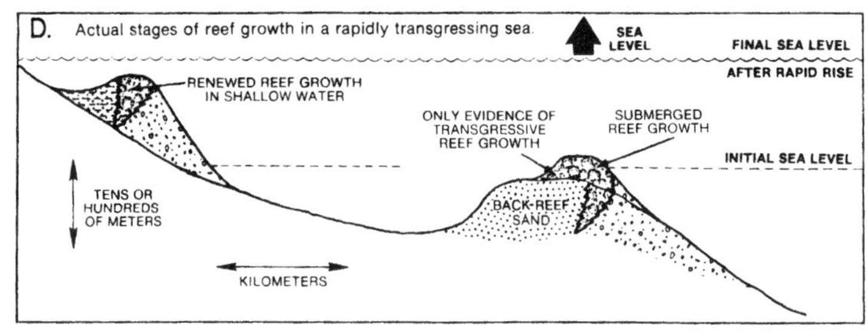

Figure 2–7. Responses of sea level to **(A)** sea-level fall or tectonic uplift; **(B)** stable sea level; **(C)** rising sea level or tectonic subsidence with the reef keeping up; and **(D)** rising sea level with the submerged reef failing to keep up and backstepping (from Longman, 1981).

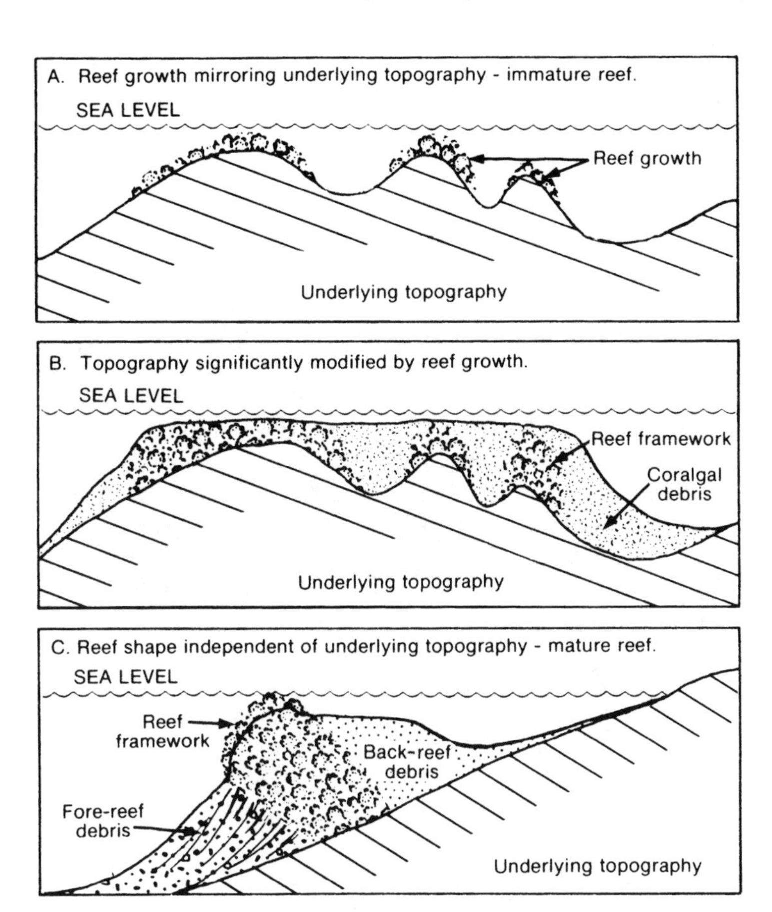

Figure 2–8. Relationships between reef growth and underlying topography: **(A)** colonizing reef-growth mirrors underlying topography; **(B)** reef represents an antecedent topographic high, but features of the underlying topography are modified by reef growth; **(C)** reef shape is independent of the underlying topography (from Longman, 1981).

Did climate or ocean geochemistry play a role? Or did sea-level rise simply slow sufficiently that reef development could keep pace from that time until the present?

Rising sea level does not always promote reef growth (Neumann and Macintyre, 1985). Approximately 4,000 years ago, sea level was about 3 m below the present level (Shinn, 1988). In southern Florida, Florida Bay was still swampland, instead of the vast restricted lagoon it is today. As sea level rose and Florida Bay developed, tides flushed water inconducive to reef growth from Florida Bay onto the reef tract; cold water alternated seasonally with warm, salty water. As a result, today Florida's reefs are best developed where protected from Florida Bay by elongate islands in the upper and lower keys. Middle key reefs, which began to develop about 6,600 years ago, have mostly failed to keep up over the past 4,000 years.

2.10. Conclusion

The history of reefs is a fascinating and ongoing study, full of paradoxes and unexpected feedback mechanisms. One of the most difficult concepts to understand is that limestones are both storehouses and sources of CO_2, depending on the time scale being considered. Through geologic time, limestones have tied up CO_2 and are therefore responsible for a livable planet. Yet on the shorter time scales of oceanic circulation and Ice Ages, interglacial reef building may move as much as 80 ppm of HCO_3^- from the bicarbonate pool in the oceans to CO_2 in the atmosphere. Relatively rapid erosion of young, predominantly aragonitic-coral limestones and sediments occurs when sea level falls. During a glacial advance, atmospheric CO_2 shifts back to HCO_3^- in the oceans. On human time scales, reefs are probably irrelevant to the CO_2 story. Fossil fuels, like ancient limestones, represent long-term storage of CO_2 in the Earth's crust. Human activities, including the burning of fossil fuels, have added more CO_2 to the atmosphere in 100 years than reef building did in 15,000 years.

The major lesson to be learned from the history of reefs is that they are constructed by complex and specialized biological communities. Because these systems are highly specialized to a limited range of environmental conditions, they are extremely sensitive to local, regional, and global environmental perturbations. Environmental crises, culminating in global mass extinction events, have repeatedly caused the extinctions of whole communities of reef-building organisms and the myriad of organisms that depended on the reef structures for habitat. Subsequent recovery of reef-building communities requires millions of years.

It can be argued that humans are simply part of nature; that exploding human populations are naturally generating another episode of mass extinction from which the Earth will recover in 20 or 30 million years. But from a human perspective there is a difference between a mass extinction event caused by a meteor impact and the ongoing one caused by human activities. Human intelligence has reduced biological limitations on both the growth rate of human populations and the environmental damage that any individual human can cause. Is it too much to hope that human intelligence can be utilized to bring an end to the current human-generated mass extinction event before the most specialized communities, like reefs and rain forests, are lost to future human generations? After all, many so-called "primitive" human cultures, including Micronesian and Polynesian inhabitants of atolls and coral pinnacles, thrived in resource-limited habitats. Can "modern" humans develop a sustainable global society based on recognition of globally limited resources? Or do humans represent the latest in the series of "disaster" species that proliferate globally at mass extinction events?

3

Reefs as Dynamic Systems

Dennis K. Hubbard

Rising concerns over the decline of "reef health" have caused scientists to critically reevaluate what we do and do not know about reefs. In the 1960s, and again in the 1980s, a sudden outbreak of *Acanthaster planci* devastated large sections of the northern Great Barrier Reef in Australia (Endean, 1973; Moran, 1986). In 1983, the long-spined sea urchin *Diadema antillarum* underwent an unprecedented population crash throughout the Caribbean. In the 1970s and 1980s, the most important shallow-water coral in the Caribbean and western Atlantic, *Acropora palmata,* was nearly irradicated over large areas by an unknown pathogen simply known as "white band disease" (Gladfelter, 1982). Most recently, the identification of widespread bleaching of Caribbean corals has led to congressional hearings on the state of the world's coral reefs. These sudden events over the past decade have highlighted the complexity of factors involved as well as our fundamental inability to separate human-induced change from the effects of natural cycles, the duration of which may exceed the history of modern reef science.

As discussed in Chapter 2, reefs have evolved throughout geologic time, largely in response to natural changes on the planet that have either placed the existing community at risk or have favored one that was significantly different—the process of natural selection. An underlying tenet of geology is that the present is the key to the past; the best way to understand how ancient organisms or systems functioned is to examine their modern counterparts—uniformitarianism. While this approach is fraught with peril, it has repeatedly provided valuable insights into the workings of ancient biological systems.

This chapter turns the uniformitarian approach around by using past history to place recent changes into a larger spatial and temporal context. Despite the difficulties in this exercise, the examination of reefs from a geological perspective offers two advantages. First, geological studies encompass a larger spatial scale than is addressed by most biological investigations. A more holistic approach integrates the details provided by biologists into a larger picture, more on the scale of events that concern us today. Second, and probably more important, present-day reefs are the end result of a process that cannot be addressed by even the most careful of biological studies—time. While individual organisms respond to local processes that can be measured with great precision, the reef itself—its location, its shape, and its very existence—is the end result of a myriad of changing processes integrated over time. These have produced the solid mass beneath the present veneer of living coral, algae, and other organisms that more often come to mind. The following discussion takes advantage of this larger perspective in an attempt to understand the factors that have made modern reefs what they are, and by extension, might be driving the changes that we are now seeing.

The remainder of this chapter is divided into three parts. The first outlines the dominant factors controlling the location and character of modern coral reefs, and by extension, their fossil counterparts. Having established these basic controls, we will turn to how corals have built the massive structures beneath them. The third and final section attempts to integrate all this information into a consideration of reefs under the influence of global warming. The discussion is necessarily circular, because reefs serve to some extent as mediators of global climate while being strongly impacted as it changes.

3.1. Process: The Ultimate Control of Reef Development

We start by examining the factors that are generally thought to be most important in controlling reef location and type. For simplicity, these have been broken down into three groups, based on the scale at which they are most important, but keeping in mind that their boundaries often overlap. Macroscale processes are those that exert control either globally or over a very large area (i.e., thousands of kilometers) and over long periods of time. Primary among these are worldwide changes in sea level and large-scale movements of the Earth's crust (tectonics). Mesoscale processes are generally physical-oceanographic in nature and operate within individual oceans or basins (e.g., latitudinal changes in mean water temperature, wave-energy variations across an ocean basin) over a variety of time periods. Microscale processes are those that affect organisms at a reef-wide level. Conspicuously absent from this discussion, although no less important, are the organism-level processes that are discussed elsewhere in this volume.

3.1.1. Macroscale Controls

Tectonics

Our earliest studies of reefs focused on the role of tectonics. Workers in the Italian Alps during the nineteenth century (e.g., Mojsisovics, 1879) concluded that great upheavals of the Earth's crust must have been responsible for fossil reefs that occurred so far above sea level. It was also known before Darwin that, while modern reef corals do not generally exist in water any deeper than roughly 100 meters, many fossil reefs have attained thicknesses of hundreds or even thousands of meters. It was correctly presumed that this reflected the upward building of the reef by organisms as the foundation beneath them sank gradually—the process of subsidence.

Throughout the 1800s, the primary focus of the geological community was on subsidence and how it controlled the morphology of modern reefs. Darwin, in 1874, argued that the worldwide distribution of reefs could be explained by an understanding of the tectonics involved. He proposed that fringing reefs, barrier reefs, and atolls represented an evolutionary continuum from the former to the latter, with the underlying control being subsidence (Fig. 3–1). As a volcanic island slowly sinks, the fringing reef builds vertically as space is made available. The process continues, with the island progressively disappearing and

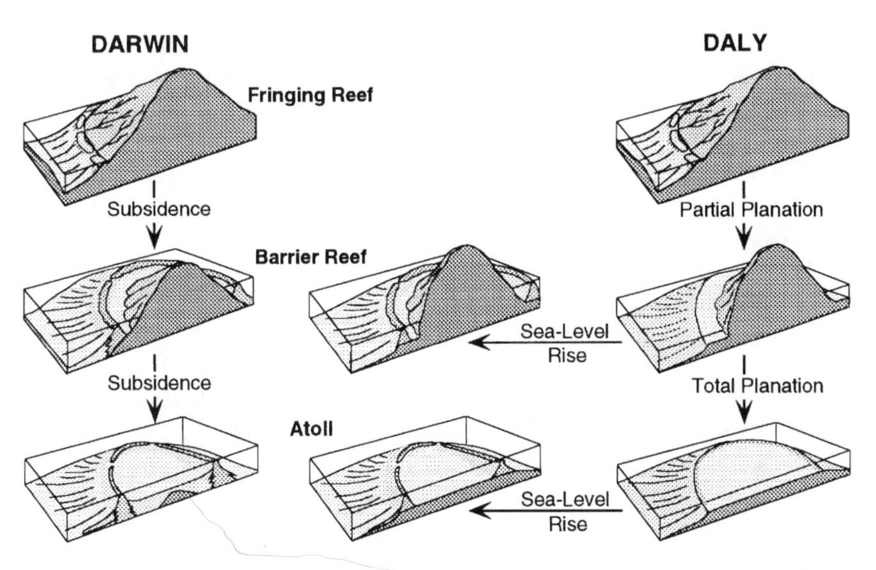

Figure 3–1. Two contrasting theories on the origins of fringing reefs, barrier reefs and atolls. Darwin (left) felt that the three were part of an evolutionary sequence he observed during his Pacific voyage on the *H.M.S. Beagle.* As the volcano subsides and the reef accretes vertically to stay near sea level, the fringing reef gradually evolves into a barrier reef and, eventually, an atoll. In contrast, Daly (right) envisioned changing sea level as the sole determinant of reef character. During a lowstand, waves plane off a platform. During the ensuing sea-level rise, reefs form near the edge of the new shelf. Whether a barrier reef or an atoll will form depends on the completeness of the planation process. In map view, the end results of the two mechanisms would be identical. The truth can be revealed only by examining the resulting accretionary sequence in cross section. Deep cores through Enewetak Atoll eventually supported Darwin's hypothesis.

the lagoon between it and the reef growing wider and deeper, until the island eventually sinks out of sight to form an atoll. From this, Darwin further argued that fringing reefs formed in areas of tectonic stability and that barrier reefs were reflective of subsidence.

While Darwin's ideas about the genetic sequence of fringing reefs to atolls were subsequently confirmed by deep cores through Enewetak and Bikini atolls, (Ladd and Schlanger, 1960), our growing knowledge of global tectonics has shown that Darwin's ideas about subsidence as a universal and sole control of reef type were flawed. Contrary to his prediction, the fringing reefs of the eastern Caribbean exist in an extremely volatile tectonic setting. Conversely, barrier reefs often occur in areas that are tectonically stable (e.g., the Great Barrier Reef of Australia). Nevertheless, the work of Darwin and his contemporaries provided many of the fundamental truths about reefs that form the foundation of our science today.

The relatively recent recognition of plate tectonics has provided a unifying theory to help explain the large-scale distribution of reefs on a global scale. The constant motion of giant fragments of oceanic and continental crust creates dynamic zones of seafloor spreading and crustal collision. Although the rates of movement involved will never affect transoceanic airfares, predictable patterns of uplift, subsidence, or stability have emerged as important, long-term controls of reef development. For example, the tropical ocean formed a continuous girdle around the equator during Jurassic time. The subsequent closing of this "Tethys Seaway" broke the continuum and led to a substantial separation between Atlantic and Pacific reef fauna.

The apparent decline of reefs in recent years is nothing new. The present Caribbean community is but a remnant of a cosmopolitan assemblage that existed during Oligocene time (37 to 24 million years ago). In relatively recent geologic time (Pliocene), the Isthmus of Panama separated the Pacific and Caribbean provinces (Vermeij, 1978, 1982). As a result, the diversity of Caribbean and eastern Pacific reefs has dropped continually (Frost, 1977). A better understanding of the community dynamics of these and other changes in reefs through time holds important information that might allow us to separate natural from human-induced change in reefs today.

Sea Level

Periodic changes in the spatial relationship between Earth and the Sun have resulted in predictable and recurring changes in global climate. As our planet moves closer to the Sun, climate warms. Freshwater is added to the oceans from melting glaciers, and sea level rises. During cooler periods, sea level falls. Climate-driven fluctuations in sea level have occurred throughout geological time, but appear to have become more pronounced after the breakup of the single supercontinent, Pangaea, 250 to 200 million years ago. With each waxing and waning of sea level, reefs that lived near the upper limit of each sea-level rise were alternately flooded and exposed. The result is a series of reefs, each built upon the remains of its predecessor, much like the ancient cities of Greece and Rome stacked one upon another. Present-day reefs, which have formed over the past 10,000 years, are simply the latest participants in this cycle of colonization and abandonment.

In 1915, Reginald Daly proposed an alternative to Darwin's subsidence theory. It centered around the formation of a wave-cut bench during each drop in sea level (Fig. 3–1), followed by a subsequent sea-level rise and reef colonization on top of the resulting terrace. This idea was born from his observations that, in most of the areas he examined, the lagoons were of similar depth. He argued that this reflected wave cut platforms at similar depths from site to site, and proposed that the only plausible mechanism to explain this was a worldwide drop in sea level.

The main blow to Daly's ideas came in 1923, when William Morris Davis pointed out the lack of cliffed shorelines that should be commonplace behind the barrier reefs of Pacific islands were Daly's hypothesis correct. Without these logical remnants of the wave-planation process, the sea-level hypothesis of Daly became untenable.

Despite the demise of Daly's theory as a universal explanation for barrier reefs and atolls, geologists have come to appreciate the importance of sea level as a major control of reef development, both today and in the geologic past (see Davies and Montaggioni, 1985, for a summary). Equally important is the recognition that sea level can operate in conjunction with local tectonic motions to exert a complex control over the ability of reefs to form in the first place as well as the character and history of the structures that they build.

Based on thousands of kilometers of seismic profiles (records that show the stacking of geologic strata in much the same way that echo sounders show the shape of the present seafloor), it has become possible to match recurring depositional patterns with episodes of either shelf flooding or sea-level fall (Vail et al., 1977). The common characteristics of these events around the globe argue for a control by worldwide sea level (glacial eustasy). By identifying the patterns common to depositional sequences from many different ocean basins, seismic stratigraphy has been used to construct a record of global sea-level change through time (Haq et al., 1987). Careful analyses of the variability of this pattern from site to site have allowed us to understand the superimposed effects of local tectonics on a site-to-site basis. The term relative sea level refers to the change in sea level at any one locality resulting from the combined effect of glacio-eustatic sea-level rise and fall (that related to melting and freezing of polar ice) and local tectonic motions.

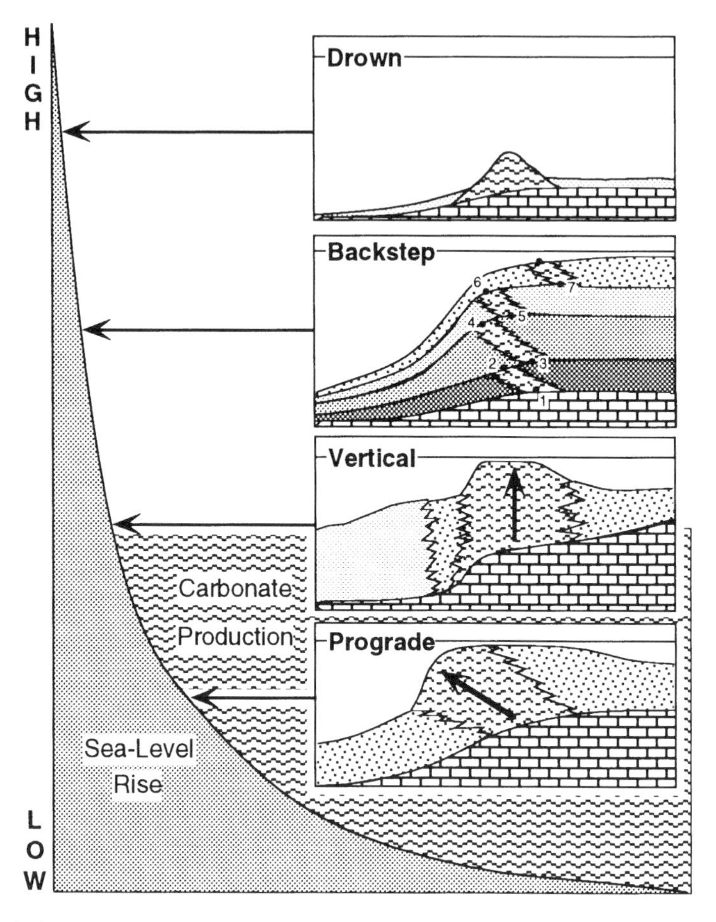

Figure 3–2. Diagram illustrating the accretion of a reef under varying rates of sea-level rise. If sea level is rising slowly relative to the reef's ability to produce carbonate, then the reef will prograde seaward as it aggrades upward. A near balance between the accommodation space created by rising sea level and the carbonate being produced within the reef will result in vertical accretion. As sea level outpaces the accreting reef, it will either backstep to higher ground or "drown."

Reef Accretion Under the Influence of Relative Sea-Level Change

Each time sea level rises over the edge of a platform in the tropics, reefs will usually form. The site of colonization will be dictated by factors discussed later in this chapter. The general pattern of reef development will depend on (1) the rate of relative sea-level rise compared to the upward accretion of the reef, and (2) how long sea level rises at a rate that exceeds the ability of the reef to keep up. In areas where carbonate production is high compared to the rate of sea-level rise, the reef produces more material than is needed to keep pace and the reef will prograde (Fig. 3–2). Under conditions of near balance, the carbonate produced in the reef just offsets sea-level rise, and reef accretion is vertical. When carbonate production cannot keep pace with rising sea level, the reef will either retrograde, backstep, or drown.

In 1985, Neumann and Macintyre proposed a tripartite classification for reefs viewed in cores or outcrop; reefs are inclined to either "keep up," "catch up," or "give up" (Fig. 3–3). "Keep-up" reefs were able to maintain their crests at or

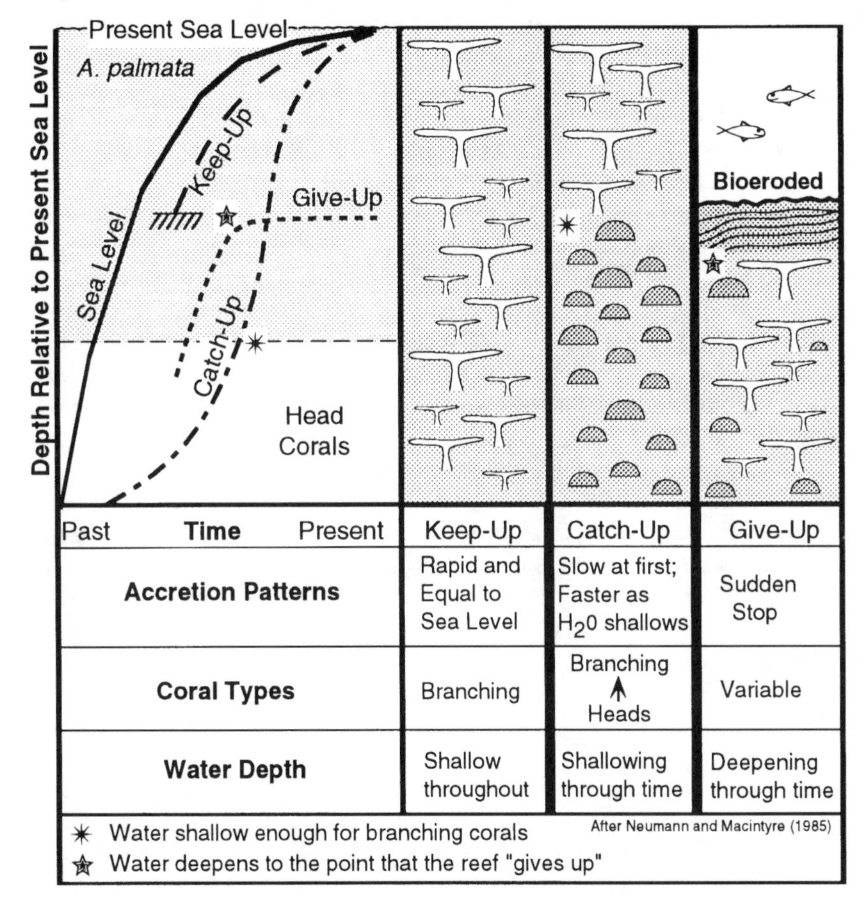

Figure 3–3. Reef sequences as indicators of sea-level rise. A "keep-up" reef is capable of matching the rate of sea-level rise. It will be comprised almost exclusively of shallow-water corals. "Catch-up" reefs generally started in deeper water (head corals) and gradually accreted closer to sea level (branching corals), often after sea level slowed in its rise. "Give-up" reefs can be highly variable because their character will reflect the conditions before they gave up as well as those responsible for their demise.

near sea level throughout their history. As such, the preserved corals (or other carbonate producers) exhibit growth forms typically found in shallow water (e.g., shallow-water branching corals). "Catch-up" reefs started in deeper water (ca. 10–20 m), but later caught up, usually after the rate of sea-level rise slowed down. These reefs are characterized by a lower (older) section dominated by deeper-water organisms, topped by younger biota that formed in progressively shallower water as the reef "caught up." "Give-up" reefs are those that, for whatever reason, simply stopped accreting. The two most common reasons for this sudden cessation are (1) a sudden rise in sea level that increases water depth to the point where net carbonate production can no longer keep up, and (2) a sudden change of oceanographic conditions such that the environment is no longer conducive to rapid carbonate production. Possible contributors to the latter inimical conditions (Fig. 3–4) include sediment stripped from an updrift shelf (Adey et al., 1977), elevated nutrients (Hallock and Schlager, 1986), and sudden changes in temperature (Hudson, 1981; Walker et al., 1982).

The recognition of these kinds of relationships allows us to better understand the large-scale factors that have shaped reefs throughout geologic time. Where sea-level history can be derived from some evidence other than that contained in the reef itself, patterns of accretion contrary to those that might be expected based on sea level alone can provide valuable information about other factors that may have come into play.

It has recently been proposed that worldwide sea level will rise at an accelerated rate by the end of the century, due to increasing levels of "greenhouse gases" in the atmosphere and the resulting global warming. Concerns over this possibility underscore the importance of understanding the factors that influence carbonate production in a regime of changing sea level. Will present-day reefs be able to keep pace with this rise or will they be left behind? Answers to such questions are of great importance culturally and economically as well as scientifically.

3.1.2. Mesoscale Controls

Temperature

Coral reefs are generally restricted to water between 18° and 36°C, with an optimal range of 26–28°C. This is expressed in latitudinal patterns of coral-reef distribution and diversity (Fig. 3–5). Within this range, certain corals will grow faster or slower, depending on temperature (Weber and White, 1974; Glynn and Stewart, 1973). Drastic thermal shifts can result in reduced coral vitality (e.g., bleaching, reproductive inhibition) or, in extreme instances, total destruction of entire reef systems.

It is important to note that most corals exist near their upper thermal limits. Therefore, even a slight increase in tropical temperatures in the future could have significant impact on the distribution of corals in the tropics. Glynn (1984)

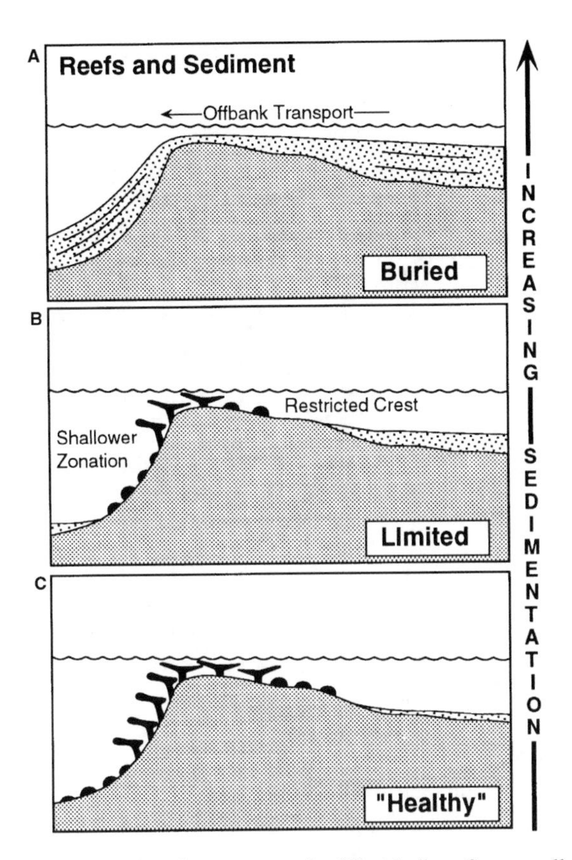

Figure 3–4. The effects of sediment on reefs. "Healthy" reefs generally exhibit well-defined zonation, with factors such as wave energy exerting the dominant control on coral type **(C).** In areas of intermediate sedimentation **(B)** the border between forereef zones often becomes shallower, a response to reduced light levels. Also, backreef zones will become more depauperate in response to increased sediment loading. In extreme instances, an updrift sediment supply can actually bury a reef **(A).** This usually occurs along the downwind flanks of broad platforms and is often a response to natural causes. As sea level floods the adjacent shelf, sediment is swept off the shelf, covering reefs that flourished when the banktop was exposed.

emphasized the importance of unusually high water temperatures in the widespread mortality of corals along the western coast of Panama. Recent episodes of coral bleaching, where corals expel their zooxanthellae (described in Chapter 5), may be related to higher-than-normal maximum temperatures over the past several years. At the other end of the scale, Walker et al. (1982) proposed that well-developed stands of *Montastrea annularis* in the northern Florida Keys are limited to areas shielded from the periodic influx of cold water pushed out from Florida Bay during the passage of major cold fronts.

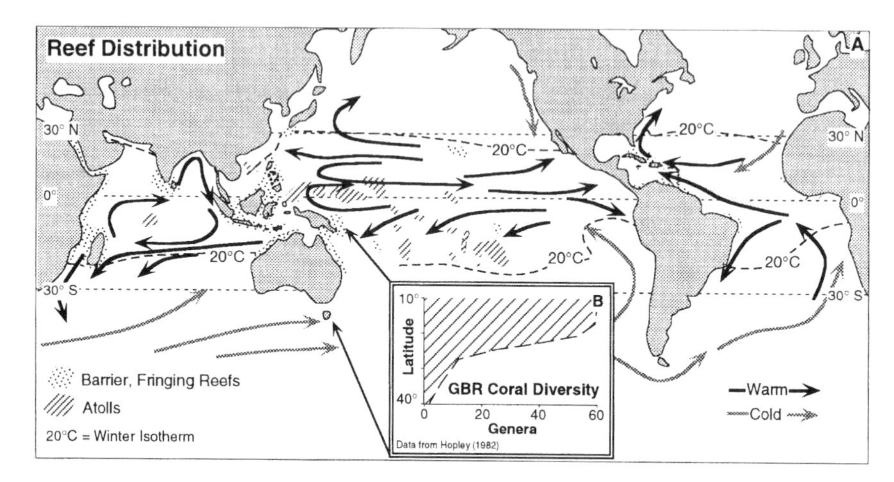

Figure 3–5. **(A)** Worldwide distribution of reefs. Reefs are confined to areas washed by warm currents (>20°C). **(B)** Latitudinal variation in species diversity along the GBR. Note the increase in the number of genera toward the equator.

On a geological scale, long-term patterns of reef development can be explained as a response to temperature changes, often in conjunction with tectonic effects. Grigg (1982) proposed that north of roughly 29°N in the Hawaiian Archipelago, lower temperatures have depressed carbonate production to the point where reefs cannot keep up with subsidence (Fig. 3–6). South of this "Darwin Point," barrier reefs and atolls are maintained. To the north, reefs are in various stages of "giving up." Cores through the Great Barrier Reef of Australia record a gradual shift from subtropical seas to a more tropical climate over the past 30 million years as the Australasian plate slowly moved toward the equator (Davies et al., 1987).

Salinity

Coral reefs are limited to areas of reasonably normal marine salinity (3.3–3.6%). Below these levels, carbonate buildups are progressively dominated by vermetids, oysters, serpulids, and blue-green algae (Teichert, 1958; Heckel, 1974). Reefs do not generally occur above this range. Low salinity (and turbidity) is a primary reason why extensive coral reefs do not occur opposite the mouths of major rivers (e.g., the Amazon and Orinoco rivers of northern South America empty into seas that are otherwise suitable for reef development). On a smaller scale, the passes through many nearshore reefs can be controlled by the present or past locations of streams.

Wave Energy

The oceanographic regime in which a reef occurs is among the most important determinants of its character. Breaking waves generate currents that move nutri-

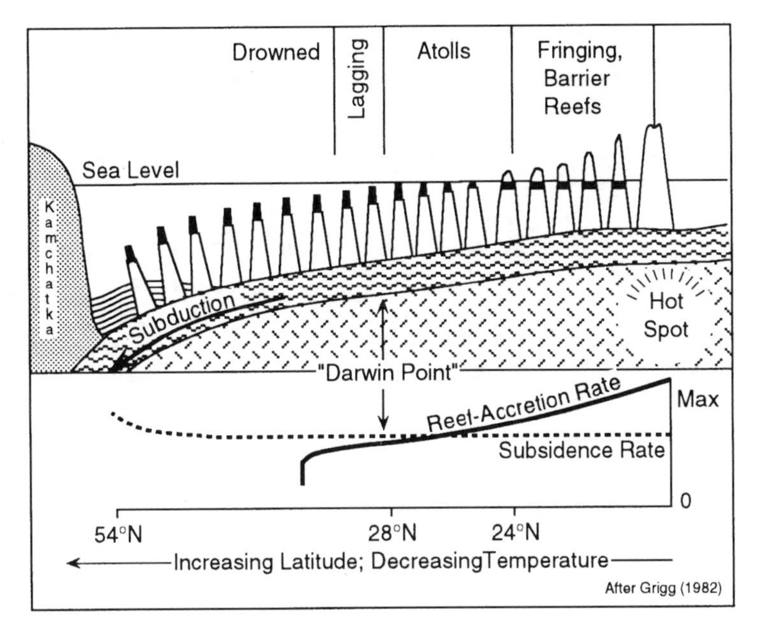

Figure 3–6. The combined effects of subsidence and temperature on reef development in the Hawaiian Archipelago. As the Pacific crust beneath the Hawaiian islands moves to the northwest, it cools and subsides. South of roughly 28°N, the reefs can produce carbonate at a rate fast enough to offset subsidence. North of this parallel, however, cooler water inhibits coral growth to the point that the reefs can no longer keep up. As a result, they are gradually drowned.

ents and waste products through the system. Benthic zonation within a single reef is to a large extent determined by the relative abilities of various organisms to either survive the turbulent conditions of breaking waves or to take advantage of the vigorous water motion caused by passing waves.

Prevailing wave energy has been shown to exert a primary control on the character of the shallow-water reef crest and the zonation along the front of Caribbean reefs (Adey and Burke, 1977; Geister, 1977; Fig. 3–7). In addition, storms can play an important role in determining reef character. In the Caribbean, three primary hurricane tracks emerge from data compiled by Neumann et al. (1981) over the past century (Hubbard, 1989; Fig. 3–8); reef type is quite sensitive to this pattern. In the Pacific, the location of Guam in the western monsoon trough correlates highly with the "topographically dull" reefs offshore and the architectural preponderance of "square cement bunkers with small windows and large shutters" on land (C. Birkeland, pers. commun.).

Caribbean reefs can be roughly divided into three types (Hubbard, 1989) that are responding to (1) storm frequency and (2) the regional distribution of total wave energy computed from synoptic wave observations for the region (U.S.

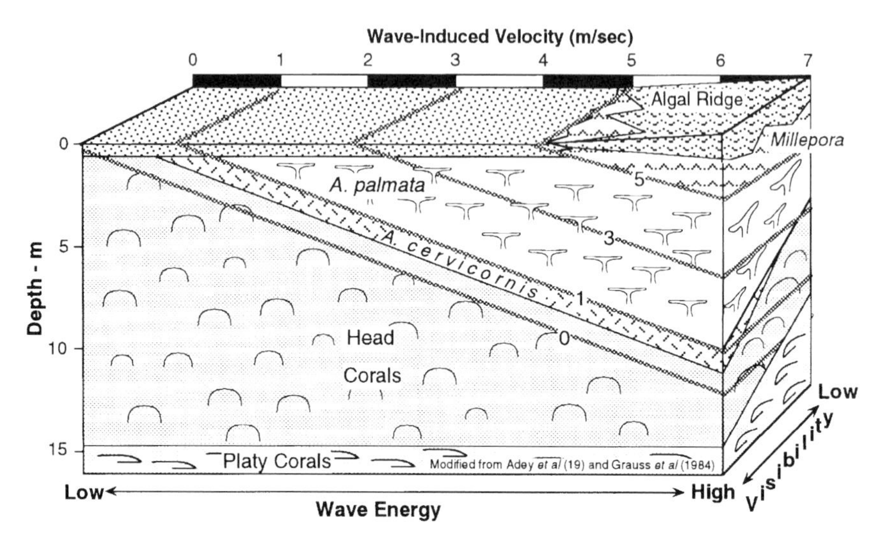

Figure 3–7. Diagram summarizing the effects of wave energy and water clarity on the zonation of Caribbean reefs (after Adey and Burke, 1977). With increasing wave energy, the boundaries between the *A. palmata,* the *A. cervicornis,* and the head-coral zones are progressively deepened. This is closely tied to magnitude of the wave-induced velocity that increases with larger waves (boundaries from Grauss et al., 1984). Turbidity can also affect coral zonation. Increased sediment in the water drives corals into ever-shallower water, seeking light. The shallowest (algal ridges, *Millepora*) are less affected by sediment. They occur in such shallow water that light attenuation is unimportant. Platy corals are less sensitive to changes in wave energy because of the greatly attenuated surge at depth.

Naval Weather Service Command, 1979). These correspond strongly to a pattern originally identified by Adey and Burke (1977).

Areas of high prevailing wave energy and moderate-to-frequent hurricane disturbance (Type I) are characterized by algal ridges (Fig. 3–8), reefs dominated along their crests not by coral but by coralline algae. Frequent storm disruption breaks down branching *Acropora palmata* colonies and compacts them into piles upon which coralline algae can recruit. High wave energy on a day-to-day basis discourages grazing by fishes that would otherwise inhibit the accumulation of thick algal crusts under calmer conditions. The best examples of these types of reefs occur along the eastern shores of the Windward Islands (Adey and Burke, 1977), exposed to open-ocean swell and directly in the path of most hurricanes that pass into the Caribbean along the southern storm route. Algal ridges can also be found along the southern shore of St. Croix. While somewhat removed from the southern storm track, this area receives long-period swell (ht ~ 5 m; T > 10 sec) every time a major storm passes within 100 km of the island, frequently disrupting the resident *A. palmata* community (Rogers et al., 1982).

Areas of high prevailing wave energy but less frequent disruption by storms

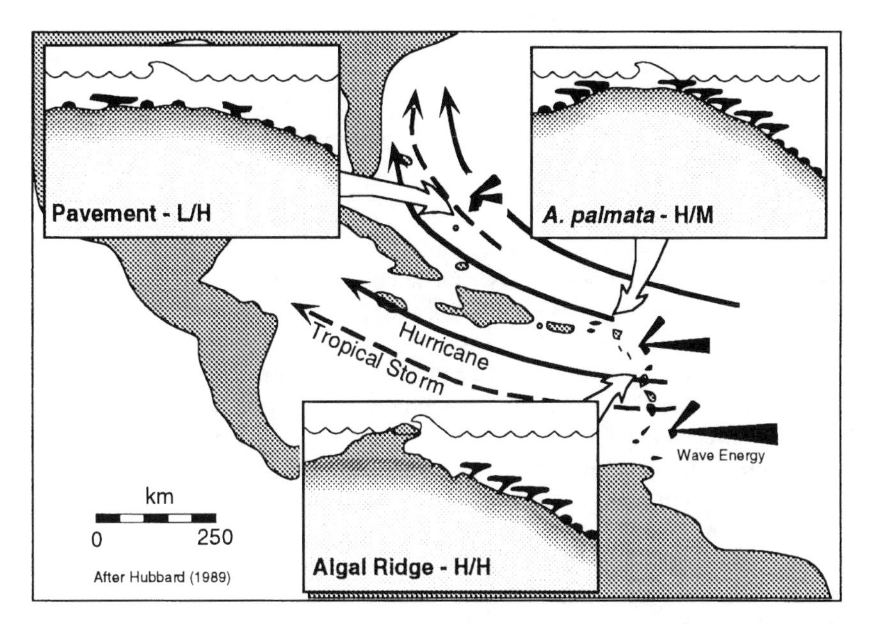

Figure 3–8. Wave energy as a control of Caribbean reef type. Major storm tracks are shown by the solid and broken arrows. Wave energy is shown graphically by the rose diagrams at individual sites. The importance of ambient wave energy and major storms (e.g., L/H = Low ambient wave conditions/High hurricane frequency) and the resulting reef types are shown in the insets. Areas where both ambient wave energy and hurricane frequency are high are characterized by algal ridges (e.g., the Windward Islands). Where day-to-day wave energy is still high but hurricanes are less frequent, well-zoned and more diverse reefs occur (e.g., St. Croix). Areas that are generally less energetic except for the frequent passage of major storms (e.g., the Bahamas) are characterized by broad wave-swept pavements.

(Type II) are dominated by branching *A. palmata.* From this, it would appear that the important distinction is the lack of frequent destruction by passing storms. Excellent examples of this type of reef occur along the north shore of St. Croix, as well as in more protected sites in the Florida Keys and throughout the western Caribbean.

Areas of low prevailing wave energy and frequent storm disruption (Type III) are dominated by open pavements with only scattered coral cover. As with Type I reefs, frequent disruption by storms discourages the establishment of a diverse and abundant reef-crest community. Low wave-energy conditions in the interim permit grazing, which effectively discourages the thick algal crusts seen in Type I reefs and probably reduces the success rate of coral recruitment as well. Type III reefs are best developed in the northern Bahamas (e.g., southern Little Bahama Bank: Hubbard et al., 1974). Reefs dominated by *A. palmata* (Type II) can be found in these areas, but they generally occur as refugia on reefs protected from the brunt of passing storms.

3.1.3. Microscale Controls

Light

The importance of light intensity and spectral character has been generally understood since the work of Quoy and Gaimard in 1825. Darwin (1874) clearly recognized its importance in controlling the maximum depths to which he observed corals throughout the Pacific.

More recent studies have quantified this long-recognized relationship. In clear water, light intensity decreases exponentially with water depth and the light spectrum shifts rapidly toward its blue end. As a result, photosynthesis and $CaCO_3$ production (which is partly dependent on it) drop off as well. In addition to changes in growth rate, reduced light at greater depths can cause polymorphic corals to change their shape from mounds in shallow water to plates at greater depths. The latter is more efficient for light gathering as it places all the photoreceptors (symbiotic algae within the coral tissue) along vertically facing surfaces, where light intensity is greatest. Perhaps the best understood coral in this respect is *Montastrea annularis* from the Caribbean. Its growth rate drops from nearly a centimeter per year in shallow water (0–10 m) to a millimeter or less in deeper water (>12–15 m: Dustan, 1975; Hubbard and Scaturo, 1985). This is accompanied by a change in morphology from heads to plates (Grauss and Macintyre, 1982). Light dependence has also been shown for *Porites lutea,* a similarly important coral from the Pacific (Buddemeier et al., 1974; Isdale, 1981, 1984a).

Reef zonation is in large part controlled by the ability of various corals to adapt to conditions of either very high or very low levels of light (Adey and Burke, 1977; Grauss and Macintyre, 1982). The importance of light is reflected in the success of a recent model of reef accretion that depends solely on light intensity as a controlling variable (Bosscher and Schlager, 1992). While accretion in the reefs used for model verification has probably been the result of several factors that roughly covary with light intensity, the general agreement between the predictions and actual accretion patterns documented from cores is striking.

Nutrients

Until recently, nutrients have generally been considered beneficial to reefs. Early references to reefs preferring areas of upwelling or other sources of nutrients underscore this misconception. More recently, it has been recognized that high nutrient levels are actually detrimental to "reef health" (Kinsey and Davies, 1979). Hallock and Schlager (1986) proposed that elevated nutrient levels may have been responsible for widespread reef degradation in the Cretaceous (144 million to 68 million years ago), and suggested that nutrient availability has been greatly underrated as a primary control of reefs over large spatial and temporal scales. In a natural experiment in Kaneohe Bay, Hawaii, reefs were severely damaged after the installation of a sewer outfall near the reef (Johannes, 1975). Since the discharge has been eliminated, the reef has shown signs of recovery (Smith et al., 1981).

The role of nutrient inhibition in coral reefs is multifaceted. On the organism level, it has been proposed that high phosphate levels in the water may shut down the calcification mechanism (phosphate "poisoning" of Simkiss, 1964). At the community scale, higher nutrient levels tend to favor sponges (Wilkinson, 1987a, b) and algae (Steneck, 1988), which can outcompete corals for space and prevent larval settling. And once the coral dies, higher levels of nutrient availability favor heavy infestation by infaunal borers such as *Cliona* spp. (Moore and Shedd, 1977), which will progressively destroy the remaining skeleton and can remove any record of its existence (Chapter 4).

Sediment

Despite an impressive body of literature (for reviews, see Hubbard and Pocock, 1974; Hubbard, 1987), little quantitative information exists on the specific responses of reef organisms to sediment loading. Laboratory experiments have documented surprising tolerance by corals to high doses of sediment over short periods of time (Taylor and Saloman, 1978; Rogers, 1983). Nevertheless, the literature is replete with what amount to postmortem autopsies of reefs killed by sediment. The obvious factor here is time. Although corals can survive short-term loading at very high levels, even lower-level stress over an extended period of time can gradually wear down the reef's defenses.

The four most important types of sediment stress are (1) smothering, (2) abrasion, (3) shading, and (4) inhibition of recruitment. Of the three, smothering is the easiest to visualize. Under natural conditions, reefs on the downwind flanks of large carbonate platforms can be buried by sediment derived from the banktop (Hine and Neumann, 1977; Fig. 3–4). During storms, or more recently, dredging of nearby areas, the levels of suspended sediment can increase markedly, resulting in extensive damage to reef corals and other sediment-sensitive biota. Such problems have been described in Australia (Fairbridge and Teichert, 1948), Hawaii (Johannes, 1975; Maragos, 1972), Puerto Rico (Kaye, 1959), and the U.S. Virgin Islands (van Eepol and Grigg, 1970; Dubois and Towle, 1985).

During storms, physical abrasion by moving sediment can cause substantial damage to coral tissue (Hubbard, 1992b). Even under less energetic conditions, sediment scour can play a role in limiting the types of corals that can exist on the shallow reef crest. Although head corals are more resistant to physical disruption by wave action, their slow growth rates virtually guarantee that wave-induced sediment scour will severely damage or kill a young colony before it can grow above the level of frequent sediment motion. In contrast, rapidly growing branching corals can elevate themselves above this traction carpet very quickly.

While more subtle in its effects than abrasion or smothering, shading is probably the most important of all the sediment-related effects. The reduced levels of light due to suspended sediment in the water column can reduce coral growth (Hubbard and Scaturo, 1985; Hubbard et al., 1986), impact natural zonation patterns (Fig. 3–4B; Morelock et al., 1979; Hubbard et al., 1986), and induce wholesale mortality if allowed to persist for an extended period of time.

Excess sedimentation can also discourage the settlement of coral larvae. While this is generally accepted, it is difficult to quantify. Morelock et al. (1979) discussed the importance of substrate type in larval recruitment in Puerto Rico. Roy and Smith (1971) proposed that on Fanning Island, the increased vulnerability of young corals to sediment damage was a more limiting factor than sediment covering available space.

All of these effects can act together to exert a significant natural control on the distribution of coral reefs. Hubbard (1986) showed that, on a local scale, the presence or absence of an updrift source of sediment exerts perhaps the greatest control on the location and character of reefs on the north coast of St. Croix. Along a reef system off Costa Rica, Cortes and Risk (1985) proposed that coral growth (and probably cover) has been gradually reduced by increasing development pressure, and specifically widespread agriculture and logging since the late 1950s. Under the influence of humans and ever-increasing population pressure, elevated levels of suspended sediment (and nutrients) represent perhaps the greatest single threat to nearshore reefs that has been documented to date (Johannes, 1975; Hatcher et al., 1989) including possible global warming.

Antecedent Topography

Like ancient cities, many reefs are built upon the ruins of their predecessors (Fig. 3–9). Because of the fierce competition for space, and the necessity for reefs to rid themselves of excess sediment produced by bioerosion, topographically elevated areas offer significant benefits to larval recruits and well-developed reefs alike. As sea level falls and rises again, the remnants of the last generation of reefs serve as areas of preferential coral recruitment. Thus, reef sequences that are recognized in the fossil record are often not single depositional units, but rather a complex of several generations of reefs, each localized atop the remains of an earlier one. Similarly, many of the present-day reefs sit astride their Pleistocene ancestors that formed 120,000 years earlier (e.g., Fig. 3–9).

In addition to older reefs, there are many other origins of antecedent topography. These remnant highs can form along the edges of tectonically created blocks, cemented sand bars, fossilized dunes, and even ancient river deltas.

One common control of antecedent topography is a process called karsting. During episodes of lowered sea level, reactive limestone strata are dissolved by

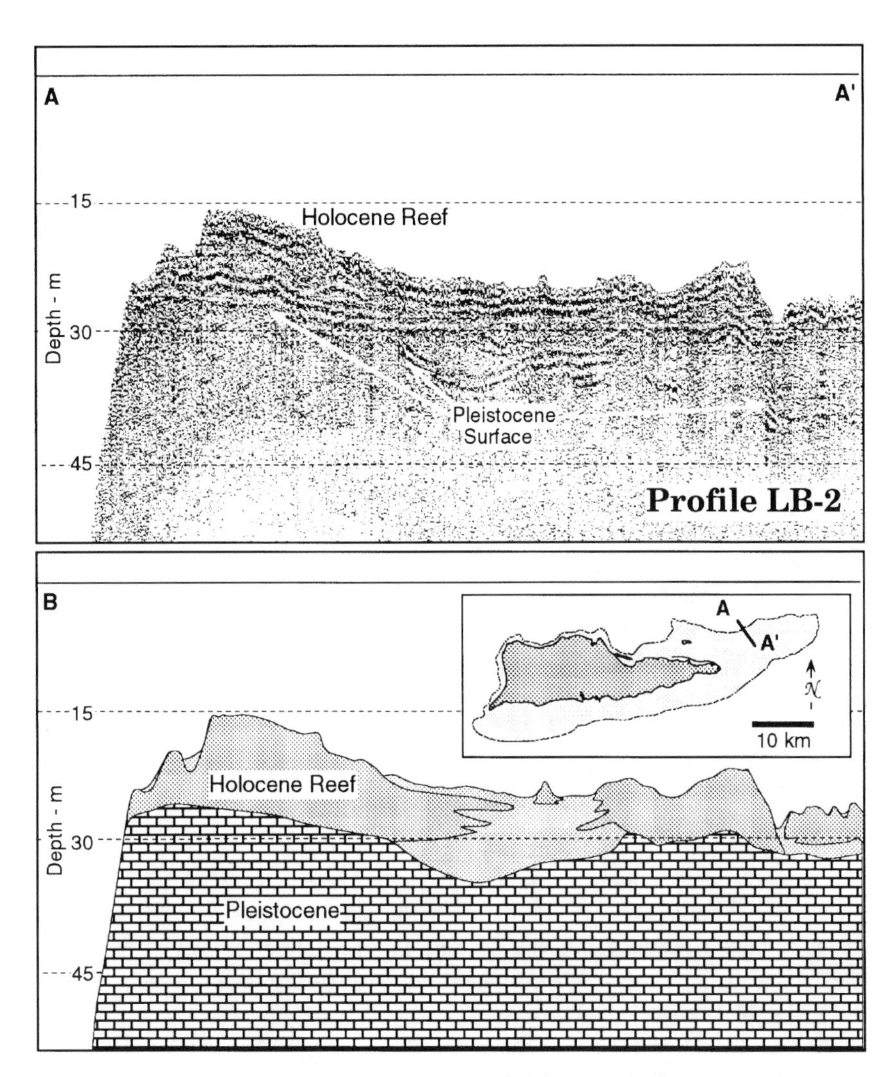

Figure 3–9. Antecedent control of present-day reefs. **(A)** Seismic line across the shelf east of St. Croix, U.S. Virgin Islands. **(B)** Interpretation of the seismic profile in **(A).** Note how the Holocene reefs sit atop highs on the underlying Pleistocene surface. The location of the profile is shown in the inset.

rainwater. Based on experimental studies and numerous field examples, Purdy (1974) proposed karsting as the primary mechanism for creating antecedent topography. Because many reefs are reinforced by aragonite and magnesian-calcite cements formed within their interstices while the reef is still accreting, the resulting mass is more resistant to karsting than are the muddier and uncemented sediments of the platform interior.

3.2. Reef Accretion: How It Works

The earliest "reefs" were defined by mariners as anything on which they could damage their keels. These included biologically produced reefs, the subject of this book, as well as piles of rocks left in shallow water by the last glaciation or the U.S. Army Corps of Engineers. Geologists largely characterize reefs based on their shape when viewed in cross section and their relation to surrounding strata. Biologists are understandably more concerned with organisms and ecological interactions among them.

For purposes of discussion, we consider the following to be the critical characteristics of reefs, both in modern tropical seas and in environments preserved in the fossil record. First, reefs owe their origin mainly to biological production. While the origin of the cements that infill voids within the reef and add to its rigidity is still a subject of debate, there is little argument that the bulk of the carbonate contained within reefs is ultimately of biological origin. Second, reefs are rigid structures that stand topographically above surrounding environments and, therefore, exert some degree of physical control over the oceanographic processes in their vicinity. This sets "true reefs" aside from isolated stands of coral that may be ephemeral in nature and shift from spot to spot. To the biologist, the key here is the feedback mechanism between the reef and oceanographic process. For the geologist, the critical distinction is that a recognizable "reef core" will be preserved as a moundlike structure relative to adjacent flat-lying beds—a bioherm.

While this seems to be a simple set of criteria, there is still considerable disagreement among geologists as to what is and what is not a reef. The focus of the argument centers around the term framework. It is a generally held principle that modern coral reefs are dominated within their interior by in-place and interlocking coral framework. Furthermore, it is assumed that the rigidity of the reef is largely a result of that interlocking internal fabric (e.g., see reviews of Heckel, 1974; Fagerstrom, 1987; Stanley and Fagerstrom, 1988). As a result, in-place and interlocking framework has become a prerequisite for a geologic assemblage to be called a true reef; or conversely, anything not containing these attributes is not a reef (Fagerstrom, 1987).

Attempts to apply this framework-based classification to the rock record have been fraught with problems, however. The vast majority of ancient reef deposits are comprised not of in-place and interlocking framework, but rather of loose assemblages of reef-building organisms usually "floating" in a matrix of reef-derived debris. This has necessitated models in which these deposits are "near to" or "related to" reefs in the strict sense, but are not the reefs themselves. Furthermore, because of the fundamental differences between so many ancient reefs and the framework model derived from their modern counterparts, an opinion has been expressed that modern reefs are in fact poor models for their ancient brethren. Possible explanations for this difference between modern and ancient reefs have included: (1) evolution of reef organisms through time, (2) oceanographic differences between shallow, epicontinental seas that fostered many ancient reefs and the more open and exposed ocean margins inhabited by modern reefs, and (3) fundamental changes in the chemistry of oceanic waters through time.

With the introduction of the submersible rock drill by Ian Macintyre in 1975, geologists have increasingly gained access to the interiors of modern reefs. Although some reefs contain a high proportion of what may be in-place "framework" (a possible example is Geleta Reef off western Panama: Macintyre and Glynn, 1976), most of the cores from modern reefs have contained what is arguably out-of-place coral debris mixed in with a high proportion of loose sediment and debris (Hubbard et al., 1986, 1990; Hubbard, 1992a). Furthermore, the highest recovery rates are often associated with deeper and "less productive" reefs, rather than their shallow-water counterparts (Fig. 3–10). It appears, therefore, that in-place and interlocking framework is unreasonable as a prerequisite for classification as a true reef. Furthermore, careful examination of many reefs reveals that their rigidity is often as dependent on secondary overgrowth by organisms such as coralline algae and bryozoans and penecontemporaneous cementation as it is on the structural continuity afforded by a meshwork of in-place framework.

The emphasis on in-place framework, although flawed, is understandable. Most of our early geologic models of modern reefs are based on surface observations extrapolated to the reef's interior. It is difficult, while swimming over the magnificant architecture of reefs in either the Caribbean or the Indo-Pacific, to envision them as being constructed of anything other than corals growing on the backs of other corals. The flaw in this approach is in assuming that carbonate production by the corals that dominate our mental picture of reefs is the only process responsible for reef accretion or that it is at least of overwhelming importance.

In addition to this initial carbonate production, there are many other processes that contribute to the ultimate character of the preserved reef. Foremost among these is bioerosion, discussed in detail in Chapter 4. It has long been recognized that once they die, corals are subjected to both physical and biological breakdown. Algae that overgrow the dead substrate are a preferred diet of many fishes, gastropods, and other marine grazers. In the process of scraping or biting away these algae, grazers also remove bits of carbonate substrate. Alongside the grazers, countless other organisms bore into the reef, seeking shelter (Fig. 3–11).

The internal fabric of a reef is determined by the relative importance of (1) initial carbonate production, (2) the type and intensity of bioerosion, (3) the patterns of sediment storage within or removal from the reef, and (4) levels of secondary cementation and encrustation. Most of the building blocks of the reef started as coral, but they have all been, to some extent, replaced by multiple generations of boring, sedimentary infill and cementation of those fills. This cycle can occur several times over a period of only a few years and can result in

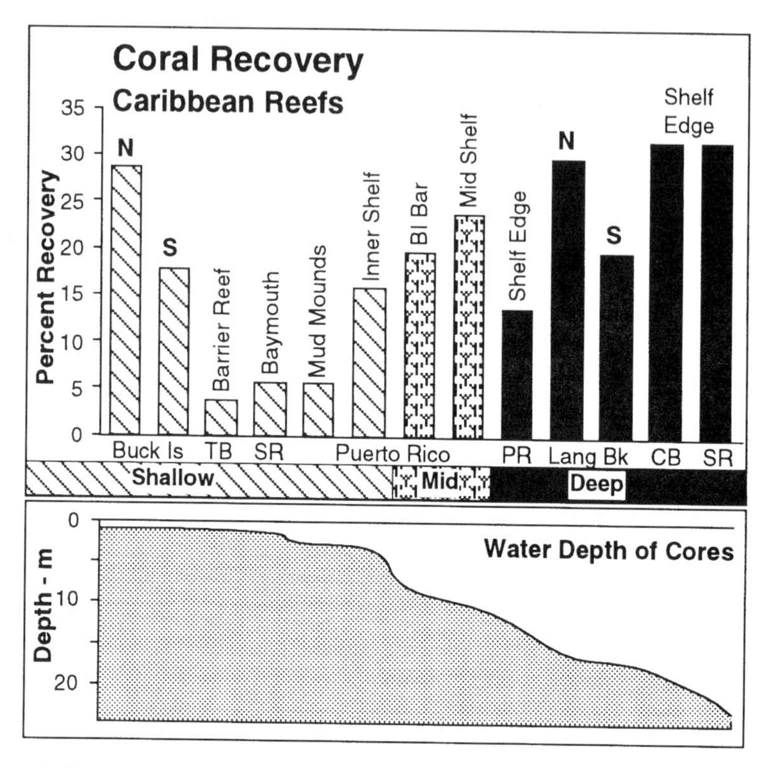

Figure 3–10. Core recovery from several Caribbean reefs. The bottom inset is a composite profile showing the depths from which each set of cores was recovered (i.e., cores are not from a single site). Note that recovery is less than 35% and the highest recovery is generally found in deeper reefs that are accreting more slowly. Also, north-facing (i.e., more exposed to ambient waves) reefs generally have better recovery, perhaps reflecting more robust coral growth. The majority of the reef fabric in all cores consists of open voids and sediment produced by bioerosion and *not* in place and interlocking framework.

sections of reef with a fabric that contains few reminders of the original corals (Schroeder and Zankl, 1977; Schroeder and Purser, 1986).

Starting in the mid-1970s, investigators attempted to quantify the relative importance of these processes in reefs. Stearn and Scoffin (1977), Land (1979), and Sadd (1984) provided estimates of the relative importance of carbonate production and bioerosion on the reefs of Barbados, Jamaica, and St. Croix, respectively. The drawbacks to these earlier efforts, however, were a lack of information on critical parts of the "carbonate budget" and the necessity of extrapolation or assumption. The most serious gap was the lack of core data that were needed to test the assumptions of the proposed budgets.

Although self-admittedly flawed, the study of Land (1979) represented a landmark in our understanding of reef development in that it provided a quantitative

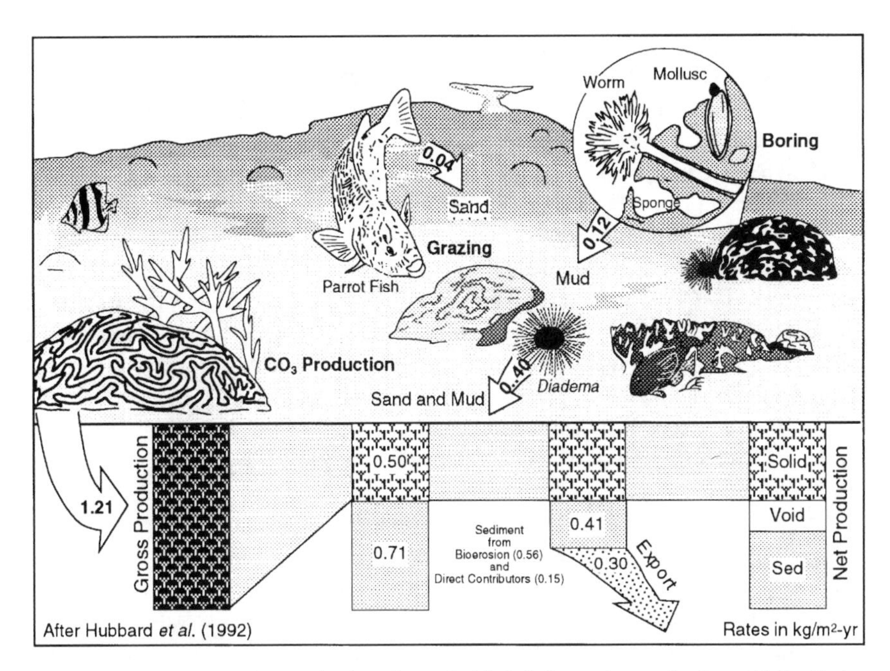

Figure 3–11. The carbonate budget for a shelf reef along the north shore of St. Croix. Various carbonate producers (mostly corals) add 1.21 kg of carbonate per square meter of bottom to the system annually (kg m^2 yr). Of that, 0.56 kg m^2 yr are reduced to sediment by grazers and borers; this is added to the 0.15 kg m^2 yr of sand derived from molluscs, forams, and other small organisms that live and die on the reef. On average, slightly over 40% of this sediment is exported from the reef each year, with most of that occurring during the passage of major storms. This is all reflected in a reef fabric comprised less of recognizable coral (42%) than of sediment and void space (58%). Furthermore, much of the recognizable coral is demonstrably moved from its original location.

model that could be subsequently tested. Land argued that what is found within the reef falls mainly into two categories: original coral skeletons and sediment derived from their physical and biological breakdown. The total volume of carbonate initially produced within the system (P_g) would be lowered by some amount due to the removal of bioeroded sediment (SED_e). The carbonate remaining within the reef (P_n) should be comprised of recognizable coral, loose sediment, and bits of debris that have been reworked by multicyclic boring, infill, and cementation. Stated mathematically:

$$P_n = P_g - SED_e \qquad (3\text{–}1)$$

where P_n is the carbonate remaining in the reef (including skeletal debris reincorporated in the reef); P_g is the carbonate produced by corals, etc.; SED_e is the bioeroded sediment exported from the reef.

The most detailed carbonate budget to date is from a reef along the north coast of St. Croix (Hubbard et al., 1990). Using quantitative data on reef cover, coral growth rates, cores, and sediment export data, all from the same locale, Hubbard et al. concluded that nearly 60% of the carbonate that was produced within the reef had been reduced to sediment by bioerosion. Of that, about 40% was removed from the reef, primarily during storms (Hubbard, 1992b), the remainder being reincorporated within the reef interior (Fig. 3–11). From this study, two important points emerged. First, recognizable coral was less important in the reef (41%) than sediment (45%) and open void space (14%). Second, it was difficult to demonstrate that the majority of the remaining coral was in place while some pieces had been demonstrably moved or toppled. From these observations and measurements, Hubbard et al. argued that not only is in-place and interlocking framework *not* necessarily an integral characteristic of a true reef, but its presence in the rock record generally reflects unusual conditions (e.g., rapid burial) that have allowed its preservation.

The lesson is that the structure of both modern reefs and those preserved in the geologic record is the result of a complex interplay among processes that produce solid substrate, those that break it down, and those that transport and redistribute the sedimentary by-products. To understand the development of reefs from either a geological or biological perspective, one must have an equal appreciation of all three. We need to redirect our interests to some extent away from just coral growth and the patterns of carbonate production with depth. This is not to say that coral growth is insignificant. To the contrary, it provides the very building blocks from which the reef is constructed. However, calcification and coral growth are not the only processes that are operating, and a model that ignores other elements of the "reef equation" is doomed to failure. We must come to understand how the factors discussed earlier in this chapter influence not only coral growth but bioerosion, sediment transport, and biological/chemical cementation as well.

3.3. Reefs and Global Warming

Having discussed the factors that control reef formation and the ways in which reefs accrete, the discussion returns to where it started—the changing reef seascape. Are reefs declining as many scientists fear? If so, what is driving these changes? How do they fit into the mechanism of climate modulation? And how might they respond to the dramatic changes in sea level that have been proposed by some researchers for the coming decades?

3.3.1. Reefs in Flux

Based on widespread opinion among both the lay and scientific communities, it seems an inescapable conclusion that the world's reefs are changing—and in the opinion of many, they are "declining." The magnitude of this change, however, is largely unknown except for anecdotal discussions based on single locales that have been the focus of study for many years. In this regard, it must be remembered that these sites were initially chosen because they were the "best developed," or the "most diverse," or whatever else, of the reefs that could be found. Areas in decline or already degraded were typically passed over as research sites. If we are going to quantify changing reefs, then we do need to take into account the likelihood that our only baseline on what constitutes a "healthy" reef is what existed at the time reef science began in earnest.

Earlier scientific wisdom held that reefs were inherently stable systems in which little changed from year to year. With further study, we have come to realize that they are very dynamic systems, subject to tremendous change over relatively short spans of time. The question now before us is how the changes that we seem to be seeing fit into the grand scheme of things. Does the presumed decline in coral cover on a worldwide basis represent a response to pressure by increased nutrients, sedimentation, and other factors related to population growth, or is it simply part of a naturally occurring boom and bust cycle that operates on a scale of decades, centuries, or even thousands of years?

3.3.2. Reefs as Mediators of Global Climate

Coral reefs lie on both sides of the global-climate equation. They both mediate climate and respond to it. We are far from the point where we can adequately model this complex feedback loop. However, we are at least starting to understand the details of certain parts of the formula.

Based on data collected over the past few decades, a continuous increase in atmospheric carbon dioxide (CO_2) has been identified as an important factor in the warming of the Earth over the past century. Among the likely causes are increased industrialization and wholesale cutting of forests that absorb CO_2. Early in the debate, reefs were characterized as possible sinks for CO_2, one of the justifications for our concern over their well-being. Examination of the processes involved in carbonate production indicate that this is not the case, however. Calcium carbonate is formed by the linking of free calcium (Ca^{++}) with carbonic acid (HCO_3^-. As part of the process, carbon dioxide is generated and not consumed:

$$Ca^{++} + 2HCO_3^- \Leftrightarrow CaCO_3 + CO_2 + H_2O \qquad (3\text{–}2)$$

Several models have been proposed in which calcium carbonate has been identified as a mediator of atmospheric CO_2 (Broecker and Peng, 1987; Martin, 1990). A recent model by Opdyke and Walker (1992) accounts for nearly all the change in global CO_2 over the past 160,000 years. Their model is based on the fact that as sea level rises, it progressively invades ever-greater areas of broad, shallow platforms. Using published data on carbonate production rates and the aerial distribution of reefs, ooid shoals, and other carbonate-producing environments, they were able to mimic the patterns of global CO_2 recorded in polar-ice cores (Barnola et al., 1987).

3.3.3. The Reef Response

Coral reefs serve many functions. In addition to their obvious biological importance, they also serve as natural breakwaters against wave energy. Countless natural harbors rely on the protection afforded by reefs, as do many centers of population. The concern over how reefs might respond to rising sea level is, therefore, an important one. Will reefs keep pace with the proposed rise in sea level or will they fall behind, exposing tropical coastlines to increasing wave energy? If the character of reefs is substantially changed as water deepens, what effect will this have on already taxed fishery stocks that are tied, at least in part, to reefs?

The best place to look for answers to questions about the response of reefs to rising sea level is within the reefs that exist today—reefs that formed under the influence of a sea-level rise that was very similar to the one proposed for the near future. These reefs initially formed 9,000–10,000 years ago as rising sea level first broke over the same shelves that had supported reefs some 110,000 years earlier. The nature of that sea-level rise is reasonably understood, as are the suite of accretionary histories of many Holocene reefs in a variety of physical settings.

While seemingly a simple matter given the existing database, the answer is unfortunately quite complex. Some reefs have managed to keep pace whereas others have been left behind. On the one hand, the reefs that are of the greatest concern are today at sea level because they were able to keep pace. We might optimistically predict that they will simply continue to do so. On the other hand, Schlager (1981) pointed out that many of the ancient reefs that were "drowned" in the geologic past were likely capable of keeping up with rising sea level, but did not. Among the mechanisms he cited were increased sediment stress, nutrient loading, or changing temperature—all induced by natural processes. Given the increased stresses placed on today's reefs, their response to a sudden increase in sea level over the coming decades is far from a certainty.

3.3.4. Evolving Methodologies

While the effects of individual environmental parameters are reasonably well known, the complex interplay that is at the heart of reef development is still poorly defined. The answers to many of the questions asked in this chapter likely lie in fossil reefs. Unfortunately, however, the record is an imprecise one that is difficult to interpret on the time scale that we are discussing. Methodologies are emerging, however, that may allow us to hindcast climatic conditions over the past few centuries. Annual density bands within many reef corals can provide a calendar—similar to growth bands in trees—upon which we can plot reef change. While there is still discussion of whether high-density bands are a response to high temperature, low temperature, changing light levels, or some as yet undetermined "stress," they nevertheless provide a chronological record of years in which corals grew faster or more slowly than normal.

In addition, recent studies have revealed potentially decipherable records of water temperature, salinity, rainfall, nutrients, and exposure to pollutants within the skeletons of growing corals. Coupled with the chronology provided by annual bands, these may become powerful tools for interpreting patterns of local change over the past two to three centuries. Spread over wide geographical areas, such studies can potentially answer many critical questions about temporal changes in the factors that drive the coral-reef system.

3.4. Where Do We Go from Here?

The questions that dominate the recent debate over "reef health" are difficult to answer with our available database. Reefs have waxed and waned throughout geological time, and whether the recent changes are simply part of a larger cycle cannot be stated with certainty. In this regard, however, there can be no doubt that reefs are being exposed to ever-increasing stress as the developed nations spread their influence and developing nations struggle to "catch up." Easily quantified increases in sedimentation and nutrient levels have contributed to some extent to what is probably a recent "decline" in reefs. For perhaps the first time in the history of our planet, a single species may be close to exerting an effect that rivals the importance of those natural controls outlined in the first part of this chapter. It is an inescapable conclusion that reducing the levels of pollutants and exploitation that are presently increasing at exponential rates cannot hurt and will likely help a great deal. Doing so seems the only sensible strategy while we are trying to address difficult questions that can be answered only after the problems of temporal scale are resolved.

4

Bioerosion and Coral-Reef Growth: A Dynamic Balance

Peter W. Glynn

> The question at once arises, how is it that even the stoutest corals, resting with broad base upon the ground, and doubly secure from their spreading proportions, become so easily a prey to the action of the same sea which they met shortly before with such effectual resistance? The solution of this enigma is to be found in the mode of growth of the corals themselves. Living in communities, death begins first at the base or centre of the group, while the surface or tips still continue to grow, so that it resembles a dying centennial tree, rotten at the heart, but still apparently green and flourishing without, till the first heavy gale of wind snaps the hollow trunk, and betrays its decay. Again, innumerable boring animals establish themselves in the lifeless stem, piercing holes in all directions into its interior, like so many augurs, dissolving its solid connexion with the ground, and even penetrating far into the living portion of these compact communities.
>
> —L. Agassiz, 1852

Coral reefs are among the Earth's most biologically diverse ecosystems, and many of the organisms contributing to the high species diversity of reefs normally weaken them and convert massive reef structures to rubble, sand, and silt. The various activities of those reef species that cause coral and coralline algal erosion are collectively termed bioerosion, a name coined by Neumann (1966). A bioeroder is any organism that, through its assorted activities, erodes and weakens the calcareous skeletons of reef-building species. Although an extensive terminology has been adopted only during the past three decades, bioerosion has been recognized as an important process in reef development and maturation for more than a century (e.g., Darwin, 1842; Agassiz, 1852). Traces of biologically induced

This chapter benefited from National Science Foundation support for bioerosion studies in the eastern Pacific. Contribution from the University of Miami, Rosenstiel School of Marine and Atmospheric Science.

erosion in ancient reef structures indicate that bioerosion has probably figured prominently in reef carbonate budgets since Precambrian and Cambrian times (Vogel, 1993).

Most bioeroder species are both small in size and secretive in living habits. Although the majority of bioeroders are not visible on coral reefs, it has been suggested that their numbers and combined mass equal or exceed that of the surface biota (Grassle, 1973; Ginsburg, 1983). For convenience, bioeroders that are usually present and visible on reef surfaces are termed external bioeroders and those living within calcareous skeletons are termed internal bioeroders (Fig. 4–1). Several studies have shown that bioeroders are important in sculpting coral-reef growth and in producing the sediments (rubble, sand, silt, and clay) that characterize coral-reef environments. Indeed, carbonate budget studies have demonstrated that constructive and destructive processes are closely balanced on many reefs with net reef accumulation barely ahead of net reef loss (Scoffin et al., 1980; Glynn, 1988a; Fig. 4–2). Bioerosion proceeds at high rates in certain zones that have high living coral cover and high rates of accretion (Kiene, 1988).

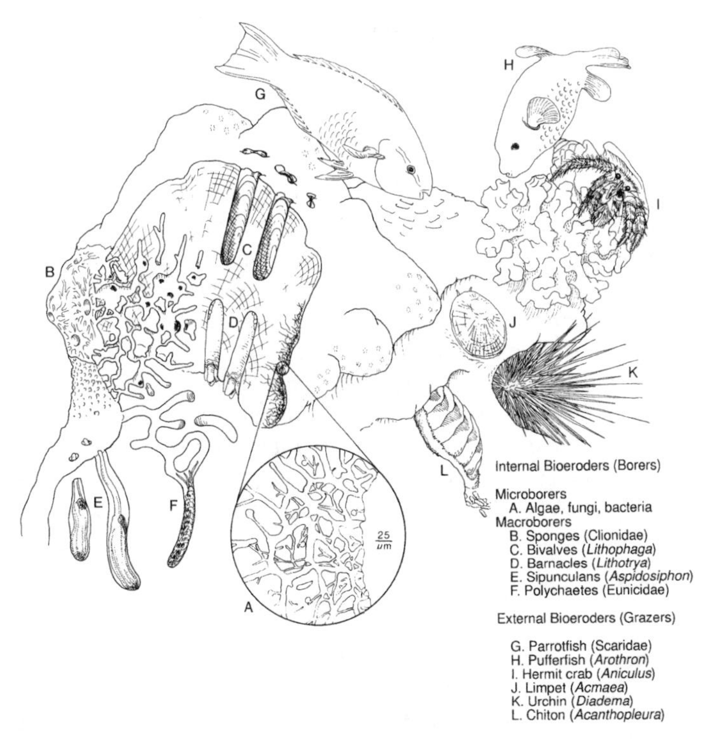

Figure 4–1. Variety of external and internal bioeroders that commonly attack coral skeletons. A legend provides identification of the taxa illustrated.

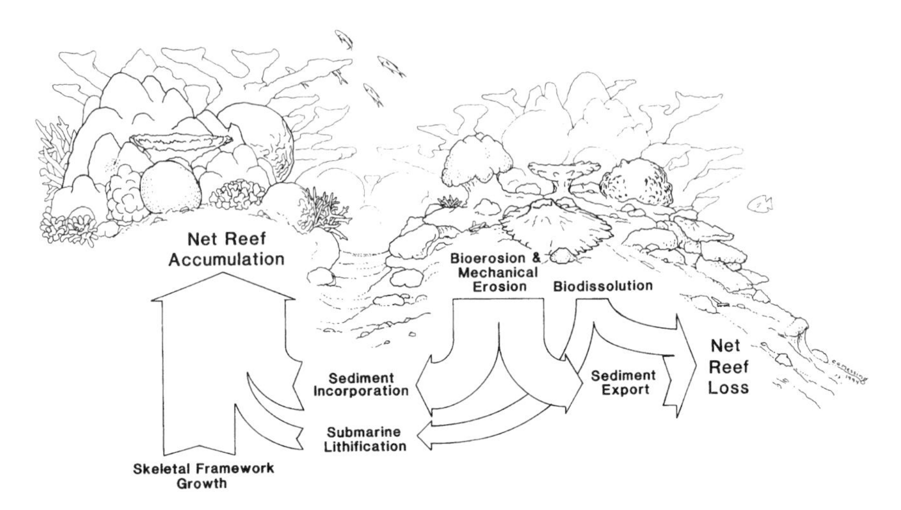

Figure 4–2. A generalized scheme illustrating the principal components of coral-reef construction and destruction. In order for reef growth to occur, rates of bioerosion and mechanical erosion must not exceed the rate of net reef accumulation.

Sometimes, however, an imbalance develops with erosional processes gaining the upper hand. When environmental conditions decline over an extended period, reef growth ceases, reef foundations are destroyed, and reef death ensues.

The aim of this chapter is to (1) illustrate the diversity of bioeroders on coral reefs, (2) identify the most destructive bioeroder groups, (3) describe the more prevalent modes of limestone destruction, and (4) highlight some case studies of intensified bioerosion on particular reef systems. Some recent evidence suggests that increasing anthropogenic stresses to coral reefs are shifting the balance in favor of coral-reef degradation on a global scale. Thus, it is timely to understand the conditions that promote bioerosion, a pivotal process affecting the growth potential of coral reefs. For more technical information on this subject, the reader may consult the articles in Carriker et al. (1969) and Barnes (1983), and the reviews by Golubic et al. (1975), Warme (1975, 1977), Risk and MacGeachy (1978), Trudgill (1983), Macintyre (1984), and Hutchings (1986).

4.1. Bioeroder Diversity

Bioeroders are abundant and diverse members of coral-reef communities, belonging to four of the five kingdoms of life on Earth, and to most animal phyla. Why have so many taxa become bioeroders? By far, the bioeroders hidden within coral skeletons, the cryptic biota, have the greatest taxonomic diversity. It is probable that intense competition and predation have led to the selection and evolution of cryptic lifestyles. Many of these secretive species are without toxins, armature, spines, and thick shells—traits that are so common to their congeners living on reef surfaces and exposed to predators.

Depending on their location on calcareous substrata, bioeroders can be classified as epiliths, chasmoliths, and endoliths (Golubic et al., 1975). Epilithic species live on exposed surfaces, chasmoliths occupy cracks and holes, and endoliths are present within skeletons. Assignment to these categories is not always clear, however, for some bioeroders may belong to more than one microhabitat or change microhabitats during development.

Bioeroders break down calcareous substrata in a variety of ways. The majority of epilithic bioeroders are herbivorous grazers that scrape and erode limestone rock while feeding on associated algae. In terms of eroding capabilities, grazers range from nondenuding and denuding herbivores that remove mainly algae and cause little or no damage to substrata to excavating species that remove relatively large amounts of algae, including calcareous algae, and the underlying limestone substrata (Steneck, 1983a, b). Most endoliths are borers that erode limestone mechanically, chemically, or by a combination of these processes. The important role of bioeroders can be appreciated when one realizes that coral reefs are predominantly sedimentary environments made up of calcareous particles that are generated in large measure by the activities of bioeroders (Chapter 2).

Many species that bioerode calcareous skeletons are minute, requiring microscopical methods for study, and are referred to as microborers or endolithic microorganisms (Golubic et al., 1975; Macintyre, 1984). To this group belong three kingdoms, namely bacteria (Prokaryotae), Fungi, and eukaryotic microorganisms such as protozoans and algae (Protoctista). The macroborers are generally more conspicuous on coral reefs, and include numerous invertebrate and vertebrate taxa in the kingdom Animalia. Most endolithic invertebrates are suspension feeders, gathering their food passively or actively from the water column.

4.1.1. Bacteria

Although our knowledge of the bioeroding potential of bacteria and the various taxa involved is very limited, preliminary observations suggest that these organisms may be important under certain conditions. A pilot study in Hawaii indicated that brownish areas inside the skeletons of massive corals contained from 10^4 to 10^5 bacteria per gram dry weight (DiSalvo, 1969). Boring sponges also were closely associated with bacteria, which could possibly have assisted the penetration of the sponges into the coral. Different workers have shown that bacteria can etch the surface of limestone crystals and dissolve the organic matrix of coral skeletons, causing internal bioerosion (DiSalvo, 1969; Risk and MacGeachy, 1978).

Several species of Cyanobacteria, formerly known as blue-green algae, are capable of eroding reef rock from the splash zone to depths of at least 75 meters. Species of *Hyella, Plectonema, Mastigocoleus,* and *Entophysalis,* for example,

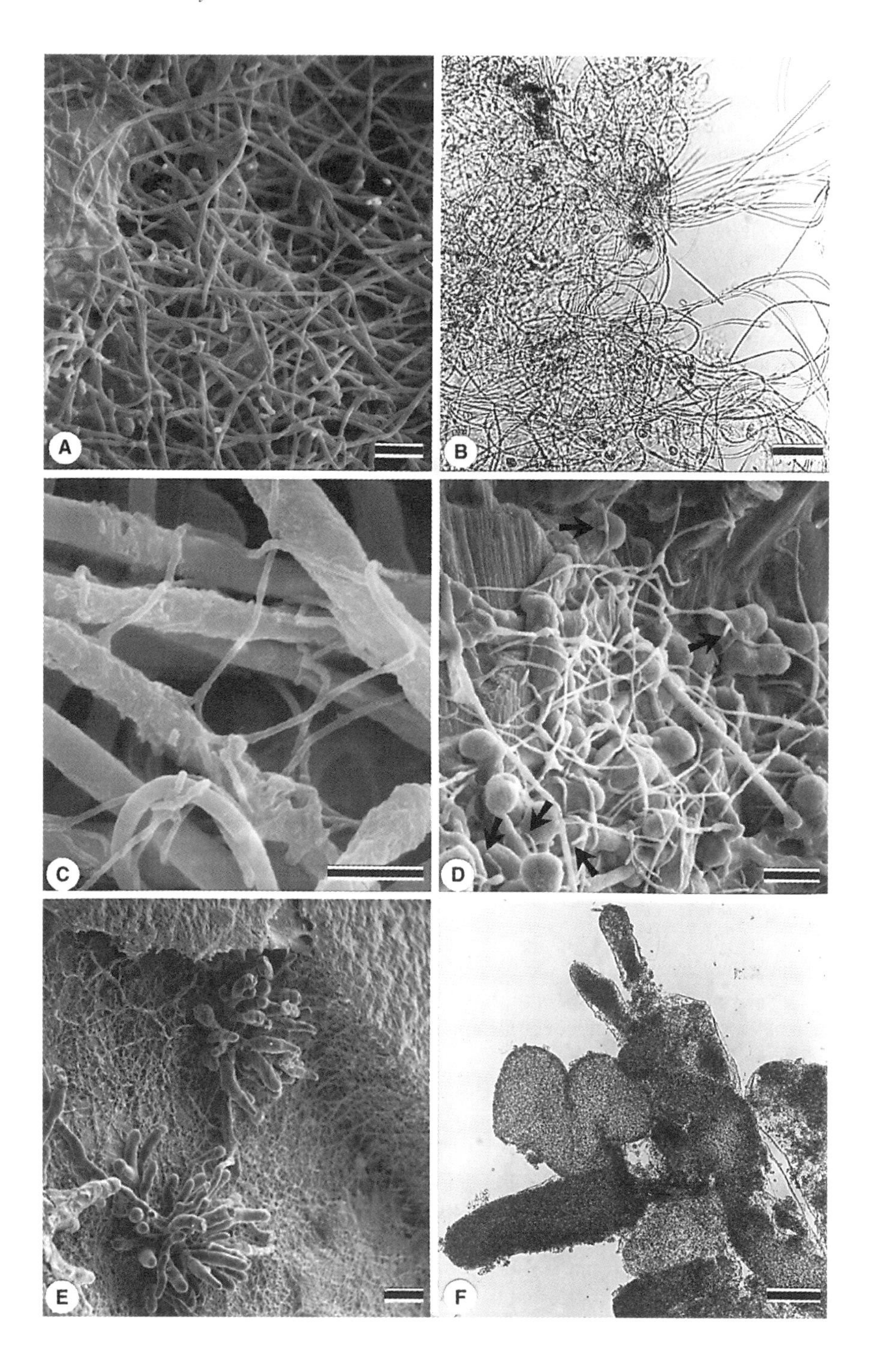
A
B
C
D
E
F

have been found on limestone surfaces, inside cavities, and penetrating reef rock (Fig. 4–3A, B). A close relative of *Hyella* has been found in Precambrian algal reefs that existed 1.7 billion years ago (Vogel, 1993). The boring is a dissolution process accomplished by the terminal cells of specialized filaments. Cyanobacteria have been implicated in the erosion of lagoon-floor sediments on the Great Barrier Reef, amounting to the dissolution of between 18% and 30% of the sediment influx rate (Tudhope and Risk, 1985; Table 4–1). (It should be stressed that most of the rates of erosion listed in Tables 4–1 and 4–2 were obtained with different methods and therefore should be compared with due caution. See Kiene [1988] for an assessment of the strengths of the methods and some problems with intercomparisons.)

4.1.2. Fungi

Boring fungi have been found in modern corals in the Caribbean, French Polynesia, and on the Great Barrier Reef (Australia). Twelve genera belonging to the Deuteromycota or Fungi imperfecti have been isolated from a variety of scleractinian corals and a hydrocoral (Kendrick et al., 1982). Fungi are capable of deep penetration into coral skeletons by chemical dissolution. The hyphae produce narrow borings and penetrate the deepest recesses of coral skeletons, probably because of their ability to utilize the organic matrix of coral skeletons (Fig. 4–3C, D). Fungi have also been implicated in the etching of calcareous surfaces, the weakening and dissolution of calcareous sediments, as well as the calcareous tube linings of various endoliths. Because of the difficulty of distinguishing between fungal and algal borings, estimates of dissolution rates due to boring fungi alone are not yet available.

4.1.3. Algae

Green (Chlorophyta) and red (Rhodophyta) algae have been implicated in the erosion of coral rock under various reef settings. Green and red algae occur on limestone surfaces, in cavities, and within coral skeletons (Fig. 4–3E, F). Freshly

◀ *Figure 4–3.* Photomicrographs of endolithic microborers in limestone substrates. Cyanobacteria: **(A)** *Plectonema terebrans* Bornet and Flahault, scanning electron micrograph (SEM) of plastic casts of filaments in an acid-etched shell; **(B)** *P. terebrans,* transmitted light micrograph (TLM) of filaments isolated by dissolution. Fungi: **(C)** SEM of plastic casts of fine fungal hyphae intertwined with the larger filaments of *P. terebrans;* **(D)** SEM of fungal borings covering and possibly feeding (arrows) on the underlying cyanobacterium. Chlorophyta: **(E)** *Ostreobium brabantium* Weber Van-Bosse, SEM of plastic cast of large radiating growth form in an acid-etched shell fragment; **(F)** *O. brabantium,* TLM of filaments isolated by dissolution. Scale bars: **A** = 50 μm, **B** = 40 μm, **C** = 5 μm, **D** = 25 μm, **E** = 200 μm, **F** = 100 μm (from May et al., 1982).

Table 4-1. Rates of Bioerosion by Internal Borers

Taxonomic Group	Erosion Rate (g $CaCO_3/m^2/yr$)	Borer Abundance	Particle Size (μm)	Habitat	Locality	Source
Cyanobacteria						
mostly cyanobacteria with some fungi	350	microborings permeated sediment grains	2–6	lagoon-floor carbonate sediments	Davies Reef, Great Barrier Reef, Australia	Tudhope and Risk (1985)
Porifera						
clionid sponges, *Cliona lampa* Laubenfels predominant	23,000	infested limestone substrates	30–80	subtidal limestone notch, 1–3 m depth	Bermuda	Neumann (1966)
Cliona and Siphonodictyon	7,000	infested limestone substrates	30–80	subtidal test blocks fringing reef	Bermuda	Rützler (1975)
	180[a]	abundant in crustose coralline algae and in dead and live corals			Barbados	Scoffin et al. (1980)
Polychaeta						
cirratulid, eunicid, sabellid, and spionid worms	690	13,000 ind. m^2		forereef slope	Lizard Island, Great Barrier Reef, Australia	Davies and Hutchings (1983)
	840	24,000 ind. m^2	10–30[b]	reef flat		
	1,800	85,000 ind. m^2		lagoonal patch reef		
Crustacea						
Lithotrya ?dorsalis Sowerby	14[a]	common	?	fringing reef	Barbados	Scoffin et al. 1980)
Lithotrya sp.	0.8 cm^3 yr ind.	common	2–4[c]	intertidal limestone shore	Aldabra Atoll, Indian Ocean	Trudgill (1976)
Sipuncula						
Phascolosoma, 3 spp. *Paraspidosiphon,* 3 spp. *Lithacrosiphon gurjanovae* Murina	8[a]	uncommon in corals	<63	fringing reef	Barbados	Scoffin et al. (1980)
Mollusca						
Lithophaga nasuta (Phillipi)	0.9 cm^3 yr ind.	common	?	intertidal limestone shore	Aldabra Atoll, Indian Ocean	Trudgill (1976)
Lithophaga laevigata (Quoy and Gaimard) *Lithophaga aristata* (Dillwyn)	9,000	1,870 ind. m^2	10–100	largely dead patch reef, 6–10 m depth	Cáno Island, Costa Rica	Scott et al. (1988)

[a]Calculated from an overall borer bioerosion rate of 200 g m^2 yr, and assuming that sponges were responsible for 89%, barnacles for 7%, and sipunculans for 4% of the total bioerosion (Scofflin et al., 1980).

[b]For an eunicid (Ebbs, 1966), and from information supplied by P. Hutchings (pers. comm).

[c]From Ahr and Stanton (1973).

fractured corals often reveal layers of green banding a few centimeters beneath the live coral surface. The green color is due to the presence of chlorophyll pigments, which intercept light passing through the coral's tissues and skeleton. This greenish layer is often referred to as the *Ostreobium* band, named after a green alga that is commonly present in coral skeletons. However, the green band may also contain a variety of different kinds of algae, for example, species of *Codiolum, Entocladia, Eugomontia,* and *Phaeophila.* The importance of boring algae as bioeroders is controversial; some workers claim that they are among the most destructive agents of reef erosion whereas others maintain that they cause only minimal damage.

4.1.4. Sponges

The most important genera of siliceous sponges known to bore into calcareous substrata are *Cliona, Anthosigmella,* and *Spheciospongia,* order Hadromerida, and *Siphonodictyon,* order Haplosclerida (Wilkinson, 1983). Clionid sponges are among the most common and destructive endolithic borers on coral reefs worldwide. Upon splitting open infested corals, clionid sponges are revealed as brown, yellow, or orange patches lining the corroded interiors of the coral skeleton (Fig. 4–4A, D). Most boring sponges form 5–15-mm-diameter chambers with smaller galleries branching off the main chambers. Their depth of penetration into the coral skeleton is usually no greater than about 2 cm. Some sponges *(Siphonodictyon),* however, can form chambers up to 100 mm in diameter that penetrate to 12 cm into coral colonies. In highly infested colonies, some boring sponges emerge from the skeleton, grow over and even kill live coral tissues on reef surfaces. On western Atlantic reefs, *Cliona caribbaea* is sometimes very abundant, forming dark brown patches several meters long that kill or overgrow dead surfaces and erode all calcifying organisms (Fig. 4–5).

Sponge boring is accomplished by amoebocytes that etch and chip minute calcareous fragments from limestone substrata (Rützler and Rieger, 1973; Pomponi, 1979). The ends of etching amoebocytes flatten against the calcareous substratum and extend fine pseudopodial (filopodia) sheets into the limestone at the cell periphery. The filopodia coalesce centrally, cutting out a hemispherical carbonate chip (Fig. 4–4E, G). This cutting is accomplished by enzymes that simultaneously dissolve calcium carbonate and the organic matter matrix of skeletons. At the end of this process, both the chip and the etching cell are transported away from the site of erosion and are expelled from the sponge. Only about 2–3% of coral skeletons are dissolved, with the remainder dispersed as silt-sized chips. These oval-shaped (faceted) chips are easily recognized in sediments and have been found to contribute up to 30–40% numerically to the fine silt fraction of sediments on Pacific and Caribbean reefs.

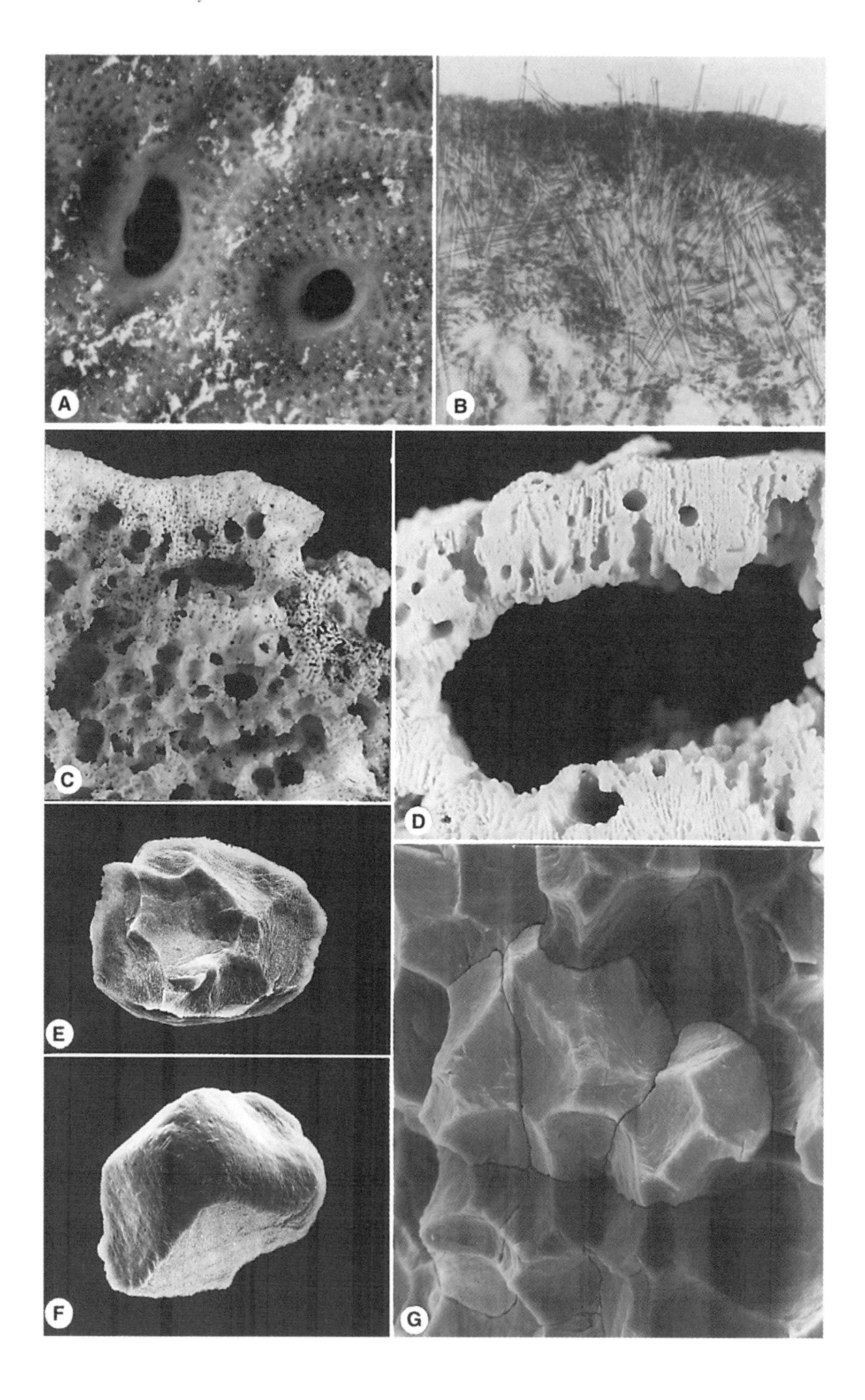
A
B
C
D
E
F
G

4.1.5. Polychaete Worms

Polychaete worms that bore into reef rock are enormously abundant in certain environments, prompting some workers to conclude that they are among the most important endolithic borers on coral reefs (Davies and Hutchings, 1983). Various species in the following families typically form circular holes 0.5–2 mm in diameter that penetrate up to 10 cm into the interiors of coral skeletons: Cirratulidae, Eunicidae, Sabellidae, and Spionidae. Eunicid holes often form a sinuous and anastomosing network (Fig. 4–1). The mechanism of boring has been reported for a few polychaete species. Some eunicids employ their mandibles to excavate. Spionids bore mainly by chemical dissolution, with some removal probably due to mechanical abrasion by setae (Haigler, 1969). Cirratulid and eunicid species are predominantly deposit feeders whereas sabellids and spionids are mainly filter feeders. The close physical association of eunicids and spionids with endolithic algae also has suggested the utilization of boring algae as a food source (Risk and MacGeachy, 1978).

A quantitative study of boring polychaetes conducted at Lizard Island, Great Barrier Reef, provides numerical abundances and bioerosion rates of a pioneer polychaete community. At various times during the study it was not uncommon to find between 27,000 and 80,000 boring polychaetes per m^2 in experimental coral blocks set out in three different reef environments (Davies and Hutchings, 1983). These worms caused erosional losses of from 0.7 kg m^{-2} yr^{-1} on the reef front to 1.8 kg m^{-2} yr^{-1} on a leeward patch reef (Table 4–1).

4.1.6. Crustacea

Barnacles , shrimp, hermit crabs, and other kinds of crustaceans can erode reef rock (Warme, 1975). Barnacles and shrimp are endolithic borers, producing cylindrical chambers, whereas hermit crabs are external bioeroders that abrade live coral surfaces.

Three groups of barnacles contain species that reside in skeletons of dead corals, namely thoracicans, acrothoracicans, and ascothoracicans. Members of the latter two taxa occupy small, millimeter-sized cavities that keep pace with

◀ *Figure 4–4.* Boring sponges in limestone substrates. **A** Two oscula of *Cliona lampa* (Laubenfels) visible on the surface of a massive coral *(Diploria).* **(B)** Vertical section through peripheral region of *Spheciospongia othella* Laubenfels revealing abundant spicules. **(C)** Chambers of *Cliona dioryssa* (Laubenfels) in porous coral rock. **(D)** A large tunnel running below the surface of coral rock excavated by *S. othella.* **(E)** Upper scalloped and **(F)** lower convex surfaces of isolated limestone chips discharged through the osculum of *Cliona lampa.* **(G)** Group of chips etched from substratum by *Cliona lampa* but still in place. Magnification: **A, C, D** × 3; **B** × 140; **E, F** × 1,500; **G** × 600 (**A–D** from Rützler, 1984; **E, F, G** from Rützler and Rieger, 1973).

Figure 4–5. A Caribbean boring sponge *(Cliona caribbea)* covering and eroding several square meters of reef substrate, San Blas, Panama, 3 meters depth (30 June 1993). Arrows denote perimeter of sponge patch.

the growth of the host coral, that is, they become embedded within the coral skeleton without causing extensive erosion. *Lithotrya* species—of thoracican barnacles—erode 2–10-cm long oval-shaped cavities on the undersides of reef rock and beach rock in shallow, agitated waters (Fig. 4–1). The barnacle's basal plate is attached at the innermost end of the cavity and the body hangs downward toward the opening with cirri exposed to food-bearing currents. The cavities are formed apparently by mechanical abrasion effected by calcified plates that cover the barnacle's body. Unlike other invertebrate endoliths, such as polychaete worms and gastropods, adjacent tubes of boring *Lithotrya* are commonly interconnected, and heavily infested limestones are thoroughly honeycombed and subject to frequent breakage. An average of one boring per cm^2 was observed on beach rock in Puerto Rico, and up to 30% of the substratum had been removed from some of the samples examined (Ahr and Stanton, 1973). Overall, however, results from studies in the Caribbean and Indian Ocean indicate that boring barnacles cause relatively little erosion compared with other internal borers (Table 4–1).

Alpheus simus Guerin-Meneville, a pistol shrimp, bores into coral rock on Caribbean reefs and causes considerable erosion on some Costa Rican reefs (Cortés, 1985). Male/female pairs excavate 10–15-mm-diameter chambers that penetrate as deep as 15 cm into dead coral rock. Microscopical study of the chamber walls suggests that this shrimp bores mainly by chemical means. Seven pairs of shrimp were found in one 1,500 cm^2 block, and each pair occupied an average chamber volume of 20 cm^3. This is equivalent to the removal of about 950 cm^3 of calcium carbonate m^{-2}. The life span of the shrimp is about 2 years, but since succeeding generations of shrimp probably occupy the same chambers, it is not possible to calculate annual erosion rates.

Two species of hermit crabs that feed on live coral produce large amounts of calcareous sediment when they scrape corals to remove soft tissues (Fig. 4–1). The average mass of coral abraded by a small hermit crab (*Trizopagurus magnificus* [Bouvier]) was about 10 mg $ind.^{-1}$ day^{-1}, and for a large hermit crab (*Aniculus elegans* Stimpson) about 1 g $ind.^{-1}$ day^{-1} (Glynn et al., 1972). Relating hermit crab population densities and erosion rates, it was found that *Trizopagurus* and *Aniculus* respectively were responsible for the generation of about 1 and 0.1 metric tons of coral sediment per hectare per year on a fringing reef in Panama (Table 4–2). Since this rate of coral abrasion by hermit crabs has not been reported elsewhere, it is possible that these high levels of erosion are unique to the eastern Pacific.

4.1.7. Sipuncula

Although it is well known that species in several genera of sipunculans (peanut worms) penetrate coral skeletons, there is no general agreement on the overall importance of this group in the bioerosion of coral reefs. Perhaps this is due to their great variation in abundance from reef to reef and across reef zones (Macintyre, 1984).

Sipunculan borings are cylindrical and pencil sized or slightly smaller, ranging from straight to sinuous and from near surface to several centimeters deep in coral skeletons, depending on the species (Fig. 4–1). Sipunculans are abundant on some reefs: nearly 800 ind. m^{-2} were present in reef crest substrata, and 1,200 ind. m^{-2} in *Porites coral* skeletons in Belize (Rice and Macintyre, 1982). Even at 30 m depth, 40 ind. m^{-2} were found. While feeding, sipunculans extend their introverts outside of their cavities and appear to ingest debris, sand, and algae. The exact manner of boring is not known, but may involve both chemical dissolution and mechanical abrasion (Rice and Macintyre, 1972). An estimated sipunculan erosion rate on a Barbados reef indicated only minor carbonate loss (Table 4–1).

4.1.8. Mollusca

Most bioeroding molluscs are external grazers that abrade reef rock while feeding on algae and associated organisms residing on and within limestone substrata. The eroding capacity of surface-enmeshed and endolithic algae, important components of the diet of grazing molluscs, also weakens the substratum and thus

Table 4-2. Rates of Bioerosion by External Grazers

Taxonomic Group	Erosion Rate (g $CaCO_3$ m^2 yr)	Borer Abundance (ind. m^2)	Particle Size (mm)	Habitat	Locality	Source
Crustacea (hermit crabs)						
Trizopagurus magnificus (Bouvier)	103	27.5	0.12–0.5	pocilloporid patchreef	Pearl Islands, Panama	Glynn et al. 1972)
Aniculus elegans Stimpson	8.5	0.02	0.25–3.0			
Mollusca						
Chitons						
Acanthopleura granulata Gmelin	227	5.5	0.03–1.0	intertidal limestone rock	San Salvador Island, Bahamas	Rassmussen and Frankenberg (1990)
Chiton tuberculatus Linné	394	22	?	lower intertidal coral rubble	La Parguera, Puerto Rico	Glynn (1970)
Gastropods						
Acmaea sp.	19.2	8	0.03–1.0	intertidal limestone rock	Andros Island, Bahamas	Donn and Boardman (1988)
Nerita tesselata	154	220	0.03–1.0	intertidal limestone rock	Andros Island, Bahamas	McLean (1967)
Echinodernata (sea urchins)						
Diadema antillarum Phillipi	4,600	9	?	patch reef	St. Croix, U.S. Virgin Islands	Ogden (1977)
Diadema antillarum	5,300	23	0.05–0.5	fringing reef	Barbados	Scoffin et al. (1980)
Diadema mexicanum A. Agassiz	139–277 3,470–10,400	2–4 50–150	0.5–2.0	lower seaward slope	Gulf of Chiriquí, Panama	Glynn (1988)
Diadema savignyi Michelin	3,400	4.8	sand	reef lagoon	Moorea, French Polynesia	Bak (1990)
Echinometra lucunter (Linnaeus)	3,900	100	?	algal ridge	St. Croix, U.S. Virgin Islands	Ogden (1977)
Echinometra mashei (Blainville)	70–260	2–7	?	limestone rock	Enewetak Atoll	Russo (1980)
Echinothrix diadema (Linnaeus)	803	0.6	sand	reef lagoon	Moorea, French Polynesia	Bak (1990)
Eucidaris thouarsii (Valenciennes)	3,320 22,300	4.6 30.8	0.05–3.0	reef flat, pre-1982 reef flat, post-1983	Floreana Island, Galápagos Islands	Glynn (1988)

Continued

Table 4-2. Continued

Taxonomic Group	Erosion Rate (g $CaCO_3$ m^2 yr)	Borer Abundance (ind. m^2)	Particle Size (mm)	Habitat	Locality	Source
Pisces						
Scarus iserti	490	0.6	0.015–0.25	patch reef	Barbados	Frydl and Stearn (1978)
Sparisoma viride (Bonnaterre)	61	0.01	silt-sand	fringing reef	Llewellyn reef, Australia	Kiene (1988)
Parrotfishes (dominantly)	<100–9,000	?	?	reef flat, slope, and lagoon habitats	Saipan, Mariana Islands	Cloud (1959)
Grazing and browsing fishes	400–600	0.04–0.06	fine sand–gravel	coral reef	Bermuda	Bardach (1959, 1961)
	110	0.01	?	patch reef		
Arothron meleagris (Bloch and Schneider)	30	0.004	2–8	pocilloporid reef	Pearl Islands, Panama	Glynn et al. (1972)

facilitates erosion during feeding. A group of mussel-like endolithic borers also is prominent on many reefs worldwide.

Molluscan bioeroders are generally most abundant in the intertidal zone, with some species extending their ranges into supratidal and subtidal habitats (Fig. 4–6a). Species abundances also change horizontally with chitons, which are often most plentiful in areas protected from strong wave assault, and with limpets, certain snails, and echinoids, which are more common in wave-swept habitats (Fig. 4–6B). Under quiet to rough water conditions, grazing molluscs are largely responsible for producing the notches and nips on tropical limestone shores. Under extremely rough conditions, however, many bioeroders either disappear or their activities are greatly reduced. Calcifying taxa, such as coralline algae and vermetid molluscs, increase in abundance with increasing exposure, probably because of ecologic requirements for high-energy habitats and a lower abundance of fish consumers in rough-water areas (Fig. 4–6C). Vermetid/coralline algal buildups help protect the underlying limestone, thus limiting bioerosion and the development of intertidal notches and nips in such areas (Focke, 1978).

Several species of chitons (class Amphineura), for example, members of *Acanthopleura* and *Chiton,* erode chiefly intertidal limestone substrata while grazing on algae. The grazing is achieved with a magnetite (Fe_3O_4) or other mineral-enriched radula, a tooth-bearing strap of chitinous material that effectively abrades the substratum (Lowenstam and Weiner, 1989). Some erosion also occurs at homing sites, rock depressions that are occupied by chitons when not foraging. As many as 50–100 sausage-shaped, 1–3-mm-long fecal pellets are voided daily by individual chitons (Rasmussen and Frankenberg, 1990). Erosion rates vary greatly among sites as they are influenced by local differences in rock type and condition, and ecological factors affecting chiton abundances and feeding activities (Table 4–2).

Limpets and snails (class Gastropoda) often occur with chitons on intertidal carbonate substrata. *Acmaea, Cellana,* and *Patelloida* are common limpet genera, and *Cittarium, Littorina, Nerita,* and *Nodilittorina* are some common snail genera. Like chitons, limpets and snails utilize a radula to scrape rock surfaces. The radula of patellacean limpets is an especially effective excavating organ with opal ($SiO_2 \cdot nH_2O$) or goethite ($HFeO_2$)-sheathed radular teeth (Lowenstam and Weiner, 1989). The radula of snails contains proteinaceous teeth, but these grazers are still capable of erosion because of the often weakened condition of the rock substratum upon which they feed (Table 4–2).

Species of *Lithophaga* and *Gastrochaena* (class Pelecypoda) bore into dead and live corals, and are most abundant subtidally, with some of these bivalves attacking reef corals to their lower-depth limits. *Fungiacava* spp. penetrate live mushroom corals, but their activities are relatively minor. The siphonal openings of *Lithophaga* typically have a keyhole-like appearance on coral surfaces and the circular holes penetrate vertically into the skeleton, from 1 to 10 cm deep depending on the species (Fig. 4–1). The lithophagines are deposit and suspension

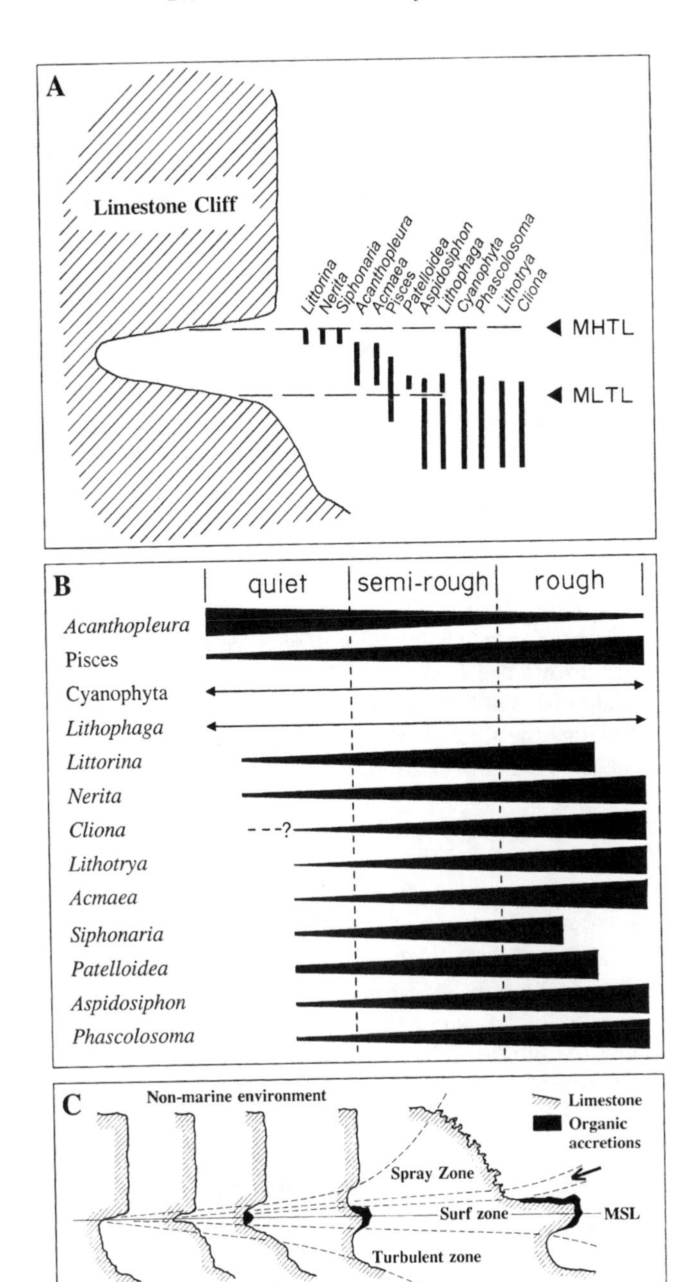

Figure 4–6. Vertical **(A)** and horizontal **(B)** distributions of bioeroding molluscs and other bioeroder taxa on a limestone shore at Palau, Caroline Islands. Theoretical relationship **(C)** of coastal profile morphology to water turbulence at Curaçao, Netherlands Antilles. An arrow locates a "transition zone" between the "spray" and "surf zones" (**A** and **B** after Lowenstam, 1974; **C** after Focke, 1978).

feeders, often most abundant in areas of high productivity. The mantle glands of *Lithophaga* secrete acid that dissolves and weakens the limestone substratum. The vertical and rotational movements of the shell also assist in boring, resulting in the production of silt/sand-sized sediment. Population densities in productive equatorial eastern Pacific waters range from 500 to 10,000 ind. m^{-2} (Scott et al., 1988), which can lead to rapid reef erosion (Table 4–1).

4.1.9. Echinoidea

Sea urchins (Echinoidea) are the only echinoderms capable of significant bioerosion. Several species in the following genera abrade large amounts of reef rock while feeding and excavating burrows: *Diadema, Echinometra, Echinostrephus,* and *Eucidaris.* Sea urchins possess a highly evolved jaw apparatus (Aristotle's lantern), a flexible and protrusible mastigatory organ consisting of five radially arranged, calcified teeth. The teeth are mineralized, and must be harder than the corroded surfaces they scrape. Sea urchin spines also assist in bioerosion when they are employed in the enlargement of burrows. Sea urchins graze on algae growing on dead coral substrata, but in some areas they also attack live coral. On seaward reef platforms where water flow is vigorous, sea urchins usually remain in their burrows and feed predominantly on drift algae. Sea urchins can cause substantial erosion at low and moderate population densities (Table 4–2); at high densities, their destruction of reef substrata rivals clionid sponge erosion and can lead to rapid framework loss.

4.1.10. Fishes

Numerous fish species erode reef substrata while grazing on algae, and also fragment colonies while feeding on live coral tissues or when extracting invertebrates from coral colonies (Randall, 1974). Surgeonfishes (Acanthuridae) and parrotfishes (Scaridae) are the principal grazing groups with many fishes in the latter family capable of scraping and extensive excavation. On western Pacific reefs, excavating parrotfishes primarily bite convex surfaces, thus reducing the topographic complexity of reefs (Bellwood and Choat, 1990). Some Atlantic and Pacific parrotfishes occasionally scrape and ingest live coral tissues (Bellwood and Choat, 1990; Glynn, 1990b). Triggerfishes (Balistidae), filefishes (Monacanthidae), and puffers (Tetraodontidae, Canthigasteridae) are largely carnivorous in feeding habits and are responsible for fragmenting live coral colonies. The jaw muscles and tooth armature are well developed in all of these families. Parrotfishes also have a pharyngeal mill, a gizzardlike organ that further reduces the size of ingested sediment. Fish teeth are composed of dahllite [$Ca_5(PO_4.CO_3)_3(OH)$] or francolite (the fluorinated form), which are apatite minerals that are harder than $CaCO_3$ (Lowenstam and Weiner, 1989).

Parrotfish grazers can produce large amounts of sediment on reefs, especially when their population densities are high. For example, a species of *Scarus* generated nearly 0.5 kg $CaCO_3$ m^{-2} yr^{-1} on a Caribbean reef in Panama with a high abundance of just under 1 fish per m^2 (Table 4–2). Entire grazing fish communities, comprised primarily of parrotfishes, typically erode large amounts of reef substrata. One of the highest erosion rates reported for fishes, 9 kg $CaCO_3$ m^{-2} yr^{-1}, occurred in the lagoon of an Australian reef (Table 4–2). While carnivorous fishes can cause substantial damage locally, their reef-wide effects seem to be relatively minor. For example, a pufferfish *(Arothron)* that erodes about 20 g of coral per day results in a total reef loss of only 30 g $CaCO_3$ m^{-2} yr^{-1} (Glynn et al., 1972) because of a relatively low population size of 40 individuals per hectare (Table 4–2).

Several other bioeroders known to produce traces or otherwise damage reef rock, for example, foraminifers, zoanthids, bryozoans, and brachiopods (Warme, 1975), may contribute to reef degradation under special conditions. To assess the relative importance of the various bioeroders considered in this survey, their rates of reef destruction may be compared with known carbonate production rates. Net carbonate production rates vary greatly among reefs and between reef zones, but 3,000–5,000 g $CaCO_3$ m^{-2} yr^{-1} have been reported for many of the world's coral reefs (Kinsey, 1983). Among the internal borers, clionid sponges and lithophagine bivalves can cause a comparable level of bioerosion, and of the external grazers, sea urchins are equally destructive. Reef frameworks are generally reduced to silt and fine sand by internal borers and to fine and coarse sand by external grazers. The combined effects of other bioeroders may also contribute importantly to reef erosion in particular areas or zones and at different times.

4.2. Conditions Favoring Bioerosion

Bioerosion increases under a variety of circumstances that can be classified according to (1) conditions causing coral tissue death and (2) conditions that provide a growth advantage to bioeroder compared with calcifying species' populations. Some of the more important situations that can alter the course of bioerosion are noted here in general terms. Specific examples are considered in section 4.4.

Aside from a few species that invade coral rock directly through living tissues (e.g., some boring sponges, bivalves, and possibly barnacles), the great majority of endolithic borers attack dead skeletons (Fig. 4–7). In general, any condition that causes coral tissue death will increase the probability of invasion by borers and grazers. Thus, any natural or anthropogenic disturbances that lead to the loss of live coral tissues will ultimately increase the chances of bioeroder invasion and higher rates of limestone loss. Many disturbances leading to tissue loss are obvious, including storm-generated surge that uproots and topples corals, sedi-

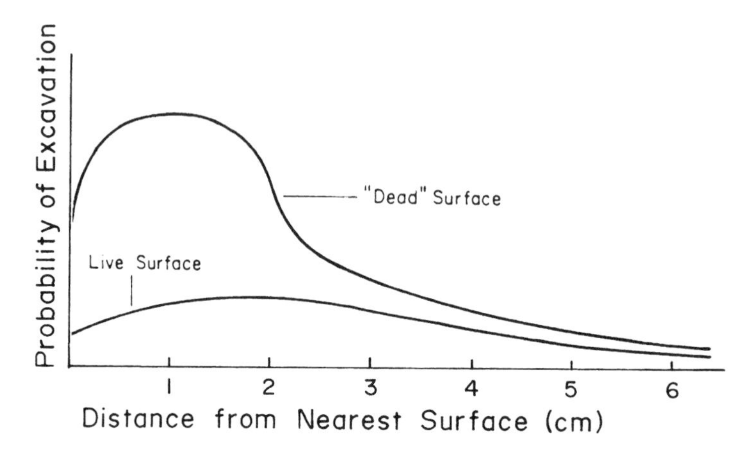

Figure 4–7. Graphic model showing the probability of excavation of endolithic bioeroders as a function of distance from a coral's surface (redrawn from Highsmith, 1981). Curves are illustrated for corals with dead and live surfaces.

ment scour and burial, tidal exposures, sudden temperature changes, freshwater dilution, sewage and eutrophication, predation, and disease outbreaks (Endean, 1976; Pearson, 1981; Grigg and Dollar, 1990).

While violent tropical storms are natural events that are known to seriously affect coral reefs, storm damage certainly must be exacerbated on reefs that have been heavily bioeroded beforehand. Sudden chilling episodes are also natural disturbances that can have devastating effects on tidally exposed or shallow coral assemblages, especially on high-latitude reefs. Numerous incidences of coral bleaching (loss of zooxanthellae and/or pigmentation) and mortality were observed worldwide in the 1980s, and many of these events occurred during periods of elevated sea temperatures coincident with El Niño–Southern Oscillation activity. Corals that were damaged or killed during these bleaching events have been subject to further damage by bioerosion. In some parts of the eastern Pacific where coral mortality was high and community recovery was slow, extensive damage by both internal and external bioeroders has been observed.

Predator outbreaks leading to high coral mortality, such as by seastar *(Acanthaster)* and snail *(Drupella)* corallivores reported from various areas of the Indo-Pacific, can set the stage for rapid bioerosion. Territorial damselfishes that colonize dead reef surfaces can cause complex responses that both increase and decrease bioerosion. Damselfishes that invade dead coral patches typically kill nearby corals while enlarging their territories. Studies in Australia have shown that the algal turf communities maintained by damselfishes favor the proliferation of internal bioeroders (Risk and Sammarco, 1982). However, the territorial defensive behavior of damselfishes also limits the bioerosive activities of external grazers such as parrotfishes and sea urchins (Glynn and Wellington, 1983; Eakin, 1993).

Coral tissue loss due to a variety of diseases can be substantial (Chapter 6; Peters, 1984). For example, "black line disease" or "black band disease," the result of a cyanobacterial infection (Rützler et al., 1983), may consume one-half of the living tissues of a coral during a single warm-season infestation. All live tissues may be sloughed from corals by white band disease, shut-down reaction, or stress-related necrosis. Though the causative agents of such diseases often remain ellusive, their occurrence seems to be influenced by increased sedimentation and turbidity.

Since the majority of endolith bioeroders are suspension or filter feeders in contrast to calcifying species, which are dominantly autotrophic, generally increases in nutrients, organic matter, and plankton biomass tend to favor increases in bioeroder compared with calcifier populations (Fig. 4–8). Because land runoff usually augments siltation and nutrient loading simultaneously (and sometimes pollutant levels), it is often difficult to distinguish between these effects.

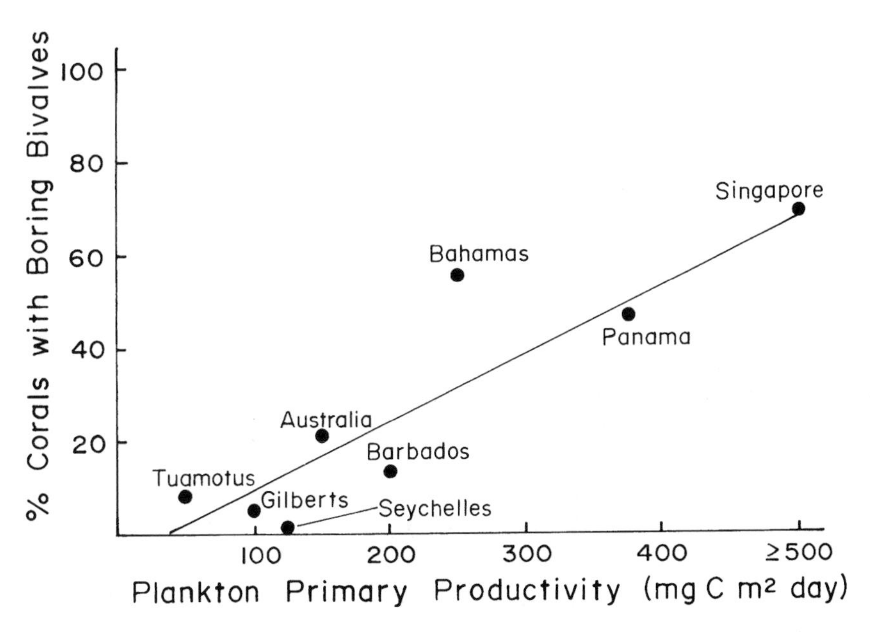

Figure 4–8. Relationship between the percentage of massive corals infested with boring bivalves and levels of phytoplankton productivity at several geographic locations (redrawn from Highsmith, 1980). Selected areas with values close to the plotted means are indicated. Each mean consists of various sampling areas and colony numbers, respectively, as follows: Tuamotu Islands—6, 212; Gilbert Islands—2, 58; Seychelle Islands—2, 12; Australia—7, 135; Barbados—7, 55; Bahama Islands—2, 64; Panama—4, 70; Singapore—5, 144.

There are at least two ways in which bioerosion is self-reinforcing. The first of these is the weakening effect of bioeroders on reef structures and the skeletons of calcifying organisms. For example, as bioerosion increases the volume of internal spaces (porosity) of coral skeletons, less mechanical force is required for breakage, toppling, and overturning (Fig. 4–9). Thus, heavily bioeroded reefs are more susceptible to damage by strong surge and projectiles accompanying violent storms. The second kind of positive feedback results from increasing levels of sediment production by bioeroders and its deleterious effects on calcifying populations.

Overfishing can also promote increased bioerosion on reefs. If natural fish predators of some bioeroder populations are eliminated, for example, triggerfishes that prey on sea urchins, then it is possible for grazing sea-urchin populations to increase in size with a devastating effect on reef limestones (see section 4.4).

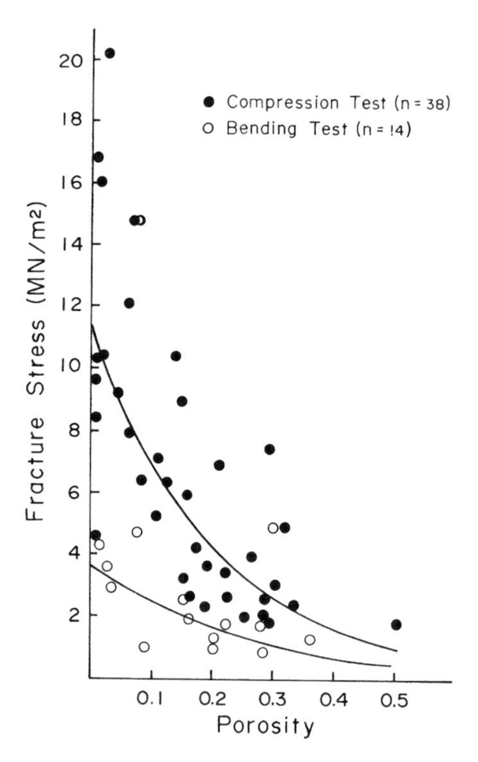

Figure 4–9. Plot of coral strength to breaking versus amount of bioerosion by *Lithophaga* (redrawn from Scott and Risk, 1988). The compression and bending tests are two measures of a coral's strength. n = newton, a unit of force; MN = 0.22481×10^6 kg m s^{-2}. Porosity indicates the percent of the skeleton removed.

4.3. Variety of Effects

The chief effect of bioerosion emphasized thus far is the mass of calcium carbonate that is reduced to sediments or is dissolved from reef substrata. The weakening of reef substrata by bioeroders that remove relatively little carbonate, but attack critical supporting structures, can be just as important in promoting reef erosion. Large massive corals may be easily toppled or overturned after their supporting bases have been weakened by endolithic borers such as *Cliona, Lithotrya,* and *Lithophaga,* or by grazers that attack bases and hollow out the interiors of colonies such as *Diadema* and *Eucidaris.* Many of the displaced corals on reefs, for example, those making up emergent, rubble ramparts or deep, forereef talus accumulations, owe their new locations in large measure to bioerosion. Large stands of *Acropora* corals that collapsed after *Acanthaster* predation on reefs in Japan, Palau, and Australia were presumably destabilized as a result of the weakening of dead skeletons by intensified bioerosion (Moran, 1986; Birkeland and Lucas, 1990).

Aside from weakening reef substrata, the cavities produced by bioeroders increase habitat complexity and thus the variety and biomass of reef-associated organisms. Numerous reef species live permanently attached to cavity walls, pass particular stages of development in cavities, and reside in cavities by day or night. Reef cavities tend to collect sediments that are produced locally or are transported to reefs from more distant sources. The microenvironmental settings of cavities promote internal cementation and the strengthening of reef substrata. Cycles of internal bioerosion, infilling of cavities, and cementation may be repeated so that eventually the reef rock appears quite different from its original condition.

The sediments generated by bioeroders accumulate around reefs and eventually infill and bury frame-building species (Fig. 4–10). This effect leads to the shoaling of reef waters and influences the development of reef zonation. Under moderate regimes of bioerosion, sediment accumulation does not overwhelm reef framework growth; however, excessive bioerosion can lead to premature burial and widespread coral death.

When bioerosion is excessive it can reduce the topographic complexity of reefs. The reefs noted above in the western Pacific, which were subjected to intense predation by *Acanthaster* and then bioeroded, lost much of their three-dimensional structure with the collapse of the *Acropora* canopies. The loss of these erect corals would eliminate important microhabitats for fishes. The topographic complexity of eastern Pacific reefs can also be reduced by echinoid bioerosion following El Niño disturbances. Coral reefs in the eastern Pacific, particularly in the Galápagos Islands, are currently being bioeroded to rubble and sediment following high coral mortality and low recruitment, respectively, during and after the 1982–1983 El Niño event. Erect, branching coral frameworks

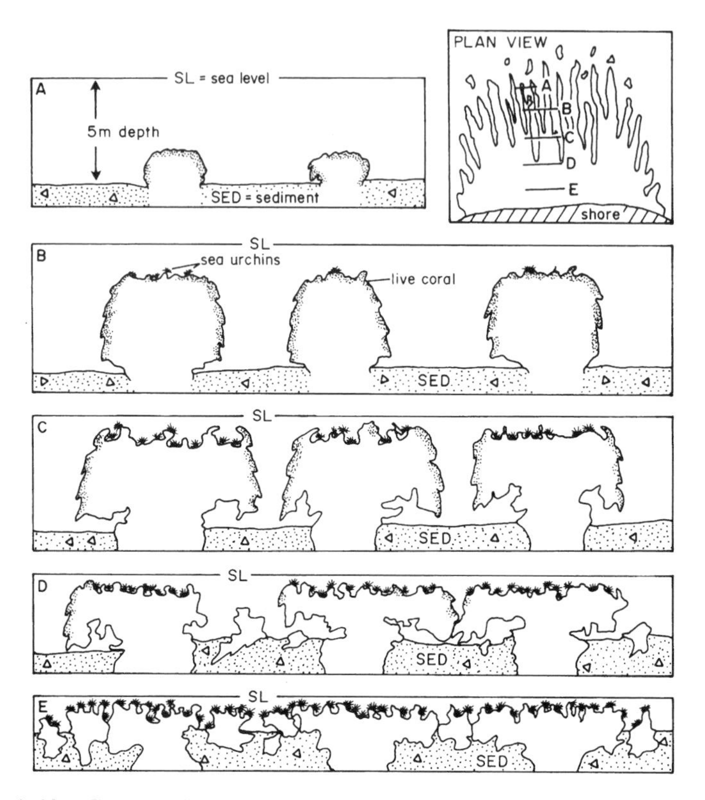

Figure 4–10. Cross-section views of a fringing reef off the west coast of Barbados showing coral framework growth, bioerosion, and infilling by bioeroded sediments. Panels **A** to **E** illustrate seaward (deep) to shoreward (shallow) reef sections. The inset plan view shows the location of the panels (after Scoffin et al., 1980).

are collapsing and massive corals are being detached from the substratum and crumbling. If present trends continue, coral recruitment could be suppressed altogether with the establishment of alternative community types.

Like many kinds of plants that spread from cuttings, it seems that some corals may actually benefit from increased breakage facilitated by bioerosion. A common mode of reproduction in many branching coral species is by asexual fragmentation (Tunnicliffe, 1979; Highsmith, 1982). It has been argued that propagation by this means, which usually results in local rather than distant dispersal, is advantageous to populations that are well adapted to particular environmental settings. Asexual reproduction occurs most commonly among branching, platy, and other such colonies of delicate morphology, with bioerosion aiding breakage by mechanical and biotic agents. Large clones of corals that dominate certain reef zones have arisen by this means (Highsmith, 1982).

4.4. Case Studies

Four documented cases of environmental alterations that have affected reef-building corals and led to extensive bioerosion are now examined. The first two examples, disturbances caused by El Niño–Southern Oscillation and predator outbreaks, are natural events. Runoff and overfishing effects are then examined, representing two examples caused by humankind.

4.4.1. El Niño–Southern Oscillation

Elevated seawater temperatures that accompanied the 1982–1983 El Niño–Southern Oscillation caused high coral mortality on reefs in the equatorial eastern Pacific. Mortality ranged from 50% to 99%, resulting in the virtual elimination of coral cover on many reefs. Coral recruitment has been low to nonexistent on the affected reefs, which have shown little signs of recovery after 10 years.

Sea-urchin abundances have increased dramatically on dead reef patches. In Panama, *Diadema* population densities have increased from 3 ind. m^{-2} before 1983 to 80 ind. m^{-2} after 1983 (Glynn, 1988). Similarly, in the Galápagos Islands, *Eucidaris* population densities increased from 5 to 30 ind m^{-2} from before to after 1983. The grazing activities of these sea urchins are very destructive (Table 4–2) and their sudden increases in population size, combined with low coral recruitment, have resulted in severe bioerosion of coral-reef frameworks. Post El Niño bioerosion rates for *Diadema* in Panama amounted to 10–30 g dry wt $CaCO_3$ m^{-2} day^{-1}, and for *Eucidaris* in the Galápagos 50–100 g dry wt $CaCO_3$ m^{-2} day^{-1}. Carbonate breakdown caused by other external and internal bioeroders was about equal to that caused by sea urchins in Panama, but only about one-fifth of the erosion caused by sea urchins in the Galápagos Islands. Total bioerosion ranged from 10 to 20 kg $CaCO_3$ m^{-2} yr^{-1} in Panama and from 20 to 40 kg $CaCO_3$ m^{-2} yr^{-1} in the Galápagos Islands. Both of these rates exceed net carbonate production of 10 kg $CaCO_3$ m^{-2} yr^{-1}, estimated for reefs in these areas before 1983. If bioerosion continues at this pace, without an increase in coral recruitment, it is highly likely that many reef formations in the eastern Pacific will disappear.

4.4.2. Crown-of-Thorns Seastar (Acanthaster)

This example is instructive because it reveals some of the long-term consequences of coral death and bioerosion at the community level. Between 1981 and 1982, the corallivore *Acanthaster planci* increased greatly in abundance at Iriomote Island, southern Japan, and by the end of 1982, it had killed virtually all the corals on a large study reef (Sano et al., 1987). This sudden loss of live coral precipitated major changes in the physical and biological character of the coral reef.

About 2 years following the *Acanthaster* outbreak, most of the erect coral *(Acropora)* canopy had collapsed, a result of bioerosion and water movement. Compared with the live reef, the dead reef exhibited low structural complexity. By 1986 all of the corals were broken apart and the reef formation had been converted into a flat plain of unstructured coral rubble. The degradation of the reef was correlated with marked changes in the fish community. As the topographic complexity of the reef decreased, the numbers of associated fish species and their abundances also declined. Fishes that fed exclusively on live coral tissues disappeared completely from the dead reefs. The declines in fishes with other diets, for example, planktivores, herbivores, and omnivores, were believed to be due in large measure to the loss of living space and to overall declines in prey on the degraded reef.

4.4.3. Runoff (Eutrophication, Sedimentation, Freshwater, and Pollutants)

One of the best examples of reef degradation caused by runoff is that reported for the Kaneohe Bay, Hawaii, coral reef ecosystem (Smith et al., 1981; Jokiel et al., 1993). Because human mismanagement of the Kaneohe Bay watershed has led to multiple effects, for example, sewage pollution, agricultural runoff, increased sedimentation, and freshwater dilution, it is not always possible to identify individual or combined stressor effects. However, the occurrence of coral-reef mass mortalities during storm floods and a general decline in coral cover during a period of increasing sewage stress implicate these stressors in the degradation of Kaneohe Bay coral reefs over the past several decades.

During the first half of the twentieth century the coral reefs of Kaneohe Bay were in a healthy state, supporting local artisanal fisheries and offering one of the best underwater vistas of "coral gardens" in the Hawaiian Islands. In 1963, a large sewage outfall was installed in the bay, which had an increasing effect on corals until 1978, when the outfall was moved to the deep ocean outside the bay. The eutrophication caused by increasing sewage loads favored the growth of a bubble alga *(Dictyosphaeria)* and suspension-feeding and bioeroding species that combined to degrade the reef communities over a 15-year period (Fig. 4–11). Following the sewage diversion, clear signs of renewed coral growth, reduced bioerosion, and reef community recovery were evident by 1983. Severe storm flooding in 1987 caused extensive coral mortality, but surviving corals quickly resumed rapid growth and the condition of reef communities (as of 1993) has remained favorable. This case history illustrates a degree of resiliency to a disturbance that might have led to reef community collapse in a sewage-stressed environment.

4.4.4. Overfishing

Several studies in the Caribbean and off the Kenyan coast in the Indian Ocean have presented evidence suggesting that sea-urchin abundances are controlled

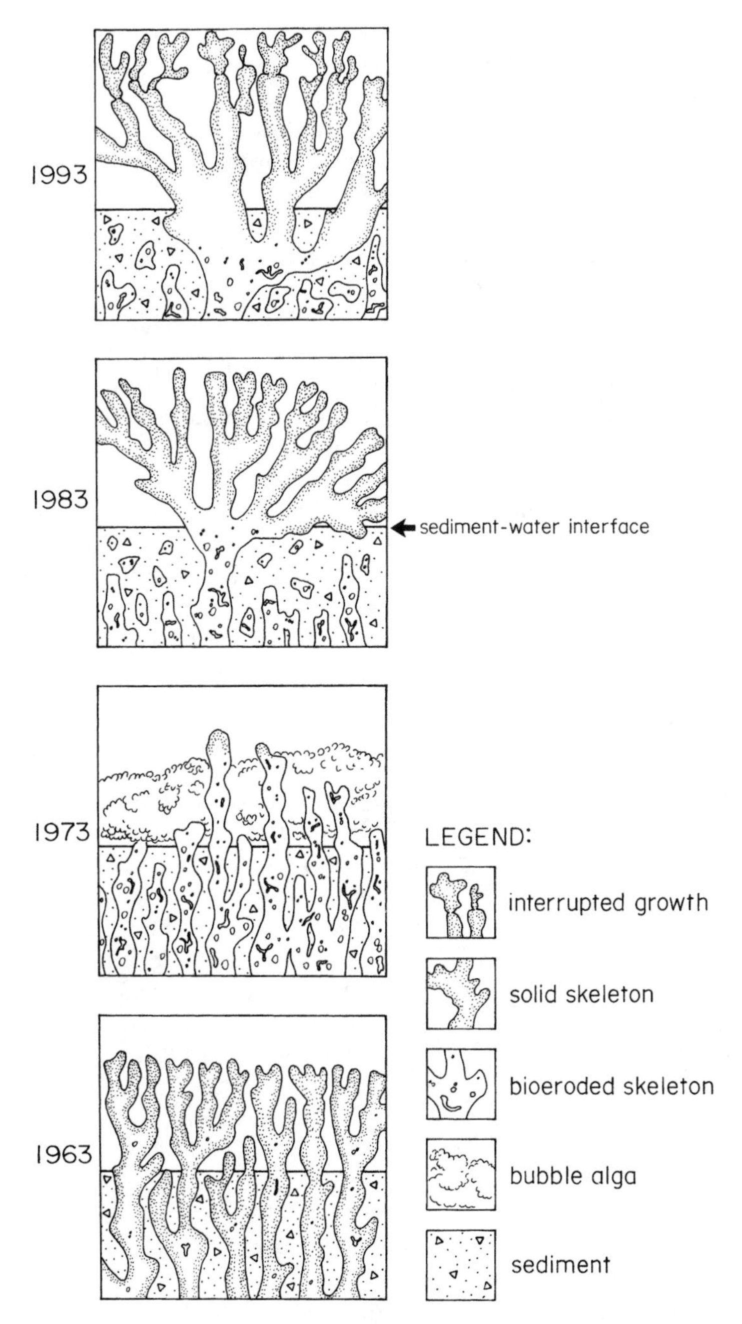

Figure 4–11. Cross-section sketches of *Porites compressa,* the predominant frame-building coral of the Coconut Island fringing reef. Prepollution (1963), pollution (1973), and postpollution (1983 and 1993) periods are shown (modified after Jokiel, 1986).

by finfish predators. When fish predators of sea urchins are abundant, urchin abundances tend to be low, but when fishing pressure is high, leading to the disappearance of urchin predators, then urchins can become exceedingly abundant. A study of protected (nonfished) and overfished Kenyan coral-reef lagoons indicates that the removal of top, invertebrate-eating, fish carnivores can have cascading effects on coral-reef community structure and function (McClanahan and Shafir, 1990).

Triggerfish predators of sea urchins were relatively abundant in protected coral-reef lagoons, but rare in comparable unprotected environments. The removal of the natural predators of sea urchins by overfishing resulted in several direct effects on the urchin prey and several indirect effects on the condition of the coral-reef community. Overfished reefs demonstrated high sea-urchin abundances, high urchin survival, and high urchin diversity compared with nonfished reefs. Correlated with the dominance of sea urchins on overfished reefs were declines in (1) live coral cover, (2) calcareous and coralline algal cover, (3) substratum diversity, and (4) topographic complexity. These changes were caused by increased substratum bioerosion, especially by *Echinometra mathaei* (de Blainville), the competitively dominant sea urchin in unprotected Kenyan reef lagoons. The end result of overfishing is accelerated bioerosion, a reef surface dominated by algal turf, and likely a decline in the reef's fisheries productivity.

4.5. Conclusion

Except for obvious reef destruction by large populations of sea urchins, bioerosion per se as a possible threat to coral reefs is seldom considered explicitly. This is probably because of the large amount of "cryptic" bioerosion caused by endoliths and the often delayed effects of bioerosion on coral-reef communities. For example, descriptions of reef damage caused by violent storms are numerous in the literature, but the contributory effects of bioerosion are seldom mentioned. The prior weakening of reef structures by bioerosion or the accumulation of sediments causing scour and burial during a storm are effects that may have been initiated years before an acute disturbance event resulting in reef devastation.

What are some of the measures that can be taken to limit bioerosion? The most obvious is to reduce coral mortality because numerous bioeroders increase their activities and abundances on dead reef substrata. Direct damage to calcifying organisms can be reduced significantly by several practices already adopted within protected coral-reef parks. For example, the use of mooring buoys and navigational markers, the prohibition of destructive fishing techniques, and the banning of coral collecting or touching live corals have all alleviated damage to coral reefs in many areas. The possibility of indirect effects, such as overfishing causing increases in bioerosion, should also be considered in coral-reef management plans.

Another method of limiting coral mortality after severe physical damage, for example, by a ship grounding, involves restoration techniques to stabilize damaged corals and reef substrata (Hudson and Diaz, 1988). Hard and soft corals may be transplanted and cemented to stable reef substrata, fractured frameworks may be secured, and the rebuilding of reef topography may be accomplished by replacing and cementing dislodged corals and sections of framework.

Numerous effects that can accelerate bioerosion are often far removed from coral reefs and therefore sometimes difficult to link with reef decline. Deforestation, land-clearing and mining activities lead to increased sedimentation, freshwater dilution, and nutrient loading around reefs that may be situated hundreds of kilometers from the affected sites. These sorts of activities may alter reef environments such that certain types of bioeroders could increase in number and possibly accelerate destructive processes. The potential damage of such anthropogenic stresses to coral reefs also may be augmented by natural disturbances such as violent storms, extreme temperature changes, diseases, and predator outbreaks. For example, most corals may tolerate low salinities for a few hours or days, but salinity stress in combination with a pathogen could precipitate high coral mortality. Many kinds of runoff include combinations of several pollutants, for example, sewage, detergents, heavy metals, fertilizers, pesticides, and oil, that may act synergistically to reduce live coral cover.

In summary, the dynamic balance between reef growth and bioerosion depends on the vitality of numerous calcifying species. If humankind's activities can be limited to nonintrusive pursuits such as observing and filming reef organisms, and if reef water quality and natural circulation patterns can be safeguarded, then one of the world's most exquisite ecosystems can be enjoyed by posterity.

5

Interactions Between Corals and Their Symbiotic Algae

Gisèle Muller-Parker and Christopher F. D'Elia

The very words "coral reef" suggest the obvious adaptive and evolutionary success of hermatypic (or reef-building) corals in becoming conspicuous and abundant organisms of many of the shallow, tropical benthic communities that gird the Earth's surface. There is little doubt that the mutualistic relationship between the coral animal and its intracellular algae (the zooxanthellae) enables hermatypic corals to be significant contributors to the organic productivity and carbonate framework of coral reefs.

This chapter discusses the coral-zooxanthella symbiosis from the perspective of the nutrient dynamics and energetics of the association and in the context of the coral-reef ecosystem in which they are found. We examine aspects of the structure and function of the symbiosis that are believed to account for the high rates of calcification and productivity exhibited by reef corals. We discuss some of the factors that are believed to influence the density of symbionts and hence the physiological balance between the symbiotic partners. We consider the possible effects of both natural and anthropogenic events on coral-reef ecosystems and how they might be expected to affect the stability and survival of the symbiosis. The costs and benefits associated with the symbiotic condition are presented. We conclude by speculating about the value of using the coral symbiotic association as a measure of the "health" of coral-reef ecosystems.

5.1. Description of the Symbiosis

5.1.1. Coral Anatomy and Location of Zooxanthellae

Scleractinian corals are colonies of polyps linked by a common gastrovascular system (Fig. 5–1). Polyps are small fleshy extensions of the coral cover (typically millimeters in diameter) compared to the often massive structure of the colony, which can be meters in diameter. In spite of the large size they may attain, corals

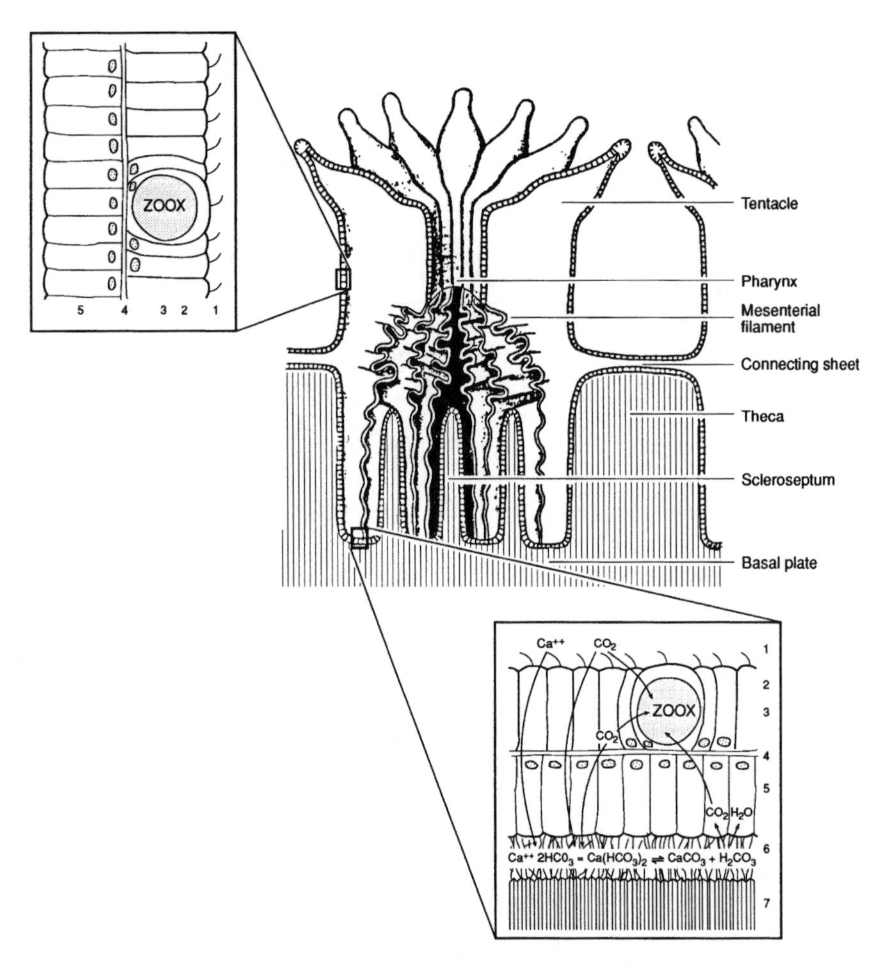

Figure 5–1. Schematic diagram of coral cross-section (modified from Barnes, 1980), with boxed insets showing coral tissue layers in contact with the seawater and the skeleton (insets from Kuhlmann, 1988, Figure 1). Numbers represent the following: (1) seawater in the gastrovascular cavity; (2) gastrodermis; (3) zooxanthella in gastrodermis; (4) mesoglea; (5) epidermis; (6) matrix with minute crystals of calcium carbonate; (7) skeleton. The equation for coral calcification is provided in (6).

are simply composed of two cell layers: the epidermis (sometimes referred to as ectodermis) and the gastrodermis (sometimes referred to as endodermis). These cell layers are separated by a thin connective-tissue layer, the mesoglea, that is composed of collagen, mucopolysaccharides, and cells (Fig. 5–1). The calcareous external skeleton is secreted by the lower epidermal layer (calicoblastic epidermis; Fig. 5–1). The upper layer of epidermis (oral or free epidermis) is in contact with seawater.

Many coral polyps are biradially symmetrical, with the central gut cavity lined by gastrodermis (Fig. 5–1). Tentacles surrounding the mouth are used for capture of zooplankton. Food consumed by one polyp is shared with neighboring polyps via the gastrovascular system that functions in the circulation and digestion of food particles. Polyp mouths also provide direct exchange of water and particulate food and wastes between the gastrovascular system and the external seawater. There is also general diffusion of gases and soluble wastes across the body surface.

The arrangement, number, and size of polyps are characteristic for each coral species. However, coral species may exhibit different morphologies depending on environmental conditions such as water motion and light, and this causes problems in taxonomy of corals (Veron and Pichon, 1976).

Zooxanthellae reside exclusively in membrane-bound vacuoles in the gastrodermal cells (Fig. 5–1). Areal densities of zooxanthellae normally exceed one million per square centimeter of coral surface, although algal density in corals may be highly variable on both temporal and spatial scales. There is a lack of information about seasonal and spatial variability of zooxanthellae densities within colonies and among coral species. This information is important given the critical role of zooxanthellae in coral nutrition, and our search for causes of coral bleaching, a phenomenon in which the host loses its symbionts.

5.1.2. Zooxanthellae

"Zooxanthella" is a general descriptive term for dinoflagellate symbiotic algae that live in animals, including corals, sea anemones, molluscs, and other taxa. Most zooxanthellae are currently placed in the dinoflagellate genus *Symbiodinium*, although zooxanthellae belonging to the genus *Amphidinium* are also represented. The true taxonomic affiliation of zooxanthellae from different animal hosts has not been investigated very thoroughly, and it is likely that other taxa will be derived in the future (Blank and Trench, 1985; Rowan and Powers, 1992).

The dinoflagellates comprise a diverse group of mostly free-swimming single-celled microscopic planktonic algae that exhibit a variety of feeding modes ranging from photoautotrophy (photosynthetic carbon fixation) to heterotrophy (dissolved organic carbon uptake or feeding on particulate food). Zooxanthellae are able to photosynthesize and contain characteristic dinoflagellate pigments (diadinoxanthin, peridinin) in addition to chlorophylls *a* and *c*. They are brown or yellow-brown in color. There is some evidence that zooxanthellae take up dissolved organic carbon from host sea anemones, although photosynthesis is likely to contribute most of their carbon requirement.

Zooxanthellae can live independently of their animal host. Zooxanthellae living in animal cells are usually found in the coccoid stage (nonmotile, lacking flagella); this differs from the free-living motile (dinomastigote) stage that possesses two flagella and exhibits a characteristic swimming pattern. In culture, zooxanthellae alternate between the coccoid and dinomastigote stages. The dominant dinoflagellate feature evident on the ultrastructural level is the nucleus with permanently condensed chromosomes (dinokaryon). Sexual reproduction has not been documented for these algae.

Although numerous animals in the coral-reef community are hosts to zooxanthellae, for this chapter we restrict our discussion to zooxanthellae in symbiosis with scleractinian corals. To the casual observer, zooxanthellae from different coral species look the same. However, detailed studies of the ultrastructure and genetic composition of zooxanthellae have shown that zooxanthellae from different hosts are highly diverse and distinct (reviewed by Trench, 1993; Rowan, 1991). Consequently, although it is possible to make some broad generalizations about the responses of symbiotic zooxanthellae from studies conducted with one or two species, it is important to remember that each zooxanthella strain or species is likely to have different adaptive capabilities and tolerances to environmental extremes (Chapter 15).

5.1.3. Acquisition of Zooxanthellae by Corals

Zooxanthellae are well established in new corals derived from both asexual and sexual reproduction. In asexually produced (clonal) coral colonies, zooxanthellae are directly transmitted in coral fragments that form the basis of new colonies. In sexually produced corals, acquisition of zooxanthellae is either directly from the parent or indirectly from the environment. When zooxanthellae are acquired and whether or not the eggs contain zooxanthellae are characteristics of coral species. A confounding factor is the frequency of sexual versus asexual reproduction in each coral. A coral that relies almost exclusively on asexual reproduction for propagation, where direct transmission of zooxanthellae is guaranteed, may not exhibit highly developed mechanisms for transmission of these algae during sexual reproduction.

During direct transmission via sexual reproduction, parental zooxanthellae are transferred to the eggs or to larvae brooded by the parent. How the algae get into the eggs has not been well investigated. Cytoplasmic extensions of the gastrodermal cells that contain zooxanthellae may invade the egg plasm, as has been described for marine hydroids (Trench, 1987). These eggs may be released and fertilized in the water, or the larvae may develop within the mother coral. If eggs do not contain zooxanthellae, larvae brooded by the parent through the early stages of development (Chapter 8) may take up algae at any time prior to release.

Corals that do not inherit parental zooxanthellae must obtain them from seawater. The concentration of zooxanthellae in seawater over the reefs is likely to be quite low under normal conditions; positive chemotaxis of motile zooxanthellae toward the coral animal may increase the probability of contact between appropriate partners. Zooxanthellae also may be supplied indirectly to the coral by ingestion of fecal material released by corallivores and of zooplankton prey containing zooxanthellae. Regardless of the mechanism, indirect acquisition of zooxanthellae provides the potential for colonization by zooxanthellae that are genetically distinct from parental symbionts. Whether or not this actually occurs depends on host animal recognition of a suitable symbiont and the chance encounter of the appropriate partners. We do not know if the specificity of host recognition and speciation in zooxanthellae are related to the modes of symbiont acquisition by different corals.

Coral "bleaching" may also provide the potential for establishment of a new population of zooxanthellae in adult corals (Buddemeier and Fautin, 1993) since zooxanthellae may be reacquired indirectly. Corals turn white (become bleached) when they lose zooxanthellae; expulsion of zooxanthellae may involve loss of the host coral cells. Corals may also appear bleached when zooxanthellae are retained but lose their photosynthetic pigments. Corals that survive a bleaching event involving the loss of zooxanthellae eventually regain normal densities of zooxanthellae (they "rebrown") when environmental conditions improve. The source of zooxanthellae for the recovery and rebrowning of a coral is unknown. Free-living zooxanthellae may invade corals after a bleaching event, residual zooxanthellae may repopulate their bleached host coral, or both may occur. If genetically distinct strains or species of zooxanthellae reside in corals, rebrowning by residual zooxanthellae in corals after bleaching may also change the genetic composition of the population of symbiotic algae within a coral.

5.2. Nutrition and Adaptations to Environmental Factors

5.2.1. Coral Nutrition

The success of corals in low-nutrient tropical waters is due largely to the variety of modes that corals utilize to obtain nutrition (Fig. 5–2). The animal has two primary feeding modes: capture of zooplankton by polyps and receipt of translocated photosynthetic products from its zooxanthellae. The amount of photosynthetic carbon translocated to the animal host is often sufficient to meet its metabolic respiratory requirements. There is some evidence that corals also take up dissolved organic compounds from seawater, a process that is aided by the extremely high surface area to volume ratio of corals. However, the nutritional importance of this uptake, and of that of other food sources such as microplankton and bacteria, is uncertain. Animal metabolic waste products derived from holozoic feeding are retained within the coral, as they are a source of inorganic nutrients required by the zooxanthellae (Fig. 5–3).

Zooxanthellae are autotrophs, and thus require only inorganic nutrients, carbon dioxide, and light for photosynthetic carbon fixation. Inorganic nutrients may be acquired from animal waste metabolites, or directly from seawater. There is some evidence that zooxanthellae can obtain organic nutrients from the animal.

The variety of coral nutritional modes suggests that corals are adaptively

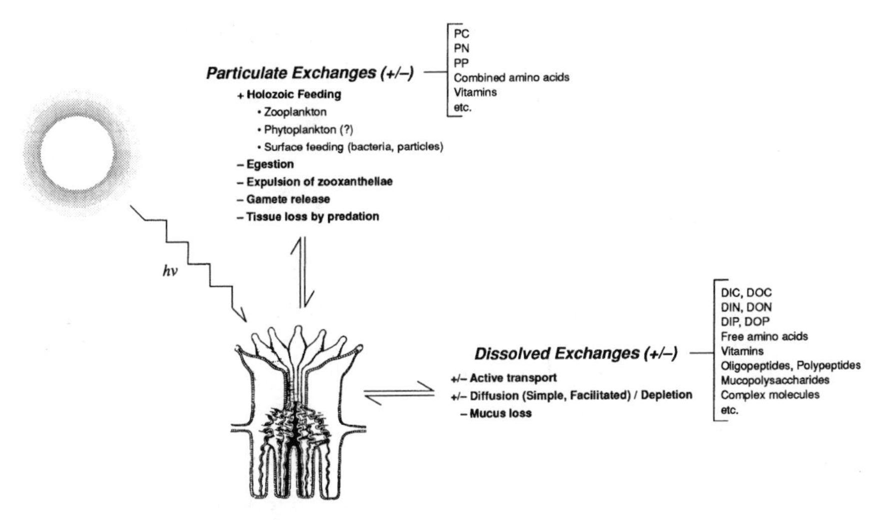

Figure 5–2. Particulate (P) and dissolved (D) exchanges of inorganic (I) and organic (O) carbon (C), nitrogen (N), and phosphorus (P) between a coral and the seawater environment. "+" exchanges represent inputs from the environment to the coral, and "–" exchanges represent losses from the coral to the environment. Internal exchanges (not shown) include uptake by symbiotic algae and translocation between the algae and the host.

polytrophic and opportunistic feeders. This polytrophism seems to account for the ability of corals to thrive in low-nutrient water (Fig. 5–3). However, there are environmental constraints and energetic costs associated with the maintenance of symbiotic algae, as discussed below, that may favor holozoic modes of nutrition under certain circumstances.

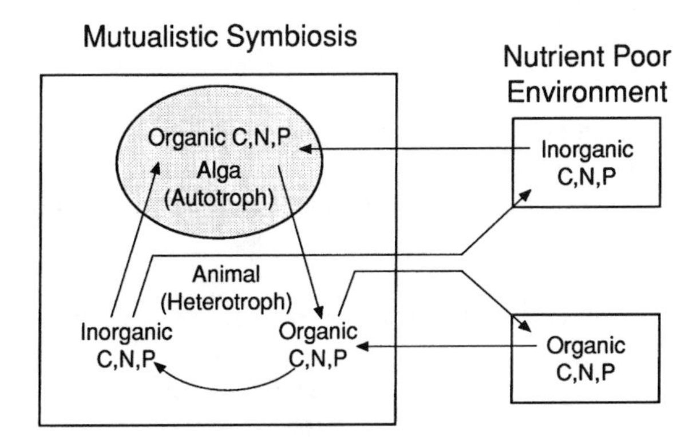

Figure 5–3. A model of the potential pathways of carbon and nutrient (N,P) flux in a symbiotic reef coral (after Lewis and Smith, 1971).

5.2.2. *Productivity of Corals and Role of Zooxanthellae in Calcification*

Photosynthetic carbon fixation by zooxanthellae ($P_{zooxanthellae}$) accounts for the high productivity of corals (Hatcher, 1988; Chapter 7). Any carbon fixed by zooxanthellae in excess of their own respiratory (R) and growth requirements is potentially available to the host coral as a carbon and energy source. If the carbon fixed by zooxanthellae meets or exceeds the combined coral and zooxanthellae respiratory carbon requirement ($P_{zooxanthellae} > R_{zooxanthellae+animal}$; ratio of $P:R > 1$), the coral is potentially photoautotrophic with respect to carbon and does not require external carbon sources. When P:R is less than one, carbon must be supplied from other nutritional sources (Fig. 5–2). P:R ratios derived from oxygen measurements generally show that shallow-water corals have a $P:R > 1$, while the P:R of deep-water corals is less than one. Therefore, deep-water corals are more likely to require external sources of carbon than are shallow-water corals.

Zooxanthellae enhance coral calcification and are responsible for much of the formation of the massive coral-reef framework. The direct relationship between coral calcification rate and light indicates the fundamental importance of photosynthesis (Barnes and Chalker, 1990), although how photosynthesis enhances calcification is still unresolved. Models that have been proposed for the mechanism of enhancement of calcification by zooxanthellae are reviewed by Johnston (1980). The models consider the alteration of the physicochemical environment by zooxanthellae to provide favorable conditions for calcification and the contribution of zooxanthella products to the process of calcification. Photosynthesis raises the pH, providing more carbonate ions for calcium carbonate precipitation (Fig. 5–1). Zooxanthellae, through the uptake of inorganic nutrients, may remove $CaCO_3$ crystal inhibitors such as phosphate from calcification sites. Photosynthesis may provide the energy for active calcification, or promote the synthesis of the organic matrix upon which $CaCO_3$ is deposited. Despite the abundance of models, the significance of each one with respect to the link between calcification and symbiosis with zooxanthellae has not been determined.

The following section describes the factors that influence the productivity of zooxanthellae and hence the amount of carbon potentially available to the coral. The balance between primary production and respiration (P:R) for a coral with a constant population density of zooxanthellae depends on environmental factors that affect both photosynthesis and respiration rates. The most important factors are light and temperature.

5.2.3. *Effect of Light and Temperature on Productivity of Zooxanthellae*

Total daily production depends on the amount of light experienced by zooxanthellae, which depends on the light incident at their depth and is transmitted through the coral animal tissues. Photosynthetic rates increase directly in response to increase in light intensity up to a certain light intensity, after which they are independent of further increases in light (Hatcher, 1988). Consequently, mechanisms for enhancing light capture and penetration through the coral could be advantageous and are evident in both the algal and animal partners. For example, the animal tissue pigmentation may filter light, transmitting photosynthetically active wavelengths and blocking harmful ultraviolet (UV) light. By their location in a thin layer of living tissue, supported by a strong skeleton, zooxanthellae maximize light capture and are highly productive. They are phytoplankton that enjoy many of the benefits of a macroscopic benthic lifestyle such as that of a seaweed. Such advantages include maintenance in the photic zone with good water exchange.

As in trees, upper layers of the coral canopy receive more light than the understory surfaces of coral branches, and zooxanthellae in shaded and unshaded portions of the colony may exhibit different degrees of photoadaptation and exist at different population densities. This suggests that algae are localized like leaves on a tree, and that there is little exchange of algae between different parts of the coral colony.

Individual zooxanthellae acclimate to changes in light by changes in their photosynthetic systems, including the light-harvesting ability of photosynthetic units (amount of pigment) and the rate of carbon fixation (enzymatic adaptations). Zooxanthellae in corals from shaded habitats usually contain more chlorophyll and thus are more efficient at light capture; the size of their light-harvesting units is large. Zooxanthellae in corals in high light environments contain less photosynthetic pigment, in smaller light-harvesting units, but have high rates of carbon fixation by containing more photosynthetic units. Light intensity also varies on a daily and seasonal basis, and zooxanthellae are likely to acclimate to these changes.

The ability of different coral species to photoadapt to light via these mechanisms may limit the distribution of coral species in different light environments. Genetic differences in the photophysiology among different species or strains of zooxanthellae are also likely to have a large effect on the optimal light regimes of individual coral species, and on that species distribution and ecological role. Clearly, proper quality and quantity of light are essential at the organismal level for (1) the overall stability of the association; (2) the ability of the symbiosis to exhibit net production ($P > R$); (3) the expression of photosynthetic pigments and the abundance of the zooxanthellae; and (4) diel behavioral aspects such as polyp expansion and contraction.

The reliance of corals on phototrophic nutrition and photosynthetically enhanced calcification favors the proliferation of corals in shallow, clear waters. Recently, attention has focused on the harmful effects of high light energy on corals. Clear, shallow, tropical waters transmit UV radiation much deeper than was previously believed. Some corals grow well in the presence of high UV radiation, whereas others are killed by exposure to high UV. The former include corals that are normally found in high light (in shallow water), while the latter include corals that live at greater depths or that are "shade-loving" species. Corals have UV-absorbing "sunscreen" compounds that protect against UV damage. The quantity of their UV-absorbing sunscreens is related to the incident UV energy, and declines in corals with depth. It is not known if the animal or the zooxanthellae (or both) produce these sunscreens; both contain UV-absorbing compounds.

Temperature affects metabolic rates of corals and their symbionts. The influence of temperature on productivity depends on how photosynthesis and respiration of both the algae and animal respond to changes in temperature. The ability of corals to acclimate to change in temperature may vary for individual species and for corals in different habitats. For example, shallow corals may tolerate a wider range of temperatures than deep-water corals. Exposure to temperatures exceeding the tolerance range of the symbiosis affects its stability, generally resulting in the loss of zooxanthellae and possibly in the death of the host. Whether the loss is due to a direct temperature effect on the animal, zooxanthellae, or both, is not known. However, the rate of the temperature change and the duration of the temperature extreme appear to be factors. As with changes in light, the ability of the coral to adapt to change in temperature or in other environmental factors depends on the acclimatory capability of one or more of the following: the animal, the zooxanthellae, and the symbiotic association as a whole.

5.2.4. Effect of Nutrient Supply on Zooxanthellae in Corals

Corals thrive in seawater where the supply of the major growth-limiting nutrients, nitrogen and phosphorus, is very low and limits net coral production. A tenet of algal-animal symbiosis is that it evolved in response to low nutrient supply, and that accordingly, such conditions provide corals with a competitive advantage over other benthic species. Corals conserve nitrogen by having low rates of protein catabolism and catabolizing translocated lipids and carbohydrates (Szmant et al., 1990).

Various sources of nitrogen and phosphorus exist for symbiotic zooxanthellae. The algae can obtain inorganic nutrients via recycling of waste products from their animal host, and "new" nutrients from the uptake of dissolved inorganic compounds from seawater and zooplankton capture by the coral (Figs. 5–2 and 5–3). In fact, early investigations of the role of symbiotic algae in corals suggested that they served as the kidneys for the animal (Yonge and Nicholls, 1931), although this is now considered highly unlikely. Although dissolved concentrations of nutrients are very low in most tropical waters, mass transport of nutrients via advection across coral surfaces may be sufficient to supply the nutrient requirements of the algae.

Although corals are undoubtedly adapted to waters containing very low levels of nutrients, corals can persist when nutrient levels around reefs become periodically elevated due to increased runoff from adjacent land, point source inputs (from sewage and industrial effluents), or periodic upwelling. Corals are unable to adapt to acute, high-level nutrient enrichments and generally compete poorly with benthic macroalgae under such conditions. Excess nutrients may decrease calcification rates, presumably because phosphate interferes with aragonite crystal formation during calcification (Simkiss, 1964).

Although there is some evidence that high phosphate levels (exceeding about 1 μM) reduce calcification and that high dissolved inorganic nitrogen levels (exceeding about 10 μM) destabilize the symbiosis by enhancing growth rates of the zooxanthellae, there is no firm evidence that even these levels, which would be considered extremely high for reef waters, directly affect the survival of the symbiosis. When growth rates of the zooxanthellae exceed coral growth or result in high population levels that are stressful to the coral animal, extra algae may be simply expelled. The expulsion of zooxanthellae may also represent a unique detoxification mechanism for the coral. For example, zooxanthellae have a high tolerance for heavy metals and accumulate them from seawater. Periodic expulsion of zooxanthellae could reduce the "body burden" of the heavy metals in the coral animal, as demonstrated for temperate sea anemones (Harland and Nganro, 1990).

Corals that survive direct effects of added nutrients may succumb to indirect effects such as reduction of light and overgrowth by fleshy macroalgae.

5.3. Stability of the Symbiosis

A stable symbiosis is defined as one in which the density of zooxanthellae in corals remains relatively constant under a given set of environmental conditions. This constancy may be important in balancing the benefits and the costs of the symbiosis (Table 5–1). It is therefore likely that the growth of zooxanthellae in corals is somehow regulated relative to the growth of the host. Either the growth rates of the zooxanthellae and the animal cells are comparable, or, if different, excess zooxanthellae are released from the coral.

It is difficult to make direct measurements of growth rates of zooxanthellae in coral tissue; doubling times have been estimated from diel measurements of the mitotic index of zooxanthellae. Doubling times for symbiotic zooxanthellae calculated by this approach show that these are at least an order of magnitude lower than doubling times of cultured zooxanthellae maintained in nutrient-enriched seawater.

Zooxanthellae densities may be self-regulating, as resource limitation may help preserve the balance between zooxanthellae and animal biomass and growth rates. The animal cell habitat indirectly slows zooxanthella division rates by limitations on space or diffusion of gases (CO_2, O_2) through animal protoplasm. As numbers of zooxanthellae increase in coral tissue, self-shading of cells will reduce available light and there will be intense competition for limited resources,

CO_2, and nutrients, potentially reducing net production and growth. In certain rapidly growing areas of the coral, animal growth rates may exceed those of the zooxanthellae, and populations of zooxanthellae are "diluted." For example, tips of branches of rapidly growing species often appear white and have reduced algal densities.

Environmental factors that are likely to affect both animal and algal growth include physical conditions and the availability of prey. Light directly affects photosynthetic productivity, while prey capture directly affects animal tissue growth and indirectly affects growth of zooxanthellae by its potential supply of nutrients. Until recently, zooxanthellae were considered to live in a nutrient-rich environment by virtue of their intracellular habitat (Fig. 5–3). However, these algae display characteristics that suggest that their growth is normally nutrient limited (Cook and D'Elia, 1987). The addition of dissolved inorganic nitrogen to seawater causes an increase in the growth rate of zooxanthellae (Hoegh-Guldberg, 1994) and in their nitrogen to phosphorus ratio (Muller-Parker et al., 1994). The opposite trends are obtained when symbiotic associations are maintained with no particulate food resources in low-nutrient seawater. It is unknown if the animal withholds nutrients from its algae, or if the supply is limited by the availability of nutrients (including animal prey and seawater as sources), or both. Nutrient limitation of the growth of zooxanthellae may favor the coral animal by creating an excess of photosynthetic carbon products that cannot be used for production of new cells and is therefore translocated to the animal host (Falkowski et al., 1993).

The stability of the symbiosis may be disrupted by changes in environmental factors that have different effects on algal and animal growth, by stressors that result in the mass expulsion of zooxanthellae resulting in coral bleaching, and ultimately, by a positive or negative shift in the balance between benefits and costs of the symbiosis (Table 5–1). For example, environmental stresses such as extreme temperatures, air exposure, or rapid change in salinity cause coral bleaching. If the coral survives the stress and regains a normal population of zooxanthellae, there may be a period during the repopulation phase when algal growth rates exceed those of the animal tissue.

5.4. Cost-Benefit Analysis of the Symbiosis

Table 5–1 presents features that we consider to represent significant benefits and costs of the symbiotic relationship between zooxanthellae and their coral animal hosts. These features are presented from the perspectives of both partners and the complete association. We do not regard this table to be all inclusive or complete, but we do suggest that this approach, albeit anthropomorphic, is a useful way to consider the symbiosis that may be helpful in framing future research directions.

Table 5-1. Benefits and Costs of the Symbiotic Relationship for the Coral Animal and for Zooxanthellae

Benefits	Costs	Indirect (+/–) Effects
A. Animal		
Supply of reduced carbon, providing low respiration costs and conservation of metabolic reserves	Regulation of algal growth and production of perialgal vacuoles	High surface area to volume ratio favors both light capture and prey capture
Increased growth and reproduction	Defenses against high oxygen tension, high light, and UV	Restriction to the photic zone
Increased calcification rate	Mechanisms for rejection of foreign or excess algae	
Conservation of nutrients	Vulnerability to environmental stresses that affect plants	
Sequestration of toxic compounds by algae		
B. Zooxanthellae		
Supply of CO_2 and nutrients from host	Translocation of a significant fraction of photosynthetic carbon to animal	Nutrient supply is regulated
Maintenance in photic zone	Regulation of growth rate; growth slower in coral than in free-living state	Protection from grazers
Protection from UV damage by animal tissues	Expulsion from host	Dispersal by predators on animal tissue
Maintenance of a high population density of a single genotype by host under uniform environmental conditions	Supply of CO_2 and nutrients limited by host?	
C. Coral Symbiotic Association		
Increased growth, more competitive for space on reef	Compounded sensitivity to environmental stresses that affect algae, animals, or both	
Resource partitioning for food and space	Restricted tolerance range of light, temperature, and sedimentation conditions for growth	
High calcification provides stronger resistance to water motion		

Notes: Factors that are not direct benefits or costs are listed as indirect effects. The relative contribution of each factor to maintaining the balance between benefit and cost of the symbiotic association is unknown.

From the animal's perspective, "sufficient" numbers of zooxanthellae must provide some input of energy toward offsetting respiratory requirements. There must be a balance between photosynthetic production and the metabolic cost of maintaining the algae. The costs include mechanisms to cope with high oxygen tension and possible regulation of the growth rate of zooxanthellae (Table 5–1). Since most corals contain 1–2 million zooxanthellae per square centimeter, it is likely that this range represents an optimal algal density that balances the benefits and costs of the symbiosis. Rapid changes in densities of zooxanthellae in corals due to environmental perturbations, for example, coral bleaching in response to high temperature and algal growth in response to increase in seawater nutrients, will upset this balance and may stress the coral by uncoupling algal and animal growth. From the alga's perspective, the coral must provide a good habitat. The "economic" benefit of the partnership may be viewed as the net return based on the relative costs of the symbiosis between zooxanthellae and the coral animal.

Although it is difficult to evaluate benefits and costs, obviously when benefits exceed costs there is a net benefit to sustaining the symbiosis, and the association might be expected to persist in a stable state. Conversely, when costs exceed benefits, net costs exist and the association might not persist. Thus, the persistence and stability of the symbiotic relationship at both ecological and evolutionary scales must depend on the net benefit of the symbiosis over relevant time scales with respect to its ability to withstand environmental stresses and its capabilities to compete for space and other resources with other benthic organisms.

In some cases, benefits or costs of the partnership have been experimentally verified. For example, the enhancement of coral calcification by zooxanthellae is documented, both from comparison of calcification rates of symbiotic and nonzooxanthellate corals and by the light-enhanced calcification rates of symbiotic corals. In other cases, the relationships are less obvious. It must be recognized that not only is our knowledge of the costs and benefits of the relationship limited, but it can be misleading to apply anthropomorphic interpretations of risks and benefits. It is entirely possible that subtle yet crucial benefits and costs exist that we cannot yet identify or quantify. Moreover, it is also possible that the cumulative effects of different costs and benefits are not simply additive. The interactive effects between factors are not likely to be easily quantified.

The diversity and number of the entries in Table 5–1 suggest that the balance between benefit and cost for the relationship is highly dynamic and varies according to both previous and current conditions. Organisms have a physiological minimum and maximum tolerance to, and an optimum value for, any given factor. Within limits, such ranges of tolerance are useful constructs for the consideration of the environmental conditions both necessary and sufficient for survival. We can as yet only speculate whether the susceptibility of the coral to given stressors will either be increased, decreased, or modulated when compared to the susceptibility of the individual partners to the same stressors.

The stability of the coral-zooxanthella association may provide a useful measure of the well-being of corals on long time scales. One might assume that a stable relationship between the symbionts indicates a net benefit to each partner and to the symbiosis as a whole. Factors that stress corals to a point where the relationship is disrupted seem to imply that the costs of maintaining a symbiosis have exceeded the benefits. This may provide for the short-term survival of the coral. When favorable conditions return, the symbiosis is reestablished because the benefits to the coral are required in the long term.

Disruption of the association by stressors may, in turn, have major consequences not only for the individual corals but also for the coral-reef ecosystem. A particularly good example can be seen in the effect of temperature-induced coral bleaching on community structure in the Eastern Tropical Pacific (Glynn, 1991). In 1982–1983, a very strong El Niño–Southern Ocean (ENSO) oscillation event resulted in severe warming and severe bleaching of corals in Costa Rica, Panama, Colombia, and Ecuador. Mass mortalities of corals occurred and reef structure changed substantially. Such severe effects notwithstanding, disruption of the symbiosis by stressors may also provide the opportunity not just to "weather a storm" but to "change partners" to other zooxanthella clones or species that can provide better benefits and lower costs for particular environmental conditions (Buddemeier and Fautin, 1993). We are only now beginning to consider such subtleties.

As we consider the factors affecting the costs and benefits of maintaining the symbiosis, which is a dynamic state in and of itself, it seems appropriate to consider three questions, namely: Is viewing symbiosis in terms of benefits and costs a practical way of assessing the ability of a symbiosis to persist? What are the known factors that shift the balance from benefit (+) to cost (–) to the symbiotic association? Are such factors interrelated? Since we are only capable of making crude determinations of relative cost or benefit of a given factor, we cannot realistically provide numbers (limits) for the quantification of benefits and costs. Although this means that the answer to the first question is "no" in most cases, consideration of the relative benefits and costs does facilitate our ability to conceptualize the response of the symbiotic association to changes in any factor.

Exposures to extremes in temperature, oxygen, and salinity are known to destabilize the symbiosis and result in the loss of zooxanthellae (coral bleaching). For each of these stressors, it is believed that the cost of sustaining the zooxanthellae is too great and that either the host actively expels them or they leave on their own accord. Nutrients, on the other hand, seem to have a different effect that may also result in the active expulsion of zooxanthellae by the host. In this case, under conditions of nutrient repletion, algal expulsion seems only to keep the host from being overgrown by its endosymbionts. A disruption of the balance between the animal host and its zooxanthellae may result in reductions in productivity and coral growth, leading to possible overgrowth by faster-growing organisms (macroalgae). Below we discuss some practical examples of how natural and anthropogenic stresses to corals affect the stability of the symbiosis.

The final question posed above asked whether factors that affect the net benefit of the symbiosis interrelate with each other. It is possible to define a set of conditions under which a symbiosis will persist, and conversely, under which it will not. Nonetheless, we presently have almost no information regarding synergistic interactions and the effects of multiple stress factors on the net benefits of maintaining the symbiosis.

5.5. Environmental Effects on the Symbiosis

Other chapters in this book review general ecological features relating to corals and coral reefs. It is appropriate here to consider the stability of coral/zooxanthellae symbioses with respect to environmental stresses. We approach this topic first from the perspectives of local and regional effects, and then from the perspective of global environmental changes and effects through the alteration of the essential factors of sedimentation, light, nutrients, and temperature.

5.5.1. Local and Regional Stresses to Symbiotic Corals

In coastal areas, human population densities are increasing at an alarming rate, as people are migrating to within a few hundred kilometers of coasts (LMER, 1993). This demographic factor is having substantial environmental effects in all coastal areas in temperate and tropical regions, but to date most attention has been paid to temperate areas where more scientific study and environmental concern occur. That situation is beginning to change. Several meetings of international authorities on coral reefs recently concluded that the cumulative effects of local coastal development are likely to pose more immediate problems than any present global effect such as ozone depletion or enhanced greenhouse effect due to the anthropogenic release of carbon dioxide. This is particularly important because most international policy concerns are focused on controlling greenhouse gases for the sake of environmental protection at the global level (see below).

Increased sedimentation and runoff are two of the most pronounced early effects of coastal development. In high precipitation areas especially, clear-cutting of forests and development of agrarian economies result in increased levels of waterborne sediments and nutrients and decreases (or increases in the seasonal variation) in salinity. These activities have been associated with a reduction in coral cover and diversity (Kühlmann, 1988). At the same time, symbiotic corals can provide a useful temporal record of environmental changes within a reef ecosystem because of the dependence of calcification on zooxanthellae; changes in calcification rate due to variation in parameters such as temperature, salinity, turbidity, and pollution are recorded in the density banding patterns of the coral skeletons.

Studies of Kaneohe Bay, Hawaii, and other places suggest that turbidity (suspended sediment in the water) is the foremost enemy of reef corals (Chapter 15). Although the probable greatest effect of sediment on corals relates to the accumulation of particles on coral surfaces and interference with feeding, the effect of turbidity on the quantity and quality of light available for photosynthesis is also important. Alteration of light quality and quantity is due directly to the higher turbidity related to sediment loads and indirectly to turbidity resulting from the stimulation of phytoplankton growth by increased nutrient loadings associated with sedimentation and agricultural land practices (increased fertilizer and pesticide application, slash and burn and deep tillage agriculture) (see sections 5.2.1 and 5.2.2 on coral nutrition and calcification).

In addition to the effects of increased runoff and sedimentation on nutrient levels, phytoplankton biomass, and turbidity, also to be considered are the effects on the trophic status of the water column overlying reefs that in turn may affect the nutrition and stability of the symbiosis. The predominant effect of elevated nutrient levels on corals and coral reefs seems to result from altered trophic structure resulting from overgrowth of corals by fleshy green algae, high bacterial biomass, increased disease (Chapter 6), and so forth, which are beyond the scope of this chapter.

5.5.2. Global Stresses on Symbiotic Corals

With respect to corals, global stresses include (1) increased UV light due to a reduction in the ozone layer; (2) temperature increases due to global warming and related changes in oceanic circulation patterns leading to variation in temperature and nutrient inputs; (3) increased nutrients and turbidity due to industrial development in other areas.

The effect of chlorofluorocarbons (CFCs) on the depletion of the ozone layer and the subsequent increase in the flux of UV light to the Earth's surface have received substantial attention with respect to coral reefs. Conditions that favor photosynthesis by zooxanthellae expose corals to UV damage. Although corals contain pigments that may afford considerable protection from UV, the effective metabolic cost of UV protection for the animal and zooxanthellae with respect to the symbiosis is unknown. If the cost to the symbiosis is greater than the benefit of light-driven photosynthesis, then the symbiosis becomes a liability.

Temperature is a crucial factor affecting the stability of the coral/zooxanthellae symbiosis at the individual level, and certainly, in a larger sense, of coral reefs (Glynn, 1991). Limits of temperature tolerance for corals and well-developed coral reefs are considered to range from a winter minimum of approximately 18°C to a summer maximum of approximately 30°C, although to be sure, there are thriving reefs found at either extreme that appear to be uniquely adapted to such conditions. While initial interest in the practical ramifications of temperature stress on corals and coral reefs developed as a result of concern about the thermal effects of electrical power generation on local biota, the present interest in this topic is related to concerns about potential increases in global temperatures due to the enhanced greenhouse effect resulting from anthropogenic emissions of infrared-absorbing greenhouse gases.

Exposure to temperature extremes may or may not affect the stability of the symbiosis. Both the length of exposure to, and the severity of, a given temperature stress or anomaly are important factors. As an example, probably the best known response that indicates a destabilization of the coral/zooxanthellae symbiosis, *bleaching*, is dependent on both of these factors. Corals bleach or actively expel their zooxanthellae most typically when temperatures increase sharply for a short period of time (3–4°C, several days) or increase moderately for a longer period of time (0.5–1.5°C, several weeks) (Glynn and D'Croz, 1990; Jokiel and Coles, 1990). Since coral calcification, and therefore reef growth, depends on the presence of zooxanthellae, a gradual rise in sea level with global warming might result in the demise of coral reefs at low latitudes and a shift to higher latitudes. The effects of low light and increased nutrient inputs from global changes in atmospheric deposition and oceanic circulation patterns are discussed in the previous section.

Whether disruption of the coral symbiotic association with zooxanthellae provides an indication of global climate change is less certain, but is an issue of debate (Miller, 1991; D'Elia et al., 1991). Bleaching may simply represent a temporary disruption of the symbiosis that allows each partner to survive the stress on its own. The potential for new and more tolerant combinations of partners after bleaching makes this issue more complicated.

5.6. Conclusion

It is clear that the symbiotic association with zooxanthellae is beneficial to corals. There is increasing evidence that the symbiotic state is accompanied by sensitivity to environmental stress, since a common response to a stress is the disruption of the symbiosis, resulting in coral bleaching. The response is complex, since zooxanthella strains (or species) and different species or genotypes of coral animals may have different adaptive capabilities and tolerances to environmental extremes. As the host animal depends on its complement of zooxanthellae for reduced carbon compounds, coral death will ensue if stresses persist for long periods of time or if they are at levels outside of the tolerance range of the coral and the zooxanthellae. Factors that induce a stress response include light (quantity and UV), temperature, sewage and runoff inputs (high nutrients, increased turbidity), salinity (freshwater runoff from land due to deforestation), and physical damage.

Disruption of the symbiotic association, in turn, has potential for use as an indicator of the health of the coral-reef ecosystem. Drastic changes in the stability of the symbiosis, evidenced by changes in the ratio of zooxanthellae to animal biomass in corals, may turn out to be a useful diagnostic indicator of stresses to coral reefs. Present research is leading to improved understanding of how and when this can occur.

6

Diseases of Coral-Reef Organisms

Esther C. Peters

Most people picture reefs, and their associated fauna and flora, as vigorous, flourishing, and healthy. Only since the mid-1970s have we realized that corals and other reef organisms are susceptible to diseases caused by pathogens and parasites as well as to those conditions caused or aggravated by exposures to anthropogenic pollutants and habitat degradation.

Disease is defined as any impairment (interruption, cessation, proliferation, or other disorder) of vital body functions, systems, or organs. Diseases are usually characterized either by (1) an identifiable group of signs (observed anomalies indicative of disease), and/or (2) a recognized etiologic or causal agent, and/or (3) consistent structural alterations (e.g., developmental disorders, changes in cellular composition or morphology, and tumors).

Biotic diseases are those in which the etiologic agent is a living organism such as a pathogen or parasite. A variety of organisms normally live in interspecific associations known as symbioses on or within the tissues of other organisms (Ahmadjian and Paracer, 1986). Such associations can range from mutualistic symbioses (beneficial to both organism and host) to parasitic symbioses where the organism derives a nutritional benefit from the host. If a parasite causes disease and death of the host, then it is known as a pathogen. Infectious agents, those that are spread from host to host, include viruses, bacteria, fungi, and protozoans (also known as microparasites), and metazoans such as helminths and arthropods (macroparasites).

Many interactions between organisms and their hosts occur without clinical signs of disease (Scott, 1988). These symbioses enable their hosts to live in potentially toxic environments or to subsist on nutritionally limited diets, for example. Other associations induce changes in host behavior that may enhance transmission of the parasite. Shorebirds prey on clams infected with trematode sporocysts and metacercaria that either cannot burrow or that surface and produce conspicuous tracks at low tide (Sousa, 1991). Parasites may damage the host's reproductive capabilities (parasitic castration) or seriously affect the functioning of vital organs. Similarly, infectious host-specific diseases caused by pathogens may weaken or disable individuals so they are only more susceptible to predation or stressful environmental conditions. However, such diseases may also occur as epizootics (similar to epidemics in humans), causing disease and mortalities in large numbers of organisms of a single species.

Abiotic diseases are those structural and functional body impairments that result only from exposure to abiotic environmental stresses such as changes in physical conditions (salinity, temperature, light intensity or wavelength, sedimentation, oxygen concentrations, currents) or exposures to toxic chemicals (such as heavy metals and organics like oils or pesticides). The effects of environmental pollution and physical alterations in the reef habitat on corals are discussed in Chapter 15; only biotic diseases will be discussed in this chapter. Although the causal agent of a disease in a tropical marine organism may appear to be either biotic or abiotic, both types of diseases are often closely interrelated. Therefore, determining the primary cause may be difficult. In some cases a pathogen or true parasite may not harm its host unless the host is stressed by some other biotic or abiotic disease factor (a "stress-provoked latent infection"). Conversely, an abiotic disease can become complicated by secondary infections from normally harmless microorganisms. Usually, however, biotic diseases develop in specific hosts.

As also noted in Chapter 15, corals and other tropical marine organisms possess a variety of defense mechanisms to protect themselves from invasion by pathogens. These include the secretion of mucus, production of antibiotic compounds or noxious chemicals that repel parasites, a protective epidermis, and a variety of amoeboid cells that engulf pathogens or surround parasites and produce toxicants to destroy them. The degree of vulnerability or susceptibility of an animal to penetration by a pathogen and successful establishment of the pathogen may vary between and within species and individuals or may be altered as a result of changes in environmental conditions, nutritional state, developmental stage, and other factors. Resistance to infection is characterized by those physiological alterations or responses that occur naturally or develop in the course of exposure of an animal to invasion by pathogens or parasites.

In the tropics, opportunistic pathogens may replicate rapidly and reach the peak of their growth curve in only a few hours. Susceptibility and the relative resistance of the host can also change with the size of the population and the genetic constitution of the microorganisms present, but little is known about the regulation of symbiont populations by the host. Furthermore, as pathogens and parasites influence the abundance of host populations, they exert strong selective pressures on the genetically based variability of an individual host's resistance or its ability to recover from infection within the population. Thus, the nature of the association may be altered over time (Anderson, 1986).

Organisms also have developed certain generalized behavioral, physiological, and biochemical responses that may be invoked over the short or long term, allowing them to adapt to a range of changing conditions while maintaining a preferred state, level (homeostasis), or rate of some process (homeorhesis) (Stebbing, 1981; Sindermann, 1990). The counteractive capacity of these adaptive responses will allow an organism to maintain its health while being subjected to changing conditions, leading to resistance to the stressor(s). In order to adapt, the organism will expend energy for survival, growth, and reproduction. However, as the number of stressors and/or their level of intensity increases, energy expenditures will increase but growth and reproduction will slow or cease. The ability of the organism to deal with stress decreases or disappears as the result of exhaustion of critical biochemical and physiological functions, until finally, disease appears. Death of the organism will result if the condition is irreversible.

The mechanisms by which changing environmental conditions, toxins, or pathogens cause disease appear varied and will also differ with the species and individual (for reviews see Sparks, 1985; Sindermann, 1990). **Epizootics** or dramatic increases in disease prevalence may result from the introduction of a new pathogen into a susceptible population, increases in pathogen numbers or virulence, or lowered resistance of the population. Furthermore, although motile organisms may be able to avoid or limit their contact with pathogens, toxic agents, or adverse physicochemical conditions, sedentary invertebrates generally cannot. (They may produce planktonic larvae to escape, however.)

In summary, diseases occur as the result of interactions between a susceptible host, a virulent pathogen, and prevailing environmental conditions. As illustrated in Figure 6–1, the circles representing each of these factors will increasingly overlap as they become more significant, with the result that disease is more likely to occur. Diseases caused by infectious microorganisms, parasites, and noninfectious (nutritional, environmental, or genetic) disorders have been reported from most phyla of marine plants and animals. However, most of our information on diseases of marine organisms has come from studies of commercially important temperate fish and shellfish species. These studies have received extensive funding and were investigated by multidisciplinary pathobiology teams. For tropical species, many reports in the literature are descriptions of "parasites" where the true nature of the organism's association with the host has not been experimentally determined. There are also a number of reports where the etiologic agent of mass mortalities has not been identified because the disease was not recognized until most of the population was affected and there were few survivors available for study (e.g., the *Diadema antillarum* mass mortalities of 1983).

6.1. Diseases of Reef Plants

Little is known about biotic and abiotic diseases of marine algae and seagrasses in tropical waters. There have been some observations of viruses, bacteria, and

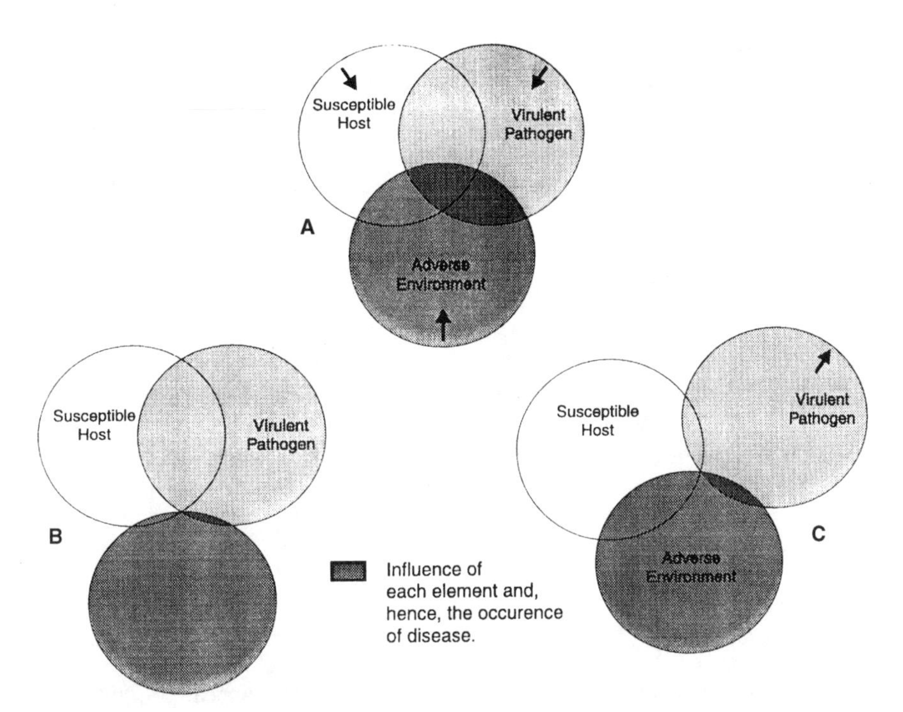

Figure 6–1. Host-pathogen-environment interrelationship with disease. **(A)** The interaction of adverse environmental conditions with a susceptible host and virulent pathogen produces disease; the relative contribution of each variable influences the incidence and severity of the disease. **(B)** With improved environmental conditions, disease is reduced. **(C)** If no virulent pathogens are present, the host is affected by environmental conditions only (reproduced from Warren, 1991).

other pathogens infecting temperate species (Andrews, 1979). One tropical seagrass, *Thalassia testudinum,* in the environmentally stressed Florida Bay has been affected by a marine slime mold, genus *Labyrinthula.* This pathogen caused blackened, necrotic lesions on the seagrass blades, reduced the photosynthetic production of oxygen in the plant (Durako and Kuss, 1994), and resulted in massive die-offs of this important species (Robblee et al., 1991; Thayer et al., 1994).

Littler and Littler (1994, 1995) reported the appearance, first in 1992, of a bacterial pathogen of encrusting coralline algae on reefs in the Cook Islands, Fiji, Solomon Islands, and Papua New Guinea. The pathogen was bright orange and spread across the algal surface, leaving behind the bleached skeletal carbonate remains of the coralline algae. When the pathogen reached the margin of the algal thallus, it formed upright filaments and globules, similar to those formed by terrestrial slime molds. Microscopic examination revealed a motile gliding rods of a colonial bacterium in a mucilagenous matrix. Experimental studies confirmed that the pathogen globules were highly infectious to a variety of coralline algal species. The recent appearance of this new disease, termed coralline lethal orange disease (CLOD), could potentially affect the structure and function of many reef sites, since the dead corallines no longer contribute to productivity and carbonate accretion processes, and fleshy algae overgrow the dead coralline algae and inhibit the settlement and growth of reef-building corals.

6.2. Diseases of Reef Invertebrates

Sponges, scleractinian or stony corals, soft corals (alcyonaceans), sea fans and sea whips (gorgonaceans), polychaete worms, a wide variety of bivalve and gastropod molluscs, octopus and squid, spiny lobsters and crabs, sea urchins, seastars, sea cucumbers, crinoids, and brittlestars are some of the more prominent members of the coral-reef community and associated tropical marine habitats. While we understand much about their ecological roles, studies of the nature and effects of diseases on these organisms are relatively recent and far from completion.

6.2.1. Corals

The first reports of a disease affecting scleractinian corals appeared in the mid-1970s. Black-band disease (BBD) was first reported from reefs off Belize and Bermuda, but has since been found throughout the Caribbean as well as the Indo-Pacific (Rützler et al., 1983; Antonius, 1985). BBD has also been reported on milleporinids (fire corals) and gorgonaceans. Not all coral species appear to be susceptible to this disease. Massive brain corals (*Diploria* spp., *Colpophyllia* spp.) and star corals (*Montastraea* spp.) are the most commonly affected members of the family Faviidae, while elkhorn, staghorn, and pillar corals resist natural infections.

Figure 6–2 shows the characteristic appearance of this disease. BBD results from the invasion of coral tissue by a cyanobacterium, *Phormidium corallyticum.* The disease line appears as a black mat a few millimeters wide composed of fine cyanobacterial filaments that also contains sulfate-reducing bacteria, sulfide-oxidizing bacteria, and sometimes fungi and protozoans. These microorganisms produce anoxia deep in the band next to the tissue and hydrogen sulfide, which kills the coral tissue and allows the microorganisms to use the organic compounds released by the dying coral cells for their own growth and reproduction (Carlton and Richardson, 1995). This band or mat moves across the surface of the coral at the rate of a few millimeters per day, leaving behind bare coral skeleton that is eventually colonized by filamentous algae.

Healthy corals can become infected with BBD when in contact with an infected colony, but injured colonies are most susceptible. Most studies have found that less than 2% of Caribbean corals are infected with BBD on any given reef area,

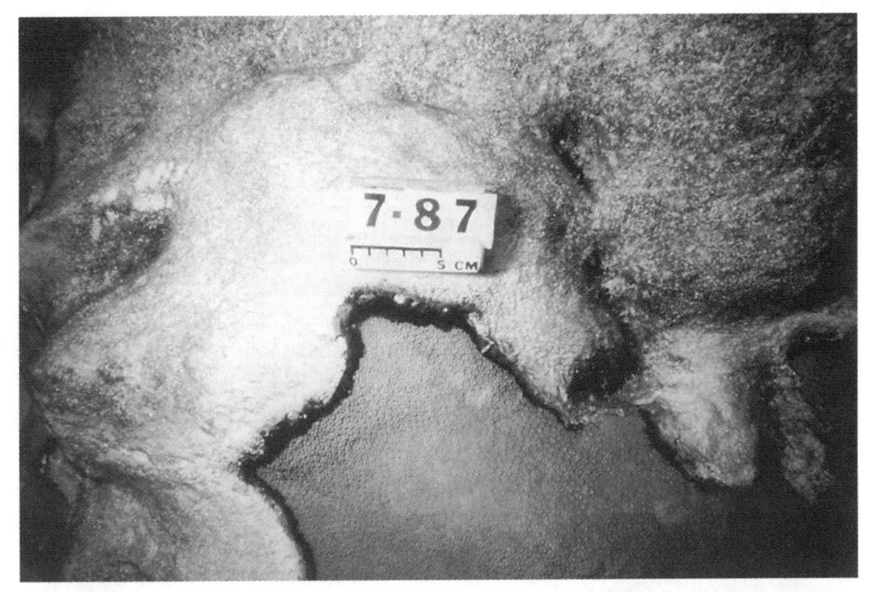

A

B

Figure 6–2. **(A)** Diseases of scleractinian corals. Black-band disease destroying living coral tissue on a colony of star coral, *Montastraea annularis,* at Looe Key, Florida Keys National Marine Sanctuary. The skeleton in the upper portion of the picture is being colonized by filamentous algae (photo: H. H. Hudson). **(B)** Diseases of scleractinian corals. Black band disease destroying living coral tissue on a colony of star coral, *Montastraea annularis,* at Looe Key, Florida Keys National Marine Sanctuary. The diver on the left is using a suction device to remove the cyanobacterium from the coral (photo: H. H. Hudson).

although there have been recent outbreaks at several locations, including Looe Key in Florida. Edmunds (1991) found that 58% of BBD-infected brain corals *(Diploria strigosa)* on reefs off St. John, U.S. Virgin Islands, lost over 75% of their tissues within 7 months.

About the same time that BBD was reported, acroporid (elkhorn and staghorn) corals off St. Croix, U.S. Virgin Islands, exhibited tissue sloughing, which started at the base of the branches and moved toward the branch tip at the rate of a few millimeters per day. In contrast to BBD, however, no consistent assemblage of microorganisms could be found at the junction separating the sloughing tissue from bare coral skeleton. This disease was termed white-band disease (WBD), and has also been referred to as "white plague" and "white death," because the sloughing left a broad band of bare skeleton several centimeters wide on the colony that was also eventually colonized by filamentous algae (Fig. 6–3). These disease signs can be distinguished from predator damage (e.g., fish, gastropod, or worm feeding scars) and have since been observed on acroporid species throughout the Caribbean, the Red Sea, and off the Philippines (Peters, 1993).

The etiology of white-band disease is unknown. Unusual aggregates of Gram-negative rod-shaped bacteria were found scattered in the calicoblastic (skeleton-producing) epidermis that lined the gastrovascular canals of the porous skeleton

Figure 6–3. Diseases of scleractinian corals. Elkhorn coral, *Acropora palmata,* afflicted with the characteristic basal tissue sloughing of white band disease at Grecian Rocks, Florida Keys National Marine Sanctuary (photo: J. C. Halas). Reproduced with permission from Couch and Fournie, 1993.

in affected acroporids from St. Croix and Bonaire, Netherlands Antilles (Peters, 1984). The bacterial aggregates were also found in apparently healthy colonies at St. Croix. Five years later, up to 95% of the elkhorn corals there had died. Although these bacteria have not been found in apparently healthy acroporids from other sites, the role of this microorganism in the development of disease has not been determined. No studies have been conducted on afflicted Indo-Pacific acroporids to demonstrate the presence of such bacterial aggregates there.

Basal tissue sloughing has also been observed in acroporids and other species of scleractinian corals from the field as well as in captivity, but microscopic studies have not found observable microorganisms within their tissues. Because there was an association with adverse environmental changes in some of these cases, Peters (1984) proposed the label "stress-related necrosis" for cases in which degenerative changes in cell structure are observed in the absence of obvious pathogens, particularly bacterial aggregates, as determined by microscopic examination of fixed, embedded tissues. Much remains to be learned about the nature of tissue sloughing in corals and how many conditions caused by different pathogens or environmental stresses may actually be represented by the same disease signs.

Corals also harbor a variety of protozoan and metazoan microorganisms, some of which may be parasites. Gregarine protozoans have been found in Caribbean coral tissues, causing localized adverse host tissue reactions, including loss of zooxanthellae and necrosis (Peters, 1993). Another such relationship has been examined in Hawaiian corals *(Porites compressa)* that contained the metacercarial stage of a digenetic trematode, *Plagioporus* sp. The host for the final stage of this parasite is probably a coral-feeding fish. Aeby (1991) examined the effects of the parasite on the coral polyps, which became pink with swollen nodules and altered the ability of the polyps to retract into their calices. Parasite encystment resulted in reduced growth rates of parasitized corals. Fish fed preferentially on infected polyps, and as a result the altered polyp appearance provided both an enhancement of the parasite's transmission rate and parasite removal from the coral. Healthy polyps then grew back over the feeding scars. Thus, this phenomenon may act as a host strategy of parasite defense.

Anomalous calcification patterns in scleractinian corals may be caused by parasites or commensals. Other examples of enlarged corallites or tumors in the exoskeleton have been attributed to cellular proliferative disorders, including neoplasia (reviewed in Peters et al., 1986). Whitened protuberant calcified tumors have been found infrequently on branching acroporid corals in the Caribbean and Indo-Pacific. These skeletal masses have proliferating gastrovascular canals and associated calicoblastic epidermis (the calicoblast cells lay down the exoskeleton of the coral). As the calicoblastic epithelioma grows, porous skeleton is laid down more rapidly than the surrounding tissue, resulting in degeneration of normal polyp structures and loss of zooxanthellae from gastrodermal cells. Mucus secretory cells normally in the epidermis of the coral tissue disappear from the epidermis covering the tumor as the tumor mass grows larger. Having lost the mucus secretory capabilities of the epidermis, the coral is unable to shed sediments and the tissue becomes ulcerated and invaded with filamentous algae. Branches having tumors also exhibit reduced skeletal accretion and growth. Both genetic and environmental factors appear to affect the distribution of tumor-bearing colonies (Peters et al., 1986).

Another sign of disease in corals is the loss of zooxanthellae, which normally give the coral tissue a brownish coloration, and/or to the loss of photosynthetic pigments from the zooxanthellae. The phenomenon is called coral bleaching. Bleaching of corals, gorgonaceans, alcyonaceans, and anemones has been attributed to exposure to high light levels, increased solar ultraviolet radiation, high turbidity and sedimentation resulting in reduced light levels, temperature and salinity extremes, and other factors. The nature and extent of bleaching vary between individuals and among species at the same location during a bleaching event and have been attributed to different physiological tolerances of the strains (or species) of zooxanthellae and the coral hosts. Chronic partial or widespread loss of zooxanthellae, for whatever reason, signals a disturbance in the normal metabolism of the coral host and can lead to delayed or reduced reproduction, tissue degeneration, reduced growth, and death of the affected tissue (Chapter 15; Williams and Bunkley-Williams, 1990).

6.2.2. Sponges

While scleractinian corals are usually the most noticeable members of the reef community, at least in size if not numbers, members of several other invertebrate phyla are also important components of reef habitats. Yet observations on diseases affecting these organisms have been limited. Bleaching has also occurred in tropical marine sponges that contain photosynthesizing symbionts in their tissues, particularly in the Caribbean during the recent coral bleaching events. There have been only a few reports of diseases in sponges, primarily the commercial species of the genera *Spongia* and *Hippospongia.* In the late 1930s, widespread mortalities occurred among these species in the Caribbean. The timing and distribution of these mortalities followed the major current patterns. Commercial sponge fisheries were effectively eliminated, although some sponges did recover. Affected sponges exhibited "bald patches," followed by "rotting" of tissue beneath the patches, with the entire sponge degenerating within one week. The lesions always contained long slender aseptate (without interior walls) filaments that were believed to be a fungus. Studies suggested that bacteria and changes in water temperature might be responsible, but these observations were never confirmed (Peters, 1993).

Healthy sponges contain a variety of mutualistic symbiotic bacteria (Wilkinson, 1987c) that may provide nutrition for their sponge hosts or that use metabolic wastes produced by the host. Thus, investigations of the causal agent(s) of diseases

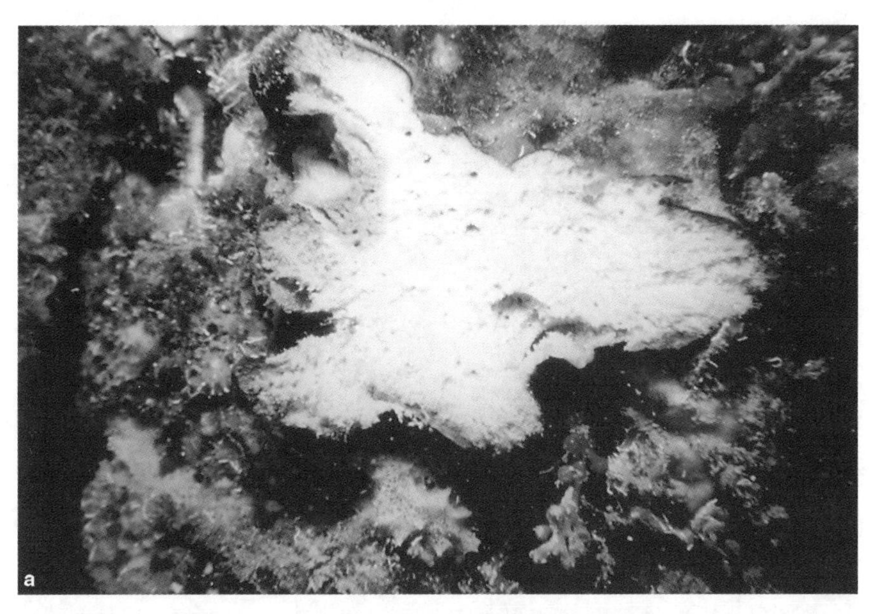

Figure 6–4. Diseased sponge. **(A)** Cut surface of *Geodia papyracea,* Belize. Pale gray areas just inside the outer margin are infected with cyanobacteria; these areas appear yellow in color and are decaying. **(B)** Bacterial flora in a thin section of affected tissues examined by transmission electron microscopy. The large circular to ovoid cells with concentric rings are the pathogenic cyanobacteria (solid arrows), approximately 2.2×1.3 µm in size; other smaller bacterial symbionts, 1.4×0.8 µm, are in the area to the upper left (clear arrows) (photo: K. Rützler).

in sponges may be complicated by the presence of these microorganisms or by secondary infestations from seawater populations of microorganisms. Sponges possess a variety of cellular defense mechanisms and many sponges can also produce antimicrobial compounds to control pathogenic microorganisms; however, the relationships of sponge-dwelling bacteria and other micro- and macroorganisms with host metabolism and health are poorly understood.

Although a recent epizootic among commercial sponges in southern Mediterranean countries is suspected to be the result of a bacterial disease, reports of diseases in other tropical marine sponges are lacking. Rützler (1988) described a disease in the mangrove demosponge *Geodia papyracea* from Belize. Apparently, the normal cyanobacterial symbionts of this sponge multiplied out of control, resulting in the destruction of the host sponge tissue (Fig. 6–4).

6.2.3. Molluscs

Diverse species of molluscs live on coral reefs or in adjacent seagrass beds, but few studies have been conducted on these animals, except for the commercially

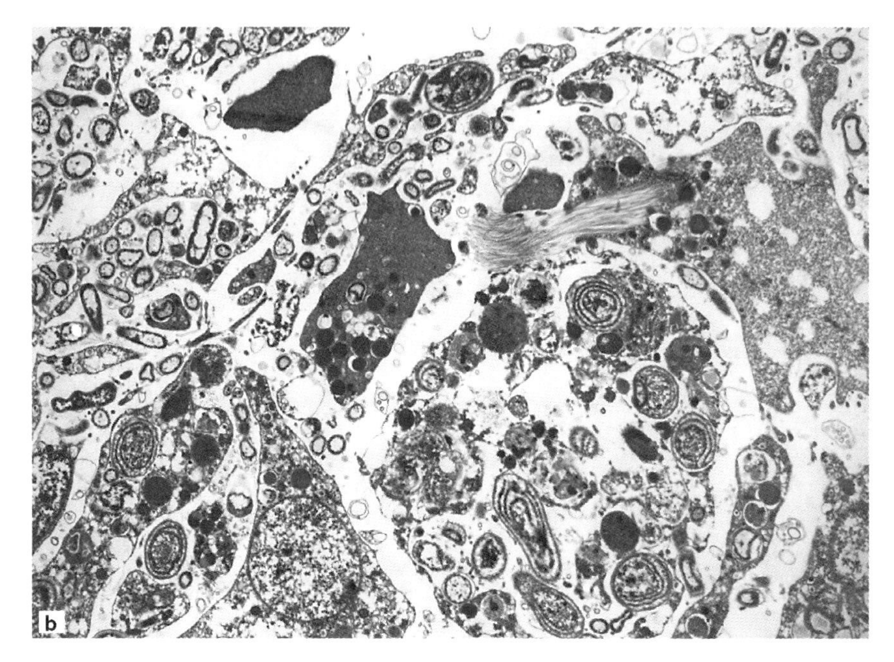

Figure 6-4. Continued

important queen conch *(Strombus gigas)* and giant and fluted clams (*Tridacna* spp.). Giant clams from the Great Barrier Reef were found to contain large numbers of the protozoan *Perkinsus* sp. (Fig. 6–5) It has been suggested that this microorganism, in conjunction with cooler water temperatures, was responsible for observed mortalities affecting up to a third of the giant clams at Lizard Island from 1984 to 1987. However, the protozoan was also found at low levels of infection and not associated with mortalities in other bivalves on the reef (84 species from the families Spondylidae, Arcidae, and Chamidae, as well as the Tridacnidae were examined). These results indicate that there may be species-specific host susceptibilities among the bivalves, several different species of *Perkinsus,* or variations in the prevalence of pathogenic strains of *Perkinsus* (Goggin and Lester, 1987).

A temperate relative of this protozoan, *Perkinsus marinus,* has been responsible for extensive mortalities of oysters along the east coast of the United States. The relationship of this pathogen to changes in environmental conditions, including salinity and temperature, has been investigated, and careful monitoring and control measures have been undertaken to protect these food resources (Fisher, 1988). Further histological studies of the giant clam mortalities off Lizard Island revealed the presence of an unidentified unicellular organism in some of the clams (Alder and Braley, 1989), but many questions remain about the nature of this epizootic.

Other diseases of temperate molluscs are caused by viruses, bacteria, fungi,

A

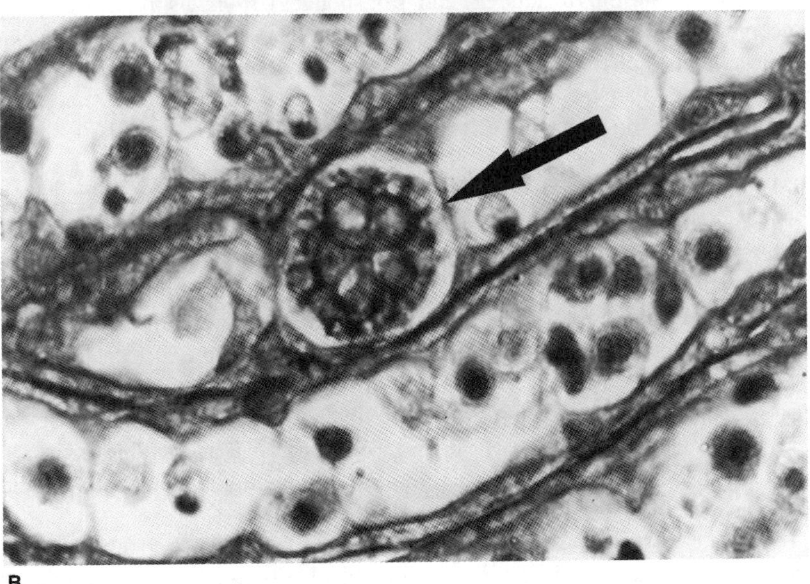

B

Figure 6–5. Protozoan parasite of the giant clam. (**A**) Divers sampling a dying Tridacna gigas from North Direction Reef, Great Barrier Reef, Australia (photo N. Quinn). (**B**) Perkinsus sp. schizont from Tridacna crocea gill from Lizard Island, Great Barrier Reef, Australia. Diameter of schizont is approximately 12 microns (photo L. Goggin).

and protozoan and metazoan parasites; nutritional, developmental, and neoplastic disorders are also known. The commercial market for mother-of-pearl oysters in French Polynesia was severely damaged in the late 1980s by a virus that killed up to a million oysters. Larval trematodes of the family Bucephalidae cause parasitic castration or destruction of gonadal tissue in marine bivalves; such an infection has also been observed in a burrowing tridacnid clam from the Great Barrier Reef (Shelley et al., 1988). Molluscs possess cellular and humoral defense mechanisms that help to control pathogens, but investigations of these defense mechanisms have not been performed on their counterparts in subtropical and tropical marine habitats. An interesting relationship has been discovered in the *Lucina* clams of seagrass beds, where bacteria enable them to live in the anoxic sediment layer (Fisher and Hand, 1984). These bacteria are contained within vacuoles of gill cells and apparently assist in oxidizing toxic hydrogen sulfide present in the sediments and provide nutrients to the clam.

6.2.4. Crustaceans

Most reef crustaceans are small and inconspicuous or cryptic. Examples include banded coral shrimps (*Stenopus* spp.), cleaner shrimps (*Periclimenes* spp.) associated with anemones, burrowing mantis shrimps (*Gonodactylus* spp., *Callianassa* spp.), snapping shrimps that hide in corals or sponges (*Alpheus* spp.), and a variety of decorator crabs, coral crabs, hermit crabs, and the arrow crab *(Stenorhynchus seticornis).* The spiny lobsters (*Panulirus* spp.) and slipper or Spanish lobsters (*Scyllarides* spp.) are the objects of important subsistence and commercial fisheries in tropical marine waters. However, reports of disease in all of these species are lacking. Temperate lobsters (*Homarus* spp.), penaeid shrimp, and various edible crabs are known to be susceptible to a variety of pathogens and parasites, as well as abiotic diseases related to poor nutrition and water quality (Sparks, 1985; Sindermann, 1990). Again, most of the research has been performed on commercial species, particularly those held in mariculture facilities.

Crustaceans also possess fixed and mobile phagocytic cells in the gills, the pericardial sinus, and at the bases of appendages. They also produce bactericidins, agglutinins, and lysins to deal with pathogens and parasites. Bactericidins of the West Indian spiny lobster *(Panulirus argus)* have been examined (see review in Sindermann, 1990) and found to be partially nonspecific. Bactericidin activity was enhanced against other Gram-negative bacteria following injections of formalin-killed bacteria.

Temperate crustaceans are affected by exposure to pollutants and other environmental stresses, resulting in damage to gills (black gill disease) and exoskeletons (shell disease), and reducing the quality of lobster and crab fisheries. Both of these diseases involve ulcerations of tissue with necrosis and bacterial invasion. In shell disease, chitinoclastic microorganisms are responsible for eroding the shell, which may have been damaged by mechanical, chemical, or microbial action, followed by secondary infection of the underlying tissue by facultative pathogens.

6.2.5. *Echinoderms*

The echinoderms are represented by such diverse animals as sea cucumbers, seastars, sea urchins, and brittlestars, and all are found in tropical marine habitats. While bacterial and protozoal diseases of temperate sea urchins and seastars have received much attention from invertebrate pathobiologists, there have been few studies on the etiologies of diseases and mass mortalities in tropical echinoderms, including the most extensive epizootic ever reported for a marine invertebrate.

Mortalities of the long-spined sea urchin *Diadema antillarum* were first observed on reefs off the Caribbean coast of Panama in January 1983. Mortalities of only this species subsequently occurred at other sites around the Caribbean and Bermuda for one year, in a pattern that followed major water currents from west to east, with a few exceptions (Lessios, 1988). A species-specific waterborne pathogen, perhaps introduced to the Caribbean from the ballast water of ships traversing the Panama Canal and discharged at the Caribbean entrance and at Barbados, was suspected to be the causal agent of this epizootic, since no other species were affected and no adverse changes in environmental conditions were noted at any of the sites. Overall, adult populations of this urchin were reduced by 85–100%, with juveniles rarely affected.

Diseased urchins were initially recognized by an accumulation of sediment on their spines and sloughing of the spine epidermis. Pigment in the skin covering the spine muscles, peristome, and anal cone then disappeared, and the spines broke off. The tube feet that normally hold the urchin to the seabed weakened and could not fully retract. Finally, patches of skin and spines sloughed off and the test disintegrated (Fig. 6–6). Diseased urchins died within 4 days to 6 weeks, depending on locality, although some urchins apparently survived the disease and recovered from the broken spines and skin lesions. Affected urchins also exhibited unusual behavior, moving out from their normal hiding places in the reef into the open during daylight, where they were preyed upon by fish.

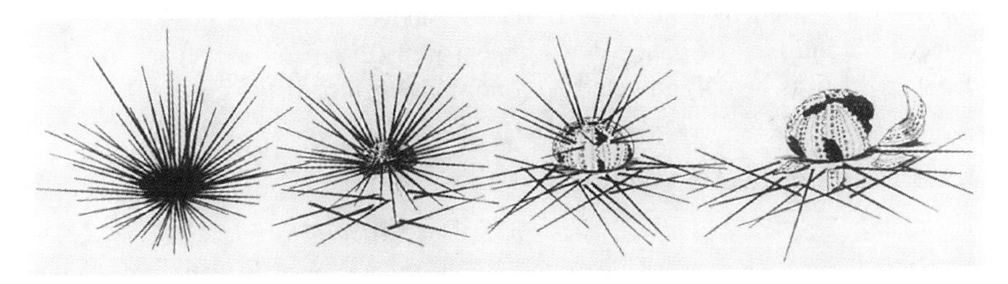

Figure 6–6. Progressive stages in the death of tropical western Atlantic long-spined sea urchins, *Diadema antillarum,* during the mass mortalities. (Reproduced with permission from Williams and Sindermann, 1992.)

Although numerous studies were conducted on the ecology of the urchin die-offs in 1983, and on later isolated mortalities of remaining populations of *D. antillarum* off St. Croix, Grand Cayman, and Jamaica, few samples were obtained for histological or microbiological investigations. Gram-positive anaerobic spore-forming rods of the bacterial genus *Clostridium* were isolated from two urchins showing similar signs of the disease that died while in a flow-through seawater aquarium in Miami in 1983. Laboratory experiments with cultures of these bacteria caused death of healthy urchins in 10 hours to 6 days, depending on water temperature. Microscopic examination of fixed tissues from St. Croix urchins revealed Gram-positive micrococci in mucoid cells of the glandular crypts of the esophagus and in connective tissue and muscle bundles of the peristome, spines, and ampullae. However, bacterial samples were not taken in this study (see Peters, 1993, for review). A mass mortality of D. antillarum that occurred in the Florida Keys in 1990 provided samples for bacteriological, virological, and histopathological studies, but results proved inconclusive and differed from the earlier observations. Localized mass mortalities of unknown causes of two species of diadematid urchins were observed off Hawaii and in the species *Astropyga magnifica* and *Eucidaris tribuloides* off Puerto Rico in the 1980s.

Echinoderms possess cellular and humoral defense mechanisms in the coelomic fluid and associated tissues that can protect them from invasion by potential pathogens. However, they also usually contain mutualistic bacteria in the gut or other organs. The coral-eating Indo-Pacific crown-of-thorns seastar *Acanthaster planci* has an unusual bacterium living in its body wall, mucus secretions, and pyloric caecum. When the seastar is held in aquaria, however, these bacteria become facultative pathogens and leave the animal vulnerable to secondary infections by *Vibrios* and other bacteria. Mass mortalities of juvenile *A. planci* that occurred near Fiji over 3 years were not caused by bacteria, however, but apparently resulted from sporozoan parasites infecting the digestive tract (reviewed in Birkeland and Lucas, 1990).

Similarly, a wide variety of metazoans have been observed to associate with echinoderms, usually without harming their hosts. In tropical marine species, however, Emson et al. (1985) reported severe damage to brittlestars that were heavily parasitized by copepods. Williams and Wolfe-Waters (1990) observed emaciation and lack of gonad development in a Caribbean basketstar in which the stomach was heavily infected by a normally ectoparasitic copepod.

6.3. Diseases of Reef Vertebrates

Although there are numerous examples of viral and bacterial diseases and epizootics in temperate marine fishes, particularly commercially gathered species and those species used in aquaculture, few studies have examined their counterparts in tropical species. Besides fishes, sea turtles are globally important tropical marine vertebrates and sea snakes may also be encountered on Indo-Pacific reefs. Some information on the normal physiology, biochemistry, behavior, and diseases has accumulated from research on those tropical marine species popular in aquariums and oceanariums (Stoskopf, 1993) and in aquaculture (Glazebrook and Campbell, 1990a). Diseases that occur under these artificial conditions tell us little about what occurs on coral reefs and there have been few published field observations. Studies on captive fishes and turtles have confirmed their sensitivity to adverse changes in environmental conditions and the importance of appropriate water quality and proper nutrition in maintaining their health (Roberts, 1989).

6.3.1. Fishes

Injured or weakened reef species, such as damselfishes, squirrelfishes, soldierfishes, and angelfishes, may become infected with marine bacteria. Gram-negative *Vibrio damsela, Vibrio alginolyticus, Vibrio anguillarum,* and Gram-positive Streptococcus spp. have been isolated from fish suffering from skin ulcerations, septicemias, exophthalmias or popeye, and other lesions. Captive tropical marine fishes are also susceptible to fungal infections such as *Ichthyophonus hoferi.* Lymphocystis disease, caused by an iridovirus, is characterized by giant cell "tumors" and has been reported from Australia, Hawaii, the Pacific coast of Panama, Indochina, the South Pacific, and the Caribbean. Other viral diseases, including viral erythrocytic necrosis (VEN) caused by another apparent iridovirus, an infectious pancreatic necrosislike virus (IPN), and a rhabdovirus infection, have been found in captive tropical marine species such as angelfishes, wrasses, and blennies. However, little work has been done on viruses isolated from these fishes (Stoskopf, 1993).

A wide variety of protozoan and metazoan parasites is known from examination of field-caught and captive tropical marine fishes. They include ectoparasitic flagellates (*Amyloodinium* and *Crepidoodinium* pathogens of gills and skin), trypanosomes, hemoflagellates of the genus *Trypanoplasma,* various genera of ciliates, sporozoans (phylum Apicomplexa), microsporidians, myxosporidians (genera *Ceratomyxa, Myxidium,* or *Leptotheca* have been found in gallbladders of marine tropical fish), turbellarians, nematodes, digenetic and monogenetic trematodes, aspidogastrids, cestodes, leeches, copepods, and isopods. The trematodes, cestodes, and nematodes require one or more intermediate hosts to complete their life cycle. Many of these parasites do not cause overt disease or mortalities among wild fish, although they may be found in potentially pathogenic numbers during mass mortalities; however, losses of captive fish have been attributed to parasitic infestations, especially coral fish disease or velvet disease caused by *Amyloodinium* spp. The cleaner fishes and shrimps in tropical marine habitats apparently keep the levels of most pathogens and parasites quite low in wild fish populations, although this has not been demonstrated experimentally (Chapter 11).

A few tumors (neoplasms) have been reported in tropical fish. Of particular interest are the neurofibromas and malignant schwannomas of the bicolor damselfish, *Pomacentrus partitus* (Schmale, 1991). Damselfish neurofibromatosis (DNF), which is similar to the disease known as von Recklinghausen neurofibromatosis in humans, is characterized by the appearance of conspicuous hyperpigmented spots on the skin and fins that arise from multicentric peripheral nerve-sheath tumors that vary in degree of pigmentation (Fig. 6–7). The tumors are highly malignant, spreading throughout the animal and eventually causing its death as the result of destruction of vital organs or as secondary infections develop. Affected bicolor damselfish have been found only on reefs off South Florida at prevalences of up to 23.8%. Disease rates remained relatively stable for each reef site examined over a 9-year period.

The distribution of tumorous fish and the results of laboratory transmission experiments suggest that an infectious agent, such as a virus, is involved. The virus could be transmitted during the frequent aggressive interactions that occur among neighboring individuals defending their territories. Other species of damselfish and other tropical reef species such as snappers in the Florida Keys have also been found to be afflicted with a similar disease. Damselfish neurofibromatosis is being developed as an animal model to study neurofibromatosis and possible treatments for humans.

Figure 6–7. Neurofibromas developing on a bicolor damselfish, *Pomacentrus partitus,* Molasses Reef, Florida. A single, unpigmented, nodular tumor appears on the upper back under the dorsal fin; several small pigmented tumors are apparent near the eyes, on the head (photo: M. C. Schmale, from Schmale et al., 1986).

The limited observations of disease in fishes in the wild may be the result of predation. Any condition that weakens a fish or changes its normal behavior, or causes its death, would make it susceptible prey. Isolated cases of diseased fish may thus be overlooked. However, there have been mass mortalities of tropical marine fish. Mass mortalities of mullet in the mid-1970s near Miami, Florida, were believed to result from infection of their brains with a newly described bacterium that caused twirling and other signs of neurologic disease (Udey et al., 1977). A Caribbean-wide massive fish kill occurred in August and September of 1980, following Hurricane Allen (Williams and Bunkley-Williams, 1987). Tons of dead and dying fishes washed onto beaches. Wild and captive fish exhibited odd behavior suggesting that fishes surviving the mortalities were "sick" for several months. The cause was never identified. Millions of herrings (*Harengula* spp.) died at eight locations around the Caribbean during the 1980s (Williams and Bunkley-Williams, 1990). These mass mortalities were disjunct over time and geographic location, and examinations of moribund fish did not reveal any bacterial infections or other conditions that may have been responsible.

6.3.2. Sea Snakes

Little is known about diseases of the poisonous sea snakes. External parasites include ticks on the skin of laticaudids and a turbellarian on *Pelamis* sp. from the Gulf of Panama. Foraminiferans, hydrozoans, serpulid polychaetes, bivalve molluscs, bryozoans, and a small stalked (gooseneck) barnacle have been recorded as fouling organisms on sea snakes. Most external symbionts are probably dislodged, however, as the result of frequent shedding of the skin and knotting behavior in these organisms (Zann et al., 1975). Endoparasites include chigger mites in the lungs of the semiterrestrial *Laticauda* spp., nematodes, and trematodes (Heatwole, 1987; see also Culotta and Pickwell, 1993).

6.3.3. Sea Turtles

The green *(Chelonia mydas)*, loggerhead *(Caretta caretta)*, and hawksbill *(Eretmochelys inbricata)* turtles are frequently found on coral reefs and associated habitats, where they feed on seagrasses, macroinvertebrates, and sponges, respectively. Most observations of diseases in sea turtles have also been conducted on oceanarium- and aquaculture-reared animals, with reports consisting primarily of systemic bacterial diseases, metazoan parasites, nutritional disorders, and skin tumors.

Glazebrook and Campbell (1990b) examined 22 wild turtles obtained from the Torres Strait off Townsville, Queensland, Australia. The wild turtles had the greatest number and diversity of parasites. Two species of flukes (Digenea and Spirorchiidae) were found in the heart and major associated arteries and their eggs were found in other organs and tissues. The presence of the flukes was

Figure 6–8. A green sea turtle, *Chelonia mydas,* with large fibropapillomas on its neck and flippers. The turtle was collected at El Tuque, near Ponce, Puerto Rico (photo: L. Bunkley-Williams).

associated with clinical signs of disease, and muscle wasting, bronchopneumonia, and septicemia-toxemia were present in some of the afflicted turtles. Cardiovascular flukes have also been found in wild turtles in the United States, India, Puerto Rico, and elsewhere in Australia. Seven species of flukes were found in the gastrointestinal tracts of nine wild turtles, but there were no signs of pathological changes.

One neoplastic disease is now commonly reported in wild green turtles from the tropical Atlantic and Pacific oceans, and has also been found in hawksbill and loggerhead turtles (Balazs and Pooley, 1991; Williams et al., 1994). The green turtle fibropapilloma (Fig. 6–8) appears as irregular lobulated tumors, up to 30 cm or more in diameter, on the skin, scales, scutes, eyes, and surrounding tissues, and may also appear internally. The tumors may interfere with vision, breathing, feeding, and swimming. The condition seems to be afflicting increasing numbers of turtles and is also spreading geographically. The eggs of parasitic trematode worms occur often, but not always, in dermal capillaries within the tumors. A herpesvirus is now suspected to be the etiologic agent of these tumors (Jacobsen et al., 1991).

6.4. Unanswered Questions

Etiologies remain to be investigated for white band disease in the scleractinia, mass mortalities of sea urchins and fishes, and suspected diseases in many other organisms. New disease conditions are being discovered and their relationship to environmental stresses examined. For example, field and laboratory research indicate that sponge health is influenced in large part by environmental conditions, particularly water flow rates and temperature (Peters, 1993). Vicente (1989) proposed that Caribbean commercial sponges evolved under slightly cooler climatic conditions. Rising seawater temperatures during the first half of the 20th century apparently caused partial or total extinctions of these species, as indicated by analyses of species richness patterns. Whether the demise of commercial sponge species occurred in direct response to the change in environmental conditions or as the indirect result of diseases, such as the suspected fungal epizootic of the 1930s, is still uncertain, however.

Large patches of acroporid corals have disappeared from reefs in the Caribbean in the 1980s, apparently as a result of a white band–type disease. The branches themselves did not appear physically damaged, as would have occurred during storms. The occurrence of WBD on broken branches of acroporids following hurricanes or physical damage needs to be investigated, and the pathogen(s) or stresses causing the condition need to be identified. Presence of WBD has not been correlated with any adverse human-induced or natural environmental factors.

The increased incidence and prevalence of black band disease (BBD) at some sites during certain times of the year, however, does appear to be related to adverse environmental conditions, including warmer than normal temperatures, nutrient loading, increased sedimentation and turbidity, predation resulting in colony injury, and toxics (Peters, 1993). In the laboratory, antibiotics can control bacterial colonization of mucus on the surface of corals in BBD. Similar results were obtained for corals exposed to elevated concentrations of crude oil, copper sulfate, potassium phosphate, dextrose, or sedimentation (Mitchell and Chet, 1975; Hodgson, 1990). Increased mucus production, while an important defense mechanism for corals and other tropical marine organisms, requires high energy expenditures. Available energy can be limited when nutritional sources are scarce or metabolic processes are altered by exposure to temperature extremes or pollutants, leaving the animal susceptible to attack by pathogens.

A study of corals and octocorals from Biscayne National Park, off southeast Florida, in the 1970s revealed high levels of organochlorine pesticides and heavy metals, similar to those levels used in toxicity tests that led to bleaching and mortality of the same reef-building species of corals in the laboratory. Approximately one-third of coral colonies sampled from this site exhibited lesions and possibly pathogenic microorganisms were found in their tissues, although the presence of the lesions could not be linked to contaminants in this study (Glynn et al., 1989). High tissue burdens of such chemicals resulting from chronic exposures may increase the susceptibility of the organisms to disease when additional physical or chemical stresses are encountered. Similar studies of corals off Australia and elsewhere have detected uptake of pesticides and heavy metals into tissues; however, the presence of biotic or abiotic diseases in contaminated corals from these sites has not been documented.

While few diseases in tropical reef fish have been linked to anthropogenic pollutants, a number of studies have found that temperate fish exposed to various chemicals in degraded habitats exhibit increased incidences of fin erosion, ulcerations, epidermal papillomas, gill hyperplasia, exopthalmia, and lymphocystis (Sindermann, 1990). Organic loadings from domestic sewage or agricultural runoff into estuarine and coastal waters may stimulate the growth of large populations of facultative pathogens of the genera *Vibrio, Pseudomonas,* and *Aeromonas.* These bacteria then invade the skin and produce surficial or penetrating ulcers on fish that are stressed under low dissolved oxygen levels, abnormal temperatures, or the presence of other pollutants.

Mucus secretion, the principal external defense mechanism for fish (and other organisms), is suppressed following prolonged exposure to pollutants, leaving the fish susceptible to viral (e.g., lymphocystis), bacterial, or fungal infections and fin erosion. Cellular and humoral immune responses, metabolism, production of detoxification enzymes, and other physiological and biochemical components are also altered by exposures to anthropogenic pollutants and changes in water quality (Adams, 1990), resulting in structural and functional changes that may lead to morbidity and mortality of resident fish populations. Of particular concern is the possibility that environmental stresses may activate latent infections, where pathogens normally exist in a "carrier state" within the host, causing no harm until the fish are stressed.

Observations on commercial catches of fish in Biscayne Bay, off Miami, Florida, over 10 years (1970–1982) revealed a variety of surficial lesions and abnormalities, including ulcerations, fin and integumental hemorrhages, fin erosion, eye abnormalities, scoliosis, scale disorientation, parasitoses, emaciation, and tumors. Bottom-feeding fish were most seriously affected by both infectious and noninfectious diseases. Excessive nutrients from sewage; petroleum hydrocarbons from marinas, shipping, port facilities, and industries; and toxic chemicals, pesticides, and polychlorinated biphenyls leaching from landfills, rivers, and canals have contributed to the decline in water quality of Biscayne Bay. Fishes from the mouths of canals that empty into the southern part of Biscayne Bay exhibit higher prevalences of gill parasites and histopathological changes in the gills than those from (relatively) less polluted areas of the bay (Skinner and Kandrashoff, 1988). Because many young stages of reef fishes utilize inshore waters that may contain toxic chemicals, high levels of nutrients from sewage disposal, heavy sedimentation and turbidity, they may also be susceptible to diseases caused by microbial pathogens and parasites. The role of these diseases in limiting reef fish populations, however, is unknown because appropriate studies have not been conducted.

Similarly, defense mechanisms of higher invertebrate phyla have been shown to be impaired by exposures to anthropogenic pollutants, leading to greater susceptibility to infections by pathogens and parasites (Sparks, 1985; Fisher, 1988; Sindermann, 1990; Couch and Fournie, 1993). Long-term changes in water quality, such as elevated levels of nutrients, are of particular concern for attached or sedentary invertebrates, since they cannot escape. Loss of important members of a food web will ultimately affect higher predators, as will the loss of those organisms that perform important functions in an ecosystem, such as decomposition, bioturbation, or nutrient recycling, and the loss of organisms that provide protection and specialized habitats. Elevated water temperature, the recent local decline in frequency of hurricanes, elevated salinity, and chronic sediment hypoxia were noted as possible causative factors associated with the "wasting" disease of *Thalassia testudinum* off Florida, although a pathogenic protozoan was also found in diseased seagrass blades (Roblee et al., 1991).

In addition to the above research needs, scientists in tropical marine environments should be aware of the possibility that diseases and mortalities observed locally may actually signal major epizootics or mass mortalities that cover large areas of coral reefs or coastal habitats or even regions. Williams and Bunkley-Williams (1990) discussed the problem of major marine ecological disasters (MMEDs) and the difficulties in communicating reports and alerting others around the region or globally. Experience thus far suggests that networks of scientists need to be developed to track such events and to locate expertise to investigate the role of biotic and abiotic diseases in MMEDs. Both observers in the field and specialists in laboratories will be required to respond quickly and gather samples and essential data. Multidisciplinary studies will be most important, drawing on information from ecological, chemical, physiological, microbiological, histopathological, and other types of research. Information from these studies would then be compiled to determine the shared causes, interrelationships, and significance of MMEDs.

The Caribbean Aquatic Animal Health Project at the University of Puerto Rico has established one such network, the Marine Ecological Disturbance Information Center (MEDIC), to track reports of diseases and mass mortalities in tropical aquatic organisms and to alert appropriate experts when field and laboratory studies are required. Scientists in temperate regions are currently developing important biomarkers or biological indicators to assess the stresses that may be encountered by aquatic organisms in their natural habitats. Biomarkers are defined as "molecular, biochemical, and cellular changes caused by pollutant chemicals which are measurable in biological media such as cells, tissues and body fluids" (McCarthy and Shugart, 1990, p. 457). These biomarkers can provide information on current and potential adverse ecological effects of changes in environmental conditions. They may also help to identify causal mechanisms underlying observed effects at the population or community levels. Biomarkers permit the simultaneous monitoring of sensitivity to stress, specific responses to exposure to toxicants or other changes in water quality, effects on the health of the organism, and pertinent ecological factors.

Biomarkers that have been examined include molecular and biochemical responses such as DNA (deoxyribonucleic acid) adduct formation and cytochrome

P-450 system enzyme induction that occurs during exposure to certain pollutants. Physiological and behavioral responses such as respiration rate and changes in swimming patterns, alterations in the immune system as a result of exposure to chronic stress, histopathological changes in cells and tissues, and food consumption or bioenergetic studies that may be indicative of changes in growth and fecundity have also been examined. Similar biomarkers may be useful in monitoring the condition and health of coral-reef organisms; however, the normal ranges for the various parameters must be carefully examined first (Adams, 1990; Huggett et al., 1992). Knowledge of the variability of normal and abnormal structure and function in most tropical marine organisms, both within and between individuals and species, is still lacking. Comparisons that would be useful indicators of environmental stress are few. The study of bleaching in corals and other organisms containing photosynthetic symbionts has increased to the point that algal densities and photosynthetic pigment fluorescence may be useful indicators.

6.5. Ecological Implications

Previous investigations of the structure and function of coral reefs and associated soft-bottom habitats (mangrove and seagrass communities) generally have failed to consider the role of pathogens and parasites in population and community development and alterations. Not only may these organisms affect host responses to environmental stresses, thereby altering population size and geographic distribution, but they can also influence intra- and interspecific interactions of species (Scott, 1988; Sousa, 1991; Rohde, 1993).

Otto Kinne (1980, p. 1) aptly summarized the role of diseases in the marine environment:

> Diseases affect basic phenomena of life in oceans and coastal waters: for example, life span, life cycle, abundance, distribution, metabolic performance, nutritional requirements, growth, reproduction, competition, evolution, as well as organismic tolerances to natural and manmade environmental stress. In short, diseases are a major denominator of population dynamics.

Other chapters in this book examine the changes occurring when populations of reef organisms are altered as a result of predation or exposure to adverse environmental conditions. The same principles and effects apply when populations are affected by biotic diseases.

Disease can have effects that become manifested at the community level. Disease may either produce mortalities of individuals over varying periods of time (acute to chronic progressive mortalities) or alter the structure or function of the individual in such a manner as to make it more susceptible to predation or environmental stresses. Diseases affecting the gonads, sexual maturation, associated tissues, and mating or spawning behavior may also lead to reduced fecundity of individuals with subsequent reductions in population size or changes in gender and age composition. There have been few studies addressing the role of disease in altering populations and communities of tropical marine organisms.

As discussed in other chapters, the demise of even one species of coral-reef organism may have serious repercussions for the structure and function of a particular community and the reef ecosystem (e.g., *Diadema antillarum,* Chapter 9 and section 15.3.4 in Chapter 15). Diseases of scleractinian corals are an important factor in changing the structure and function of coral-reef communities because loss of live tissue cover not only reduces the number of polyps producing gametes for potential new recruits, but also opens up new hard substratum space for settlement of benthic organisms. From studies at seven sites in the U.S. Virgin Islands, Edmunds (1991) estimated that chronic BBD infections in brain corals *(Diploria strigosa)* could increase bare substratum by 3.9% per year, allowing colonization by a variety of organisms (not necessarily new corals). He noted that the distribution of infected corals was not clumped, suggesting that the disease is not highly infectious between colonies, but could have been picked up by chance encounters with drifting filaments of *Phormidium corallyticum* when the colony had been previously injured.

Edmunds further noted that coral colonies killed by BBD did not show any scleractinian recruits after 2 years of observation. This study was undertaken in a bay with a fully protected watershed and relatively pristine reefs, and the only increase in infection rates and loss of tissue occurred during late summer when seawater temperatures were at their peak. Thus, this disease may prove even more damaging in altering the growth and development of the massive reef-framework-building corals on reefs chronically exposed to higher seawater temperatures and anthropogenic stresses. The outbreaks of BBD observed off the Florida Keys (Looe Key and Key West) and other sites in the Caribbean in the late 1980s may be due to adverse environmental changes and human interactions.

The enigmatic white band disease has also affected large areas of reef throughout the Caribbean and Florida Keys. The loss of branching acroporid corals as a result of this disease, with bioerosion of the remaining exoskeletons, has changed reef structure (Gladfelter, 1982). As coral tissue dies and sloughs off the skeleton, endolithic boring sponges, algae, and other organisms also invade the carbonate substratum, breaking it down and destroying reef topography, with eventual collapse of the reef. Extensive loss of topographic relief has affected fish populations as protective niches and important habitats have been altered. The decline of coral-reef fisheries among western Atlantic coral reefs may be related to such structural changes, although overfishing has probably contributed to losses of fish species diversity and numbers as well (Rogers, 1985). Acroporid corals damaged by Hurricane Allen off Jamaica initially survived as fragments, then apparently succumbed to white band disease, changing the pattern of predation by coral snails and shifting the roles of other predators there.

Although community alterations have been observed more often on coral reefs, diseases may also affect a variety of relationships and processes in soft-bottom benthic communities on reefs and in seagrass beds and mangroves. In addition to predation and competition, bioturbation, sedimentation, primary and secondary production, and other phenomena may be changed with the loss of one or more micro- or macroorganisms to biotic or abiotic diseases. During the recent mass mortality of turtlegrass *Thalassia testudinum* in Florida Bay, more than 23,000 ha were affected, with 4,000 ha of seagrass beds completely lost, especially in protected basins. Many reef organisms are associated with seagrass habitats. They are used by larvae and juveniles as refugia from predators. Soft-bottom invertebrates used as prey may suffer from the loss of the seagrass and they can also be affected by pathogens and parasites, as documented in a recent review by Sousa (1991). As mentioned above, there is evidence that fish diseases are increasing in tropical bays and mangrove areas where man-made solid and other wastes are disposed.

Coral reefs in the Caribbean and Indo-Pacific differ in a number of respects, as discussed in Chapter 12. There are some differences in the pathogens and parasites, their hosts, and the environmental stresses found in these regions, although the same general principles apply. White band and black band disease in scleractinian corals, coral bleaching, turtle fibropapillomas, and other diseases are now reported from reefs all over the world. Many diseases may be increasing as the result of new stresses in the coral-reef environment from human activities, and/or are being noticed because of increased research by pathobiologists in the tropics.

Another concern is that organisms are being transported out of their home habitat and introduced into similar or new habitats many miles away, often in different regions or even oceans, as the result of global shipping, oceanarium, aquaculture, marine laboratory, and tropical aquarium trade activities. In addition to attaching to the hull of a vessel, organisms, particularly their larvae, may be picked up with ballast water and deposited offshore or in another harbor. Specimens for research and commercially desirable organisms may be flown to new oceanariums or aquaculture facilities in a matter of hours. Shipments contain not only the desired (or target) organisms or their eggs and larvae, but also any organisms in the water or seaweed in which they were originally packed and pathogens or parasites within the target organisms. If the new tanks are operated as flow-through systems, without proper quarantine or treatment of the discharged water, these animals will be released to the sea. Similarly, releases of unwanted fish or other animals by tropical aquarists have also led to introductions in foreign lands. If conditions are suitable for survival and reproduction, the species may become established at the new site.

Some introduced species may have positive effects in their new environment. But the vast majority of documented introductions have adversely affected existing commercial and recreational fisheries as the result of competition, predation, and the introduction of pathogens and parasites (Sindermann, 1990; Carlton,

1992). Again, studies of this phenomenon in the tropics are limited compared to research in temperate areas. However, many species of tropical marine fish, penaeid shrimp, and tridacnid clams are now being transferred to new sites, some even into new ocean basins. These organisms may carry symbionts that could be pathogenic to susceptible species when introduced into new habitats. A number of countries are now developing regulations and procedures to restrict such introductions, including permit and certification requirements (DeVoe, 1992), such as the following guidelines:

- A species should not be transferred out of its known natural range.
- If a species is to be transferred within its natural range, the organisms should be cultured in filtered seawater and/or maintained in filtered, UV-irradiated, recirculating seawater for one month preceding the transfer.
- The receiving institution should quarantine the organisms in special tanks or raceways using filtered seawater for at least 6 months and the used seawater must not be disposed of in the sea unless properly treated.
- If disease, pathogens, parasites, or predators are detected during the quarantine period, the stock must be destroyed and all equipment sterilized.

In summary, investigations on the nature and role of diseases in coral-reef organisms are only just beginning. Based on observations of commercially important invertebrates from temperate seas, more diseases and diverse parasitic infestations no doubt remain to be discovered (Sparks, 1985; Sindermann, 1990). Population losses thought to be tied to overfishing or natural predation may have been the result of disease. Pathogens and parasites exert tremendous pressures on individuals, populations, and communities in their interactions with the host and environmental conditions (Scott, 1988), and must be taken into consideration in any discussion of the dynamics of coral reefs and associated tropical marine habitats.

7

Organic Production and Decomposition

Bruce G. Hatcher

Coral reefs are gigantic structures of limestone with a thin veneer of living organic material—but what a veneer! Everything that is useful about reefs (to humans and to the rest of nature) is produced by this organic film, which is approximately equivalent (in terms of biomass or carbon) to a large jar of peanut butter (or vegemite) spread over each square meter of reef. The production of inorganic materials that accumulate over geological time to form reefs (Chapter 2) is the major component of reef growth (Chapter 3), the balance between production and destruction (accretion or erosion) being modulated by the interaction of physical and biotic controls (Chapter 4). A major biotic control of the production of calcium carbonate is the rate and form of photosynthetic (primary) production, which enhances calcification (Chapter 5). The metabolic destruction of this organic production by respiration serves, conversely, to dissolve calcium carbonate by altering the pH of the seawater-carbonate medium. Thus, the balance of massive production-destruction of inorganic material is intimately linked to that of much lower production-decomposition of organic material across the broad range of spatial and temporal scales characterizing coral-reef ecosystems. The contributors to, and controls of, the balance between organic matter production and its decomposition within coral-reef ecosystems are the subjects of this chapter.

One way that scientists have adapted to the bewildering diversity of coral-reef communities (Chapter 14) is to treat whole sections of reefs, or whole reef ecosystems, as single organisms that process (metabolize) energy and materials according to the laws of physics and thermodynamics. An economy of the reef is derived in terms of pools of materials and the transport of materials (i.e., fluxes) into and out of those pools. Indeed, the trophodynamic, or so-called "systems," approach to ecology had some of its first and most successful applications on coral reefs (Sargent and Austin, 1949; Odum and Odum, 1955). These seminal papers and the many they spawned (see reviews by Kinsey, 1985a;

Smith, 1988; Hatcher, 1988, 1990) still provide valuable insights and provocative questions.

Much of our current understanding of the processes, rates, and patterns of organic production and consumption in coral-reef communities is based on inference from the changes in ratios of concentrations of chemicals (i.e., the stoichiometry) in the water bathing the living veneer, rather than from direct measurements of changes in the mass of individual organisms due to the processes of growth and decay. For example, attempting to calculate the production of plant tissue on a 100-meter-wide reef flat by collecting samples of each of the hundreds of species of algae, measuring their metabolism or growth, determining their relative abundance on the reef, and then multiplying through in order to scale up the errors of integration would be horrendous (Smith and Kinsey, 1988). How much more accurate (and elegant) to simply measure the change in oxygen concentration of the water overlying the benthic community as it flowed across the reef flat! This change represents the net result of photosynthesis and respiration by all the organisms of the reef flat (i.e., the community metabolism). The temporal and spatial scales of the measurements of flux are perfectly matched to the scales of the community and its production processes (Hatcher et al., 1987). We are then in a position to answer the question, How much food can this reef flat provide to animals living in the adjacent lagoon?

Of course, measures of community metabolism (often summarized as the ratio of photosynthesis to respiration) tell us nothing about how a particular species is performing. For that we must make the species-specific measurements. Many such measurements have been made of the metabolism of individual reef organisms as well, using both direct measures (e.g., assimilation of radioactive carbon) and inference from changes in enclosed water volumes. The choice of approach depends on the question being asked, and we get into real trouble when we mismatch the scales of the process and the measurement (Smith and Kinsey, 1988). For example, a clear understanding of the processes and accurate measures of rates by which areal primary production is converted into total fish biomass on a reef will provide a poor answer to the question, How many coral trout will I find on this reef? We might as well expect to learn how a television works by burning it in a bomb calorimeter (May, 1988)!

The question most often asked about coral-reef production is, How can there be so much of it in the nutrient deserts of the tropical oceans? The apparent paradox is a bit of a disappointment in that it arises largely from semantics and oversimplification. When properly scaled, the net organic productivity of entire atolls is actually little different than that of the surrounding oligotrophic ocean (i.e., low; see Crossland et al., 1991), and there is plenty of nutrient in that water to support the net ecosystem production, despite its low concentration (Atkinson, 1988, 1992). This trite dismissal of the reef paradox ignores some profound and complex realities however, the understanding and measurement of which are keys to sustainable use of the organic resources of coral reefs.

NITROGEN
ADVECTIVE INPUTS
$N_{inorg.}$
$N_{org.}$
PLANKTON
WALL OF MOUTHS
LAGRANGIAN CONTROL VOLUME
T'
T"
COMMUNITY PRODUCTION
$P_{net} = O_2'' - O_2'$
$P_{gross} = P_{net} + R_{night}$
$E = P_{gross} - R_{24hr}$
CO_2
RESP
FAECES
$N_{org.}$
SECONDARY PRODUCTION
ORGANIC NITROGEN
NITROGEN FIXATION
R E E F F L A T
REGENERATIVE SPACES
CO_2' O_2' O_2'' CO_2''
SPECIFIC PRIMARY PRODUCTION
P.A.R.
CO_2
PP_{gross}
CO_2
RESP.
Biomass
N_{inorg}
GROWTH
PP_{net}
FRAGMENTATION
EULARIAN CONTROL VOLUME
PLANKTON $N_{org.}$
MUCUS $N_{org.}$
CO_2
$N_{inorg.}$
CO_2
CO_2
$N_{org.}$
$N_{inorg.}$
HARVEST EXPORT
ADVECTIVE EXPORT
$N_{inorg.}$
$N_{org.}$
L A G O O N
DETRITIS
MIGRATION EXPORT
CO_2 RESP.
$N_{inorg.}$ REMIN.
SEDIMENT MICROBIAL COMMUNITY
KITTIWAKE ILLUSTRATIONS

▲
Figure 7–1. Methods and measures of coral-reef production. Nutrients (N) in both inorganic (dissolved) and organic (dissolved and particulate) forms are advected into and out of a cartoon reef ecosystem by the movement of water. A small fraction of this total flux is captured by the reef community and incorporated into living tissue. Plankton-feeding assemblages of corals and fish on the outer margins of the reef (wall of mouths) intercept the particulate input and metabolize it, so that a portion is subsequently available for uptake by plants (primary producers). Symbiotic and free-living plants combine assimilated nutrients (including inorganic carbon, CO_2) to form new tissue (growth) using energy in photosynthetically active radiation (P.A.R.). The tissue may accumulate as biomass, or be lost through fragmentation or grazing. In the latter case, the plant-specific primary production directly supports secondary production. Some algae are able to take up dissolved gaseous nitrogen and convert it to a reduced, and ultimately inorganic, form (nitrogen fixation). The decomposition of dead and dissolved organic material (detritus, feces, mucus) takes place through microbial metabolism in the water column, the regenerative spaces in the reef matrix, and the sediment microbial community of the lagoon. The organic nutrients (inorganic N) thus released (REMINeralization) are again available in dissolved form to contribute to primary production (PP_{gross}). Within the coral-algal symbiosis, the recycling of nutrients between organic and inorganic phase is largely internal, with both organic (plankton) and inorganic nutrients taken up, and organic material released as mucus. Primary production at the plant-specific (PP_{gross}) and community (P_{gross}) levels results in the uptake of CO_2 and the release of oxygen during the daylight hours. RESPiration by plants and animals and microbes in the reef community produces opposite fluxes to and from the surrounding water. During the night, these are the only fluxes, but RESPiration during the day is offset by the photosynthetic fluxes, the net result being a measure of net production (PP_{net} and P_{net}). Gross production (PP_{gross} and P_{gross}) is never measured directly; it is calculated by adding dark RESPiration to net production (PP_{net} and P_{net}). At the level of the community, the excess production *(E)* is calculated by subtracting the total diel RESPiration of the entire plant and animal community ($R_{24\ hr}$) from the gross production during the daylight hours (P_{gross}). Changes in dissolved oxygen (O_2) in the water surrounding a plant, coral, or portion of a habitat are used to measure the rates of primary production and respiration. Often the volume of water is fixed in an enclosure (Eulerian control volume), which restricts cross-boundary fluxes. In uniform flow conditions, dye may be used to mark a parcel of water (Lagrangian control volume) that is followed downstream and sampled at time intervals (T', T'') as it traverses the reef community. In addition to advective exports, secondary production may leave the reef ecosystem through migration and harvest.

The production-decomposition functions of coral reefs (e.g., the plant growth, herbivory, remineralization) are exquisitely tied to their structures at all levels of organization, in the physiological, physiographic, and community senses (Chapter 12). Adaptations at all these levels enhance efficiency of capture and recycling of energy and nutrients within the ecosystem, thereby reducing the one-way losses or need for inputs across system boundaries (e.g., detritus and nutrients that are carried off and on the reef by water flows). As a result, entire coral-reef ecosystems place little demand on, and contribute little to, adjacent ecosystems, relative to the total pools and turnover of materials within the reef system (Crossland and Barnes, 1983; Smith, 1988; Crossland et al., 1991). However, the pools (i.e., solar energy, dissolved nutrients, organic material) are so large that the relatively small differences between inputs and outputs (i.e., the net fluxes) can still be large in absolute terms. If one of these is the export of secondary production from the reef system to the people on the adjacent terrestrial system, then the determinants of the magnitude of the flux may be a matter of life and death, for both the people and the reef (Chapter 18)!

The problem of determining the available production of a coral reef then becomes a nontrivial one of understanding controls on relatively small deviations from zero net flux in a system with massive total flux. A positive deviation indicates a reef accumulating or exporting organic material, a negative deviation a reef decomposing or importing organics (Fig. 7–1). This chapter will present what we know about this balance and how our activities might tip it. Our knowledge of the positive (production) side of the equation is far greater than that of the negative (decomposition) side (Fig. 7–2).

7.1. Definitions and Methodologies of Production

Terms and measures used to describe and quantify organic production are confusing in their use and application. A simplistic portrayal of those relevant to coral reefs appears in Figure 7–1. More detailed definitions are given in Kinsey, (1978) and Hatcher (1988).

The essential concepts are those of currencies, pools, fluxes, turnover, and scales. In the simplest sense, the balance between organic production and decomposition is the rate of biogenic flux from inorganic to organic pools divided by the opposite flux. In a closed euphotic system, this ratio is well approximated by the photosynthetic (primary) production divided by the total respiration during equal time periods—the P/R ratio. If it is greater than unity, the system is producing; if less than unity, the system is decomposing. The value of the ratio is highly dependent on the pools of materials measured and the spatial and temporal scales of measurement. The trick in using this powerful metric is to match form and scales of the measurements to those of the question being asked.

REDUCTION	DECOMPOSITION	POOL	PRODUCTION	ACCUMULATION
of Biomass	← (Negative FLUX)	of Biomass	(Positive FLUX) →	of Biomass
Smaller individuals	Tissue LOSS due to Herbivory, Predation, Growth, Overfishing, Erosion, etc.	ORGANISM	Tissue GROWTH (1° & 2° Production)	Larger individuals
Smaller colonies	Polyp LOSS due to Abraison, Grazing, Bleaching, Disease, Erosion, etc.	SYMBIOSIS, COLONY	Colony GROWTH (2° Production)	Larger colonies
Fewer individuals	MORTALITY & EMIGRATION	POPULATION	REPRODUCTION & IMMIGRATION	More individuals
Fewer large individuals	Selective MORTALITY (storms, fishing, etc.)	ASSEMBLAGE & COMMUNITY	SUCCESSION	More large individuals
Less reef area	ALL of ABOVE, plus EXPORT	ECOSYSTEM	ALL of ABOVE, plus ACCRETION	More reef area

Figure 7–2. The balance of production and decomposition of reef biomass. The mass of organic material in a coral-reef ecosystem is the result of opposing processes of production and decomposition that operate at most levels of organization. At the highest levels of biological integration (the community and ecosystem), coral reefs exhibit relatively low net rates of biomass accretion or erosion (i.e., are homeostatic or self-regulating), even though large imbalances may concurrently exist at lower levels of organization (e.g., within a population or organism). Major disturbances such as catastrophic storms or changes in water quality can produce large deviations from modal flux rates, leading to rapid reef destruction or accretion (growth).

7.1.1. Currencies of Production

The currencies used to quantify metabolism are not easily interchangeable because they are stored and processed differently by different components of the community. For example, nitrogen and phosphorus are not immediately required in the process of photosynthesis (they are, however, essential to plant growth and to maintain the photosynthetic machinery). Carbon fixation can proceed apace in the absence of these nutrients if light and carbon dioxide are abundant (Raven, 1974), but most of the production is rapidly lost from the plant if the structures for carbohydrate storage do not exist (Hellebust, 1974). The small plants and photosynthetic bacteria that characterize primary production in coral-reef communities have little capacity for the bulk storage of organic carbon or inorganic nutrients (Borowitzka and Larkum, 1986; Ducklow, 1990). They depend on the surrounding seawater to supply the inorganic nutrients they require and to disperse the organic compounds they leak or excrete.

Carbon is the most common currency for production measures because it is a dominant component of all organic material, is universal to all organic production and decomposition processes, and can be accurately converted into units of energy. Carbon occurs in many phases in both organic and inorganic form, which are related by a complex chemistry (i.e., stoichiometry) in seawater (Millero, 1979).

The balance of carbon phases in coral-reef waters is influenced directly and strongly by calcification as well as by organic production and decomposition, so simple measures of inorganic carbon phase changes (e.g., CO_2 flux) cannot be converted to metabolic rates without concurrent, independent measures of calcification (Kinsey, 1978). In practice, change in dissolved oxygen (O_2) concentration in seawater (O_2 has a simpler stoichiometry than CO_2 and is not directly affected by calcification) is more commonly used to measure organic production and decomposition on reefs, although the rates are usually expressed in the carbon currency.

Virtually all organic production includes materials containing other nutrients besides carbon (e.g., proteins and lipids, as well as carbohydrates), so production is often measured and expressed in terms of the most important of these other constituents (i.e., the macronutrients: nitrogen and phosphorus). These elements (especially nitrogen) also have complex cycles of transformation in reef systems (D'Elia and Wiebe, 1990), such that simple measures of their fluxes may provide ambiguous estimates of production.

Inorganic carbon concentration or supply has never been shown to limit primary production in reef environments, but the growth of individual reef plants and assemblages has been shown to be limited by low concentrations of inorganic nutrients (e.g., Hatcher and Larkum, 1983; Lapointe et al., 1987). Using different nutrient currencies for production measures can suggest different controls on production rates (Smith, 1984). The conversion from macronutrient to carbon (or energy) currencies is confounded by the great variability in the ratios of these elements in different types of organic material, both within individuals of the same species and within communities containing many species (Atkinson and Smith, 1983). Despite these caveats, measurements in nutrient currencies (especially phosphorus, because it has the simplest cycle of transformations) provide our best insights and estimates of net production for entire reef ecosystems (e.g., Smith and Jokiel, 1978; Atkinson, 1992).

7.1.2. Pools of Organic Materials in Coral Reefs and Control Volumes

The smaller the organism, the smaller its pool of materials, the lower its capacity for storage and buffering from variability in supply rate. Coral-reef communities are dominated by small plants and animals living symbiotically or freely on or in dead coral rock, internal cavities, and sediments (Fagerstrom, 1987). The complexity of reef structures creates habitat across a range of spatial scales

(Reichelt, 1988), which partially isolate volumes of seawater that exchange slowly with the bulk water mass (Andrews and Muller, 1983). Lagoon sediments isolate pore water in interstitial spaces rich in microbes (Ducklow, 1990). Coral-reef structures enclose lagoons and moats that integrate metabolic signals at much larger scales.

The pools of materials within these different volumes are chemically altered by the metabolism of the organisms they enclose. As a result, the concentrations of materials within reef water masses may differ markedly from those in the surrounding ocean. In volumes that are small relative to the biomass, there is strong potential for feedback because many of the metabolic processes related to production and decomposition (e.g., nutrient uptake kinetics) are concentration dependent. These water masses can thus be treated as control volumes (Hatcher et al., 1987), in which measured changes in the contained pools of different materials (e.g., CO_2, NO_3, PO_4) reflect, and sometimes regulate, the metabolism of the enclosed community.

The concept of the control volume is fundamental to the measurements of coral-reef production (Fig. 7–1). It links organism-scale metabolism to system-scale hydrodynamics. Control volumes may be considered as fixed in space (i.e., Eulerian), such as a tide pool or lagoon, or mobile (i.e., Lagrangian), such as a mass of water flowing across a reef flat. Because control volumes are never completely isolated from the surrounding system, changes in their contained pools of materials (concentrations) result from fluxes across their boundaries, as well as from the fluxes into and out of the organisms they contain.

If the magnitude of the boundary flux is of similar order as the biogenic flux, then it must be accurately measured or inhibited if useful estimates of metabolism are to be made. The most common method is to artificially enclose a water volume in a bottle, dome, or plastic bag that limits advective and diffusive exchange across the control volume boundaries. This is fine for short measurements of individual organisms or small portions of communities, but it is impractical for the large and long scale measurements required for integrated production rates of habitats and ecosystems. Any useful discussion of production must specify the pool size and control volume if rate measures (e.g., turnover) are to be meaningful.

7.1.3. The Concepts of Turnover in Coral-Reef Ecology

The turnover of the materials in a pool is the essence of production. A single-celled phytoplankter or zooxanthella produces carbon equivalent to that contained in its tiny body pool (turns over) once every couple of days when it is photosynthesizing at peak capacity (Platt et al., 1984; Muscatine and Weis, 1991). An entire coral-reef community takes months to years to turn over its combined pools of organic carbon (Fig. 7–3). Other nutrients (currencies) cycle at different rates.

Time scale: Period, turnover or generation time (s)	Class of process: Physical → Biological		
	Geophysical and hydrodynamical processes	Geological, biogeological and biogeochemical processes	Ecological, population and organismal processes
(10^6 years)		Formation of island arcs and guyots	
10^{13} (10^5 years)		Turnover of reef provinces	
10^{12} (10^4 years)	Renewal of ocean water Isostatic sea-level oscillations	Turnover of individual reefs Upward growth of individual reefs Lagoon infilling	Evolution of species Bioerosion of massive coral structures
10^{11} (10^3 years)		Growth of lagoon patch reefs	
10^{10} (10^2 years)	Mixing of ocean water Recurrence of catastrophic physical disturbances	Turnover of nutrients in deep sediments	Regeneration of massive coral colonies
10^9 (10 years)	Recurrence of major storms		Regeneration of large reef vertebrates
10^8 (1 year)	Exchange of water in reef provinces		Regeneration of large reef invertebrates
10^7 (1 month)	Exchange of water in closed-reef lagoons Sedimentation down mixed layer	Turnover of C in benthic algae and corals Turnover of C in lagoon water column Turnover of C, N, and P in interactive sediments	Regeneration of benthic macroalgae
10^6 (1 week)	Exchange in open-reef lagoons Vertical mixing in photic zone	Turnover of C in plankton communities	Regeneration of small reef invertebrates Development of larvae Regeneration of benthic microalgae and planktonic invertebrates
10^5 (1 day)	Mixing of water in reef lagoons Lunar tides and internal waves	Turnover of N and P in lagoon water column	Regeneration of planktonic algae
10^4 (1 h)	Exchange of water on reef flats	Turnover of C in microbial communities	Regeneration of marine bacteria
10^3	Dissipation of energy in small-scale turbulence		
10^2 (1 min)			
10^1	Ocean swell Wind waves		
10^0 (1 s)			

Figure 7–3. The time scales of coral-reef processes. Most processes span a range of periods constrained by their spatial scale; generalized time scales are accurate only to an order of magnitude. The time scales of relevance to organic production and decomposition are generally on the order of one month or less (from Hatcher et al., 1987).

Turnover rate (flux/pool: units = T^{-1}) introduces the temporal dimension to production-decomposition processes (Fig. 7–2).

The rate of organic production (productivity) is a flux that is poorly related to pool size (biomass) on an areal basis. For example, forests and coral reefs have similar primary productivities, but vastly different biomasses, and hence turnover rates (e.g., compare the figures in Whittaker, 1970, with those in Lewis, 1981). It is a common error to equate biomass with production. Compared to terrestrial systems of similar biomass (e.g., grasslands; Whitaker, 1970), the organic pools of coral reefs turn over quickly. This does not necessarily equate to short life spans or generation times of reef organisms: turnover is not the same as exchange. A 500-year-old coral colony has a production to biomass (P/B) ratio of about 0.05 d^{-1} (i.e., a turnover time of about a month), while small benthic algae have turnover and generation times of the same order (Jokiel and Morrissey, 1986). The alga may exchange every carbon molecule in its tissue with a newly produced one during the turnover period; the coral certainly does not.

Turnover is a relative parameter that, when applied to individual organisms, provides a measure of production (or consumption, respiration, or some other flux) per unit of organism mass in the same currency. It is sometimes referred to as specific productivity. For primary producers (i.e., the autotrophs: all plants and some protozoa), the relevant parameter is the rate of inorganic carbon fixation (usually photosynthesis) divided by the carbon mass of the autotroph (i.e., specific production; Fig. 7–1). The growth of heterotrophic organisms (i.e., most protozoans and all animals) is sometimes termed *secondary production*, and is essentially the conversion of one form of organic material into another (e.g., algal tissue into parrotfish tissue and gametes). It does not involve the production of new organic material from inorganic pools (i.e., primary production), just different organic material. The cost of conversion is high (most of the assimilated food is metabolized in the same day), so that growth turnover is slow relative to specific turnover rates calculated on the basis of ingestion or respiration.

The combined areal biomass of all heterotrophs (from bacteria to whale sharks) in reef communities approximates that of the autotrophs (pyramids of biomass for reef communities are steep sided: Odum and Odum, 1955; Polovina et al., 1984), and the body sizes of the organisms comprising most this biomass are small (no herds of wildebeest). As a result, heterotrophic respiration is the major pathway by which organic material is turned over (recycled back to inorganic form) in coral-reef ecosystems.

Turnover can be calculated in both gross and net terms. Gross turnover is the total flux through the pool divided by pool size, while net turnover is flux into the pool minus flux out divided by pool size. The distinction is particularly relevant to primary production because it is the ultimate source of new organic material within an ecosystem. Gross primary production is the total organic fixation of carbon by autotrophs, while net primary production is the amount of this organic fixation of carbon remaining after the immediate costs of the metabolism of autotrophs have been paid.

Turnover of a biomass pool by net production (e.g., the growth of plant tissue) will clearly be much slower than that by gross production and respiration. Regardless of how it is calculated, the primary productivity of coral-reef autotrophs is very high in terms of photosynthetic carbon fixation (Fig. 7–4) because of the large photosynthetically active area of reef surfaces, the great abundance of light and inorganic carbon, and the small size of most of the plants (i.e., they have high specific productivities: Littler et al., 1983a).

Plant growth is not precisely equivalent to net primary production of carbon, although the terms are often used interchangeably. The production of tissue

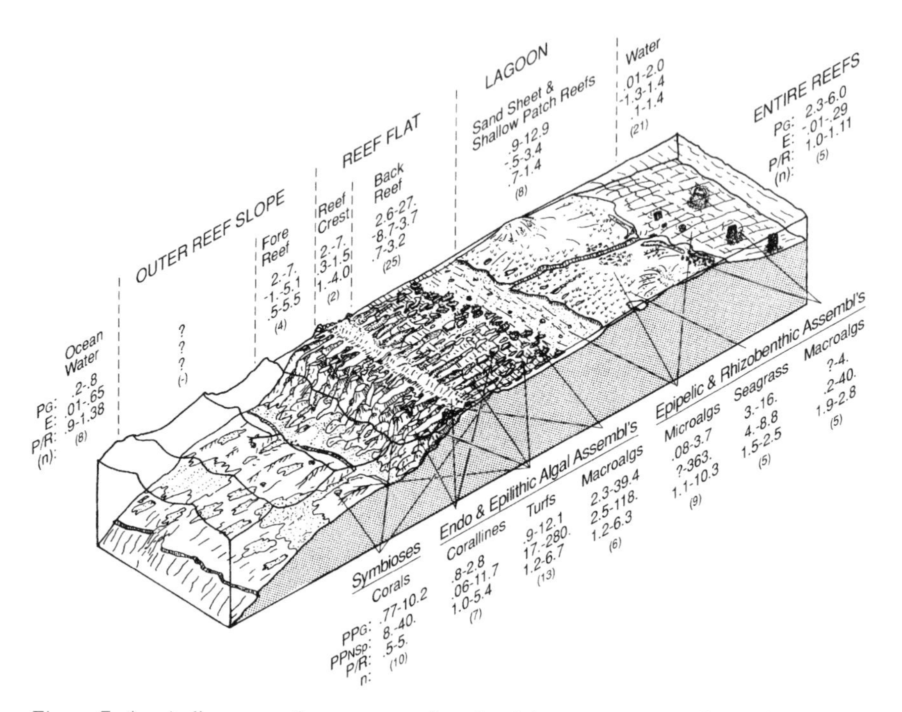

Figure 7–4. A diagrammatic summary of productivity measurements in coral-reef ecosystems. The physiographic zones of an idealized reef are illustrated from the seaward margin to the lagoon. Minimum and maximum diel (24 hr) gross community production (P_G) and excess production *(E)* per unit of vertically projected surface area are expressed in units of gC m^{-2} d^{-1}. The diel gross production to 24 hr community respiration ratio (P/R) is also given for the major reef zones and entire reef ecosystems. Minimum and maximum gross primary production (PP_G) per unit of actual benthic surface area (gC m^{-2} d^{-1}), diel net weight–specific primary productivity (PP_{NSp}) in mgC per g Dry Wt. d^{-1}, and diel gross photosynthesis to respiration ratio are shown for each major component of reef primary producer biomass along the lower panel. The number of species populations, assemblages, communities, or ecosystems is given in parentheses (from Hatcher, 1988).

for growth does not include organic exudates or reproductive products. The discrepancy is extreme in symbiotic zooxanthellae, which have a carbon turnover rate of less than 2 days, but a tissue turnover rate of over 50 days (Muscatine and Weis, 1991). Tissue growth also requires the uptake of inorganic nutrients in addition to photosynthetically fixed carbon. Inorganic nutrients are not always abundantly available to autotrophs on coral reefs, and so they have the potential to regulate primary productivity at both organism and community levels (Atkinson, 1988).

When a causal relationship exists between nutrient supply and productivity, nutrient turnover in a pool provides a useful measure of organic production.

Inorganic nutrients fixed into primary production can originate from two sources: within the boundaries of the system (recycled within the pool) and outside the system (imported from another pool). Organic material produced by plants using imported nutrients that are new to a pool is called **new production** (Dugdale and Goering, 1967; Platt et al., 1991), while that produced using old nutrients regenerated within the pool may be termed **recycled production.** The distinction between new versus recycled production is entirely dependent on boundaries of the system and the integration time used to calculate turnover.

At the scales of the individual plant on a reef, virtually all net primary production is new production because inorganic nutrients for growth are taken up from the water outside the plant wall. But at higher levels of integration (e.g., the coral-algal symbiosis, the coral-reef lagoon, or the entire reef ecosystem: Fig. 7–2), the distinction becomes crucial as more and more of the production depends on the turnover of nutrients already within the system (recycling). In the simplest sense, the faster and more efficient the recycling of materials required for production within a system, the lower the demands a given level of production will place on external sources. Coral reefs abound with adaptations that enhance recycling at all levels of organization (Chapters 5 and 12).

7.1.4. Scaling Up

The concepts of production and turnover highlighted above can be applied at all levels of organization in coral-reef ecosystems, but there are important differences in the definitions of terms and methodologies employed at the progressively larger spatial and temporal scales (Figs. 7–1 and 7–3). The most significant difference concerns net production at the scales of habitats or entire communities of organisms.

Productivity is generally expressed per unit of vertically projected surface area at these scales. (Note that the actual biogenic surface area may be as much as 15 times the projected area [Dahl, 1973]). The community biomass pool includes both the autotrophs and heterotrophs on and in the reef substratum, as well as in the overlying water. The respiratory destruction of organic matter, which is divided into the gross production by the autotrophs to calculate community P/R ratios, is due to all living biomass in an area, not just the plant metabolism. Net community production is thus much lower than net primary production in any reef habitat because the latter measure only deducts autotroph respiration from the gross primary production, while the net community is a measure of what remains after the total metabolic needs of the whole community are met. It is only this remaining material that is available to increase the community biomass (through either the increase in individual size or the numerical density of its organisms), or for export from the community (e.g., emigration, harvest; see Fig. 7–2).

A 24-hour (diurnal) integration time scale is appropriate for the calculation of net community production because of the destruction of organic material that occurs during the night hours when respiration continues in the absence of photosynthetic production. The 24-hour net community production integral is termed **excess production,** or E (Kinsey, 1983), which avoids confusion with net primary production. An excess production value of $E = 0$ is equivalent to a community P/R ratio of 1, indicating that all of the local primary production is consumed within the community during the day-night cycle. Production and decomposition are perfectly balanced when $E = 1$, and the reef community cannot export biomass without shrinking unless there is an equivalent import of organics. Calculation of this single parameter of coral-reef metabolism characterizes the trophic status of the community. If E exceeds zero, the community is said to be **net autotrophic;** if E is less than zero, it is **net heterotrophic** (Smith, 1988).

The metabolism of coral-reef communities is calculated from the change in concentrations of dissolved materials it induces in an adjacent control volume of water at spatial scales ranging from an organism to a reef lagoon (Fig. 7–1). Many assumptions and interpolations are required to convert these indirect measures into estimates of reef production and decomposition, and the approach cannot be used in all habitats (Marsh and Smith, 1978; Kinsey, 1978). We have virtually no data from windward reef slopes and very few estimates for entire reefs (Fig. 7–4).

The fact that we measure the small net signal rather than the gross flux (gross production is calculated by adding measures of daytime net photosynthetic fixation to nighttime respiration) means that differences in excess production (E) are often lost in the measurement error. Furthermore, the many processes that interact to produce the net signal confound the interpretation of the components and controls of metabolism. For example, a community showing a negative E ($P/R < 1$) for many days could be suffering a collapse of primary production (e.g., coral death in a storm) and could be losing biomass, or it could be receiving a large subsidy of organic material from upstream (e.g., a sewage pipe), with no reduction in primary production or biomass. Extra measures of the fluxes of various materials across the boundaries of the control volume are required to resolve these matters. Budgets of materials in open-reef systems experiencing rapid water flows are problematic because of measurement errors and mismatching of the spatial-temporal scales of the components and processes (Smith and Kinsey, 1988).

The processes of organic production and decomposition in coral reefs are clustered at several different time scales (Fig. 7–3), and our measures and interpretations of reef metabolism are strongly scale dependent (Hatcher et al., 1987). Photosynthesis, assimilation, respiration, and growth proceed and are mediated at the physiological scales of the cell within the organism. Processes crucial to the accretion of entire reef ecosystems, such as light-enhanced calcification (Chapter 5) and nutrient limitation of algal growth in symbioses, occur at the millimeter and minute scales characteristic of the small pool sizes and control volumes (Hinde, 1988).

The integration of cellular processes over measurement scales characteristic of reef communities and their turnover (meters and days up) is not always straightforward. For example, low local nutrient concentration has been shown experimentally to limit instantaneous productivity of individual coral-reef algae (Lapointe et al., 1987), but long-term additions of nutrients to coral-reef communities have had ambiguous effects on gross productivity and biomass accumulation (Kinsey and Davies, 1979; Hatcher and Larkum, 1983). Measurements and predictions of coral-reef metabolism must always be made in the context of clearly defined spatial and temporal scales.

7.2. Producers and Consumers

7.2.1. Microbial Communities Rule

The smaller the size of an organism, the greater its importance in the processes of production and decomposition in coral-reef communities, and the less we know about it. The microbial biomass of reef communities approximates that of other shallow marine communities (Millis, 1981; Ducklow, 1990), but virtually all of it is concentrated in the benthos because of the abundance of interstitial habitats and the longer residence times of organic inputs (Ducklow, 1990). The high conversion efficiencies and specific growth rates of bacteria (carbon turnover rates often less than 1.0 d^{-1}) mean that their assemblages generally require more than the in-situ primary production to support their metabolism in many reef habitats, despite their diminutive biomass (Fig. 7–5).

Indeed, metabolism by both aerobic and anaerobic, heterotrophic bacteria in the upper few centimeters of sediments is responsible for the decomposition of virtually all of the organic material reaching coral-reef surfaces, leaving little to be buried as organic sediment (Ducklow, 1990). The fast generation times and large surface to volume ratios of microbes allow their populations to respond rapidly in numerical and metabolic terms to changes in the supply and quality of organic material (Sherr and Sherr, 1984). As a result, microbial biomass and metabolism vary greatly through time and between reef habitats (e.g., Moriarty et al., 1985; Hansen et al., 1987). The sediments of back reefs, moats, and lagoons are major sites of microbial activity, making these habitats net heterotrophic in most conditions (but see Kinsey, 1985b, who discusses net autotrophic lagoons). Organic material must be suppled to these consumption zones from adjacent, net-autotrophic (production) zones, primarily in the form of detritus (Alongi, 1988).

Communities of microbes form virtually isolated food webs in the water column over coral reefs—the **microbial loop** (Hopkinson et al., 1987; Ducklow, 1990). They recycle dissolved organic material efficiently, leaking from primary produc-

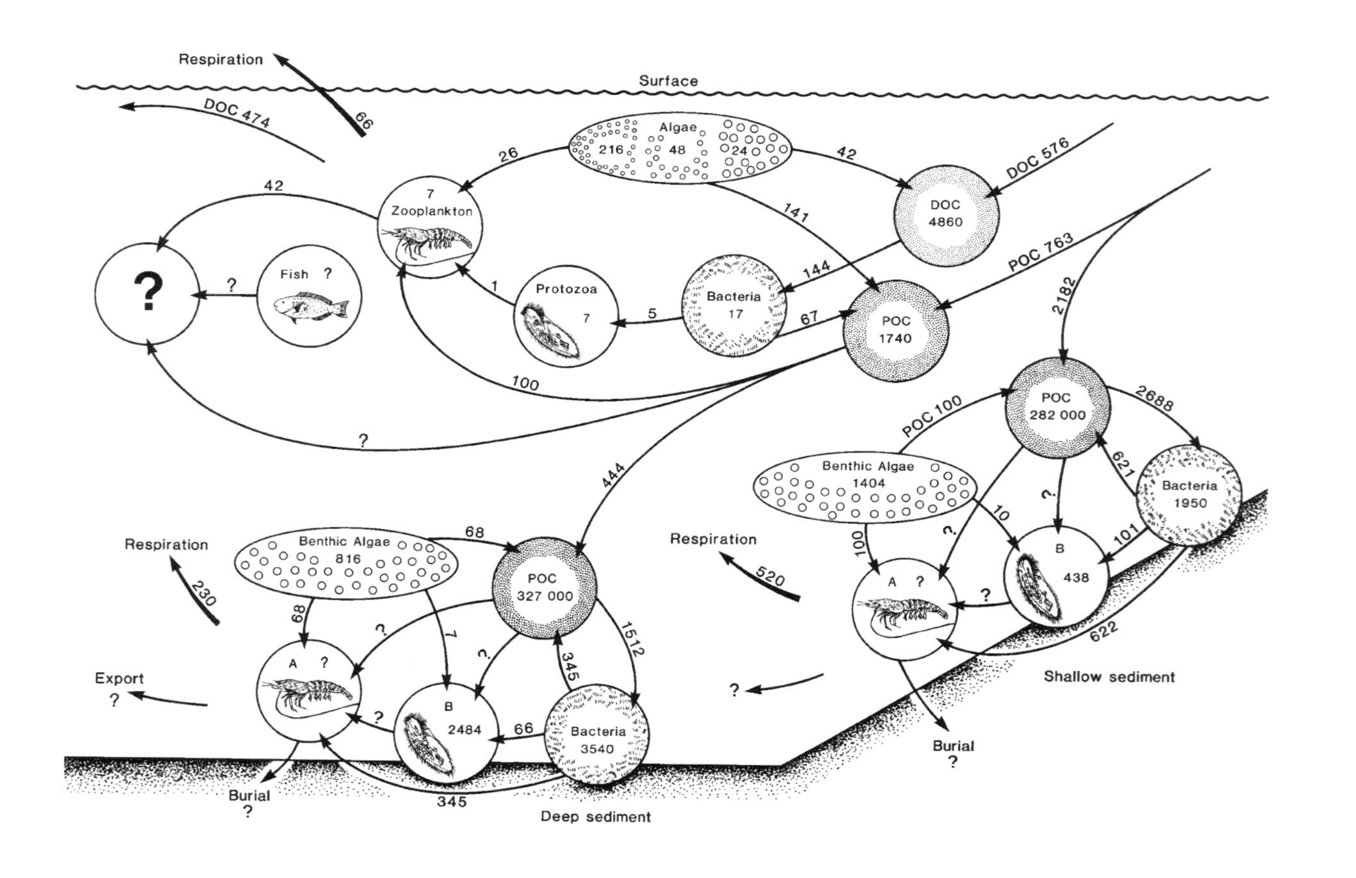
Respiration
66
Surface
DOC 474
Algae
216
48
24
26
42
DOC 576
DOC
4860
42
7
Zooplankton
141
Fish ?
?
?
Protozoa
7
1
Bacteria
17
5
144
67
POC 763
POC
1740
2182
100
POC
282 000
2688
POC 100
?
444
Benthic Algae
1404
621
Bacteria
1950
?
10
?
B
438
101
Respiration
Benthic Algae
816
68
POC
327 000
Respiration
520
100
A ?
?
230
68
?
7
?
1512
345
622
Export
?
A ?
?
B
2484
66
Bacteria
3540
?
Shallow sediment
Burial
?
Burial
?
345
Deep sediment

ers (algae and photosynthetic bacteria) and corals through heterotrophic bacteria and microscopic bacteriovores (protozoans and meiofauna). As a result, relatively little of this organic production moves up to larger members of the coral-reef food web. The isolation is less pronounced in benthic communities where most of the organic supply is particulate detritus, and grazers and infauna (e.g., holothuroids) can feed effectively on microbe-rich sediments (Hammond, 1983).

Even in sediments, however, coupled aerobic and anaerobic microbial communities produce, and decompose and recycle, vast amounts of organic material without involving macroscopic organisms (Koop and Larkum, 1987). For example, in the most detailed study of microbial processes on a reef to date (Wilkinson, 1987), it was possible to account for only about 20% of the microbial production in detrital food webs (Fig. 7–5). In the context of nutrient cycling it is important to note that bacteria actually require (take up) inorganic nutrients as they grow, because most of their food is carbon rich and nutrient poor (Alongi, 1988). The crucial role of nutrient regeneration often ascribed to bacteria usually requires their consumption and metabolism by other members of the microbial community (Ducklow, 1990).

Our ignorance of the controls and patterns of microbial metabolism highlights the most significant unanswered question about coral-reef production and decomposition, the role of detritus pathways. As long as the microbial community remains the blackest of the many black boxes in reef ecosystems, understanding and quantification of the flux of materials through detrital food webs (which dissipate the majority of reef primary production) will remain elusive (Alongi, 1988; Ducklow, 1990). This ignorance may cost reefs, and therefore us, dearly. It is the microbes that respond first to anthropogenic inputs of organic materials, with the potential to drive community metabolism into the strongly heterotrophic modes that characterize polluted reefs (e.g., Smith et al., 1981).

7.2.2. Algae: The Engines of the Reef

Sharing the physiological advantages at the microscopic end of the size spectrum are the single-celled algae: the phytoplankton, the zooxanthellae, and the sedi-

---

◀ *Figure 7–5.* Carbon flow in the inner lagoon of Davies Reef (GBR) in winter. Three locations are represented: the water column to 15 m depth, shallow sediments on the lagoon slope, and deep sediments in the middle of the lagoon. Numbers in ovals and circles show biomass in mgC m^{-2}. Numbers on arrows are fluxes in mgC m^{-2} d^{-1}. Only about 30% of the dissolved organic carbon entering the lagoon is shown; the remainder is not readily metabolized, so it travels unaltered. The consumer communities in the lagoon sediments are classified as: A—large consumers such as crustaceans, polychaetes, molluscs, holothuroids, and fishes; B—microscopic consumers such as protozoa, nematodes, copepods, and flatworms. The question marks represent major gaps in our knowledge of decomposition processes (from Wilkinson, 1986).

ment-dwelling species. The first can be discounted in many reef habitats because the shallowness, short residence time, and low nutrient concentrations of the water column prevent the accumulation of significant chlorophyll levels. In deeper lagoons, however, where limited exchange circulation or large inputs allow the accumulation of nutrients, photosynthesis by phytoplankton can make significant (but still relatively minor) contributions to areal primary production (Legendre et al., 1988; Sorokin, 1990) and water-column food webs (Ducklow, 1990).

The symbiotic dinoflagellates (now known to comprise many species or strains: Rowan and Powers, 1991a) are the most successful reef algae, for reasons detailed in Chapter 5. In the context of community metabolism, the algal-coral symbiosis is appropriately treated as a polytrophic plant. It suffices to say here that they dominate gross primary production in many reef habitats, but make less than half of that organic material available to bacterial consumers as mucus, wax esters, and dissolved organics (Muscatine, 1990). Little coral-algal tissue is consumed directly, and most of the production is respired, recycled, or accumulated within the coral colony (Chapter 5). As a result of these adaptations and attributes, a major portion of the exceptional primary productivity of reef systems is prevented from dissipating up through higher levels of secondary production in the reef food web.

The unicellular algae in the interstices and on the surfaces of reef sediments and rock are the smallest of a spectrum of free-living benthic plants extending up to seagrasses and kelps. This continuum can be divided into functional groups having progressively lower specific productivities (turnover rates) and greater structural and defensive adaptations with increasing size (Littler and Littler, 1988). Virtually every sunlit surface of a reef not covered in corals and other sessile, symbiotic organisms (e.g., sponges, giant clams) is coated in a layer of small algae (Borowitzka and Larkum, 1986). The mats of filamentous green and blue-green algae on the surface of sediments often fix enough carbon to make some shallow lagoon habitats net autotrophic for at least part of the year (Kinsey, 1985b) and, as discussed above, support a benthic microbial loop.

The hard surfaces of the reef harbor epilithic and endolithic algal assemblages of remarkable diversity (Price and Scott, 1992). These range from encrusting coralline paints to dense stands of macroalgae such as sargassum, but it is the fine turfs of filimentous and small fleshy algae that make the major contribution to the primary productivity of reefs directly available to higher consumers (Larkum, 1983).

On the basis of both areal biomass and production, it is the small, filimentous and fleshy algae that play the dominant role in coral-reef trophodynamics (Hatcher, 1988). The reasons are twofold: turf algae capitalize on the amplification of photosynthetic surface area that complex reef topography provides, and they are constantly grazed by a great diversity of invertebrate and vertebrate herbivores (Steneck, 1988; Chapters 9 and 10). The destructive process of grazing actually enhances primary productivity by keeping the individual algae in an exponential growth phase, by keeping algal assemblages in early successional stages (Birkeland et al., 1985; Carpenter, 1986a), and by recycling inorganic nutrients back to the benthic community (Williams and Carpenter, 1988). These algal-herbivore interactions are so finely balanced that epilithic algal biomass within many reef habitats shows relatively little temporal variation, despite very high rates of net primary production (e.g., Hatcher and Larkum, 1983; Klumpp and McKinnon, 1992). Between 20% and 90% of the epilithic algal production enters grazing food webs in different reef habitats (e.g., Hatcher, 1982). These are the trophic pathways that lead most directly to the organisms harvested by humans from coral-reef communties (Polovina, 1984).

When conditions of elevated nutrient concentration and/or reduced grazing pressure allow biomass to increase in epilithic algal assemblages, an increasing abundance of large algae often occurs (Chapters 9 through 12). These algal species (many of which also exist in turf-forming morphologies) have poorer nutrient uptake efficiencies and slower turnover rates than the turf varieties (Littler et al., 1983), but they are often perennial and well defended from herbivores (Steneck, 1988; Hay, 1991). Under certain conditions, they may form alternate stable states for reef benthic communities that persist for many years (Knowlton, 1992). The areal primary productivities of macroalgal stands approximate those of low biomass turfs because lower specific productivities are offset by their great biomass (Hatcher, 1988). Very little of this production, however, is consumed directly by reef herbivores because of their inaccessibility or unpalatability (Hay, 1991).

Macroalgal detritus is either advected out of the reef system or consumed in the detrital food web (e.g., Kilar and Norris, 1988). Thus, it can be seen that the form, rather than the magnitude, of primary production can have profound effects on the pathways and magnitudes of its consumption and decomposition in reef ecosystems. Small primary producers favor grazing food webs and internal recycling; large plant producers favor detrital food webs and export. The distribution of different groups of primary producers and the measured range of their productivities across characteristic reef habitats are summarized in Figure 7–4.

7.2.3. Macroconsumers: Pretty Fishes and Others

Despite their relatively insignificant contributions to the total fluxes of materials associated with the processes of organic production and decomposition, the large reef plants and animals that catch our eyes have received the vast majority of our research attention. Macroscopic animals always account for less than 25% of the total dark respiration in coral-reef communities (e.g., Polovina, 1984), but their diversity of form and function is the essence of coral reefs as humans perceive and use them. Their importance lies primarily in their control functions (e.g., grazing of biomass, modification of habitat, transport and cycling of nutrients) and protein yields to top predators (like us). There is simply not enough space here to summarize the diversity of consumer organisms and their trophic roles in coral-reef communities, and they are discussed in most of the other chapters of this book.

Two points are worth reemphasizing. First, every job that needs doing to produce, consume, and decompose organic material in reefs has many species doing it: there is considerable redundancy within functional groups and trophic levels (Done et al., 1996). It appears that the abundance, diversity, and degree of specialization of grazers and predators exceed that of detritivores. Second, coral-reef food webs are long, so total ecological efficiencies of conversion of primary production into animal flesh are low relative to many other marine ecosystems (Grigg et al., 1986; Chapter 18).

7.3. Rates of Coral-Reef Metabolism and Their Variation in Space and Time

Great variability in metabolic rates is the norm at small spatial and temporal scales in reef communities (Figs. 7–3 and 7–4). Reef morphology creates diverse microhabitats that may favor either producers or decomposers (e.g., branching corals, regenerative spaces; Fig. 7–1). Oxygen concentrations in Lagrangian control volumes vary at scales of meters as the water flows over net heterotrophic and net autotrophic "hot spots" of metabolism on a reef flat (e.g., Barnes and Devereaux, 1984). Even in the apparently uniform sediments of lagoons, incubations in domes a few meters apart often yield metabolic measures of opposite sign because of differences in the distribution of microbial and infaunal assemblages (e.g., Hansen et al., 1987).

The relevance of this small-scale variability depends on the scale of the question. Reef scientists cope with it by choosing large control volumes that integrate over large spatial and temporal scales, thereby smoothing the small-scale variability. The scales of coarse graining are generally determined by the physiography and hydrodynamics of the reef, expressed in their characteristic zonation (Chapter 3) and water residence times. For example, the reef flat with its uniform flow field and the lagoon with its enclosed water mass are hydrodynamically defined units for which integrated productivity measurements are available.

7.3.1. Habitat-Scale Variation in Community Metabolism and Export

Areal primary production and community respiration vary predictably between the distinct habitat zones of coral reefs, often by as much as an order of magnitude (Fig. 7–4). Gross production can be predicted to increase with decreasing depth up the reef slope, as light and photosynthetic surface area increase. In the shallower forereef habitats, gross production is high enough to offset respiration by the large heterotroph biomass, such that P/R ratios exceed unity and there is excess production available for export downstream (Crossland and Barnes, 1983; Kinsey,

1985a). Logistic problems of access and measurement have severely limited metabolic studies in this mare incognitum, however (Smith and Harrison, 1977); and we really have little idea of how much the outer reef slope contributes to the total productivity of reef systems.

Given their large relative area and rich benthic communities, the contribution of the outer reef slopes is probably important knowledge that we are lacking. Organic material produced on the outer reef slopes may be transferred further into the reef system by advection upslope or by the feeding migrations of consumers from the reef flat and lagoon. This production is thus available for recycling.

Organics may also be exported from the reef slope community by falling downslope (e.g., Josselyn et al., 1983), lateral advection beyond the reef (e.g., Kilar and Norris, 1988), or feeding incursions of pelagic consumers (see Birkeland, 1985). This export production (Fig. 7–6) is not available for recycling within the reef system. Given that the dominant near-bed flows are along reef slopes, rather than upslope (Hamner and Wolanski, 1988), it is likely that much of the excess production of organic material in these hydrodynamically energetic habitats is exported from the reef system. We need to know how much is retained if we are to make accurate estimates of whole-reef production.

Similar considerations apply to the wave-swept habitats of the shallow forereef and reef-crest zones, for which we also have few measurements of productivity (Fig. 7–4). The reduced rugosity and biomass of the intertidal pavements that characterize emergent reef crests, and the predominance of encrusting coral and algal morphologies there, are at least partially offset by the abundant light and nutrient supply, such that gross and excess productivities may be similar to or exceed those of the deeper forereef, although not as high as backreef rates (e.g., Smith and Harrison, 1977; Adey and Steneck, 1985; Kinsey, 1985a).

The essential difference from deeper reef-slope communities is that most of the excess production by these shallow, generally net autotrophic communities is advected further into the reef system by wave pumping (Andrews and Pickard, 1990). Besides the obvious mechanism of transport in water flowing over the shallow reef, organic products from these zones are forced into the porous matrix of the reef structure by wave energy (Buddemeier and Oberdorfer, 1986). There they accumulate in regenerative spaces and are decomposed by a diverse and metabolically active detritus-feeding community, subsequently releasing their nutrients to the pore water as it percolates through the reef matrix (Sansone et al., 1988). To the extent that the latter mechanism involves decomposition within the zone, it does not export excess production, but it does export nutrients fixed in that production, which are subsequently available to support primary production in downstream zones (Tribble et al., 1988).

The conditions on the reef flat and backreef zone are optimal for organic production and its measurement. Great photosynthetic surface area, abundant light, and a steady supply of nutrients combine to support some of the fastest rates of gross primary production anywhere on Earth (Lewis, 1981; Larkum,

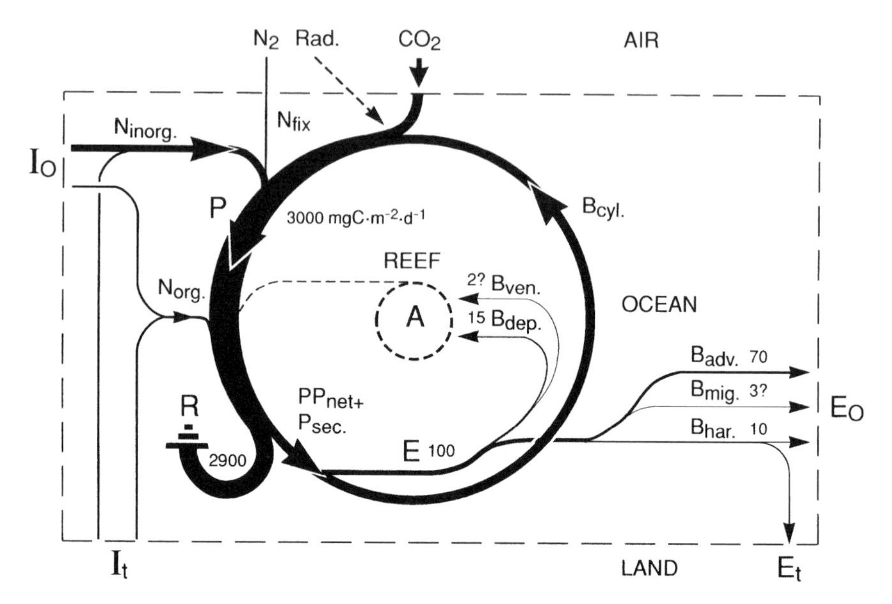

Figure 7–6. The cycling of materials in coral-reef ecosystems. Photosynthetic assimilation of CO_2 is fueled by photosynthetically active RADiation. Inorganic nutrients (N_{inorg} and N_2) are taken up and combined with the fixed carbon to produce organic plant materials (gross primary production: PP_G). Organic nutrients (N_{org}) are taken up by heterotrophs. The majority of the organic material fixed and taken up is respired within 48 hr (R). The remainder comprises the net primary production (PP_{net+}) and secondary production (P_{sec}) of the reef community. Most of this organic biomass is consumed and metabolized or decomposed within the reef system (B_{cyl}), making it again available to primary producers. The remainder is the net community, or excess, production (E). This material may contribute to the accumulation of reef biomass (A) in the living veneer (B_{ven}) or the sediments deposited (B_{dep}). Accumulated biomass may subsequently be returned to the cycle through destruction and erosion. Another component of the excess production may be exported in various types of biomass: advected dissolved and particulate forms (B_{adv}), the migration of living organisms (B_{mig}), and the harvest of living organisms (B_{har}). Imports and exports of materials may be from the adjacent ocean (I_o to E) or terrestrial systems (I_t to E_t).

1983; Hatcher, 1988). The turbulent flows of the crest have been replaced by the uniform, steady flows that favor upstream-downstream measurement of community metabolism (Marsh and Smith, 1978; Fig. 7–1).

We have a large data set for reef flats (Fig. 7–4; Kinsey, 1985a). These are also sites of high consumer biomass and metabolism that are positively correlated with the autotrophic component of the community (Kinsey, 1985a). As a result, the range of excess production measured for reef flats (ca. –4 to +4 gC m^{-2} d^{-1}) is much smaller than the range of gross production, and the P/R ratio is remarkably constant around a modal value of unity across a moderate range of sites and seasons (Kinsey, 1983). The outliers are as interesting as the modal values, which are largely derived from the broad, intertidal reef flats characteristic of the Indo-Pacific.

The highest rates of both gross and excess production are recorded from shallow reef flats having high topographic complexity that have not yet reached sea level (e.g., Adey and Steneck, 1985). The lowest rates have been measured on emergent algal pavements of low relief receiving organic pollutants (that enhance heterotrophic biomass and respiration, to the detriment of benthic primary production). When detailed measurements allow resolution at scales within zones, it becomes apparent that reef flats are divided into alternating regions of net autotrophy and net heterotrophy, connected by the advection of water and movements of consumers (e.g., Barnes and Devereaux, 1984).

Seasonal variation in rates of production can approximate the spatial range, especially in subtropical latitudes, but few records of interannual variability are available. Even reef-flat communities that oscillate minimally between net autotrophy and net heterotrophy may accumulate substantial biomass over some periods and export organics (losing biomass) in other seasons (e.g., Barnes and Devereaux, 1984). The discrepancy between the nearly balanced trophic status of the reef community (as measured in daily integrations of community metabolism) and the long-term accumulation and loss of biomass (as measured in seasonal changes in biomass) exemplifies an integration time-scale mismatch (Smith and Kinsey, 1988).

Even very small deviations from a community P/R of unity can produce large changes in excess production over long periods, given the high absolute rates of *P* and *R* involved in reef metabolism. An added complication is that biomass is controlled not only by the primary production terms, but also by the loss terms of comsumption, advection, and migration (Fig. 7–2), which may be only poorly related to community metabolism.

An essential message from all this work on the metabolism of reef-flat communities is not that all coral reefs are about the same in terms of production and decomposition rates. Rather, there are substantial differences even between these remarkably uniform reef habitats, and the measures and conclusions from a narrow range of reef flats may not apply to all morphologies and environments of reef flats, much less entire reef ecosystems.

Regardless of the trophic status of individual zones within them, the shallow perimeters of reefs generally produce more organic matter than is consumed within them. Some of this material is, of course, incorporated into the living veneer of the reef as it accretes vertically and laterally, and some is exported from the reef system (Fig. 7–6). Another portion of the excess production is transported into the deeper water behind the back reef, usually a lagoon (in the case of an atoll or barrier reef) or a moat (in the case of a fringing reef). There, reduced water motion allows the settlement of both inorganic (Chapter 3) and organic detritus to form the sediment accumulations that characterize these zones. Primary productivity is relatively low in these habitats (Fig. 7–4), and the input of organic detritus supports infaunal consumer assemblages that can drive the P/R ratio below one.

In a sense, the lagoons and moats can be seen as the stomachs of reef systems, where the excess production of the perimeter reefs is decomposed. If much of the detritus is algal tissue (with a high C:N:P ratio), then the lagoon benthos acts as a net sink for inorganic nutrients, despite being net heterotrophic in terms of carbon metabolism (Fig. 7–5). Ultimately, however, virtually all of the net organic input to, and production in, the lagoon sediments must be remineralized (recycled) in situ, because only a small fraction of the organic material accumulates in the deep sediments of reef systems. Nutrients released to the waters of the lagoon are then available again to support production in situ, or on hard reef structures bathed by lagoon waters as they flow through and out of the reef system. Scale down to the algal-coral symbiosis and see the beautiful analogy between the polyp and the atoll!

7.3.2. Reef-Scale Variation in Community Metabolism and Export

The metabolism of reef communities may thus be viewed as a mosaic of regions of net autotrophy and net heterotrophy, one feeding the other. In an ideal reef system, materials are used over and over again (endlessly recycled), and there is no need for new inputs of mass, just energy. Of course, reefs are not fully closed systems, and transformations of matter between states are never perfectly efficient. Most coral reefs are characterized by massive throughputs of energy and materials, most of which never enter the biological cycles (e.g., dissolved carbon), so the estimation of excess production at the level of the entire reef ecosystem is difficult.

One approach is to measure the metabolism of representative zones and construct a weighted average based on the area of each (e.g., Atkinson and Grigg, 1984). Another is to budget the net flux of an essential element like phosphorus, by measuring time-varying changes in its concentration in water within a reef system (e.g., Smith and Jokiel, 1978; Charpy and Charpy-Robaud, 1988). Both methods have shortcomings that limit their accuracy and generality in measurement of excess productivity.

Additive estimates suffer from compounding error (Smith and Kinsey, 1988), and mass-balance calculations exclude advectively dominated zones like outer reef slopes (Smith, 1988). Only a few numbers are available (Fig. 7–4). They suggest that the excess production by entire reef ecosystems is only slightly greater than zero (average P/R = 1.03), and is very similar in magnitude to the new production of the mixed layer of the surrounding, oligotrophic ocean (Crossland et al., 1991).

How robust this conclusion is across the broad range of geographic and oceanographic conditions in which reefs occur remains to be tested. The community metabolism of coral-reef flats does not vary greatly with latitude (Crossland,

1988), despite the large number of potentially controlling variables (e.g., light, temperature, herbivory) that do vary with latitude. Reef flats are not necessarily representative of entire reef ecosystems, however, so it is not yet possible to make clear statements about the variation in whole-reef productivity across large scales. It is the form and fate of organic production, rather than its magnitude, that varies with the large-scale differences in the physical and biotic structures of reefs.

In particular, the cycling of nutrients is sensitive to the morphology and local oceanographic environment of reefs (e.g., Hatcher and Frith, 1985; Hamner and Wolanski, 1988). If the reef system is hydrodynamically open, for example, recycling of nutrients will be inhibited because both dissolved and particulate forms are rapidly advected away from the benthic community (Fig. 7–7). The gross productivity of such a system may well be as great as that of a more closed system, because nutrients are rapidly advected into the sites of production as well. More of that organic production will be exported from the system, however, so that less will accumulate within the system for a given rate of excess production (Fig. 7–6).

Alternatively, such advectively dominated reefs may achieve higher rates of excess production than those with a large amount of recycling, especially if the advective inputs are rich in nutrients (e.g., because they are near continents or upwelling regions; see Chapter 12).

The most simplistic way to explain large-scale varation in patterns of organic production and decomposition in coral-reef ecosystems is to position them along two gradients: one of hydrodynamic closure (e.g., Smith, 1984) and the other of advective inputs of new nutrients (e.g., Hallock, 1987). The two dimensions are interactive and tend to counterbalance each other. Increasing hydrodynamic closure favors the recycling of materials within the system, allowing high rates of gross production while requiring little input of new nutrients, but leaving little excess production for accumulation or export. This is the classic reef paradigm.

Recycling of organic material is hindered as reef systems become more hydrodynamically open, so that a greater supply of new nutrients (through either greater advection or higher concentration) is reguired to support the high gross productivity, more of which is available for export or accumulation. The gross morphology of reefs and their characteristic oceanographic regimes (Chapter 12) largely determine their position across these classifying continua (Fig. 7–6).

At one extreme are the midocean atolls, their intertidal rims surrounding captive lagoons, where water circulates for as long as a year (Andrews and Pickard, 1990) before exchanging with the oligotrophic waters of the great ocean gyres, far from terrestrial inputs. The typical examples are found in the central and western Pacific, and the majority of our knowledge of coral-reef metabolism is derived from their examples (even though they account for less than 20% of the world's reef area). The low nutrient concentration around the hard substrata of the atolls is only partially offset by the advective fluxes over them. Recirculation

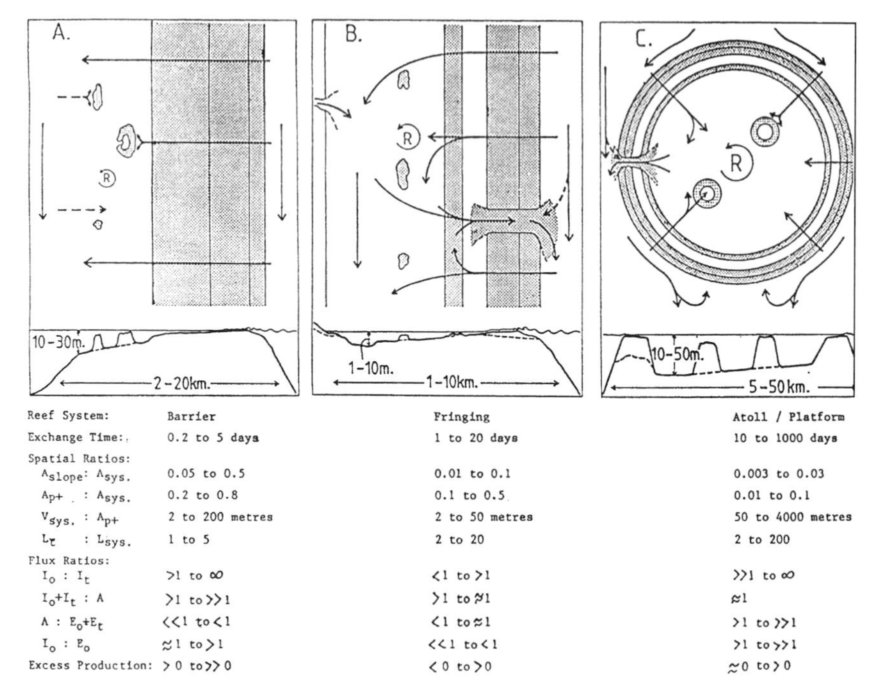

Reef System:	Barrier	Fringing	Atoll / Platform
Exchange Time:	0.2 to 5 days	1 to 20 days	10 to 1000 days
Spatial Ratios:			
A_{slope} : $A_{sys.}$	0.05 to 0.5	0.01 to 0.1	0.003 to 0.03
A_{p+} : $A_{sys.}$	0.2 to 0.8	0.1 to 0.5	0.01 to 0.1
$V_{sys.}$: A_{p+}	2 to 200 metres	2 to 50 metres	50 to 4000 metres
L_t : $L_{sys.}$	1 to 5	2 to 20	2 to 200
Flux Ratios:			
$I_o : I_t$	>1 to ∞	<1 to >1	$\gg 1$ to ∞
$I_o+I_t : A$	>1 to $\gg 1$	>1 to ≈ 1	≈ 1
$A : E_o+E_t$	$\ll 1$ to <1	<1 to ≈ 1	>1 to $\gg 1$
$I_o : E_o$	≈ 1 to >1	$\ll 1$ to <1	>1 to $\gg 1$
Excess Production:	>0 to $\gg 0$	<0 to >0	≈ 0 to >0

Figure 7–7. A schematic relationship between reef morphology, material fluxes, recycling, and excess production. The three basic reef morphologies are characterized in terms of their hydrodynamics, physiognomies, and fluxes. Abbreviations correspond their physiography, hydrodynamics and fluxes. These are captured in the spatial ratios: the area of the reef slope scaled to that of the entire reef system (Aslope:Asys); the area of net autotrophic substrata as a proportion of entire reef (Aps:Asys); the volume of water in the reef system divided by the area of net autotrophic substrata; and the average transit distance through the reef system scaled to the characteristic length scale such as the radially-symetrical diameter (Lt::Lsys). Relationships between fluxes characterize the relative importance of inputs and outputs to and from the reef system. The abbreviations used in these flux ratios correspond to Figure 7–6. Linear barrier reefs are the least closed hydrodynamically. Oceanic inputs dominate the fluxes, there is relatively little internal recycling, and excess production is high because most of the reef zones are net heterotrophic (shaded). At the other end of the continuum, emergent rim atolls are virtually closed hydrodynamically. Internal recycling rivals oceanic inputs, and excess production approximates zero because net autotrophic and heterotrophic zones alternate through the flow field. Fringing reefs experience relatively high terrestrial inputs that dominate the sign and magnitude of excess production.

of water within the system gives both water column and benthic organisms many opportunities to capture both new and regenerated nutrients, thus maintaining high gross productivity. Extensive, multizoned reef flats and lagoons with long retention times ensure that much of the excess production exported from the shallow perimeter is recycled within the system. The low excess production of the entire reef contributes to the slow accumulation of organic mass as the lagoon infills, or is exported in leakage of organics from the lagoon passes and steep outer slopes to the surrounding ocean.

At the other extreme are linear barrier reefs in regions of upwelling or terrestrial influence, where nutrient-rich water passes rapidly through the system. The typical examples are in the Caribbean, where many reefs have not yet reached sea level. The fact that the productive, hard substrata of these barrier reefs have only one chance to capture nutrients from the passing water is partially offset by the higher concentrations the water contains. There is little recirculation of water, so excess production exported from the shallow band of reef is likely to be lost downstream from the system rather than recycled within it (e.g., Kilar and Norris, 1988). Because little of the reef area is net heterotrophic, the high excess production contributes to biomass accumulated in the rapid vertical and horizontal accretion of these reef types (e.g., Adey and Steneck, 1985), and to the large exports of organic material (both living and dead) from them to adjacent ecosystems (Birkeland, 1985; Ogden, 1988).

Fringing reefs comprise an intermediate group that have variable water retention periods and recycling potentials, depending on the morphology of the moat and the complexity of the reef flat. They are also a special case because the close proximity of land allows substantial inputs of organic matter, which may drive the entire reef system net heterotrophic in extreme cases.

The vast majority of coral reefs fall at intermediate positions along this simplistic continuum. Atolls do occur in equatorial upwelling regions, and barrier reefs often occur in oligotrophic waters. Water recirculation, and hence the dynamics of nutrient uptake and recycling, are influenced by the physical forcing (i.e., wind strength, tides, and other factors) as well as the reef morphology. Defining a particular reef's position along the closure and delivery continua may give a reasonable indication of the trophic balance and sign of the excess production rate. It will not predict the rate or form of that production. The question arises: What limits coral reef production?

7.4. Controls on Coral-Reef Production

Simply stated, we do not know what factors control the excess production of entire reef ecosystems. Liebig's Law does not apply to complex systems; no single factor can be identified that is universally limiting. A number of measurements and experiments are available at lower levels of organization, from the individual plant to lagoon patch reefs, but these provide conflicting evidence of limitation. A reductionistic approach is invalid: light may limit the primary production of a plant assemblage while nutrient supply may limit gross community production. The identification of rate-controlling processes is highly dependent on the methodologies and scales of measurement.

Considerable confusion has been generated by the sloppy use of terms and extrapolations from measurements demonstrating limitation at one level of organization to another. D'Elia et al. (1986) discuss the numerous approaches to assessing limitation of marine metabolism, the most powerful of which is to experimentally alter the magnitude of a potentially limiting factor and measure system response (e.g., Kinsey and Davies, 1979).

7.4.1. Limitation of Gross Production

In terms of autotrophic or gross community production, it helps to view a reef ecosystem simply as a sessile coral. The stationary, living biomass captures transient light energy and both inorganic and organic nutrients passing over it to produce organic material. Factors that control the rates of delivery to, and uptake by, the organism have the potential to limit its growth (equivalent to net system production). At the most fundamental level, the size of the organism itself (equivalent to biomass of the reef) may limit production and decomposition rates. Reefs that have low biomass (due perhaps to mortality from a recent storm or cold-water event) may have insufficient plant material to capture much of the photosynthetic radiation or dissolved nutrients, and insufficient animal abundance to consume and recycle most of the organic material.

The range of biomass across which productivity can be expected to increase is small, however, because of negative feedbacks like self-shading and depleted nutrients in sluggish benthic boundary layers. Most reef communities appear to have sufficient biomass to maintain very high rates of organic production across the naturally occurring range of light and nutrient regimes, as evidenced by the lack of strong latitudinal gradients in gross community productivity (Crossland, 1988).

The delivery of photosynthetically active radiation sets the upper limit on areal gross productivity, and the rate of primary production in deep or turbid reef habitats may be light limited (Dustan, 1985). In the shallow, intensely lit portions of reefs, photosynthetically active radiation appears to be superabundant (Carpenter, 1986b), the amount actually captured depending on the photosynthetic biomass, its surface area, and efficiency.

The biomass of reef communities is controlled by a complex of interacting factors (Chapters 9, 10, and 11), including those limiting productivity, but also including evolutionary (Chapter 14) and anthropogenic processes (Chapters 15 and 17). Empirical evidence demonstrates that the biomass of primary producers on reefs is sensitive to both ambient nutrient concentrations and the feeding activities of herbivores, at both small and large scales (Hatcher and Larkum, 1983; Littler and Littler, 1988).

It is important to recognize that nutrient concentration may limit the specific rate of plant tissue production, and its subsequent accumulation, even when advective nutrient delivery far exceeds that required to support community gross production (Atkinson, 1988, 1992). Large changes in biomass may occur in coral-reef communities without accompanying changes in gross productivity.

7.4.2. Nutrient Limitation of Excess Production

Over ecological time scales (months to decades), the organic production by reef communities that accumulates in reef structures or is exported from the reef system must be matched by a net input of nutrients equivalent to that contained in the excess production (Fig. 7–6). Net input of nutrients requires that the delivery of nutrients from outside the system exceeds the loss from the system.

Nutrient inputs to reefs come from three sources: the surrounding ocean, the adjacent land, or, in the case of nitrogen, the atmosphere, through the process of biological nitrogen fixation (D'Elia and Wiebe, 1990). The relative importance of these three sources varies geographically and with reef morphology (Fig. 7–7), but the role of nitrogen fixation in community metabolism is particularly sensitive to the hydrodynamic regime (Capone, 1983; Hatcher and Frith, 1985). Net autotrophic reef communities experiencing strong flows of water may actually export more nitrogen than they import, because in-situ nitrogen fixation is great enough to support the advective loss (Smith, 1984, 1988).

Smith (1984) has argued that the greater the hydrodynamic closure of marine communities, the less likely is nitrogen limitation of their productivity, because fixed nitrogen remains available for use within the system. If the net input of nutrient limits excess production in reefs toward the hydrodynamically closed end of the spectrum, then it is likely to be by elements other than nitrogen, such as phosphorus.

Phosphorus cannot be made from thin air, and the carbonate mass of reefs binds most of the element in a form unavailable for uptake by plants. Low concentrations of phosphorus have been shown to limit the net primary production of individual reef algae, and the long-term average rates of net primary productivity in some reef communities can be predicted from rates of phosphorus uptake (Atkinson, 1988).

Indeed, the rate of uptake of dissolved inorganic phosphorus by constructed reef communities in large flumes has been shown to be dependent on its concentration in the overflowing water (Bilger and Atkinson, 1995). Budgets of phosphorus in relatively closed-reef systems have been used to infer rates of net system (excess) production (e.g., Smith and Jokiel, 1978; Atkinson, 1992), but these do not prove that a causal relationship exists between phosphorus supply and excess production.

The two cases of experimental manipulation of inorganic nutrients at the scale of coral reef habitats did not specifically address the question of excess production, and failed to distinguish between the effects of nitrogen and phosphorus because they were altered in tandem. A small-scale study in a lagoon-patch reef resulted in increased rates of gross production, but no apparent increase in excess production in the form of increased biomass (Kinsey and Davies, 1979). The experiment is currently being repeated with replicated additions of the nutrients separately and together (Larkum and Steven, 1994).

The only ecosystem scale experiment yielded somewhat different results when sewage inputs to a large barrier reef complex in Hawaii were terminated (Smith et al., 1981). Both net primary productivity and community biomass declined, implying that both gross and excess productivity had been nutrient limited prior to the sewage inputs. Most interestingly, the reef system shifted from being strongly net heterotrophic when sewage inputs were high to being slightly net autotrophic when the sewage inputs were reduced (which presumably reflected the presewage condition). Profound changes in community structure accompanied these perturbations, which affected the balance between autotrophic and heterotrophic organisms and processes.

Nutrient inputs from sewage apparently caused plant biomass to increase not only on the benthos (macroalgae), but also in the water column (phytoplankton), which intercepted light at the expense of benthic primary production. High concentrations of particulate organic material in the enclosed water volume also supported a large biomass of benthic filter feeders, which replaced many primary producers such as corals, again at the expense of benthic primary production. The relative proportions of organic material input directly in the sewage and derived from inorganic nutrients in the sewage are unclear, but both forms represent net inputs to the system. Despite the difficulty of conducting experiments and making measurements at whole-system scales, we need more studies like that of Smith et al. (1981) if we are to understand nutrient limitation of coral-reef production and export.

In summary, the relationship between nutrient delivery and excess production in entire reef ecosystems can be viewed from two opposing perspectives. First, excess production is constrained to a small value by biological processes, so the demand for new nutrients is correspondingly small. In this scenario, rapid consumption and decomposition of primary production by a large heterotrophic biomass leaves little excess. Efficient recycling meets most of the nutrient requirement by gross production, and advective delivery is more than adequate to meet the demands of the small excess production. In this scenario, nutrient supply does not limit excess production (although nutrient concentration may limit net primary production and biomass). Excess production is more probably set by other factors such as topography, water circulation, and photosynthetic efficiency. Increased nutrient supply affects excess production through changes in community structure (Chapter 12), not direct rate limitation (Atkinson, 1988).

From a second perspective, excess production is constrained to a small value by the slow net flux of new nutrients to the system. In this scenario, efficient recycling is also necessary to maintain high gross productivity, but the rate has the potential to increase further with enhanced nutrient delivery (concentration × advection). Heterotrophs do not have the capacity to consume and decompose the increase in net primary production, with the consequence that excess production is enhanced. Increased nutrient supply affects community stucture through direct limitation of the accumulation of biomass.

Either scheme may apply to different reefs, but the normal condition of low excess production is common to both. In the most extreme case of midgyre oceanic atolls, one might expect nutrient limitation to occur because the supply of new nutrients in surface water overtopping the reef perimeters is lowest. Yet the few measurements available suggest that these reef systems have gross and excess productivities similar to those where nutrient delivery is much higher (Lewis, 19981; Kinsey, 1985a; Crossland et al., 1991).

An ingenious explanation for this apparent paradox has been suggested by Rougerie and Waulthy (1993): **geothermal endo-upwelling.** The latent heat of volcanism under atolls creates a thermoconvective cell within the porous matrix of the reef structure, which draws nutrient-enriched water from below the permanent thermocline (ca. 100 m depth) in the surrounding ocean and moves it up to the surface layers of the reef. The patterns and rates of nutrient delivery to the productive living veneer by this process have not yet been unequivocally established, but the supply is small compared to the advective inputs and regeneration rate, so its potential to support observed reef production remains in question (Tribble et al., 1994). Clearly the mechanism does not apply to the majority of reefs that are not sited on volcanic structures, but it emphasizes the dynamic nature of regenerative spaces within reef structures. The controversy also illustrates our relative ignorance of controls on the balance between production and consumption of organic material in reef ecosystems.

7.4.3. Organic Inputs Support Excess Production

Nutrient inputs to coral reefs occur in both inorganic and organic forms, the latter being either dissolved or particulate, the particulates being either living (i.e., plankton) or dead (detritus). In oligotrophic waters, concentrations of organic nutrients in the mixed layer are generally greater than dissolved inorganic forms (Smith et al., 1986). Nutrient limitation is generally considered in terms of the supply of inorganic nutrients, because these are directly taken up by primary producers. Because of advection, relatively little of the inorganic nutrient flux through reef systems is captured by the benthic community, supporting the view that excess production is low (although it may be a large number in absolute terms).

But reef communities are eminently adapted to capture particulate material

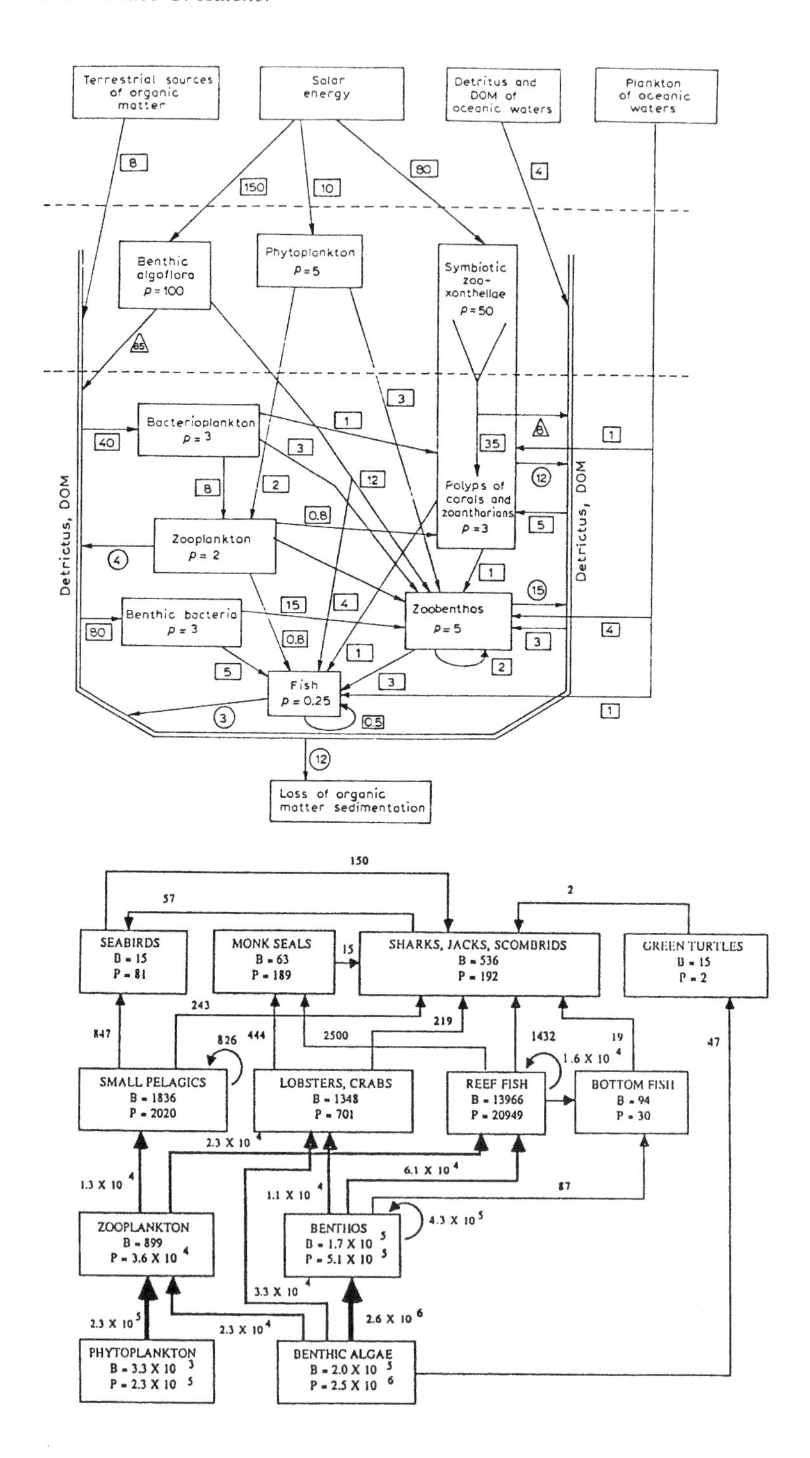
Terrestrial sources of organic matter
Solar energy
Detritus and DOM of oceanic waters
Plankton of oceanic waters
Benthic algoflora p=100
Phytoplankton p=5
Symbiotic zoo-xanthellae p=50
Bacterioplankton p=3
Polyps of corals and zoantharians p=3
Zooplankton p=2
Benthic bacteria p=3
Zoobenthos p=5
Fish p=0.25
Detrictus, DOM
Loss of organic matter sedimentation
SEABIRDS B = 15 P = 81
MONK SEALS B = 63 P = 189
SHARKS, JACKS, SCOMBRIDS B = 536 P = 192
GREEN TURTLES B = 15 P = 2
SMALL PELAGICS B = 1836 P = 2020
LOBSTERS, CRABS B = 1348 P = 701
REEF FISH B = 13966 P = 20949
BOTTOM FISH B = 94 P = 30
ZOOPLANKTON B = 899 P = 3.6 X 10 4
BENTHOS B = 1.7 X 10 5 P = 5.1 X 10 5
PHYTOPLANKTON B = 3.3 X 10 3 P = 2.3 X 10 5
BENTHIC ALGAE B = 2.0 X 10 5 P = 2.5 X 10 6

advected onto them in the form of zooplankton and larger detrital particles. Through their normal metabolism, corals and planktivorous fishes, which are abundant on the outer slopes (Hamner and Wolanski, 1988), effectively transform particulate organic nutrients into the dissolved inorganic forms taken up by zooxanthellae and free-living algae (Erez, 1990). Because they originate outside the reef system, these particulate organics represent net inputs of nutrients, a portion of which supports new reef primary production (Crossland et al., 1990). Few estimates of the magnitude of this flux are available, but it appears that it may be adequate to match the component of excess organic production exported from some reef systems (Sorokin, 1990). The use of both inorganic and organic inputs to support net community production is an important process distinguishing excess production in coral reefs from new production in the mixed layer of the surrounding ocean.

7.4.4. Secondary Production and Decomposition

The growth of individuals and populations of heterotrophic organisms that consume and decompose organic material in reef communities represents the secondary production of the ecosystem. As discussed above, some of this growth is fueled from outside the system, but the majority of the diverse and abundant reef fauna grows on organic material derived from in-situ production. The inefficiencies of metabolism and long food webs mean that secondary production will be a tiny fraction of the primary production (Fig. 7–8), even though most of that prodigious production is consumed within the system.

The question of controls on rates of secondary production in reef communities may be viewed as "bottom-up" or "top-down." Bottom-up control implies that the rate of secondary production at any given trophic level is determined by the

◀ *Figure 7–8.* Two models of coral-reef trophodynamics. Compartments represent the major functional groups; arrows quantify the fluxes between the compartments. The upper diagram (from Sorokin, 1990) is a generalized "bottom-up" model that emphasizes the dominance of microbial processing of detrital material in energy flow in reef communities. The net production of each compartment (P) may flow to subsequent trophic levels (in squares), or may flow to the detrital pool as unconsumed production (triangles) or unassimilated food (circles). All figures are in the units of kJ m^{-2} d^{-1}. Note the massive flows between benthic algae, the detritus pool, and the microbial components. The lower diagram (from Polovina, 1984) is a production/biomass model of an atoll in the Hawaiian Archipelago, from which "top-down" control of fisheries production is suggested. Mean annual biomass (B) is shown for each compartment in units of kg km^{-2}. Net production (P) and trophic fluxes between compartments (arrows) are in units of kg km^{-2} yr^{-1}. All production is assumed to be consumed (i.e., $E = 0$), with major fluxes occurring from benthic algae to the benthos (which includes microbial assemblages) to reef fish. Such models involve numerous simplifying assumptions, and reflect differences in the viewpoint of the modeller as much as differences in community structure and function.

productivity of the level below. In population terms, the population growth rate of a particular species is limited by the availability of its food. Top-down control reverses the causality. Higher tropic levels control productivity lower down; grazing and predation control plant and prey populations. The complex food webs of reef communities (Fig. 7–8) challenge the generality of such models and make the assessment of controls difficult (Chapter 11). Each trophic level is occupied by many different functional groups and species of consumers, and many reef animals feed at more than one trophic level.

The fact that as much as 100% of benthic primary production in shallow reef-slope habitats is consumed directly by grazing herbivores (Steneck, 1988) suggests that bottom-up control may operate at the second trophic level (primary consumers). Similarly, the rapid response of microbial communities to inputs of organic detritus and the low rate of accumulation of organics in reef sediments imply that bacterial production may be substrate limited in some lagoon habitats. These observations do not necessarily mean, for example, that the growth of all herbivorous fish species in reef ecosystems is food limited. They imply that if the primary productivity were to increase, the total secondary production of reef herbivores would follow suit.

Theoretical convention suggests that bottom-up control will become increasingly important at higher trophic levels (Hairston et al., 1960), the growth and abundance of top predators being most likely to be food limited. The few detailed studies available, and consideration of the high heterotrophic biomass and secondary productivity of reef communities, do not support this generalization. From a comprehensive trophodynamic analysis of an atoll ecosystem, Grigg et al. (1984) conclude that predation exerts stronger control on secondary productivity than does the rate of food production. Top-down control implies that limitation of secondary productivity is effected primarily through the limitation of biomass at lower trophic levels.

Obviously, factors other than predation-induced mortality (e.g., disease, recruitment failure, etc.) may serve to control heterotrophic biomass. In contrast, bottom-up control is effected primarily through changes in turnover, rather than biomass. Undoubtedly, both bottom-up and top-down mechanisms interact to control the balance of production and consumption in coral-reef ecosystems.

Trophodynamic models (Fig. 7–8) have little relevance below the level of the ecosystem: they tell us nothing about the dynamics of individual species populations (although they may set boundary conditions for population models). Insofar as populations are the target of human use of coral-reef resources (e.g., monospecific fisheries), trophodynamic models offer poor predictive power. When our concern is with higher-level, functional classifications, such as free-living benthic algae or multispecies demersal fish assemblages, then trophodynamic models can provide useful predictions. For example, one of the most used tools for predicting total, area-based yields from artisanal fisheries of coral reefs (ECOPATH) is based on the production/biomass model pictured in Figure 7–8 (Polovina, 1984).

7.5. Interactions Between Humans and Coral-Reef Production

What are the implications of the patterns, processes, and limitations of reef production introduced in this chapter for use of coral-reef resources by humans? How do these uses and other human activities influence the balance between production and decomposition in coral-reef ecosystems? These are crucial questions as we recognize the deterioration of reefs on a global scale (Ginsburg, 1993). Much of the discussion of human effects in Chapters 6, 8, 9, 10, 12, and 14–18 is pertinent to the topic, and the discussion here is limited to the effects of human activities on excess production and the harvestable production of coral-reef ecosystems.

7.5.1. Anthropogenic Effects on the Trophic Status of Reefs

Virtually all human activities in and adjacent to reef communities have the potential to disrupt the balance between organic production and its decomposition (Hatcher et al., 1989; Fig. 7–2). The major anthropogenic effects may be classified according to whether they perturb the reef system toward net autotrophy or net heterotrophy. Inorganic nutrient pollution, overfishing of herbivorous animals, and introduction of weed plants all enhance the primary production of organic matter in forms that are recycled less efficiently within the reef system. Phytoplankton production preempts scarce water-column resources and is advected from the system or dissipated in the microbial loop (e.g., Hopkinson et al., 1987). Macroalgae outcompete and replace the coral-algal symbiosis and are resistant to grazing. Most of their tissue production is exported or enters microbial food webs as detritus, rather than directly supporting macrofaunal production in situ. The net effect is to increase excess production by increasing organic export and decreasing heterotrophic biomass and respiration.

Organic pollution, sediment loading, and physical habitat destruction (e.g., dredging, dynamite fishing) all inhibit primary productivity and encourage the replacement of benthic autotrophs with heterotrophic biomass. Reduced topographic relief and reduced delivery of light to the benthos slow areal photosynthetic rates. Filter-feeding animals outcompete plants and corals. The net effect is to reduce excess production (even making it negative) by reducing photosynthetic production and increasing heterotrophic biomass and respiration dependent on imported organic material.

Most of the anthropogenic changes in the trophic status of reef communities outlined above are either affected by direct modification of the abundance of reef plants and animals, or result in large changes in their relative biomass. Changes in community structure do not always produce concomitant alterations in net community metabolism. For example, coral-reef flats covered in epilithic algae may exhibit rates of gross and excess organic production similar to reef-flat communities dominated by living corals. The measurements of carbon flux do not distinguish between the different forms of organic production, even though the fates of the production and the consumer communities may differ markedly in their composition and usefulness to top predators like humans. It follows that simple carbon metabolism assays of trophic status cannot be used alone as reliable metrics of coral-reef condition or "health."

More complete measurement regimes, involving other nutrients besides carbon in inorganic and organic forms, are required to qualify metrics of anthropogenic change in the trophic status of reef communities. Such ecoindicators may prove to be more sensitive (thereby providing earlier warning) than the characteristic changes in community structure that follow anthropogenic impacts, but there are too few data to judge at this time.

7.5.2. Harvestable Yields from Reefs

The net community (excess) productivity of entire reef systems sets the upper limit on sustainable yields of organic material from reefs at about 2–3% of gross productivity. This would be a huge mass (ca. 350×10^6 Mt. Fresh Wt yr^{-1}) on a global basis, but of course only a small fraction is actually available for harvest by humans. First-order estimates suggest that 75% of the excess production is exported from reefs in unusable form (e.g., dissolved organics, detritus) and 15% accumulates in reef structures, leaving only about 10% (ca. 35×10^6 Mt. yr^{-1} globally) to be fished (Crossland et al., 1991). This number is equivalent to well under 1% of the gross production by reefs, a proportion in accord with independent trophodynamic analyses (Fig. 7–8). The errors inherent in attempting to estimate reef-specific (much less species-specific) fishery yields from measures of reef metabolism and production should be apparent from this huge discrepancy between gross production and harvestable secondary production.

Current catches from coral-reef fisheries are but a few percent of this estimated sustainable yield (Marshall, 1985; Russ, 1991), suggesting that considerable potential exists for increasing the harvest from reefs. Yet there are already many examples of overfishing (Russ, 1991; Roberts and Polunin, 1993). High yields are generally obtained when the fisheries are nonselective (most species of large vertebrate and invertebrate consumer are harvested) and their long-term sustainability has yet to be demonstrated (Ludwig et al., 1993). It appears that the great diversity of reef food webs acts to limit both the harvest efficiency of multispecies fisheries and the ecological efficiency of energy transfer. The result is anomalously low fishery yields from reefs compared to other marine ecosystems having similar rates of primary production (Fig. 18–1). Given the profound effects that overfishing has already had on the structure and function of reef communities, it would be imprudent to attempt to achieve the theoretical maximum sustainable yields from them. They will cease to be coral reefs long before that point is reached.

8

Reproduction and Recruitment in Corals: Critical Links in the Persistence of Reefs

Robert H. Richmond

Each chapter in this book deals with factors that can shift the balance from processes supporting coral-reef growth and development to those that result in reef degradation. Reproduction and recruitment are among the critical processes upon which the persistence of coral reefs depends. Reproduction is the process by which new individuals are formed. Recruitment is the process by which newly formed individuals become a part of the reef community. This distinction is important, as it is possible to have successful reproduction, with healthy larvae, tissue fragments, or other types of seed material being produced, but eventual death of a coral-reef community if these new individuals are unable to find appropriate substrata for settlement, or if conditions prevent growth, maturation, and survival.

This chapter will describe the methods by which corals reproduce, how coral larvae are formed and develop, the factors that affect site selection, settlement, and metamorphosis in coral larvae, and how particular problems can affect the success of both reproduction and recruitment. By studying the biology of coral reproduction and recruitment, we gain an understanding of how it is possible to slowly degrade a reef, through the interruption of the critical processes that replenish populations of these important organisms. This understanding is of central importance to coral-reef management and preservation. Whether a reef is killed quickly by sedimentation or slowly through reproductive or recruitment failure, the result is the same: the loss of the beauty, economic and cultural value, and benefits that coral reefs provide.

Research supported by NSF grant 8813350 and NIH grant S06 GM 44796-06. This is contribution number 380 of the University of Guam Marine Laboratory, and is dedicated to Keana Avery Richmond, who recruited July 27, 1996, between the two annual coral spawning events. I hope that the information in this chapter and book will help in the protection of coral reefs for her children, as well as all future generations, to enjoy.

8.1. Coral Reproduction

Corals reproduce both asexually and sexually. Asexual reproduction in corals includes several processes by which one coral colony forms additional colonies through the separation of tissue-covered fragments, or through the shedding of tissue alone. Sexual reproduction is more complex, and requires the fusion of two gametes, egg and sperm, to form embryos that develop into free-swimming planula larvae. Asexual and sexual processes are not mutually exclusive; species and/or individuals may produce "offspring" both ways within the same time period. The products of the two types of reproduction can differ both physically and ecologically.

8.1.1. Asexual Reproduction

In discussing asexual reproduction in corals, it is helpful to separate colony growth from the formation of new colonies. Most reef-building stony corals are true colonies, made up of hundreds to thousands of interconnected polyps. Colonies grow through the asexual process of budding, during which new polyps form. Additional polyps can form when one polyp divides into two (intratentacular budding), or sometimes a new mouth with tentacles can simply form in the space between two adjacent polyps (extratentacular budding). If the polyps and tissue formed by these asexual processes remain attached to the parent colony, the result is considered growth and is seen as an increase in colony size. If polyps or buds become detached from the parent colony and give rise to new colonies, we consider this to be asexual reproduction, that is, the direct formation of new individuals from prior stock.

New coral colonies can be formed asexually in several ways. Fragmentation is common among finely branched or relatively thin plating corals (Highsmith, 1982). Coral fragments, including the underlying skeleton, may become detached from parent colonies as a result of wave action, storm surge, fish predation on associated animals, or other sources of physical impact. If a fragment lands on a solid bottom, it may fuse to the surface and continue to grow through budding. Many fragments generated by storms roll around, eventually losing their thin covering of coral tissue, and do not succeed in becoming new colonies, that is, they do not recruit (Knowlton et al., 1981).

Pieces of living tissue may leave the underlying coral skeleton, and through the use of cilia that cover the outer surface, swim and drift in the water column until coming into contact with an appropriate surface for settlement and attachment. This process has been referred to as polyp bailout (Sammarco, 1982b), as polyps appear to actively leave their skeletons. In a similar process, balls of coral tissue may remain on an otherwise dead skeleton, or may ooze out of the coral calices and later differentiate into coral polyps and begin secreting a calcareous skeleton (Rosen and Taylor, 1969; Highsmith, 1982; Krupp et al., 1993). Sections taken through coral colonies, particularly massive forms, often reveal periods of growth, diebacks, and regrowth over a previous skeletal base. For this reason, colony diameter may not always be a reliable measure of age (Hughes and Jackson, 1980).

It is also possible that coral larvae may arise from unfertilized eggs, through a process known as parthenogenesis (Stoddart, 1983). While eggs are produced, they are not fertilized by sperm, but develop directly. This asexual mechanism for production of embryos has been observed in plants and many clonal organisms.

Asexual reproduction results in the production of offspring that are genetically identical to the parent. As long as conditions remain the same, the offspring will enjoy the same level of success that the parent colonies had. In reality, the physical and biological aspects of coral-reef communities vary. Seawater temperatures may change because of El Niño events, predators may evolve new feeding habits, a new disease may appear, or a new competitor may immigrate to the reef. A population with no genetic variability is vulnerable to changes in the physical or biological components of the environment. Another disadvantage of asexual reproduction in corals is that fragments, residual tissues, and some shedded tissues have limited dispersal abilities. The distribution of offspring is important to the success of coral populations and coral reefs.

8.1.2. Sexual Reproduction

Unlike asexual reproduction, which produces exact copies of the parent, sexual reproduction offers two opportunities for new genetic combinations to occur: (1) crossing over during meiosis in the formation of eggs and sperm, and (2) the genetic contribution of two different parents when an egg is fertilized by a sperm. These serve to add genetic variation to populations, which may lead to enhanced survival of a species. In corals, the resulting embryo develops into a ciliated planula larva. Planulae are particularly well adapted for dispersal and can seed the reef of origin, nearby reefs, or reefs hundreds of kilometers away (Richmond, 1987, 1990).

8.1.3. Gonochorism Versus Hermaphroditism

If a species has separate males producing sperm and females producing eggs, it is said to be gonochoric. The term *dioecious* is also used, but it is more appropriate for plants. If, however, a single individual of a species is capable of producing both eggs and sperm, it is said to be hermaphroditic. This term originates from Greek mythology, in which Hermes was the male messenger god, and Aphrodite was the goddess of beauty. The joining of these two names is used to describe organisms that have both male and female function within the same individual.

Approximately 25% of the coral species studied to date (e.g., species of *Porites* and *Galaxea*) are gonochoric (Harrison and Wallace, 1990). The identification of separate sexes in corals is sometimes confused by the fact that it takes eggs longer to develop than sperm; hence a study early in the gametogenic cycle may lead to the conclusion that a coral is female, since no sperm would be seen until later in the year. Additionally, individual colonies of some species are distinctly male or female, while other colonies of the same species may be hermaphroditic (Chornesky and Peters, 1987; Harrison and Wallace, 1990). These cases may represent reproductive plasticity, or in some cases, differences at the species level.

If an organism has both ovaries for producing eggs and testes for producing sperm at the same time, it is called a simultaneous hermaphrodite. If, on the other hand, an individual is a functional male first, then develops into a female later (protandry), or is initially female, eventually changing into a male (protogyny), it is a sequential hermaphrodite. Corals display the full range of sexual characteristics, with the majority of species studied so far identified as simultaneous hermaphrodites (e.g., most acroporids, faviids, and some pocilloporids). A few species have been found to be sequential hermaphrodites (e.g. *Stylophora pistillata* and *Goniastrea favulus;* Rinkevich and Loya, 1979b; Kojis and Quinn, 1981), while others are gonochoric.

Hermaphroditism is particularly favorable in small populations, as it ensures sexual partners will be present if there are more than two individuals. If self-fertilization is possible, a single simultaneous hermaphrodite is capable of sexual reproduction and at least taking advantage of the genetic variation introduced during the crossing-over phase of meiosis.

8.1.4. Brooding Versus Spawning

Corals display two distinct modes of reproduction that differ in the way the gametes come into contact with each other. In brooding species, eggs are fertilized internally, with the embryo developing to the planula stage inside the coral polyp. Alternatively, spawning species release eggs and sperm into the water column, where subsequent external fertilization and development take place. The differences between the two modes of reproduction influence many aspects of coral ecology, including the transfer of symbiotic algae to the larvae, larval competency (the period during which larvae possess the ability to successfully settle and metamorphose), dispersal of larvae, biogeographic distribution patterns, genetic variability, and even rates of speciation and evolution (Richmond, 1990). Also, spawned gametes that float to the surface may be more vulnerable to the effects of pollutants often found in the surface layer in coastal waters.

Planula larvae released from brooding corals are immediately competent, that is, capable of settlement and metamorphosis. Brooded larvae are generally larger than spawned larvae, and in hermatypic (reef-building) corals, contain a full complement of symbiotic zooxanthellae from the parent colony (Fig. 8–1). It has been demonstrated that zooxanthellae contribute metabolites to the larvae, giving them additional energy sources to promote long-distance dispersal

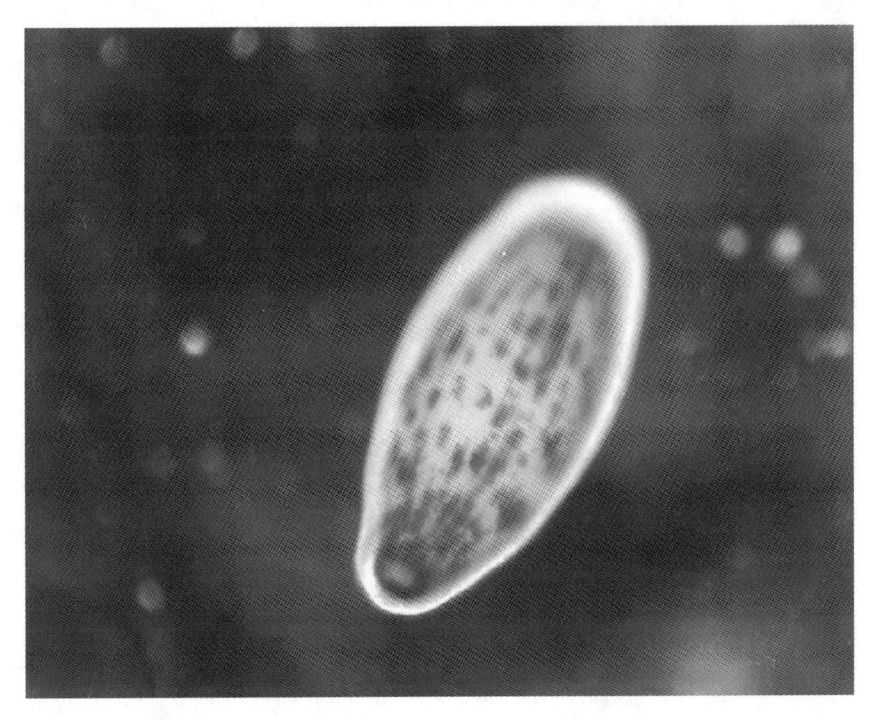

Figure 8–1. Planula larva of *Pocillopora damicornis* with bands of symbiotic zooxanthellae.

(Richmond, 1987, 1988). Nevertheless, the ability to immediately settle results in some brooded planulae attaching to the substrata within centimeters of the parent colony.

Brooders, sometimes referred to as planulators, represent only a small portion of the corals studied, perhaps 15%. The pan-Pacific coral *Pocillopora damicornis* releases brooded planulae on a lunar cycle monthly, throughout the year, on the reefs of Micronesia and Hawaii (Richmond and Jokiel, 1984; Jokiel, 1985). This same species releases larvae only a few months per year in Okinawa and in western Australia (Stoddart and Black, 1985; Richmond and Hunter, 1990). *Pocillopora damicornis* has also been reported to spawn in the eastern Pacific and western Australia, raising questions about the taxonomy of corals that look alike but demonstrate differences in reproductive characteristics (Glynn et al., 1991; Ward, 1992).

For over 250 species of corals already studied, the majority (perhaps 85%) are spawners (Table 8–1), many of which participate in multispecies mass-spawning events during limited periods each year (Fadlallah, 1983; Harrison et al., 1984; Harrison and Wallace, 1990; Richmond and Hunter, 1990). In Okinawa, most spawning species release their gametes over a 5–8-day period commencing

Table 8-1. Reproductive Characteristics of Corals from the Caribbean Sea, Hawaii, Central Pacific, Eastern Pacific, Indo-West Pacific, Great Barrier Reef, and Red Sea

Caribbean Coral	Sex	Mode	Season	Source
Acropora cervicornis	H(pg)	S	sr	1,6,29
Acropora palmata	H(pg)	S	sr	6,29
Agaricia agaricites	H	B	sp-yr	1,10
Agaricia fragilis	x	B	sr	13
Agaricia crassa	x	B	sp	13
Dendrogyra cylindris	G(pg)	S	sr	6,29
Diploria strigosa	H(pg)	S	sr	6,7
Favia fragum	H	B	sp-yr	1,6,10
Isophyllia sp.	G?	B	sp	1
Manicina areolata	H	B	sp	1
Meandrina sp.	x	B	sp	1
Montastrea annularis	H(pg)	S	sr	6,29
Montastrea cavernosa	G(pg)	S	sr	6,29
Mycetophyllia ferox	H(pg)	B	w-sp	6,29
Porites astreoides	H(s)	B	sp-yr	1,6,10
Porites porites	G?	B	w-sp	1
Siderastrea radians	G	B	yr	1,6,10
Siderastrea siderea	G	S	sr	6
	6G:9H:3?	11B:7S		
Pacific Corals Hawaii				
Acropora cytherea	H	*	*	12
Acropora humilis	H	*	*	12
Acropora valida	H	*	*	12
Cyphastrea ocellina	x	B	yr	13
Dendrophyllia manni	x	B	sr-f	1
Fungia scutaria	G	S	sr	15
Montipora verrucosa	H	S	sr	10
Pocillopora damicornis	H	B	yr	16,17
Porites lutea	G	S	sr	10
Tubastrea aurea	x	B	sr-f	1,10
	2G:5H:3?	4B:3S		
Central Pacific (Guam, Marshall Islands, Palau)				
Acrhelia horrescens	x	B	yr	20
Acropora bruggemanni	x	B	x	18
Acropora cerealis	H	x	sr	14
Acropora corymbosa	x	B	sr	13
Acropora humilis	x	B	sr	13
Acropora hystrix	H	x	sr	14
Acropora irregularis	H	S	sr	10
Acropora nasuta	H	S	sr	10
Acropora palawensis	x	B	x	19
Acropora smithi	H	x	sr	14
Acropora sp.	H	S	sr	10
Acropora striata	x	B	w	13
Acropora tenuis	H	S	sr	10
Acropora valida	H	x	sr	14

Continued

Table 8-1. Continued

	Sex	Mode	Season	Source
Central Pacific (Guam, Marshall Islands, Palau) *Continued*				
Euphyllia glabrescens	x	B	x	20
Favia mathaii	H	x	sr	14
Favites abdita	H	x	sr	14
Favites flexuosa	H	x	sr	14
Fungia actiniformis	H?	B	yr	21
Fungia fungites	G	x	sr	14
Galaxea apera	x	B	yr	23
Goniastrea aspera	x	B	f	21
Goniastrea edwardsi	H	x	sr	14
Gonioporaqueenslandiae	G	B	sr	1
Leptoria phrygia	H	x	sr	14
Platygyra pini	H	x	sr	14
Pocillopora damicornis	H	B	yr	13,17
Pocilloporaelegans	x	B	sr	13
Pocillopora verrucosa	x	B	yr	13
Porites lutea	G	x	sr	14
Seriatopora hystrix	x	B	yr	13
Stylophora pistillata	x	B	yr	13,22
16H:3G:13?	16B :4S:12?			
Philippines				
Acanthastrea hillae	x	x	sp	30
Acropora austera	x	x	sp	30
Acropora cytherea	x	x	sp-sr	30
Acropora florida	x	x	sp	30
Acropora humilis	x	x	sp	30
Acropora hyacinthus	x	x	sp	30
Acropora loripes	x	x	sp	30
Acropora pulchra	x	x	sp	30
Acropora selago	x	x	sp	30
Acropora tenuis	x	x	sp	30
Acropora valida	x	x	sp	30
Astreopora myriophthalma	x	x	sp	30
Echinophyllia aspera	x	x	sp	30
Echinopora gemmacea	x	x	sr	30
Favia pallida	x	x	sp	30
Favia helianthoides	x	x	sp	30
Galaxea fasicularis	x	x	sp	30
Goniastrea edwardsi	x	x	sp	30
Goniastrea favulus	x	x	sp	30
Montipora sp.	x	x	sp	30
Oulophyllia bennettae	x	x	sp	30
Pectinia lactuca	x	x	sp	30
Platygyra daedalea	x	x	sp	30
Platygyra sinensis	x	x	sp	30

Note: While it may be assumed that coral species identified as spawners elsewhere also spawn in the Philippines, this table lists sex and mode as unknown unless an actual observation was recorded. The corals listed as tentatively spawning in the spring were observed to contain ripe gonads in April and May. The data summarized here are based on preliminary observations reported in Bermas et al. 1993.

Continued

Table 8-1. Continued

Taiwan	Sex	Mode	Season N	Season S	Source
Acanthastrea echinata	x	x		May	31
Acropora austera	x	x		May	31
Acropora cerealis	x	x	Jun	Apr, May	31
Acropora cytherea	x	x		May	31
Acropora danai	x	x		May	31
Acropora digitifera	x	S		Apr, May	31
Acropora divaricata	x	x		Jun	31
Acropora formosa	x	S		Jun	31
Acropora humilis	x	S		Apr, May	31
Acropora hyacinthus	x	x		May	31
Acropora monticulosa	x	S		May	31
Acropora nana	x	x		Apr, Jun	31
Acropora nasuta	x	x		Jun	31
Acropora nobilis	x	S		May	31
Acropora palmerae	x	S		May	31
Acropora spicifera	x	S		Jun	31
Acropora valida	x	x	Jun		31
Acropora spp.	x	x		Apr, May	31
Astreopora gracilis	x	x		May	31
Astreopora listeri	x	S		May	31
Cyphastrea chalcidicum	x	x	Jul	Apr, May	31
Cyphastrea microphthalma	x	x		May	31
Cyphastrea serailia	x	x	Jul	May	31
Echinophyllia aspera	x	x	Jul	May	31
Echinopora lamellosa	x	x	Jul	Sep	31
Euphyllia ancora	x	S		Apr, May	31
Favia laxa	x	x	Jul	May	31
Favia pallida	x	x	Jul	May	31
Favia sp.	x	S		Apr	31
Favia speciosa	x	S	Jun	Apr, May	31
Favites abdita	x	x		Apr, May	31
Favites chinensis	x	x	Jul	Apr	31
Favites complanata	x	x	Jun	May	31
Favites russelli	x	x	Jul		31
Galexea astreata	x	S		Apr, May	31
Galaxea fascicularis	x	S		Apr, May, Jun	31
Goniastrea aspera	x	S		Apr, May	31
Goniastrea edwardsi	x	S	Jul	May	31
Goniastrea retiformis	x	S	Jul	Apr, May	31
Leptoria phrygia	x	x		Apr, May	31
Merulina ampliata	x	x		Sep	31
Montastrea curta	x	x	Jul		31
Montastrea valenciennesi	x	x	Jun	Jun	31
Montipora aequituberculata	x	S		Jun	31
Montipora digitata	x	S		Apr, May	31
Montipora efflorescens	x	S		Apr, May	31
Montipora foliosa	x	S		Jun	31

Continued

Table 8-1. Continued

Taiwan–*Continued*	Sex	Mode	Season N	Season S	Source
Montipora informis	x	S		Apr, May	31
Montipora spp.	x	S		Apr, May	31
Montipora tuberculosa	x	S		Apr, May	31
Montipora venosa	x	S		Apr, May	31
Mycedium elephantotus	x	x	Jul	May	31
Pachyseris rugosa	x	x		May	31
Pachyseris speciosa	x	x		May	31
Platygyra daedalea	x	x		Apr, May	31
Platgyra lamellina	x	S		Apr, May	31
Platgyra pini	x	x	Jul	Apr, May	31
Platygyra sinensis	x	S	Jul	Apr, May	31
Plesiastrea versipora	x	S		Apr	31
Pocillopora damicornis	x	B		Nov–Mar	31
Porites annae	x	x		May	31
Porites lobata	x	x	Jul		31
Seriatopora hystrix	x	B		yr	31
Stylophora pistillata	x	B		yr	31

N = North Taiwan, S = South Taiwan

Eastern Pacific	Sex	Mode	Season	Source
Pocillopora damicornis	H(s)	S(?)	sr	32
Pocillopora elegans	H(s)	S(?)	sr	32
Tubastrea aurea	x	B	Jun–Nov	33
Great Barrier Reef				
Acropora aspera	H(s)	S	seasonal	4
Acropora cuneata	H(s)	B	x	4
Acropora digitifera	H(s)	S	sp-sr	4
Acropora formosa	H	S	sp-sr	5
Acropora humilis	H(s)	S	sp-sr	4
Acropora hyacinthus	H(s)	S	sp-sr	4
Acropora millepora	H(s)	S	sp-sr	4
Acropora palifera	H(s)	B	x	4
Acropora pulchra	H(s)	S	sp-sr	4
Acropora robusta	H(s)	S	sp-sr	4
Acropora variablis	H(s)	S	sp-sr	4
Favia abdita	H(s)	S	sp-sr	3
Favia favus	H(s)	S	sr	5
Favia pallida	H(s)	S	sp-sr	1
Goniastrea aspera	H(pg)	S	x	9,11
Goniastrea australensis	H(pa)	S	sp-sr	2,9
Leptoria phrygia	H(s)	S	sp-sr	3
Lobophyllia corymbosa	H(s)	S	sr	1,5
Montipora ramosa	H	S	x	5
Pavona cactus	G	S	x	25
Platygyra sinensis	x	S	x	9

Continued

Table 8-1. Continued

Great Barrier Reef–*Continued*	Sex	Mode	Season	Source
Pocillopora damicornis	H	B	yr	1,25
Porites andrewsi	G	S	sp-sr	24
Porites australiensis	G	S	sp-sr-f	5
Porites haddoni	x	B	sr-f	25
Porites lobata	G	S	sp-sr	24
Porites lutea	G	S	sp-sr	5,24
Porites murrayensis	G	B	sp-sr-f	24
Seriatopora hystrix	x	B	sp-sr	26
Symphyllia recta	H	S	sp-sr	25
6G:21H:3?	6B:24S			

Note: A total of 133 species out of 356 have been observed to mass spawn during the week following the full moon in October (see Willis et al., 1985, for details).

Red Sea	Sex	Mode	Season	Source
Acropora eurystoma	H(pg)	S	sp	27
Acropora hemprichii	H	x	x	28
Acropora humilis	H(pg)	S	sp	27
Acropora hyacinthus	H(pg)	S	sr	27
Acropora scandens	H(pg)	S	sr	27
Alveopora daedalea	H(pg)	B	f-w	27
Astreopora myriophthalma	H(pg)	S	sr	27
Favia favus	H(pg)	S	sr	27
Galaxea fascicularis	H(pg)	S	sr	27
Goniastrea retiformis	H(pg)	S	sr	27
Platgyra lamellina	H(pg)	S	sr	27,28
Pocillopora verrucosa	H(pg)	S	sr	27
Seriatopora caliendrum	H(pg)	B	sp-sr-f	27,28
Stylophora pistillata	H(pg)	B	w-sp-wr	27,28
OG:14H	3B:10S:1?			

Sources: 1. Fadlallah, 1983; 2. Kojis and Quinn, 1981a; 3. Kojis and Quinn, 1982; 4. Bothwell, 1982; 5. Harriot, 1983a; 6. Szmant-Froelich, 1984; 7. Wyers, 1985; 8. Van Moorsel, 1983; 9. Babcock, 1984; 10. Richmond, pers. obs.; 11. Babcock, 1984; 12. Grigg et al., 1981; 13. Stimson, 1978; 14. Heyward, 1989; 15. Krupp, 1983; 16. Harrigan, 1972; 17. Richmond and Jokiel, 1984; 18. Atoda, 1951a; 19. Kawaguti, 1940; 20. Kawaguti, 1941; 21. Abe, 1937; 22. Atoda, 1947b; 23. Atoda, 1951b; 24. Kojis and Quinn, 1982a; 25. Marshall and Stephenson, 1933; 26. Sammarco, 1982b; 27. Shlesinger and Loya, 1985; 28. Rinkevich and Loya, 1979b; 29. Szmant, 1986; 30. Bermas et al., 1993; 31. Dai et al., 1993; 32. Glynn et al., 1991; 33. Richmond, pers. obs. *Sex:* H = hermaphroditic; G = gonochoric; pg = protogynous; pa = protandrous; s = simultaneous; x = unknown. *Mode:* S = spawner; B = brooder. *Season:* w = winter; sp = spring; sr = summer; f = fall; yr = year-round; x = unknown. Abbreviations for months are used when available and appropriate. (Updated from Richmond and Hunter, 1990.)

around the night of the May and June full moons each summer (Hayashibara et al., 1993). In Guam, Micronesia, peak spawning occurs 7–10 days after the July full moon (Richmond and Hunter, 1990). In the nearby islands of Palau, coral spawning appears to occur several months per year, including March, April, and May (Kenyon, 1995). In Australia, mass spawning events occur during November (Harrison et al., 1984).

Why are there differences in timing of coral spawning among sites, and yet so many species have a high degree of synchronization at a particular location? A critical aspect of spawning is synchronization among members of a species in the production and release of sperm and eggs. If eggs are ripe while the sperm are not, reproduction will be unsuccessful.

Corals have the ability to respond to several environmental factors. Water temperature is one signal that determines the time of year when spawning will occur (Shlesinger and Loya, 1985; Oliver et al., 1988). Many invertebrates in polar, temperate, subtropical, and tropical habitats reproduce during the times of maximum water temperatures. The "fine-tuning" seems to be in response to lunar phase. Since tides are affected by the moon, these may also affect timing, but studies have shown that nocturnal illumination plays a key role in reproductive timing in corals (Jokiel, 1985; Jokiel et al., 1985).

Corals are sensitive to chemical compounds that may also facilitate synchronized reproduction on a particular reef (Coll et al., 1989, 1990; Atkinson and Atkinson, 1992). In simultaneous comparisons among reefs in Japan separated by distances of over 50 km, different species were found to spawn on different nights during the period following the June full moon; but by the end of the week, all of the same species had released their gametes. Contagious spawning events occur as the gametes from one coral colony stimulate other colonies of the same species downcurrent to release their eggs and sperm upon contact with the gamete cloud. These observations support the notion that chemical signals are a likely cause of synchronized spawning within a reef.

Spawning species that are simultaneous hermaphrodites typically release combined egg-sperm packets (Fig. 8–2), with egg size and number of eggs per cluster varying among species. Gamete bundles may consist of between 9 and 180 eggs surrounding or embedded within a mass of sperm. Sections taken through a coral polyp prior to gamete release reveal the eggs lined up vertically or in clusters along mesenteries (Fig. 8–3). Sperm-filled packets have also been observed within the same polyp, but attached to different mesenteries (Harrison and Wallace, 1990). On the night of spawning, sperm packages are moved up from within the colony to a position near the mouth of the polyp and are rotated as eggs are either attached to the outer surface (many acroporids) or embedded within the sperm packet (e.g., *Favites*). The exact sequence of events may differ among species. In some, the transparent expanded polyps appear white as the sperm packets are moved up toward the mouth, but later become orange, pink, or red, as the colored eggs are attached (Fig. 8–4). Eventually the mouth of the polyp

Figure 8–2. Egg-sperm clusters of *Acropora* sp. Each cluster contains 9–16 eggs surrounding a central sperm packet. Each cluster is approximately 1 mm in diameter.

Figure 8–3. A cross-section of a branch of an *Acropora* containing pigmented eggs and white sperm packets.

Figure 8–4. *Acropora* colony ready to release egg-sperm clusters.

expands and the gamete clusters are released (Fig. 8–5). In the field, these events are visible to the naked eye as the outer surface of the coral colony takes on color and as the colored gamete clusters are released.

The high lipid content of eggs makes the clusters positively buoyant. Sperm are neutrally buoyant and would otherwise have to swim to the surface in order to fertilize the eggs. Combining the eggs and sperm as a cohesive unit guarantees the sperm will reach the ocean's surface within moments of their release at no energetic cost, and will be in the proximity of appropriate eggs if conspecifics are nearby.

Once the combined egg-sperm packets reach the surface, there is a delay of approximately 10–40 minutes before they break apart and fertilization can take place (Fig. 8–6). Experiments have shown that eggs will not become fertilized until after this breakup occurs. Whatever the mechanism, the time delay reduces the chance of self-fertilization among eggs and sperm from the same colony, and increases the chances of fertilization among gametes from different individuals (outcrossing). However, this observed characteristic also increases the period during which gametes will be exposed to pollutants, like oil and contaminants carried in freshwater runoff, that are found at highest concentrations at the ocean surface.

8.1.5. Self-Fertilization Versus Outcrossing

A number of interesting questions arise from the observations of multispecies mass-spawning events. Does self-fertilization occur, and if so, at what frequency

Figure 8–5. *Acropora* polyps releasing egg-sperm clusters.

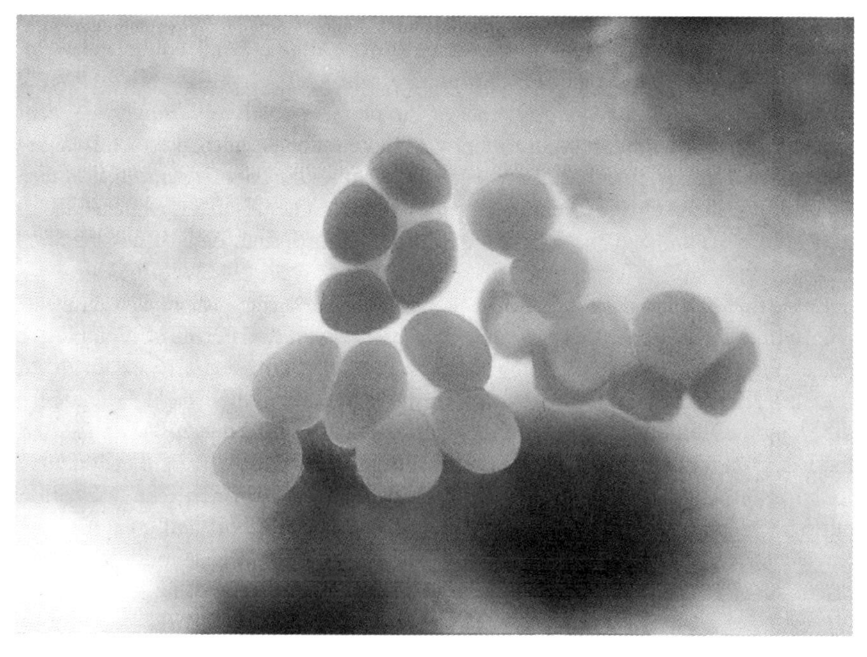

Figure 8–6. Egg-sperm clusters breaking apart approximately 20 minutes after release. The eggs are now ready to be fertilized by sperm from different colonies.

compared to outcrossing? Does hybridization occur among closely related species? Why does multispecies synchronization occur?

Barriers to self-fertilization do exist in corals that promote outcrossing, but these barriers may be time dependent and break down after several hours (Heyward and Babcock, 1986; Richmond, 1993a). In some experiments, it has been shown that sperm do not fertilize eggs from the same colony until 6 hours after release, and even then, observed rates of self-fertilization are less than 10%. The same eggs treated with sperm from another colony of the same species demonstrated fertilization rates of 70–100% within 2 hours of gamete release. This warns us that fertilization rates and reproductive success of corals may be reduced by pollution with chemical contaminants that can interfere with chemical recognition between gametes.

8.1.6. Hybridization

During a multispecies mass-spawning event, sizable slicks of coral eggs and sperm can be observed above reefs, extending hundreds to thousands of meters in some cases (Oliver and Willis, 1987). With so many eggs and sperm from a wide variety of species intermingling, the question arises, Can eggs of one species become fertilized by sperm from another?

Laboratory experiments have demonstrated that hybridization does occur among corals. This has been observed not only among closely related species within the same genus, but across genera (Richmond, 1993a, 1995; Willis et al., 1993; Kenyon, 1993). In one case, crosses of *Acropora digitifera* and *Acropora gemmifera* were unidirectional, meaning eggs of *A. digitifera* were fertilized by sperm from *A. gemmifera,* but eggs of *A. gemmifera* were not fertilized by *A. digitifera* sperm (Richmond, 1995). This type of unidirectional hybridization has been observed in echinoderms and fruit flies (Kaneshiro, 1987; Uehara et al., 1990).

Hybridization among species spawning on the same evening may be deterred by differences in the timing of gamete release. Some species usually release their eggs and sperm around 8 P.M. (e.g., *Acropora tenuis* on Guam), whereas others consistently release gametes at 10 P.M. (*Acropora humilis* on Guam), 11 P.M. (*Acropora valida* on Guam), and 11:45 P.M. (*Acropora irregularis* on Guam). Corals may also spawn during the day, as demonstrated by certain pocilloporids (Kinzie, 1993).

8.1.7. Larval Development

The planula larvae that develop from brooding and spawning corals are similar in some aspects. Both are ciliated, rich in lipid, and have chemoreceptors used for detecting the appropriate substrata for settlement and metamorphosis. But brooded larvae tend to be larger than larvae that develop from spawned gametes, possess zooxanthellae, and are capable of settling immediately upon release from the parent colony. The smaller larvae that develop from spawning corals require time to reach a stage capable of settlement and metamorphosis. Smaller eggs of the spawning coral genera *Leptoria, Goniastrea,* and *Montastrea,* averaging 350–400 μm in diameter, require 18 hours before they become ciliated and capable of settlement, whereas the larger eggs of many spawning *Acropora* species take nearly 72 hours to reach the same stage of development.

Most planula larvae that develop from spawned gametes do not contain symbiotic zooxanthellae, and do not appear to obtain them until after settlement and metamorphosis. Calculated competency periods indicate such larvae retain their ability to recruit for only 3–4 weeks (Richmond, 1988). After that, they may remain alive but lack the energy reserves needed to make the transformation to a benthic calcified coral. Corals in the genus *Montipora* are an exception, as the spawned eggs contain zooxanthellae. All brooded planulae studied to date possess zooxanthellae, and in the case of *Pocillopora damicornis,* the competency period has been found to exceed 100 days, attributable in part to the contribution of the symbiotic algae to the nutritional needs of the larva (Richmond, 1987a, 1988).

The biogeographic distribution patterns of specific groups of corals suggest that possession of zooxanthellae during the larval stage enhances dispersal ability. Widely distributed species have a decreased risk of extinction from local events. Some corals endemic to Japan may be threatened with extinction due to anthropogenic (human-induced) disturbance (Chapter 14; Veron, 1992).

8.2. Larval Recruitment

Successful reproduction is only the first step in the replenishment of corals on the reef. In order for coral populations to be maintained, dead corals must be replaced, either from larvae or asexually produced products. Recruitment depends on the ability of larvae to identify an appropriate site for settlement and metamorphosis. These two distinct and critical processes are often dependent on specific chemical signals for marine invertebrates (Pawlik and Hadfield, 1990).

8.2.1. Settlement

Settlement of coral larvae is a change from a planktonic existence to a benthic lifestyle, and usually includes attachment to the reef. In order for settled larvae to survive and develop into young corals, they must settle on an appropriate site. The criteria for appropriate sites include substratum type, water motion, salinity (generally above 32‰), adequate sunlight for zooxanthellae, limited sediment deposition, and sometimes specific algal species or biological films of diatoms and bacteria.

Site selection by coral planulae may be made on the basis of chemical signals that are affecting receptors located on the outer surface of the larvae. Coral planulae react to biological films, and in the case of species of *Agaricia*, species-specific chemical signals from particular types of crustose coralline algae (Morse and Morse, 1993).

In the laboratory, competent planula larvae can be observed to swim downward until they contact a surface. If the substratum has the proper texture and biochemical coating, the planula forms an attachment with the aboral surface, begins to contract, and lays down an organic matrix layer, followed by the deposition of the stony carbonate skeleton (Vandermuelen, 1974).

Substratum type as well as orientation can affect recruitment, growth, and survival rates. In most laboratory settlement experiments, planulae would not settle on loose sediment, especially if solid substrata were available. When settlement did occur on sediment, coral planulae had poor survival rates. In field experiments, Birkeland (1977) found recruits had faster growth rates on upper surfaces of artificial substrata, but survivorship was greater for those larvae settling on vertical surfaces. The same study reported faster growth rates of recruits at shallower depths, but higher survivorship with intermediate depth and at lower nutrient levels. Survival of young corals depends not only on the attributes of the settlement surface, but on competition with other organisms including algae and encrusting invertebrates such as sponges.

8.2.2. Metamorphosis

Metamorphosis is a developmental process during which a larva undergoes a series of morphological and biochemical changes while completing the transformation to the benthic juvenile stage. A planula larva is quite different from a coral polyp in that it does not have tentacles for feeding, the mouth has not yet opened to the gastrovascular cavity, digestive enzymes have not yet been produced for heterotrophic feeding, and no calcification has taken place.

During metamorphosis, a commitment is made to the settlement site. A calcified basal plate is secreted along with the first skeletal cup, and tentacles complete with stinging cells known as nematocysts form surrounding the mouth. A new coral colony will develop from this first or primary polyp through growth, budding, and continued calcification. Larvae that develop from spawned eggs, and that did not acquire zooxanthellae from the parent colony, will incorporate these algal symbionts from external sources. Observations made on a variety of *Acropora* species found that these corals obtain their zooxanthellae only after settlement and metamorphosis, and that recruits that did not pick up their algal symbionts within 2 weeks were often overgrown by crustose coralline or other red algae.

Larvae that settle in unfavorable sites may get a second chance. If a newly metamorphosed coral is stressed within days of settlement and development, it may be able to retract its tissue from the freshly secreted skeleton and return to the plankton until another suitable site is encountered (Richmond, 1985). This has been observed in *Pocillopora damicornis,* but only from the single-polyp stage and within 3 days of initial settlement and metamorphosis.

The process of settlement does not guarantee metamorphosis will follow. For many types of invertebrate larvae, metamorphosis is a complex chain of reactions that commences only in the presence of a chemical inducer (Hadfield and Pennington, 1990). The inducer of metamorphosis can be highly specific as seen in coral planulae that will only settle on a single species of coralline algae, or more general in nature, as with short-chain peptides or general biological/diatomaceous films (Pawlik and Hadfield, 1990; Morse and Morse, 1991). Observed differences in species distribution patterns are more likely the result of larval selection than colony survivorship (Morse and Morse, 1991; Morse et al., 1994).

An important consideration in studies of recruitment is that coral larvae may be sensitive to chemical signals at levels below the detectable limits of human technology. Bioassays, biological tests using sensitive organisms, are the appropriate tools to determine if environmental contamination is at levels that can interfere with critical biological processes like reproduction and recruitment (Connell and Miller, 1984).

However, the choice of bioassays is also important (Cairns et al., 1978). Accepted standardized protocols, such as a 96-hour LC_{50} (the concentration at which 50% of the test organisms die within 96 hours of exposure), are not useful to the reef manager, as sublethal effects to organisms can be as damaging to a reef over time as lethal effects. For example, an environmental toxin at a level that allows 100% survival of larvae over 96 hours, but prevents them from detecting the appropriate settlement cue and hence prevents recruitment, has the same overall effect on the reef as causing 100% mortality of the larvae. Recent experiments demonstrated that the pesticide Chlorpyrifos, a chemical often used on golf courses, can be taken up by crustose coralline algae and that coral larvae had statistically lower recruitment rates on substrata exposed to the chemical (5 parts per billion) than on untreated controls.

By understanding the biology of coral reproduction and recruitment, it is easy to see how environmental quality can affect these two processes that are responsible for the persistence of reefs. While corals may not represent the greatest biomass on the reef, they do serve as primary framework builders and as an initial link in reef food chains as the host for symbiotic zooxanthellae.

8.3. Reproductive and Recruitment Failure of Corals

Coral reefs are diverse and productive ecosystems with complex interactions at the cellular, organismal, and community levels. Many interactions, including those between adjacent corals, between their gametes, and between larvae and their settlement substrata, are chemically mediated (Coll et al., 1990; Atkinson and Atkinson, 1992; Richmond, 1993a, b; Morse et al., 1994). Changes in water quality that affect transmission of these chemical signals can have negative effects on reproductive timing, synchronization, egg-sperm interactions, settlement, metamorphosis, and the incorporation of zooxanthellae. Jokiel (1985) observed that changes in salinity, water temperature, and light availability affected planula production in the coral *Pocillopora damicornis.* Kojis and Quinn (1984) found a negative correlation between fecundity and depth, turbidity, and sedimentation for the coral *Acropora palifera.* They also found that allocation of energy to tissue repair in *Goniastrea favulus* resulted in a decrease in reproductive ability (Kojis and Quinn, 1985).

Colony size has been found to be a determinant of fecundity. Among small-polyped coral colonies of the same size, age can also affect overall reproductive output, with older corals being more fecund (Kojis and Quinn, 1985). For large-polyped forms, such as *Lobophyllia corymbosa,* polyp size was found to be more important than colony size for both maturity and fecundity (Harriott, 1983). For branching corals, including representatives from the genera *Pocillopora* and *Acropora,* it appears to take at least 2–3 years to attain reproductive maturity and produce the first gametes and/or larvae. Massive corals (e.g., species of *Porites*) may require a longer period of growth and development, ranging from 4 to 7 years (Rinkevich and Loya, 1979a; Szmant-Froelich, 1985; Szmant, 1986; Babcock, 1988a). For species that exhibit a direct relationship between colony size and reproduction (both onset of sexual maturity and reproductive output), reduced growth from "stress" (Brown and Howard, 1985) will also lead to a depression of reproductive potential.

Normal seawater salinity for thriving coral reefs is near 35‰. Coral colonies can survive higher and lower salinities for periods of time, and if exposed by an extremely low tide, can secrete a layer of mucus to act as a barrier against desiccation. Coral gametes, however, are more sensitive to conditions of altered salinity than adult colonies. Since peak coral spawning occurs during the rainy season in Micronesia and Okinawa, gametes released may end up in a surface layer of reduced salinity. Field samples have found surface salinities over reefs adjacent to streams and storm drains may be decreased by more than 25% to 26‰ or lower. Bioassays designed to test the effects of salinity on fertilization and larval development in corals found a 20% drop in salinity from 35‰ to 28‰ caused a corresponding 86% decrease in fertilization rate (Richmond, 1993b, 1994; Fig. 8–7). If red clay soil is included in the runoff, the same drop in fertilization was observed to accompany a smaller decrease in salinity (to 33‰), demonstrating synergistic effects. Additionally, larvae formed in uncontaminated waters showed decreased settlement rates in areas of lowered salinity.

8.3.1. Terrestrial Runoff and Water Clarity

Water clarity is an important factor affecting coral growth, and has also been observed to affect reproduction and recruitment (Jokiel, 1985; Tomascik and

Figure 8–7. Coral eggs from a reduced salinity (28‰) fertilization bioassay showing only 4 of 35 eggs fertilized. The control (34‰) displayed an 88% fertilization rate versus 25% for the experimental treatment.

Sander, 1987b). The symbiotic association between corals and zooxanthellae (Chapter 5) allows reefs to flourish in nutrient-poor waters and also helps fulfill the energetic demands of coral reproduction. Photosynthetic products of zooxanthellae contribute to the production of eggs and larvae (Rinkevich, 1989). Since coral reefs are predominantly coastal, shallow-water features, they are vulnerable to the influence of land-based activities that result in runoff and increased turbidity (Chapter 15).

Sedimentation continues to be the most persistent problem affecting coastal reefs and those surrounding high islands (Johannes, 1975; Rogers, 1990; Richmond, 1993b). Sediments may exact an energetic cost to the coral that must cleanse its surface. The results are lower growth rates and less energy available for reproduction (Tomascik and Sander, 1987b). Sediment can also be a barrier preventing coral larvae from detecting the chemical signals from preferred settlement substrata like coralline algae.

Nutrient enrichment or eutrophication can be a problem affecting reproduction and recruitment of corals (Tomascik, 1991) and can originate from several sources including agricultural runoff and sewage outfalls. Eutrophication can lead to reduced light levels by increasing turbidity. Furthermore, elevated nutrient levels give fast-growing organisms like algae, sponges, tunicates, and bryozoans a competitive advantage over slower-growing corals (Birkeland, 1977, 1988a).

Such competitors can overgrow corals and dominate available substrata, preventing coral larvae from settling, and may lead to alternate stable states (Hatcher, 1984; Tomascik, 1991; Done, 1992; Hughes, 1994).

8.3.2. Water Pollution

A general consideration for islands and coastal communities is that substances used on land today often end up in the coastal zone tomorrow. The distance between sources of chemical contamination and coral reefs may be small. Common toxins in street runoff, including oil and petroleum products, cadmium from automobile tire wear, and miscellaneous heavy metals, enter the coastal zone every time it rains. If industrial wastes and toxins are released into residential drainage and sewer systems, they too can reach the reef. These problems all point to the need for sound watershed management and serve as examples of how land-based activities must be regulated to protect the marine environment.

Substances adhering to soil particles and contained in runoff water can be toxic and/or interfere with chemical signals (Ingersoll, 1995). Pesticides and other chemicals may bind to soil particles on land, but due to the difference in the pH of seawater, they can be released when these particles reach the ocean (Connell and Miller, 1984). The pesticide Chlorpyrifos was found to decrease levels of larval settlement and metamorphosis on settlement substrata that had been incubated in the presence of the chemical at the level of 5 parts per billion. The behavior of pesticides and toxic substances can change over time and space. Breakdown products may be more toxic to cells than the original chemical form, and processes including photoisomerization and methylation may increase stability, toxicity, and biological activity (Connell and Miller, 1984).

Corals in an area affected by an oil spill showed decreased gonad size compared to colonies from unaffected reefs years after the spill occurred (Guzman and Holst, 1993). Oil pollution was found to abort the formation of viable larvae in a brooding species of coral (Loya and Rinkevich, 1979). The presence of contaminants in coral-reef waters can also interfere with chemical signals that allow reproductive synchrony among coral colonies as well as interactions between egg and sperm (Richmond, 1994).

8.3.3. Population Depletion

Coral planula competency and recruitment patterns suggest some reefs may depend on distant coral communities for their supply of planulae larvae (Richmond, 1987; Babcock, 1988). If source reefs are disturbed, the effects may also be felt on downcurrent reefs. This issue is important, as it points to the need for interisland and regional cooperation if reefs are to be protected. This also has implications for planning coral-reef reserves, which should include consideration of dispersal patterns (Williams, et al., 1984).

Even if reproduction is successful, replenishment of reef populations is not guaranteed until larvae and asexual products successfully recruit. Planulae produced from corals living on healthy reefs will not recruit onto reefs where substrata or water quality are unacceptable. For example sedimentation and runoff may be sublethal to adult corals, yet can prevent larvae from settling (Richmond, 1993b). Living coral cover alone (abundance and diversity) does not reliably reflect the health of a reef. Such values only describe the state of the reef at that moment in time. Recruitment patterns are important in predicting what the future of a reef may be. Adult corals can survive in areas where reproduction is failing and larvae are unable to settle.

Commercially valuable sea cucumbers in Micronesia have provided an example of how populations of reef organisms may be affected by reproductive and recruitment failure. In the late 1930s, prior to World War II, large numbers of edible sea cucumbers were exported from Chuuk (Truk) to Japan, with records reflecting quantities in the hundreds to thousands of tons. Surveys performed in 1988 found only two individuals of the valuable species *Holothuria nobilis* from over eight sites in Chuuk Lagoon. Based on interviews with local residents and fishermen, it appears that populations of several species of sea cucumbers have not recovered from the severe exploitation levels prior to and during the Second World War. It is reasonable that for spawning species, once populations are reduced below a certain level, chances of successful reproductive events are low because of gamete dilution (**Allee effect**; Levitan et al., 1992). If an island is beyond the dispersal range of larvae from other source areas, immigration is not going to occur, and populations will not rebound. The effect of population reductions on future numbers is clear when considering the decrease in reproductive success.

8.3.4. Prevention and Mitigation

Corals can die as a result of both natural occurrences and human activities. If these corals are not replaced through the processes of reproduction and recruitment, the reef will eventually degenerate. Corals provide a primary link in the food chain as the sites of photosynthetic fixation of light energy through their zooxanthellae. They also provide a habitat for numerous associated fish and invertebrates. If the coral populations on the reef go into decline, so will the rest of the community. Whether a reef is killed quickly by catastrophe or slowly by attrition (no population replenishment), the result is the same: the loss of the reef and all it has to offer.

Our present knowledge of factors affecting reproduction and recruitment enables us to better understand how to manage activities that affect reefs, and also allows us to examine methods for applying this knowledge to reseeding and rehabilitating damaged reefs once suitable conditions return. Experiments on the cultivation of coral larvae for reseeding have proven successful. Larvae raised from mass-spawning events have been introduced onto suitable substrata in three

Figure 8–8. *Acropora* recruits, approximately 24 hours after settlement and metamorphosis.

areas damaged by crown-of-thorns starfish predation and by sediment. Numerous recruits were found in the seeded areas (Fig. 8–8), while no recruits of the species used were found in adjacent reference sites. These results indicate that if the environment is appropriate for recovery, reseeding can enhance recruitment rates above natural levels. Unfortunately, a 50-year-old coral cannot be replaced by reseeding in less than the 50 years it took to grow to a particular size. Prevention of human-induced damage and protection of water and substratum quality are the most effective means of supporting successful reproduction and recruitment of corals, and hence, the persistence of coral reefs.

9

Invertebrate Predators and Grazers

Robert C. Carpenter

Coral reefs are among the most productive and diverse biological communities on earth. Some of the diversity of coral reefs is associated with the invertebrate organisms that are the primary builders of reefs, the scleractinian corals. While sessile invertebrates, such as stony corals, soft corals, gorgonaceans, anemones, and sponges, and algae are the dominant occupiers of primary space in coral-reef communities, their relative abundances are often determined by the activities of mobile invertebrate and vertebrate predators and grazers. Hixon (Chapter 10) reviews the direct effects of fishes on coral-reef community structure and function and Glynn (1990b) has provided an excellent review of the feeding ecology of many coral-reef consumers. This chapter reviews the different types of mobile invertebrate predators and grazers on coral reefs, concentrating on those that have disproportionate effects on coral-reef communities and are intimately involved with the life and death of coral reefs.

The sheer number and diversity of mobile invertebrates associated with coral reefs are especially daunting with species from several major phyla including the Annelida, Arthropoda, Mollusca, and Echinodermata. Numerous species of minor phyla are also represented in reef communities, but their abundance and importance have not been well studied. As a result, our understanding of the effects of predation and grazing by invertebrates in coral-reef environments is based on studies of a few representatives from the major groups of mobile invertebrates.

Predators may be generalists or specialists in choosing their prey, and this may determine the effects of their feeding on community-level patterns of prey abundance (Paine, 1966). Feeding preferences are determined by a suite of factors

I appreciate the support of the University of Puerto Rico Sea Grant Program and the National Science Foundation for some of my work with coral-reef herbivores.

that include predator and prey vagility, morphological structures for prey capture and processing, relative availabilities of different prey species, and the relative sizes of predator and prey. Each of these factors is important in both an ecological and evolutionary context.

Relative constancy of prey abundance often can lead to specialization by a predator on a single prey type (Emlen, 1973). Some of the more spectacular examples of feeding specialization have been found in coral-reef communities. Examples include opisthobranchs feeding on algae (Taylor, 1971), fishes that specialize on cleaning parasites from other fishes (Losey, 1974), and butterflyfishes that specialize on corals (Reese, 1977). However, some invertebrate predators and grazers are more generalized, and while their prey are restricted to certain groups of organisms (e.g., corals, algae), they exhibit little preference or discrimination between different species within these groups.

Many invertebrate predators and grazers appear to exhibit limited preferences within major groups of prey that likely are the result of prey morphology relative to the handling capability, feeding mode, and feeding apparatus of the predator. For example, the crown-of-thorns seastar *Acanthaster planci* feeds preferentially on branching and tabulate acroporid corals, leaving poritid, fungiid, and faviid corals behind (Moran, 1986; Birkeland, 1989a). However, depending on the relative abundances of coral species, *Acanthaster* may feed on a variety of other prey species (Glynn, 1990b). Many herbivorous invertebrates, such as majid crabs and echinoids, feed preferentially on filamentous algal turfs, avoiding macroalgae species, but do not discriminate between species within the algal turf functional group (Steneck, 1988). Again, depending on prey availability, some primarily herbivorous echinoids may feed on live coral (Bak and van Eys, 1975; Carpenter, 1981). It appears that dietary plasticity and the ability to exploit a wide range of prey contribute to the success of several species of invertebrate predators and grazers associated with reefs (Birkeland, 1989a).

In general, the two most common types of benthic prey in many coral-reef environments are scleractinian corals and algae. On most shallow reefs (<20 m), these two components cover in excess of 80% of the substratum. Deeper reefs may have a significant cover of sponges, and soft corals are common on some Pacific reefs (Wilkinson and Cheshire, 1989). Nevertheless, corals and algae are prey for a large number of invertebrates, and it is the relative abundances of these two components and how this balance is regulated by some species of predators, that underlies the transition between a thriving, growing, coral-dominated reef and an algal-dominated community that ceases to expand laterally and to accrete vertically (Chapter 3; Adey, 1978).

The direct effect of invertebrate predators and grazers on their prey is reduction of their abundance. In the simplest terms, the extent to which predators limit their prey depends on predator abundance and efficiency of prey capture and the availability to the prey of spatial and/or temporal refugia from predators. Several combinations of these conditions can lead to stable predator-prey cycles (Pianka,

1988). However, predator-prey relationships in most environments probably are not at equilibrium. Most invertebrate predators and grazers on coral reefs are intermediate-level consumers and are themselves the prey for higher-level predators. Likewise, the abundance of prey species for invertebrate predators and grazers may be controlled in part by nutrient and/or light availability (Hatcher, 1990; Muscatine, 1990). As a result, both top-down and bottom-up processes, together with stochastic processes, such as recruitment and physical disturbance, regulate the relative abundances of invertebrate predators and their prey and set the stage for numerous indirect effects that cascade throughout the coral reef trophic web (Chapter 11).

The list of invertebrate predators and grazers on corals and algae given in Table 9–1, while not exhaustive, includes the majority of organisms known to feed on these prey items on coral reefs. They have been divided into two categories (those with major and those with minor effects) based on their typical abundances and their demonstrated effects on coral-reef community structure and function, and their ability to cause transitions between different community types.

9.1. Invertebrates Having Minor Effects

9.1.1. Corallivores

A variety of invertebrates have been reported to feed on live coral, including polychaetes (Robertson, 1970; Ott and Lewis, 1972; Witman, 1988), gastropods (Robertson, 1970; Moyer et al., 1982; Hayes, 1990a, b; Turner, 1994a), and crabs (Glynn, 1983a; Gilchrist, 1985; Stimson, 1990). Under normal circumstances, species within these groups are not exceedingly abundant, and coupled with their small body size and limited mobility, they appear to have only minor and localized effects on coral-reef communities.

For example, Witman (1988) reported that the polychaete *Hermodice carunculata* was responsible for removing 0.13% per day of the hydrocoral *Millepora* spp. on one Caribbean reef, resulting in a change in the relative abundance of this species and species that colonized the damaged colonies. However, the abundance of *H. carunculata* on most reefs is low and their influence in changing live coral biomass is minimal (Ott and Lewis, 1972; Knowlton et al., 1990). Likewise, brachyuran crabs and hermit crabs have limited direct influence as coral predators, and the association of some crabs, such as *Trapezia* spp., with corals is much more important because of their deterrence of other coral predators (Glynn 1983a, b; Stimson, 1990).

Some gastropod corallivores can have more significant effects on corals. In the Caribbean, *Corallophilia abbreviata* was found on up to 64% of the coral colonies examined on a reef in Panama and the gastropods were associated with coral mortality in experiments where corals were stressed (Hayes, 1990). Similarly, Knowlton et al. (1990) examined the effects of *C. abbreviata* on the

Table 9-1. Taxa of Invertebrate Corallivores and Herbivores That Have Minor and Major Effects on Coral Reefs

Corallivores		Herbivores	
Taxa	References	Taxa	References
Groups Having Minor Impacts		Groups Having Minor Impacts	
Annelida		Annelida	
Hermodice carunulata	Robertson, 1970	eunicid polychaetes	Fauchald and Jumars, 1979
	Witman, 1988	syllid polychaetes	Fauchald and Jumars, 1979
	Knowlton et al., 1990		
Mollusca		Mollusca	
Corallophilia abbreviata	Robertson, 1970	limpets	Steneck and Watling, 1982
	Hayes, 1990a,b		
	Knowlton et al., 1990	chitons	Glynn, 1970
			Steneck and Watling, 1982
Drupella cornus	Wilson and Stoddart, 1987		
	Turner, 1994a	Arthropoda	
D. elata	Moyer et al., 1982		
D. fragum	Moyer et al., 1982	amphipods	Brawley and Adey, 1981
D. rugosa	Moyer et al., 1982		Brostoff, 1988
Jennaria pustulata	Glynn, 1982		Brawley, 1992
other gastropods	Robertson, 1970	majid crabs	Coen, 1988a, b
Echinodermata		Echinodermata	
Asteroidea		Echinoidea	
Culcita novaeguineae	Endean, 1971	*Astropyga radiata*	Bak and Nojima, 1980
	Glynn and Krupp, 1986	*Colobocentrotus* spp.	see Birkeland, 1989a, for review
C. schmideliana	Thomassin, 1976	*Echinometra lucunter*	Ogden, 1977
Nidorellia armata	Glynn and Wellington, 1983	*Echinometra viridis*	Williams, 1981
Pharia pyramidata	Glynn and Wellington, 1983	*Echinostrephus* spp.	Campbell et al., 1973
Linckia laevigata	Thomassin, 1976	*Eucidaris tribuloides*	Sammarco, 1977
Nardoa variolata	Thomassin, 1976	*Heterocentrotus* spp.	Dart, 1972
Echinaster purpureus	Thomassin, 1976	*Lytechinus variegata*	Camp et al., 1973
Asterina sp.	Yamaguchi, 1975	*Tripneustes* spp.	see Birkeland, 1989a, for review

Continued

Table 9-1. Continued

Corallivores		Herbivores	
Taxa	References	Taxa	References
Groups Having Minor Impacts		Groups Having Minor Impacts	
Echinoidea			
Astropyga radiata	Herring, 1972		
Diadema antillarum	Bak and Van Eys, 1975		
	Carpenter, 1981		
D. setosum	Herring, 1972		
Echinothrix calamaris	Herring, 1972		
Groups Having Major Impacts		Groups Having Major Impacts	
Echinodermata		Echinodermata	
Asteroidea		Echinoidea	
Acanthaster planci	Goreau, 1964	*Diadema antillarum*	Randall et al., 1964
	Moran, 1986		Ogden et al., 1973
	Birkeland and Lucas, 1990		Sammarco, 1982
			Carpenter, 1986, 1990a, b
Echinoidea			Hughes et al., 1987
		D. setosum/savignyi	Mathias and Langham, 1978
			Bauer, 1980
Eucidaris thouarsii	Glynn et al., 1979		
	Glynn and Wellington, 1983		
		Echinometra mathaei	Downing and El-Zahr, 1987
			McClanahan and Muthiga, 1988
			McClanahan and Shafir, 1990
		Echinothrix spp.	Ormond and Campbell, 1971

coral *Acropora cervicornis* in Jamaica where approximately 10% of the coral colonies had snails associated with them. They found that recent coral mortality was related to the presence of *C. abbreviata* and that coral predation, combined with other stresses, had significant effects on coral mortality. Because the changes in the coral-reef community at this site were also related to the mass mortality of the echinoid *Diadema antillarum,* it was not possible to attribute which of these changes might have been caused directly by *C. abbreviata.* However, it appears that predation by this species may be important locally where snails are abundant and when other environmental conditions result in additional stresses to corals. This is in contrast to the conclusions reached by Ott and Lewis (1972), who reported that *C. abbreviata* had minimal effects on corals on the reefs of Barbados.

Other gastropods that have been demonstrated to have significant effects on coral mortality on some Pacific reefs are the ovulid snail *Jenneria pustulata* (Glynn, 1982) and muricid snails in the genus *Drupella* (Moyer et al., 1982; Turner, 1994a). Glynn (1982) found that *J. pustulata* consumed the coral *Pocillopora* spp. on reefs in Panama. At least four species of *Drupella (D. cornus, D. elata, D. fragum,* and *D. rugosa*) have been shown to cause extensive reef damage in areas where their densities have reached outbreak dimensions. Reported densities of *Drupella* spp. include abundances of <1 ind./m^2 in the Ryukyu Islands, Japan (Fujoika and Yamazato, 1983), 1–20 ind./m^2 on the Great Barrier Reef, Australia (Ayling and Ayling, 1992), and 5–19 ind./m^2 on Ningaloo Reef in western Australia (Ayling and Ayling, 1987; Turner, 1994b). Outbreak aggregations with densities of 1,500 ind./0.5 m^2 have been reported from some reefs in the Philippines (Moyer et al., 1982). Aggregations tended to be found in areas with high cover of acroporid corals. *Drupella* spp. appear to be generalist corallivores, although some studies have suggested preferences for acroporid, poritid, and pocilloporid corals (see Turner, 1994a, for a review).

The effect of coral predation by *Drupella* spp. when they reached outbreak densities was removal of up to 95% of live corals in the patchily distributed areas of high snail density. Therefore, at this scale, effects were quantitatively and qualitatively similar to the effects of predation by the seastar *Acanthaster planci.* In several cases, reefs have been simultaneously infested with high densities of both *Acanthaster* and *Drupella* spp., making it difficult to interpret the effects of *Drupella* spp. In both cases, dead coral skeletons within the patches where predation occurred were quickly colonized by filamentous algae and fouling organisms, and as a result, portions of the reef or sometimes the entire reef was transformed from a coral-dominated system to an algal-dominated one. Increased algal resources can result in higher abundances of grazing fishes and decreased abundances of obligate coral feeders such as chaetodontids (Ayling and Ayling, 1987).

The causes underlying the outbreaks of *Drupella* spp. on some reefs are not well understood. Whether outbreaks are the result of natural fluctuations in the population sizes of these species, perhaps because of stochastic processes that determine recruitment success, or anthropogenic influences that either reduce the mortality of *Drupella* spp. because of overfishing of snail predators or through modification of the coastal environment as a result of increased runoff and siltation, has not been established (Turner, 1994a). Although the effects of predation by *Drupella* spp. have been significant on a few reefs, their abundances and influences on most reefs in the Pacific are minor. However, the capability of forming outbreak populations and subsequently altering the community structure of the reef should make studies of the causes of population increases in *Drupella* spp. a priority.

Several species of echinoderms have been shown to feed primarily, or facultatively, on live corals. Birkeland (1989a) lists nine species of asteroids that prey on corals in the Pacific. With one exception *(Acanthaster planci),* these species are thought to have only minor effects on coral populations. For example, *Culcita novaeguineae,* because of its smaller extruded stomach area (compared to *A. planci*), preys mainly on smaller colonies of acroporid and pocilloporid corals and has only minor effects on reef community structure (Glynn and Krupp, 1986).

Most echinoids associated with coral reefs are herbivorous or omnivorous (Lawrence, 1975). However, several echinoid species have been found to feed, at least occassionally, on live corals. Herring (1972) reported that coral was an important food item for three species of diadematid sea urchins in Zanzibar (*Diadema setosum, Echinothrix calamaris,* and *Astropyga radiata*). Bak and van Eys (1975) reported that *Diadema antillarum* fed on live corals in the Netherlands Antilles. Carpenter (1981) found that predation on corals by *D. antillarum* varied seasonally on reefs in St. Croix, U.S. Virgin Islands, suggesting that fluctuating algal resources probably influenced the choice of prey for this echinoid species. The direct effects of coral predation by *D. antillarum* on coral-reef community structure were not considered significant in either of these studies.

While the diversity of this minor group of corallivores is high, it appears that given their typical abundances and feeding preferences, predation on live corals by these species does not have a dramatic influence on coral-reef community structure. Only for *Drupella* spp. has it been demonstrated that under conditions of high population density they have a significant effect on coral abundance and predation results in a phase shift (sensu Done, 1992) from one community type to another.

9.1.2. Herbivores

The diversity and abundances of herbivorous invertebrates associated with coral reefs are impressive, but studies of the effects of invertebrate herbivores on coral-reef algal communities primarily have focused on a few groups. Because of their small size and sometimes cryptic behavior, very little is known about the effects of feeding by species in several groups, including annelids, molluscs, and arthropods.

Polychaetes in the families Eunicidae and Syllidae are the most abundant herbivorous annelids associated with coral reefs (Fauchald and Jumars, 1979). Kohn and White (1977) reported densities of herbivorous polychaetes that exceeded 40,000 ind./m^2 on algal-turf-covered substrata in Guam. Carpenter (1986a) found densities of syllids associated with algal turfs to be positively related to algal turf biomass, and the highest densities (2,500 ind./m^2) were found in treatments protected from fish predators. Their overall body size and mouthpart size suggest that polychaetes are capable of ingesting only diatoms, cyanobacteria, and the smallest algal filaments, and although numerous, polychaetes do not have a significant effect on reef algal communities (Steneck, 1988). This is supported further by the positive relationship between algal abundance and polychaete density (Carpenter, 1986a) and suggests that algal turfs probably are a refuge for polychaetes from fish predators.

Several species of herbivorous molluscs are associated with coral reefs (Steneck and Watling, 1982; Steneck, 1983b). Limpets (e.g., *Acmaea, Fissurella*) are among the most common molluscan herbivores on shallow reefs where they maintain feeding ranges on algal-covered substrata. Depending on the species, limpets feed primarily on small algal filaments and crustose coralline algae within a very limited home range (Steneck and Watling, 1982). Repeated grazing within the feeding range results in the predominance of filamentous turf algae and crustose corallines that are able to persist under the intense grazing regime. Although capable of shifting algal community structure on a very limited spatial scale (centimeters), limpets are never abundant enough on most coral reefs to have a community-wide effect on species composition or affect the balance between the abundance of corals and algae. Similarly, other herbivorous gastropods and chitons (Glynn, 1970), while associated with coral reefs, have not been demonstrated to exert control of coral-reef community structure.

Arthropod mesograzers, such as amphipods, isopods, tanaids, and the larger majid crabs, commonly are found in coral-reef habitats. Amphipods, isopods, and majid crabs are primarily herbivores, while tanaids are most often detritivores. Klumpp et al. (1988) found densities of amphipods associated with algal communities on the Great Barrier Reef to range from 491 to 22,968 ind./m^2, while isopods in the same areas reached densities of 386 to 3,840 ind./m^2. Carpenter (1986a) reported densities of amphipods associated with reef algal turfs up to 1,406 ind./m^2 and densities of tanaids up to 6,875 ind./m^2. In both of these studies, higher abundances of mesograzers were associated with higher biomass (thicker) algal turfs.

As a result of their size and handling capabilities, many amphipods are restricted to feeding on microscopic unicellular and filamentous algae while isopods are capable of feeding on a range of algal types (Brawley, 1992). The effects of grazing by these mesoherbivores suggest that they often can control the abundance of epiphytic algae on larger macroalgae. In a coral-reef microcosm, Brawley and Adey (1981) found that grazing by the amphipod *Ampithoe ramondi* controlled algal community structure. When amphipods were absent, the algal community was dominated by several species of small filamentous algae. When grazing by amphipods occurred, the abundance of filamentous algae declined and a macroalgal species *(Hypnea)* persisted. Similarly, field experiments in Hawaii by Brostoff (1988) demonstrated that amphipod grazing reduced the abundance of filamentous algal epiphytes on two species of macroalgae.

Both of these studies were conducted under conditions where mesograzers were not subjected to normal predation. These results suggest that when amphipods are abundant, they may play an important role in reducing epiphytes on macroalgae. It is unclear the degree to which predators control the abundance of these mesograzers on most reefs. Data suggest that algal turfs on open-reef surfaces grazed by macroherbivores do not support large populations of mesograzers (Carpenter, 1986a). However, adjacent crevices and holes in the reef may provide refugia from predators, and the extent of nocturnal grazing by mesoherbivores may be underestimated.

Larger mesoherbivores, such as majid crabs, are capable of feeding on a variety of algal species (Coen, 1988a) and may limit the abundance of algae near their protective crevices. Coen (1988b) found in Carrie Bow Cay, Belize, that *Mithrax sculptus* was associated with 28% of the *Porites porites* colonies examined in a shallow-reef environment. Where *Mithrax* abundances were manipulated experiments demonstrated that crabs removed most of the algal epibionts from the coral colonies, perhaps preventing overgrowth of the coral. Their cryptic diurnal behavior suggests that the abundance of majid crabs such as *Mithrax* is probably controlled by predation. While it is clear that crabs can influence algal community structure over limited spatial scales, their ability to affect algal species composition and/or coral abundance on a community-wide scale has yet to be demonstrated.

Several echinoid species that are primarily herbivorous are associated with some coral reefs. The diadematid *Astropyga radiata* is found most often in sandy and seagrass habitats surrounding reefs, where it grazes seagrasses as well as attached algae. *Astropyga* is usually rare or absent on most reefs and has not been reported to have significant effects on the coral-reef community.

An echinometrid echinoid *Echinostrephus* sp. is often found inhabiting holes and crevices on Pacific reefs (Birkeland, 1989a). These sea urchins are sedentary, capturing drift algae with their spines and pedicellaria, and do not appear to graze the substratum surrounding their protective holes. As a result of this sedentary behavior, even population densities of several ind./m^2 probably have little or no effect on the open-reef community.

Several sea urchins in the genera *Heterocentrotus* and *Colobocentrotus* are found on many Pacific coral reefs (Birkeland, 1989a). While species in both genera are herbivorous, their distributions on the reef generally are restricted to shallow subtidal and intertidal habitats, respectively, both in areas of extreme water motion. Although Ebert (1971) reported an abundance of *Heterocentrotus mammillatus* from the Kona Coast, Hawaii, these taxa do not appear to maintain high population densities on most Pacific reefs. Because of their patchy distributions and relatively sedentary nature, these sea urchins have a limited influence in reef environments.

In the Caribbean, *Echinometra lucunter* and *E. viridis* occur in varying densities in some coral-reef habitats (Lessios, 1995). The rock-boring sea urchin *Echinometra lucunter* is found most commonly in very shallow and intertidal portions of the reef and on algal ridges (Ogden, 1977). In these areas, population densities of 50–100 ind./m^2 are common. This species, like *Echinostrephus* spp., is a sedentary crevice dweller that relies primarily on drift algae for food, although it does graze within its burrows, resulting in bioerosion of the substratum and enlargment of the crevice as the sea urchin grows. Therefore, the effect of grazing by *E. lucunter* is very localized, and its effect on the reef community is limited to those areas where it maintains high population densities.

The abundance of *Echinometra viridis* on reefs varies from site to site. In backreef habitats at Discovery Bay, Jamaica, Williams (1981) reported population densities of 10–15 ind./m^2. Lessios (1995) quantified the abundances of *E. viridis* on reefs in Panama for over 10 years and reported population densities of <1–4 ind./m^2. Population densities of *E. viridis* in forereef habitats in St. Croix, U.S. Virgin Islands, have been <1 ind.shm^2 (Carpenter, unpublished data). *Echinometra viridis,* although fairly sedentary, has been reported to compete for algal resources with *Diadema antillarum* in Jamaica (Williams, 1981). Because it is not abundant at other reef locations, *E. viridis* does not appear to have a strong influence on reef community structure.

Other echinoids occasionally found on coral reefs include *Eucidaris tribuloides, Lytechinus variegata,* and *L. williamsii* in the Caribbean. The former, although occasionally ingesting algae, feeds primarily on encrusting animals (Lawrence, 1975), and *L. variegata* is found commonly in seagrass beds and is rare on most coral reefs (Ogden and Lobel, 1978). *Lytechinus williamsii* is rare on most reefs.

Tripneustes spp. can be found on both Caribbean and Pacific reefs (Birkeland, 1989a). Although Ogden (1976) reported that *T. ventricosus* increased in abundance and cropped algae from the reef following the removal of *Diadema antillarum* from a patch reef, this pattern has not been repeated following the mass mortality of *D. antillarum* (Hughes et al., 1987; Lessios, 1995). *Tripneustes ventricosus* is found most commonly in seagrass habitats. Similarly, in the Pacific, the abundance of *T. gratilla* is patchy and this species does not appear to have any community-level effects on coral reefs.

9.2. Invertebrates Having Major Effects

9.2.1. Corallivores

Eucidaris thouarsii

In most coral-reef environments, cidaroid sea urchins are a minor component of the predator and grazer guild. They feed on a variety of encrusting animals and algae (Lawrence, 1975; Birkeland, 1989a). In the Galápagos Islands, the cidaroid *Eucidaris thouarsii* (Fig. 9–1) has played a much more important role in the coral-reef community. On the eastern Pacific coral reefs of Panama and Ecuador, population densities of adult *E. thouarsii* ranged between 1 and 5 ind./m^2, although abundances of juveniles occasionally exceeded 50 ind./m^2. Adult sea urchins in these localities were relatively small, with maximum test diameters between 2.7 and 3.5 cm (Glynn et al., 1979). Similar to cidaroids elsewhere, *E. thouarsii* in these coastal habitats are relatively sedentary, remaining cryptic during the day and foraging nocturnally. Adult populations of *E. thouarsii* in Panama and Ecuador appear to be kept in check by the abundance of echinoid predators, particularly fishes in the families Balistidae, Labridae, and Tetradontidae (Glynn et al., 1979; Glynn and Wellington, 1983).

In contrast, populations of *Eucidaris thouarsii* around several of the Galápagos islands (Fig. 9–1) were greater with adult densities between 10 and 50 ind./m^2,

Figure 9–1. *Eucidaris thouarsii* grazing on *Pocillopora* corals on Isla Onslow, Galápagos.

and these populations consisted of much larger individuals with maximum test diameters ranging from 4.3 to 6.2 cm (Glynn et al., 1979). Additionally, *E. thouarsii* in the Galápagos were neither sedentary nor cryptic during the day, but foraged over relatively large distances (1–3 m day) during both day and night. The main prey of *E. thouarsii* on Galápagos reefs also differed from mainland sea urchins. In addition to feeding on crustose coralline algae, a common prey item for mainland populations, sea urchins in the Galápagos preyed heavily on live corals. The primary prey were *Pocillopora damicornis, P. capitata,* and *P. elegans,* although sea urchins also were observed to feed on *Pavona clavus* (Glynn et al., 1979). Feeding surveys indicated that on some reefs over 50% of the sea urchins were feeding on live coral with the remainder feeding on crustose coralline algae covering the dead coral skeletons. Most of the coral colonies were fed on from the tips of branches, as sea urchins usually did not have access to the interior portions of the colony (Fig. 9–1).

The effects of predation on live coral by *Eucidarus thouarsii* had significant effects on the community structure and function of the reef (Glynn and Wellington, 1983). As bare space was opened by the removal of coral tissue, other organisms settled on the dead coral branches. The most common organisms to cover the branch tips were crustose coralline algae and barnacles *(Megabalanus galapaganus).* Crustose corallines often are able to persist despite intense repeated grazing by sea urchins because of their crustose morphology and calcification that lessens the amount of biomass removed by grazers (Steneck and Watling, 1982; Steneck, 1988). As a result of sea-urchin predation on live corals, the percent cover of live coral was reduced while the abundance of crustose coralline algae and barnacles increased. Glynn et al. (1979) found that grazing by *E. thouarsii* resulted in significant bioerosion of the coral framework (Chapter 4) and limited both the lateral expansion and vertical accretion of the reef framework.

The predation on live corals in the Galápagos by *Eucidaris thouarsii* was clearly a case of a predator having significant effects on the coral-reef community that resulted in a change in the long-term trajectory of the community (Glynn, 1990a). Increased abundance and size of *E. thouarsii,* combined with a behavioral shift leading to increased foraging time and a change in diet, led to a change in the community from one dominated by a high cover of live coral that accreted vertically to a community with reduced coral cover, increased abundances of algae and other organisms, and a reef framework that accreted and expanded more slowly.

The central question is, Why were sea-urchin populations higher in the Galápagos? It appears that high population densities of *E. thouarsii* in the Galápagos were not a recent phenomenon but had persisted for decades. Glynn et al. (1979) indicated that sea-urchin predators were more abundant on coral reefs in Panama and Ecuador than in the Galápagos. Although the same species of predators were present in all localities, predation pressure was reduced in the Galápagos because of the lowered abundance of predators. Glynn et al. (1979) speculated that sea-urchin predators themselves may be subject to higher predation in the Galápagos from higher-level predators (sharks).

The following scenario is suggested. Intense predation on sea urchins associated with mainland reefs resulted in not only lower population densities of *E. thouarsii,* but also smaller-sized sea urchins, diurnal crypsis, and reduced foraging movements. Under these conditions, prey for the sea urchins was abundant and they fed on a variety of encrusting organisms. Under conditions of reduced predation on sea urchins in the Galápagos, sea-urchin population sizes increased and selection for diurnal crypsis was reduced, allowing increased time for foraging over larger areas. Sea urchins switched to preying on live corals because of increased competition for prey as high-density populations depleted resources. Sea-urchin sizes were greater in these populations because mortality was reduced, or alternatively, sea urchins grew faster when feeding on corals.

However, several questions remain. Why have populations of *Eucidaris thouarsii* exhibited this pattern only on the coral reefs of the Galápagos and not at any other known coral reefs where sea-urchin predators are also rare? Was the reduction in predation pressure on sea urchins in the Galápagos a natural phenomenon or human induced? Or, did the increase in sea-urchin populations result, in part, from a stochastic event leading to high recruitment of juvenile sea urchins that swamped their predators and allowed them to reach a refuge in body-size? How long had the pattern of high sea-urchin densities and high rates of coral predation persisted? Unfortunately, continued study of these questions was prevented by the mass mortality of corals in the Galápagos during the El Niño event in 1982–1983.

Acanthaster planci

One of the most influential invertebrate predators on Pacific coral reefs is the crown-of-thorns seastar, *Acanthaster planci* (Fig. 9–2). A member of the Acanthasteridae, *A. planci* is distributed from the tropical Indian and western Pacific oceans east to Panama and portions of the Gulf of California (Birkeland and Lucas, 1990). Madsen (1955) proposed that *Acanthaster* in the eastern Pacific was a different species *(A. ellisi),* but this has not been accepted generally, and Glynn (1976) has regarded *Acanthaster* in the eastern Pacific as *A. planci.* This conclusion has been supported further by genetic evidence (Nishida and Lucas, 1988). The local distribution of *A. planci* is primarily in protected areas of the reef, including lagoons and deeper portions of forereef slopes (Moran, 1986). Shallow, wave-exposed portions of reefs are avoided.

Acanthaster planci is a free-spawning, dioecious, sexually reproducing species that is not known to reproduce asexually through arm autonomy (Moran, 1986). Periods and duration of spawning are temperature dependent and occur primarily during summer. Females have been reported to produce from 12 to as many as 60 million eggs per spawning season. From larval culturing studies, it is estimated that the feeding planktonic larvae spend from 9 to 23 days in the plankton before

Figure 9–2. *Acanthaster planci* in Fagatele Bay National Marine Sanctuary, American Samoa.

they settle, although the timing is influenced by environmental conditions of temperature, salinity, and food availability (Lucas, 1982; Olson, 1987). At the end of the planktonic period, larvae become negatively buoyant prior to settlement, which may lead to higher recruitment of juvenile seastars on deeper portions of reefs (Moran, 1986). Recruitment onto reefs in Fiji has been shown to be highly variable both spatially and temporally (Zann et al., 1990).

The juvenile stages of *A. planci* usually are cryptic and feed primarily on crustose coralline algae. They grow to a size of approximately 1 cm in total diameter within 4–5 months after settlement (Yamaguchi, 1973). The rate of growth is exponential, and at an age of approximately 6 months, the diet of the juvenile seastars switches to live coral (Lucas, 1975; Moran, 1986). Birkeland (1989b) and others have pointed out that the initial rapid growth and dietary plasticity of juvenile *A. planci* may be important characteristics explaining their potential for rapid population increase. Juveniles continue to grow rapidly and attain sexual maturity at an age of approximately 2 years. Adult growth appears to be indeterminate and is influenced by diet and spawning activity (Moran, 1986).

Adult seastars normally range in size from 25 to 35 cm in total diameter and have from 7 to 21 arms that are covered with spines (Moran, 1986). The tissue covering the spines contains one or more compounds that are toxic, and as a result, the spines are thought to function in predator deterrence (Cameron, 1977). The calcareous skeleton in *A. planci* is relatively thin and gives the seastar a flattened, elastic, prehensile morphology. This morphology has been proposed as another trait that contributes to the effectiveness of *A. planci* as a predator on corals, as it allows feeding on branching corals that are unavailable to other asteroid corallivores (Birkeland, 1989b).

Acanthaster planci feeds primarily on corals but will feed on a variety of other benthic organisms, depending on availability (Moran, 1986). *A. planci* is termed a generalist coral predator by some (Birkeland and Lucas, 1990), whereas others describe it as a strict coral specialist (Cameron and Endean, 1982). It appears that *A. planci* is a generalist with distinct preferences for branching corals, especially the acroporids. Feeding preferences are determined by a variety of factors, including the relative availability of prey and the nutritional state of the seastar (Ormond et al., 1976; Moran, 1986). Although growth is most rapid when *A. planci* feeds regularly, individuals are able to live without feeding for considerable periods of time.

Acanthaster planci feeds by everting the stomach over the live coral tissue and secreting an enzyme (with protease activity) that breaks down the coral tissue; the products are then absorbed (Birkeland and Lucas, 1990). This process requires 4–6 hours. The area of feeding is defined by the size of the stomach and therefore the size of the seastar. Birkeland (1989b) has pointed out that the everted stomach area for *A. planci* is about 2.5 times that for a same-sized corallivorous seastar, *Culcita novaeguiniae,* and suggests that this might contribute to the greater effects of corallivory by *A. planci.* Once a coral is digested, the seastar moves on to the next branch or colony.

A characteristic of *Acanthaster planci* that allows it to encounter its prey effectively is the ability to move relatively rapidly across the reef. The smallest of juveniles have been observed to move one body length per minute (Yamaguchi, 1973), while larger juveniles can move up to 4 m in an hour (Moran, 1986). Movements of adult seastars of 10 m per day have been measured (Keesing and Lucas, 1992), and there is indirect evidence that *A. planci* can move large distances between reefs separated by unsuitable habitat (Moran, 1986).

There has been debate about the longevity of *Acanthaster planci,* with some arguing that they have characteristics of a short-lived species (Ebert, 1983), whereas other studies suggest a much longer life span (Cameron and Endean, 1982). The best estimates of longevity, based on laboratory studies, are that seastars live for approximately 8 years (Lucas, 1984), but longevity in the field is not known for most populations. Zann et al. (1990) found that the longevity of different cohorts of *A. planci* in Fiji varied between 2 and 8 years.

The role of predators in controlling populations of *Acanthaster planci* has been a point of discussion since the mid-1960s. It is known that at least 12 different species can prey on juvenile and/or adult *A. planci* (Moran, 1986; Endean and Cameron, 1990; Birkeland and Lucas, 1990). These include xanthid crabs, a shrimp, a polychaete, two species of gastropods, and several species of fishes. Although some species prey on the egg and larval stages of the seastar, they do not ingest them preferentially and many potential predators actively avoid them. Yamaguchi (1973) concluded that these stages might contain chemical compounds to deter predators (Birkeland and Lucas, 1990). The gastropod *Charonia tritonis* was proposed as a major predator on *A. planci* (Endean, 1977), but the effect of predation by this species on populations of *A. planci* has been questioned (Moran, 1986). If eggs and larvae are protected somewhat from predation, and the heavily spined adults can resist all but the most effective predators, it may be the juvenile stage of *A. planci* that is most vulnerable to predation. However, Moran (1986) concluded that there is little evidence that predation controls population sizes of *A. planci.*

Although quantitative data are few, "normal" abundances of *Acanthaster planci* are considered to be on the order of 1 ind./100 m^2 or <20 ind. km of reef, although as Moran (1986) points out, there are no adequate data to define the "normal" density of a population. What has fueled interest in *A. planci* has been a series of irruptions in the population sizes of this seastar since the mid-1950s (Potts, 1981). These outbreaks have occurred fairly synchronously across a large number of localities including reefs in the Ryukyu Islands, Micronesia, Hawaii, Fiji, Tahiti, Panama, and on some reefs of the Great Barrier Reef (GBR) in Australia. The extent of the population increases varied between areas and, as a result of a lack of standardized sampling of the abundance of *A. planci* and the patchy distribution of seastars, reports of *A. planci* abundance during outbreaks varied greatly. However, outbreaking populations of *A. planci* generally have abundances that are at least 400–600% higher than normal (Birkeland, 1982; Moran, 1986; Birkeland and Lucas, 1990).

Outbreaks of *A. planci* populations occurred in the Ryukyu Islands beginning in the mid-1950s and spread southward. They were reported in Okinawa in 1969. Outbreaks were reported on the reefs near Green Island on the GBR in 1962 and spread southward to the Swain Reefs by 1975. In Micronesia, an outbreak occurred on the reefs in Guam in 1968 and again in the late 1970s. In 1979, a second outbreak occurred at Green Island and again spread southward, affecting many of the reefs of the central GBR (Moran, 1986). These outbreaks have been classified as primary or secondary, based mainly on their timing and presumed connection to other outbreaks. The initial outbreaks in the Ryukyu Islands and at Green Island are thought to be primary outbreaks where populations of *A.*

planci increased dramatically. Further outbreaks to the south of these areas were probably secondary outbreaks, resulting from the transport of seastar larvae produced by the outbreaking populations to the north (Moran, 1986; Dight et al., 1990a).

For the GBR, this interpretation has been supported by data on the genetics of outbreaking populations (Benzie and Stoddart, 1992). Clearly, some outbreaks are harder to explain (e.g., Hawaii, Panama) and probably represent additional primary outbreaks. Other hard-to-explain patterns include why some reefs within the same region have been susceptible to repeated outbreaks (midshelf reefs of the GBR), while outbreaks have never been reported on nearby, outer-shelf reefs (Moran, 1986; but see Dight et al., 1990b).

The effects of the outbreaks of *Acanthaster planci* have been dramatic in many cases. In Panama, Glynn (1973) reported that *A. planci* removed only 15% of annual coral growth. In most outbreak areas, coral mortality ranged from less than 50% to nearly 100% (Birkeland and Lucas, 1990; Glynn, 1990b). Mortality often was patchy spatially because aggregations of seastars disproportionately affected deeper portions of reefs and reefs protected from wave action (Moran, 1986). Acroporid corals were fed on preferentially at most locations and other coral species often survived, at least as small remnants (Colgan, 1987; Done, 1987). This was true particularly for massive species, such as *Porites* spp. It was estimated that a single crown-of-thorns seastar could consume 5–6 m^2 of live coral per year, and at outbreak densities, this translated to as much as 6 km^2 of reef being consumed per year (Birkeland, 1989a). However, as Moran (1986) emphasizes, outbreaks were variable in their duration and severity such that some reefs were affected more significantly than others.

The immediate ecological effects of *Acanthaster planci* outbreaks on most reefs were that the community structure of reefs was modified because abundances of some coral species were reduced drastically, or they were removed from the reef community entirely, and significant amounts of substratum were made available for settlement of other organisms (Moran, 1986; Birkeland, 1989a; Endean and Cameron, 1990; Birkeland and Lucas, 1990; Glynn, 1990b). In the short term, algae colonized most of the newly available space following coral mortality, and on some reefs they were displaced later by sessile invertebrates such as sponges and soft corals (Birkeland and Lucas, 1990).

If the reef previously was composed largely of coral species fed on preferentially by *A. planci* (i.e., acroporids), mortality was often greater than 80% and the species composition of the community was changed dramatically (second-order effects). Some reefs dominated previously by *Acropora* spp. were devastated and the framework of the reef collapsed because of increased bioerosion (Moran, 1986; Chapter 4), modifying spatial heterogeneity of the reef and further affecting community composition and recovery (Sano et al., 1987). If coral mortality was patchy, because of either the previous presence of mixed coral species stands or patchy distributions of *A. planci,* reef communities were modified into mosaics of different successional stages (Fig. 9–3).

Third-order changes in coral-reef communities also occurred at some locations (Chapter 11). Increases in algal abundance led to increases in the population sizes of herbivorous fishes at some locations (Wass, 1987; Birkeland and Lucas, 1990), while the abundances of obligate coral-feeding fishes, such as chaetodontids, decreased (Endean and Stablum, 1973; Bouchon-Navaro et al., 1985; Williams, 1986). These examples illustrate how the influence of a single species can have effects that cascade throughout the coral-reef community.

Longer-term changes in coral communities subjected to outbreaking populations of *Acanthaster planci* were related to patterns of coral recruitment and regeneration of coral fragments. Colgan (1987) identified five stages of recovery on corals reefs in Guam: (1) dominance of crustose and filamentous algae, (2) recruitment of coral planulae, (3) differential success of different coral growth forms, (4) coral colony expansion, and (5) competition between corals (Moran, 1986). The reported time frame of recovery of affected reefs varied between sites and according to the interpretation of the term *recovery.* Relevant components of the coral-reef community to consider when evaluating the degree of reef recovery were the percent cover of live coral, coral species composition, and the community structure of the noncoral-covered benthos (Done, 1985, 1987, 1988; Done et al., 1991).

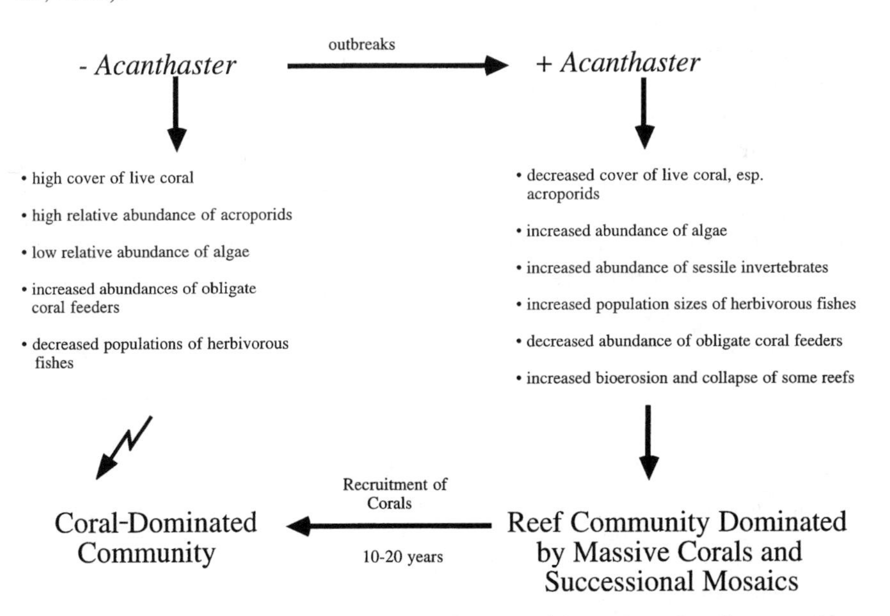

Figure 9–3. Diagrammatic representation of the transitions of coral-reef communities resulting from the presence and absence of the seastar *Acanthaster planci.*

Several investigators found that affected reefs recovered in terms of the percent cover of live coral within a 10–15-year period (Pearson, 1981; Colgan, 1982; Done, 1985). The time required for coral species diversity to approach previous levels usually was longer and depended in part on previous species diversity of corals, which coral species were killed, and the extent of coral mortality. Cameron et al. (1991) reported long-term changes in the abundance of massive corals on reefs of the GBR, with the massive corals on outbreak reefs dominated by poritids while the massive corals on nonoutbreak reefs in the same region were dominated by faviids. Differential mortality of corals because of feeding preferences or distributions of *A. planci,* together with differential recruitment of corals following the mortalities, gave rise to different trajectories of coral-reef community development following outbreaks.

Done et al. (1991) compared reef recovery from *A. planci* predation between and among reefs in French Polynesia and the GBR. They found evidence of significant changes on reefs that had been colonized by macroalgae such as *Turbinaria ornata* and *Sargassum* sp., suggesting much slower recovery rates on these reefs or the possibility that these were new alternate community states (Hatcher, 1984; Knowlton, 1992). Done et al. (1991) concluded that recovery of reefs appeared to be governed by local rather than regional processes. As Moran (1986) states, the paucity of data available for many sites prior to seastar outbreaks has made generalizations about reef community recovery difficult.

A key issue underlying the *Acanthaster planci* phenomenon is the cause of the outbreaks. Moran (1986) reviewed six hypotheses that have been proposed to explain outbreaks of *A. planci* (Table 9–2). The hypotheses can be categorized into two groups: hypotheses that view outbreaks as natural phenomena that have occurred repeatedly in the past, and hypotheses that are grounded in the premise that outbreaks are recent and novel events. Support for the view that outbreaks of *A. planci* have occurred previous to the 1950s comes from several sources. Birkeland and Randall (1979) and Birkeland (1981) concluded that outbreaks had occurred earlier in this century, based on interviews with fishermen in Samoa

Table 9-2. Hypotheses Proposed to Explain the Outbreaks of Populations of the Seastar Acanthaster planci

Outbreaks as Natural Phenomena	
Adult Aggregation Hypothesis	Dana et al., 1972
Larval Recruitment Hypothesis	Lucas, 1972
Terrestrial Runoff Hypothesis	Birkeland, 1982
Outbreaks as Unnatural Phenomena	
Reef Destruction Hypothesis	Chesher, 1969
Pollution Hypothesis	Fisher, 1969 Randall, 1972
Predator Removal Hypothesis	Endean, 1969, 1977

who remembered extremely high densities of *A. planci,* and based on the fact that the seastar is included in several cultures of human populations in Micronesia. Birkeland and Lucas (1990) also reported extensive documentation of outbreaks in Japanese waters by those using *A. planci* commercially as a source of fertilizer.

A second line of evidence for outbreaks in the past comes from geological evidence that skeletal elements of *Acanthaster planci* are found in abundance periodically in reef sediments dated over the past 7,000 years (Frankel, 1977, 1978; Walbran et al., 1989; Henderson, 1992; Henderson and Walbran, 1992). These authors suggest that predation by *A. planci* has been a recurring influence on the development of reefs on the GBR. Other reef scientists have questioned this stratigraphic evidence because of assumptions about the dispersion of skeletal ossicles and dating methodology (Moran et al., 1986; Fabricius and Fabricius, 1992; Keesing et al., 1992). The issue of whether accumulated skeletal elements represent reliable evidence of past outbreaks of *A. planci* remains unresolved.

On other grounds, Chesher (1969) and Randall (1972) have argued that outbreaks could not have been regular occurrences on reefs in the past or these reefs would be composed entirely of corals that *Acanthaster planci* avoids, such as *Porites* spp. As Moran (1986) points out, however, recovery of acroporids can occur relatively rapidly (10–15 years) after an outbreak.

The hypotheses proposed under the premise that outbreaks are natural phenomena include the adult aggregation hypothesis forwarded by Dana et al. (1972), which states that *Acanthaster planci* are more common on reefs than once thought, but that the majority of the population lives in deeper portions of the reef where they have escaped previous attention. Following a disturbance such as a storm that causes coral mortality and a disruption in food availability, seastars aggregate in areas with healthy corals and an outbreak ensues. Although this hypothesis explains why all the seastars in an outbreak are of adult size, and is consistent with the correlation for some sites between large storms and subsequent outbreaks, it has been suggested that the amount of reef damage required to result in food limitation is much greater than that observed for most storms.

Reichelt et al. (1990a, b) and Bradbury (1991) have used data on *Acanthaster planci* outbreaks on the GBR with theoretical models of predator-prey interactions to conclude that outbreaks are the result of a typical predator-prey cycle that is driven by prey (coral) abundance and recruitment success and survival of the predator (seastar). When corals are reduced in abundance, reefs cannot support large populations of *A. planci* and the outbreak ceases until coral abundance increases again. They have used this reasoning to predict that another series of outbreaks should occur on the GBR in the late 1990s.

Lucas (1975) proposed that outbreaks of *A. planci* could occur if environmental conditions were such that larval survival and recruitment were higher than normal, arguing that for such a fecund species, a small-percentage increase in survival could lead to massive increases in recruitment. In larval culturing experiments, Lucas (1973) demonstrated that larval survival increased under conditions of low salinity and high temperature, and Pearson (1975) suggested that such conditions could occur off the Australian coast, particularly in areas near large rivers. This hypothesis is consistent with the pattern that the majority of outbreaks have occurred on reefs surrounding high islands or near continental landmasses where freshwater runoff occurs, at least seasonally.

In a similar vein, Birkeland (1982) proposed the terrestrial runoff hypothesis stating that nutrients in freshwater runoff, especially from large storms, cause a phytoplankton bloom that increases the food supply for *Acanthaster planci* larvae. Increased survival results in increased recruitment of juveniles and a subsequent outbreak. Like the previous hypothesis, the terrestrial runoff hypothesis explains why outbreaks have been most common near high islands and continents and is consistent with the correlation between large storms and subsequent outbreaks that exists for some sites. This hypothesis assumes that most *A. planci* larvae starve from food limitation under "normal" conditions. Using an in-situ culturing chamber, Olson (1987) tested this hypothesis and concluded that the development times of larvae raised under conditions simulating a phytoplankton bloom were less than those for larvae in ambient seawater, but that the difference was not great enough to explain the sudden increases in *A. planci* populations. Nevertheless, a combination of temperature, salinity, and food conditions might lead to increased larval survival and recruitment and explain primary outbreaks in some reef locations (Moran, 1986).

Observations that outbreaks of *Acanthaster planci* occurred near centers of human populations led several investigators to hypothesize that outbreaks were not natural events, but instead were human-induced phenomena. Both Fisher (1969) and Randall (1972) proposed that increased pollutant concentrations in nearshore waters resulting from human activities reduced the abundances of predators of *A. planci,* allowing a population irruption. However, no studies in reef environments have shown that pesticides or other pollutants are in significantly higher concentrations in animals near human populations or that predation controls population sizes of *A. planci.* Chesher (1969) proposed that destruction of reefs as a result of human activity, including dredging, blasting, and poor land use practices leading to increased siltation, caused extensive mortality of corals. He presumed that corals were the primary predators of *A. planci* larvae and decreasing coral abundance subsequently led to outbreaks of seastars. Endean (1977) concluded that this did not explain outbreaks in areas not subject to human activities and that, in general, this hypothesis was not supported.

Of the hypotheses proposing that outbreaks of *Acanthaster planci* are human-induced events, the predator removal hypothesis of Endean (1969) has received the most attention (although the prior two hypotheses relate to predator removal as well). This hypothesis states that seastar outbreaks are the result of the reduction of predators on juvenile and adult *A. planci* by humans. The initial hypothesis referred to the triton snail, *Charonia tritonis,* which has been collected heavily in many locations. It was known previously that *C. tritonis* was a predator of adult *A. planci,* but caging experiments to estimate the rate of predation by tritons on *A. planci* suggested tritons did not prefer *A. planci* if there was a choice, and only consumed, at most, one seastar per week (Potts, 1981). This suggested that it was unlikely that *C. tritonis* controlled the population sizes of *A. planci.* Endean (1977, 1982) later incorporated other potential predators into the hypothesis, including several species of fishes.

This hypothesis—that predators are a major factor controlling the populations of *A. planci*—has received limited support from the results of theoretical models that incorporate the role of predation on seastars in the dynamics of seastar-coral interactions (Bradbury, 1991). However, it is unlikely that a reduction in predator abundance could result in sudden increases in the population sizes of *A. planci* in selected years.

The current ideas on the causes of outbreaks of *Acanthaster planci* are that it is unlikely that a single cause underlies the outbreaks at all locations. None of the proposed hypotheses fully explain the variation observed between outbreak locations or in some cases the timing of outbreaks (i.e., why outbreaks have occurred nearly simultaneously in widely separated areas). Given the complexity of the physical and biological interactions between coral-reef organisms and their environment, it is more likely that outbreaks of *A. planci* and their spread to other reefs are caused by a complex suite of interacting processes that include some aspects of many of the proposed hypotheses. As Moran (1986) states, "It is worth pondering whether our understanding of the *Acanthaster* phenomenon is hamstrung because there is a tendency to rely on hypotheses which may provide simplistic answers to what may be a far more complex question. Perhaps the real answer may lie in a collage of the main hypotheses proposed earlier."

From a management perspective, the question of outbreak frequency is acute (Kenchington and Kelleher, 1992). Birkeland (1989b) and others have argued that outbreaks have become more frequent and *Acanthaster planci* are more chronically abundant in Guam, Fiji, the Ryukyu Islands, and on the GBR (Birkeland, 1982; Yamaguchi, 1986; Reichelt et al., 1990a; Zann et al., 1990; Birkeland and Lucas, 1990). For at least some sites, the answer to the question of increasing frequency appears to be a convincing affirmative. It is tempting to correlate the increase in frequency and duration of *A. planci* outbreaks with the continuing (and accelerating) development of coastal zones throughout the Indo-Pacific that lead inevitably to increased erosion, siltation, coastal eutrophication, and increased stress on coral-reef communities.

Since some of the hypotheses to explain the cause of *A. planci* outbreaks would predict increasing frequency of outbreaks under these conditions, resource managers are left to decide a course of action in the absence of clear-cut recommendations based on expert opinions (Kenchington and Kelleher, 1992). While the economic costs of resource protection through more prudent land use and development and conservative resource management practices would be significant, the potential ecological and long-term economic costs of inaction may be much higher. The *Acanthaster* problem is one that has prompted interdisciplinary scientific approaches and will require further innovative efforts to provide resource managers with the information necessary to make informed decisions about maintaining coral reefs as highly productive and diverse ecosystems with sustainable economic benefits (Chapter 17).

9.2.2. Herbivores

Echinometra mathaei

Another invertebrate species that has been demonstrated to affect some coral-reef communities disproportionately when it is abundant is the echinoid *Echinometra mathaei* (Dart, 1972; Muthiga and McClanahan, 1987; McClanahan and Muthiga, 1988). *Echinometra mathaei* burrows into rock and coral substrata and is distributed throughout the Indo-Pacific from eastern Africa and the Red Sea to Hawaii (Clark and Rowe, 1971). The habitats in which *E. mathaei* normally are found are shallow-reef crests and lagoonal coral heads, and when present in low densities, they inhabit holes and crevices in the substratum that are enlarged by erosion of the rock or coral by the spines and grazing activity (Dart, 1972). Sheltering behavior and defense of crevices against conspecifics by *E. mathaei* are thought to be the result of predation pressure from fishes. Under conditions of low population density, *E. mathaei* feed on encrusting and upright algae in and near their protective crevices. When conditions permit high population sizes, *E. mathaei* inhabit open areas of the reef, show less agonistic behavior toward conspecifics, and graze attached algae over wide expanses of the reef (McClanahan and Muthiga, 1988; McClanahan and Shafir, 1990).

In several locations on the reefs of Kenya, McClanahan and colleagues have identified a situation where the removal of echinoid predators (primarily balistids and wrasses) has led to increases in the population densities of *Echinometra mathaei* by 2–3 orders of magnitude (McClanahan and Shafir, 1990). Reef sites vary in the abundances of echinoid predators by a factor of 5, with high abundances in reef park areas protected from fishing, and lowered abundances on unprotected, fished reefs. As a result of reduced predation, population densities of *E. mathaei* reach 13 ind./m^2 in some areas, where they outcompete other echinoid species (McClanahan and Kurtis, 1991).

Elevated population sizes of *Echinometra mathaei* have both first- and second-order effects on the coral-reef community. First-order effects include lowered algal biomass, increased cover of algal turfs, lowered cover by live corals, increased bioerosion, and lowered spatial heterogeneity on heavily grazed reefs (McClanahan and Shafir, 1990). The overall species diversity of the benthic community is significantly lower in areas of high *E. mathaei* abundance. Second-order effects are increased cover by sponges on heavily grazed reefs and decreases in the population sizes of herbivorous fishes (McClanahan and Shafir, 1990; McClanahan, 1992).

These results suggest that when released from predation, populations of *Echinometra mathaei* increase dramatically and that the resulting increase in grazing pressure reduces live coral cover. It is unclear whether the reduction in coral cover is a direct result of predation by *E. mathaei* or the indirect result of preventing recruitment of corals because of intense, generalized grazing of the reef surface. When population densities are high, *E. mathaei* apparently is capable of outcompeting other herbivores, including other echinoid species and herbivorous fishes, through more efficient utilization of algal resources and exploitative competition (McClanahan, 1992). This provides another example of a species that, under conditions of reduced population regulation by predators, can mediate coral-reef community structure.

Diadema antillarum

The long-spined sea urchin *Diadema antillarum* (Fig. 9–4) is distributed from the Gulf of Mexico, throughout the Caribbean to Surinam in South America, northward to Bermuda and eastward to the Azores, Madeira, and Cape Verde Islands and the Gulf of Guinea in western Africa (Ogden and Carpenter, 1987). *D. antillarum* is a free-spawning echinoid with populations having a female-biased sex ratio of two to one (Lessios, 1988). Spawning occurs year-round in small aggregations, and males and females must be within approximately 20 cm to effect successful fertilization (Levitan, 1991). As a result, successful reproduction is population-size dependent.

Figure 9–4. *Diadema antillarum* foraging at night at St. John in the U.S. Virgin Islands.

Planktotrophic larvae are produced that are thought to spend at least one month in the plankton, and therefore dispersal potential is great. Longevity of larvae is dependent on food quality and abundance, and under some conditions, larvae can delay metamorphorsis for up to 90 days (Carpenter, unpublished data).

Juveniles recruit into small crevices in the reef and recruitment is dependent on algal biomass, which may be a cue for the presence of conspecifics (Bak, 1985). Juveniles grow rapidly (5 mm per month in test diameter) and attain a size of 2.5–3 cm in the first year (Randall et al., 1964; Lewis, 1966). *D. antillarum* exhibits early sexual maturation after approximately one year and continues to grow to a maximum size of greater than 10 cm. Longevity in the field is not known, but 4 years is probably a minimum estimate (Ogden and Carpenter, 1987).

The population densities of Diadema antillarum have varied widely between locations throughout the western Atlantic. In some locations, such as Barbados, Curaçao, Jamaica, the Virgin Islands, Cozumel, Belize, Puerto Rico, Grand Cayman, and the San Blas Islands in Panama, population densities of 3–73 ind./m^2 have been reported (Bauer, 1980). Population densities at other sites, such as Honduras, Bermuda, and some sites in the Bahamas and in Belize, have been much lower (<1 ind./m^2; Bauer, 1980). Hay (1984b) and Levitan (1992) have correlated elevated population densities of *D. antillarum* with the overfishing of sea-urchin predators (see below) and competitors (herbivorous fishes) in locations near large human populations. While human activities appear to have influenced population sizes of *D. antillarum* at some locations in the past, other processes such as recruitment limitation may be important in determining population densities on a local scale. Presently, population densities of *D. antillarum* are reduced drastically throughout the western Atlantic because of the mass mortality event in 1983–1984 (see below).

In many reef environments, *Diadema antillarum* inhabits holes and crevices diurnally, emerging at dusk to forage all night (Fig. 9–4) and return to their protective crevices in the morning. Individuals have been demonstrated to exhibit crevice fidelity for extended periods of time and crevice fidelity is negatively correlated with sea-urchin population densities and positively correlated with predator abundances (Carpenter, 1984).

Diadema antillarum is an herbivore primarily, preferring to feed on algal turfs, but will prey on other items such as live coral (Bak and van Eys, 1975; Carpenter, 1981). Under food-limiting conditions, *D. antillarum* can increase the size of certain elements of Aristole's lantern, thereby increasing feeding efficiency (Ebert, 1980; Levitan, 1991). This may be one characteristic underlying the success of this species since it can optimally allocate resources to feeding and reproduction and maximize reproductive output over time. Dietary and morphological plasticity may be important characteristics underlying the widespread occurrence and effects of *D. antillarum* on coral-reef communities.

The main predators of *Diadema antillarum* are balistid, sparid, large labrid, and batrachoidid fishes (Ogden and Carpenter, 1987). Based on gut contents,

Randall (1967) found that the queen triggerfish, *Balistes vetula,* was the main predator on *D. antillarum* in the U.S. Virgin Islands. The king helmet snail *Cassis tuberosa* is also known to be a predator (Randall et al., 1964).

Many studies have been conducted to investigate the effects of grazing by *Diadema antillarum* on the coral-reef community. The direct effects of grazing by *D. antillarum* are to reduce algal biomass, shift algal community structure and diversity, and increase rates of primary productivity of algal turf communities (Ogden et al., 1973; Sammarco et al., 1974; Carpenter, 1981, 1986a; Sammarco, 1982a). In a sea-urchin removal experiment, Ogden et al. (1973) and Sammarco et al. (1974) demonstrated that in the absence of grazing by *D. antillarum* the algal community on a patch reef became dominated by macroalgae and that halos into the seagrass bed surrounding the reef disappeared. Results from manipulative experiments with algal turf substrata exposed to different combinations of reef herbivores indicated that grazing by *D. antillarum* reduced algal biomass and increased rates of biomass-specific primary productivity of algal turfs. These experiments indicated further that *D. antillarum* removes an average of 97% of algal turf biomass produced over a year (Carpenter, 1986a).

The effects of grazing by *Diadema antillarum* are not confined to the algal community. Sammarco (1980) found that grazing by *D. antillarum* influenced the survivorship and abundance of newly settled corals as a result of both the direct effects of grazing on settled coral spat and indirect effects of modifying competitive interactions between algae and corals. At high population densities of *D. antillarum,* intense grazing led to the mortality of juvenile corals, while in the absence of sea-urchin grazing, corals were overgrown by algae. As a result, at intermediate population densities of sea urchins, grazing was a regulating force controlling the relative abundances of corals and algae on the reef.

In addition to effects on sessile organisms, *Diadema antillarum* may affect the abundances of other herbivorous species, presumably as a result of exploitative competition for limited algal resources. Results of field experiments in Jamaica suggested that *D. antillarum* outcompeted another echinoid, *Echinometra viridis,* in areas not dominated by a territorial damselfish (Williams, 1981). Abundances of *E. viridis* increased when the damselfish excluded *D. antillarum.* Hay and Taylor (1985) demonstrated that the abundances of herbivorous fishes increased in reef areas from which *D. antillarum* had been removed. Because of the short-term nature of these experiments, they concluded that fishes from neighboring areas of the reef migrated into the sea-urchin removal areas to exploit increased algal resources, and that this was evidence for prior exploitative competition between the herbivore groups.

Prior to 1983, much of what was known about the effects of *Diadema antillarum* on components of coral-reef communities was based on manipulative field experiments. In 1983–1984, a natural experiment (unreplicated) occurred throughout the range of *D. antillarum* in the western Atlantic. Beginning in Panama in January 1983, a mass mortality of *D. antillarum* occurred throughout the Caribbean, Gulf of Mexico, Bahamas, and Bermuda (Lessios et al., 1984). Averaged over all sites where data were available, 93% of the individuals died and in many locations 100% mortality was common (Lessios, 1988). The only known populations that were not affected by the mass mortality were in the eastern Atlantic. No other species were affected. The cause of the mortality remains unknown, but given the direction and timing of the spread of mortality between populations, a species-specific waterborne pathogen is suspected (Lessios et al., 1984; for a review, see Lessios, 1988).

The mass mortality of *Diadema antillarum* provided a unique opportunity to evaluate the importance of a single species to the coral-reef community and allowed the testing of hypotheses regarding the effects of grazing on benthic community structure and function, as well as hypotheses about the competitive relationships between *D. antillarum* and other species. The effects of sea-urchin removal were immediate and substantial. In St. Croix, algal biomass increased by 27% five days after the mortality and continued to increase to 400–500% of premortality levels (Carpenter, 1985, 1990a). A similar pattern was observed in Jamaica, where algal cover increased from 31% to 50% within 2 weeks and increased further to 65% cover after one year (Liddell and Ohlhorst, 1986). Algal biomass also increased on the reefs of Curaçao (de Ruyter van Steveninck and Bak, 1986) and slight increases were observed on some reefs in Panama (Lessios, 1988).

The species composition of the algal community changed on many affected reefs from algal turfs comprised of many filamentous species (Hackney et al., 1989) and crustose corallines, to a community dominated by macroalgal species such as *Lobophora variegata, Sargassum* spp., and *Turbinaria turbinata* (Lessios, 1988). The increases in algal abundance occurred at the expense of other reef organisms since many of the macroalgal species are capable of abrasion and overgrowth of other benthic components (Fig. 9–5). Cover of crustose corallines, clionid sponges, live coral, zoanthids, and encrusting gorgonians decreased at several reef locations (Lessios, 1988).

Consistent with previous experimental results, rates of algal primary productivity decreased immediately following the mass mortality and increased thereafter as algal biomass increased (Carpenter, 1985, 1990a). At some sites, the longer-term increased rates of algal biomass production, rather than entering the trophic web via grazing, resulted in increased algal standing crop and export of algae from the reef community as detritus (Carpenter, 1990a).

Increased grazing by herbivorous fishes (mostly acanthurids and scarids), as measured by the number of fish bites, was observed at one reef location, suggesting that the increase in algae following the removal of *Diadema antillarum* resulted in an immediate functional response by herbivorous fishes (Carpenter, 1985). Similarly, increased fish grazing was observed on reefs in Jamaica (Morrison, 1984). These observations support the hypothesis that sea urchins and herbivorous fishes were competing prior to the mass mortality. However, increased grazing

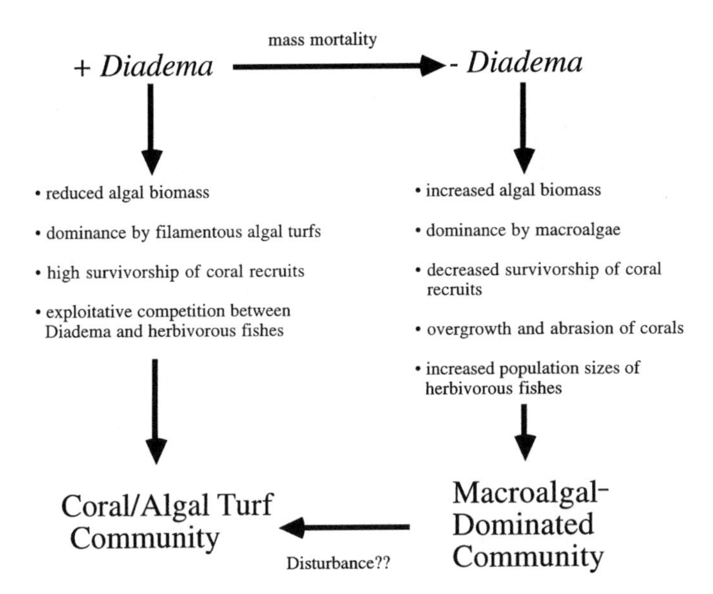

Figure 9–5. Diagrammatic representation of the transitions of the coral-reef community from one state to another as a result of the removal of *Diadema antillarum* by mass mortality. It is unclear if a transition from a macroalgal-dominated community to a coral/ algal turf community can occur or what events might facilitate this transition.

by herbivorous fishes was not sufficient to compensate for the removal of sea urchins, and algal biomass continued to increase and species composition changed until many reef areas were dominated by macroalgal species (Lessios, 1988).

The longer-term effects (several years) of the removal of *Diadema antillarum* on many coral-reef communities were a reduction in the cover of live coral and other sessile invertebrates and a shift to a reef community dominated by macroalgae (Hughes, 1994). The direction of the changes in reef communities throughout the western Atlantic following the mass mortality was consistent. However, the magnitudes of the changes were correlated with the previous population densities of *D. antillarum* (Lessios, 1988). Changes were minimal in reef community structure at reef locations where sea urchins were rare, lending support to the hypothesis that changes in community structure elsewhere were the direct result of the removal of *D. antillarum* and not other physical or biological factors.

Long-term increases in the population sizes of herbivorous fishes have been reported from two locations where data were available prior to the mass mortality. Both in St. Croix (Carpenter, 1990b) and Panama (Robertson, 1991), increased abundance of acanthurids and scarids resulted from the removal of a major competitor. It is not known if the increase in population sizes of primary consumers at these sites has led to an increase in secondary consumers. No changes have been observed in the populations of fishes that preyed previously on *D.*

antillarum, as they have switched to feeding on other prey species (Lessios, 1988, 1995).

Recruitment of *Diadema antillarum* following the mass mortality has been minimal or nonexistent on most reefs in the western Atlantic where data have been collected (Lessios, 1995). The only site where high recruitment of juveniles has been reported is Barbados (Hunte and Younglao, 1988). The most likely explanation of the lack of recruitment at most sites is the lack of larvae available to settle, which could be a function of decreased production of larvae or increased larval mortality in the plankton. Because there appears to be a minimum population size required to bring about successful fertilization in *D. antillarum* (Levitan, 1988), it is likely that the lack of recruitment is because of reproductive failure for most populations and a drastic reduction in the abundance of larvae produced (Lessios, 1995). It appears that the time required for recovery of sea-urchin populations at most locations is greater than 10 years. As populations recover, the reverse predictions of the effects of grazing by *D. antillarum* on coral-reef communities can be tested once again.

The abundance of diadematid echinoids associated with coral reefs varies greatly within the Pacific Ocean and between the Pacific and western Atlantic (Bauer, 1980; Birkeland, 1989a). Several explanations have been proposed for these patterns, including higher abundances of echinoid predators in the Pacific and the dependence of echinoid larvae on abundant phytoplankton for successful recruitment. The latter might explain the higher abundances of diadematids on inshore reefs of the GBR and the larger, high islands of the Pacific and their generally low abundances on offshore reefs of the GBR and other coral reefs in the Pacific.

An additional hypothesis that might explain the distribution and abundance of diadematids that has not received serious attention is that their abundance is related to the abundance of their prey, mainly algal turfs. In general, Pacific reefs have higher coverage of live corals than reefs in the western Atlantic. Similarly, inshore reefs and reefs surrounding high islands are subject to increased physical stresses and disturbances that decrease the cover of live coral, resulting in higher coverage of algae that can support a larger biomass of invertebrate herbivores. Recruitment of diadematids has been demonstrated to be greater under conditions where algal turfs are present (Bak, 1985), so that a combination of settlement cues and subsequent increased survivorship might explain the higher abundances of diadematids on these reefs.

9.3 Disproportionate Effects of Some Invertebrate Species

While the diversity and abundance of invertebrate predators and grazers associated with coral reefs is high, it is striking that a few species have such disproportionate effects on coral-reef community structure and function. These species are capable of changing the reef environment as a result of their direct effects and the cascade of indirect effects initiated by their activities (Chapter 11). The aggregate effect of this influence can be to shift the balance of processes, causing the reef to move from one state to another. As a result, these species are located ecologically in a position to mediate processes that involve the life and death of coral reefs. An important question is, Why do these species have the capability to alter significantly what has been regarded historically as a stable biological community?

Although the invertebrate species discussed previously that have disproportionate effects on coral reefs are all echinoderms, and therefore have some morphological similarities, other characteristics of these species must result in their important influence on reef communities. One attribute they share is the capability of rapidly attaining and maintaining large population sizes. *Acanthaster planci, Diadema antillarum, Eucidaris thouarsii,* and *Echinometra mathaei* all exhibit very high fecundity and the potential for producing extremely high numbers of larvae. Behaviors such as formation of spawning aggregations probably facilitate fertilization success and increase the number of larvae produced (Levitan, 1988; Karlson and Levitan, 1989). High recruitment into exisiting populations and establishment of new populations may be dependent on environmental conditions that promote phytoplankton abundance and contribute to increased larval survivorship, as proposed for both *A. planci* and diadematid echinoids (Birkeland, 1989a). Such conditions are more likely for reef habitats near continents and large islands where nutrient runoff is higher and often seasonal. This adds a stochastic factor that controls, in part, the population sizes of these species, and together with the longevity of adults influences whether population sizes oscillate or are more stable through time.

Acanthaster planci and *Diadema antillarum* both settle cryptically and exhibit very rapid growth of juveniles, minimizing predation on their early stages, attaining sexual maturity at an early age, and therefore contributing further to rapid increases in population size. However, much of the success of these species may be dependent on the high degree of both morphological and dietary plasticity that they exhibit (Birkeland, 1989a, b; Birkeland and Lucas, 1990). *A. planci* is unique among the seastars associated with coral reefs in having a reduced skeleton and a pliable body form that allows it access to a wide range of coral colony shapes. While it exhibits distinct preferences for some species of corals, under a variety of conditions it will readily eat other prey, effectively making it a generalist able to subsist in environments where the abundances of some prey species vary (Birkeland, 1989a).

As an early juvenile, *A. planci* feeds primarily on crustose coralline algae that are abundant in the microhabitats where its larvae settle. Later, the juvenile seastars switch to a diet of corals, which are more abundant and have higher nutritional quality. Similarly, *D. antillarum* will forage on a variety of prey, preferably filamentous algal turfs, but will feed on most algal species and even corals when conditions require it. Furthermore, *D. antillarum* allocates energy to growth of the test and feeding apparatus that allows it to forage effectively even when algal abundance is reduced (Ebert, 1980; Levitan, 1991). *Eucidaris thouarsii* and *Echinometra mathaei* also have general diets that include a variety of prey species.

On many coral reefs, predation does not appear to control the population sizes of *Acanthaster planci* and *Diadema antillarum.* This is a result of the rarity of predators in some locations, and morphologies and/or behaviors that reduce the effects of predation in general. Spines and/or toxins are used by these species to deter predators, while diel behaviors reduce the risk of predation. While a correlation exists between human fishing pressure, predator abundance, and *D. antillarum* abundance, this is confounded with environmental conditions that promote high recruitment of echinoid larvae, so a clear-cut conclusion that predation controls population sizes of *D. antillarum* is not possible. However, it appears that population sizes of *Echinometra mathaei* are controlled by predation in some locations (McClanahan and Muthiga, 1989). For *Acanthaster planci,* it is unlikely that a reduction in predator abundance could result in the rapid population irruptions that have been observed.

When population sizes of these important invertebrate species become large, the direct effects of their feeding are to reduce the abundance of their prey: corals in the case of *Acanthaster planci* and *Eucidaris thouarsii,* and algae in the case of *Diadema antillarum* and *Echinomentra mathaei.* Because these species are the main agents of biological disturbance in their respective coral-reef communities, the effects of their feeding are community-wide. Feeding activities by *E. thouarsii* and *E. mathaei* have effects that are important locally, but do not result in a new type of community. Drastic reduction in live coral cover by *A. planci* leads to changes in community structure as other sessile organisms occupy the opened space. Changes in benthic community structure lead to other second- and third-order effects.

However, high population densities of *A. planci* are not sustainable, and recovery of the reef community occurs over a time scale of 10–15 years as corals settle, grow, and outcompete other benthic organisms. Grazing by *D. antillarum* mediates competitive interactions between the coral and algal components of the reef, and the effects of grazing on the reef community are density dependent (Sammarco, 1980). When *D. antillarum* is extremely abundant, juvenile corals are grazed, live coral cover may be reduced, and increased bioerosion of the reef occurs (Chapter 4). However, the recovery to the original community state does not occur. In the absence of *D. antillarum,* algae outcompete corals and the reef is transformed into an algal-dominated community. It is unclear whether this transition is reversible over the short term or represents an alternate stable state (Knowlton, 1992).

While the vast majority of invertebrate predators and grazers have minor effects on coral communities, a few species exert important, community-wide effects on the reef. These species tend to be fecund, fast-growing generalists that exhibit both morphological and dietary plasticity, and their populations often are not controlled by predation. Their effects on the reef community are substantial and their presence or absence can result in transitions between different community types. As a result, these invertebrate species are involved in a complex of processes that determine the state of the coral reef and the trajectory of its development over time.

10

Effects of Reef Fishes on Corals and Algae

Mark A. Hixon

Fishes are among the most conspicuous and beautiful inhabitants of coral reefs. Their diversity is amazing. It has been estimated that nearly half of the 20,000–30,000 species of fishes worldwide inhabit such shallow tropical marine habitats (Cohen, 1970), and locally, hundreds of species can coexist on the same reef. For example, Smith and Tyler (1972) found 75 species occupying a 3-meter-diameter patch reef in the Caribbean, which is not a particularly speciose region compared to the Indo-Pacific (Chapter 14).

The variety of sizes and shapes of reef fishes is as remarkable as their species diversity. The smallest vertebrate is a goby less than 10 mm long that inhabits Indian Ocean reefs (Winterbottom and Emery, 1981), whereas at the other extreme, groupers, barracuda, and reef sharks can reach startling sizes. Fishes exploit virtually every conceivable microhabitat and food source on reefs, from incoming oceanic plankton, to a wide variety of benthic organisms, to other fishes. Moreover, they often occur in high-standing stocks, with about 2,000 kg ha^{-1} being the presumed maximum (Goldman and Talbot, 1976). Not surprisingly, reef fishes are an important food source for many tropical third-world nations (reviews by Russ, 1991; section 1.1 in Chapter 1).

It seems almost a foregone conclusion, then, to assert that fishes have strong effects on the dominant benthos of reefs: corals and macroalgae. In fact, herbivorous fishes do substantially affect the distribution and abundance of reef algae. Surprisingly, however, the evidence for major direct effects on corals is relatively scant. Nevertheless, the effects of herbivores, especially territorial damselfishes, can cascade through the system, indirectly affecting corals and a variety of other reef organisms. Moreover, there is substantial evidence that various fishes affect

---

I dedicate this chapter to my son, Sean Wolf, who was born the day I completed the manuscript. I thank the National Science Foundation and the University of Hawaii for supporting my research on herbivorous reef fishes.

the distribution and abundance of invertebrate corallivores and herbivores, thereby indirectly affecting corals and algae. This complex variety of direct and indirect effects has definite ramifications for understanding and managing reef systems (Chapter 11).

This chapter focuses on (1) the effects of herbivorous fishes on the distribution and abundance of reef algae, and indirectly, corals; (2) the effects of corallivorous fishes on the relative dominance of reef-building corals; and (3) the indirect effects of fishes consuming and competing with invertebrate herbivores and corallivores. This summary is by no means exhaustive; recent reviews detailing various effects of fishes on reef corals and algae include Hixon (1986), Hutchings (1986), Glynn (1988, 1990), Steneck (1988), Horn (1989), Hay (1991), and Jones et al. (1991).

The focus of this chapter is mostly on the one-way effects of fishes upon reefs, emphasizing the mechanisms and constraints under which fishes cause switches in the relative dominance of benthic organisms. However, it is important to realize that this limited perspective ignores most of the complex interactions between fishes and the reefs they inhabit. Indeed, the reciprocal effects of reefs upon fishes is a matter of life and death for many species; reef fishes are often obligatory denizens of this habitat and derive all their food and shelter from the reef. The demise of a reef certainly has repercussions for reef fishes. For example, Reese (1981) has proposed that obligate coral-feeding fishes can be used as bioindicators of the general health of a reef, an idea that has stirred considerable controversy (Bell et al., 1985; Bouchon-Navaro et al., 1985; Williams, 1986; Roberts et al., 1988; Sano et al., 1987; White, 1988a; Bouchon-Navaro and Bouchon, 1989). The link between fishes and corals has been further documented by Harmelin-Vivien (1989), who noted a significant linear relationship between the number of fish species and the number of coral species among reefs across the Indo-Pacific region,but no such relationship with the number of algal species.

Other potentially important interactions between fishes and reefs that will not be covered are assessment of the relative effects of fishes versus invertebrate herbivores (Hay, 1984; Carpenter, 1986a; Foster, 1987; Morrison, 1988; Klumpp and Pulfrich, 1989; see Chapter 9) and the role of fish feces fertilizing the reef (Meyer et al., 1983; Meyer and Schultz, 1985a, b; Polunin and Koike, 1987; Polunin, 1988; Harmelin-Vivien et al., 1992; Chapter 13). Finally, space limitations prevent summarizing the many fascinating and ecologically important interactions among fishes and the community structure of reef fishes per se. Fortunately, Peter Sale's (1991) edited volume on these topics is unparalleled and recommended for those desiring a detailed introduction to reef-fish ecology.

10.1. The Players: Corallivorous and Herbivorous Reef Fishes

Only a handful of families of fishes have been documented to have obvious direct effects on reef corals (Fig. 10–1). Although about 10 families of fishes

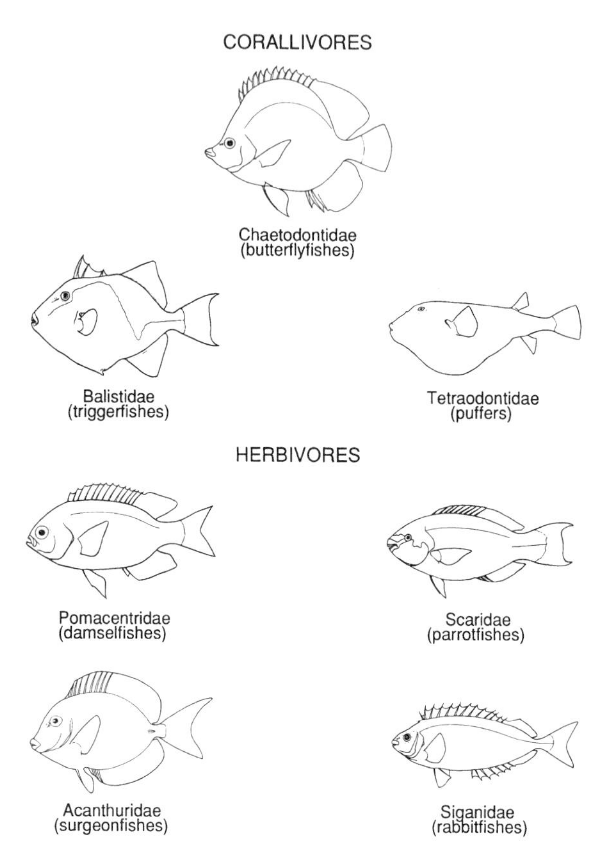

Figure 10–1. Families of larger-bodied reef fishes that include the most corallivorous and herbivorous species. Note that members of all the listed corallivore families include species that do not consume corals, and about half the species of pomacentrids are not herbivorous. Similarly, certain species in numerous other families consume corals and/or algae (drawings from Nelson, 1984).

are known to eat coral polyps, there are few groups that feed strictly on corals (reviews by Robertson, 1970; Randall, 1974). This pattern may be due to coral polyps being relatively unpalatable compared to other prey, in terms of their chemical composition, their protective nematocysts, and their calcium-carbonate skeletons. The predominant corallivores are butterflyfishes (family Chaetodontidae); about half of the over 100 species eat corals (Allen, 1981). The only other large fishes whose members regularly consume corals are some triggerfishes (Balistidae) and puffers (Tetraodontidae). Whereas the butterflyfishes delicately extract individual polyps from the coral skeleton, the triggerfishes and puffers tend to scrape and chew pieces they break off the coral colony with their beaklike mouths. Smaller corallivores include at least one goby (Gobiidae; Patton, 1974).

Among the coral eaters, the social systems of the butterflyfishes are best documented, often comprising territories defended by male-female pairs (Reese, 1975; Hourigan, 1989; Roberts and Ormond, 1992).

In contrast to the corallivores, herbivorous reef fishes are relatively speciose. As collated by Choat (1991) and Allen (1991), the predominant consumers of benthic algae are most of the approximately 75 species of surgeonfishes (Acanthuridae), all 27 species of rabbitfishes (Siganidae), all 79 species of parrotfishes (Scaridae), and over half of the approximately 320 species of damselfishes (Pomacentridae). Other herbivorous families include smaller fishes, such as the combtooth blennies (Blenniidae).

The feeding modes of these herbivores are highly variable (reviews by Ogden and Lobel, 1978; Horn, 1989; Glynn, 1990; Choat, 1991). Surgeonfishes and rabbitfishes tend to crop seaweeds in a browsing mode that leaves algal holdfasts intact. Parrotfishes, on the other hand, have highly modified jaws and teeth. With these beaks (which inspired their name) they scrape the bottom and remove algal holdfasts along with the dead coral substratum to which the algae attach. This activity makes parrotfishes the major source of bioerosion among reef fishes (reviews by Hutchings, 1986; Choat, 1991; Chapter 4). All three of these families exhibit variable social systems, from individual territories to transient foraging aggregations (e.g., Ogden and Buckman, 1973; Robertson et al., 1979; Robertson and Gaines, 1986).

Most herbivorous (actually, omnivorous) damselfishes maintain permanent individual territories, measuring about a square meter in area, which they defend vigorously against other herbivores (e.g., Low, 1971). This defense, combined with moderate browsing and even "weeding" behavior (sensu Lassuy, 1980), often maintains a distinctive mat of erect algae within the territory. By forming large schools, parrotfishes and surgeonfishes can sometimes overwhelm and denude damselfish territories (Jones, 1968; Barlow, 1974; Vine, 1974; Robertson et al., 1976; Foster, 1985; Reinthal and Lewis, 1986).

Overall, both corallivorous and herbivorous fishes display a wide variety of feeding modes and behaviors, suggesting that the ecological effects of these consumers are bound to vary widely from species to species and from reef to reef. What follows, then, are summaries of specific studies that can be generalized only with caution.

10.2. Fish Effects on Algae

10.2.1. Schooling Herbivores

Parrotfishes, surgeonfishes, and rabbitfishes often occur in dense aggregations that have obvious effects on reef macroalgae. Densities can average well over 10,000 herbivorous fish per hectare (review by Horn, 1989), standing stocks on unfished reefs in the Great Barrier Reef can reach 45 metric tons per km^2 (Williams and Hatcher, 1983), and secondary productivity can approach 3 metric tons per km^2 per year (review by Russ and St. John, 1988). In the Caribbean, parrotfishes can graze at rates of over 150,000 bites per m^2 per day (Carpenter, 1986a). In some systems, such intense grazing enhances local primary productivity by maintaining algae at an early successional stage (Montgomery, 1980; Birkeland et al., 1985; Carpenter, 1986a). Hatcher (1981) estimated that about half the net algal production on One Tree Reef, Australia, was consumed by fishes. At the same site, Hatcher and Larkum (1983) demonstrated that algal standing crops were controlled by grazing fishes all year (autumn and spring) on the reef slope (10 m depth), but only during spring in the lagoon (2 m depth). In autumn, inorganic nitrogen limited the standing crop of lagoon algae despite the continued presence of fishes (Chapter 7).

In addition to seasonal variations, an apparently general trend is that the spatial distribution of fish grazing varies inversely with tidal exposure and/or wave action (Van den Hoek et al., 1975, 1978) and directly with the availability of shelter for the herbivores from predatory fishes (Hay, 1981a; Lewis, 1986), with both turbulence and shelter decreasing with depth. Thus, as documented in Guam (Nelson and Tsutsui, 1982), the Caribbean (Hay et al., 1983; Lewis and Wainwright, 1985), and the Great Barrier Reef (Russ, 1984b), the depth distribution of herbivores and grazing intensity may often be unimodal: low in very shallow water due to limited accessibility by fishes, high at intermediate depths due to high accessibility and shelter, and low in deep reef areas (greater than about 10 m), where the abundance of coral shelter for fishes typically decreases. However, in areas where intense fishing has greatly reduced the abundance of piscivores, herbivorous fishes may be active at greater depths, with algal standing stocks consequently being lower than usual at those depths (Hay, 1984). The unimodal depth distribution of herbivorous fishes may explain the bimodal zonation of erect algal cover found on reefs such as those in Curaçao (Van den Hoek et al., 1978): high cover in the eulittoral zone (0–1 m depth), low on the upper reef slope (1–30 m), and high again on the lower slope (30–50 m).

The lack of shelter for grazing fishes probably also explains the existence of extensive algal plains occurring on sand bottoms below and between reefs, as well as high algal densities on very shallow reef flats lacking adequate shelter (Van den Hoek et al., 1978; Hay, 1981b). Overall, it appears that the risk of predation limits the grazing activities of smaller reef fishes to areas providing structural refuges (reviews by Hixon, 1991; Hixon and Beets, 1993).

At larger spatial scales, there is a trend for schooling herbivores to be more abundant on the outer Great Barrier Reef than inshore (Williams and Hatcher, 1983; Russ, 1984b). The mechanisms underlying this pattern appear to be related to between-region differences in the palatability and productivity of reef algae. (Chapter 12 gives a general review of regional variation in coral-reef processes.)

Field experiments pioneered by Stephenson and Searles (1960) and Randall

(1961), in which herbivorous fishes are excluded from reef plots by cages, have shown that these fishes strongly affect the species composition and relative abundances of algae. Typically, heavily grazed dead coral surfaces become dominated by grazer-resistant algal crusts or turfs, whereas caged but otherwise identical surfaces become covered by high-standing crops of erect algae (Vine, 1974; Wanders, 1977; Lassuy, 1980; Sammarco, 1983; Hixon and Brostoff, 1985; Carpenter, 1986b; Lewis, 1986; Morrison, 1988; Scott and Russ, 1987). Essentially, erect algae competitively exclude crusts in the absence of grazing, but crusts are more resistant to grazing (Littler et al., 1983; Steneck, 1983). Overall, the local species diversity of algae on exposed flat surfaces declines with increasing density of schooling herbivores (Day, 1977; Brock, 1979), an effect that is ameliorated on surfaces where algae can grow in crevices (Brock, 1979; Hixon and Brostoff, 1985; Hixon and Menge, 1991).

A yearlong experiment off Hawaii examined the benthic successional sequences and mechanisms that cause these general patterns (Hixon and Brostoff, 1996). Succession was followed on dead coral surfaces subjected to each of three grazing treatments: protected within grazer-exclusion cages, exposed to moderate grazing inside damselfish territories (see below), and exposed to intense parrotfish and surgeonfish grazing outside territories. The ungrazed successional sequence inside cages was an early assemblage of filamentous green and brown algae (including *Entermorpha* and *Ectocarpus*) replaced by a high-diversity assemblage of mostly red filaments (including *Centroceras* and *Ceramium*), which in turn was replaced by a low-diversity assemblage of mostly coarsely branched species (including *Hypnea* and *Tolypiocladia*).

Plotted in a multispecies ordination (detrended correspondence analysis), ungrazed succession followed a distinct left-to-right trajectory over the year (Fig. 10–2). Intense grazing by parrotfishes and surgeonfishes caused succession to follow a completely opposite path, where the early filaments were replaced immediately by grazer-resistant crustose species, including the red coralline *Hydrolithon* (Fig. 10–2). This result suggests that heavy grazing "derailed" the normal trajectory of succession (Hixon and Brostoff, 1996).

In summary, intense grazing by schooling herbivores strongly influences the standing crop, productivity, and community structure of reef algae. It also appears that selection for resistance to such grazing may compromise competitive ability among algal species (Littler and Littler, 1980; Hay, 1981b; Lewis, 1986; Morrison, 1988). Off the Caribbean coast of Panama, fishes may prevent competitively dominant (but highly palatable) sand-plain species from displacing competitively subordinate (but grazer-resistant) reef algae (Hay, 1981b; Hay et al., 1983). This dichotomy may act to maintain between-habitat diversity in algae (Hay, 1981b; see also Lewis, 1986). In any case, intense grazing appears to have selected for strong chemical defenses and morphological plasticity in some reef algae (reviews by Hay and Fenical, 1988; Steneck, 1988; Hay, 1991).

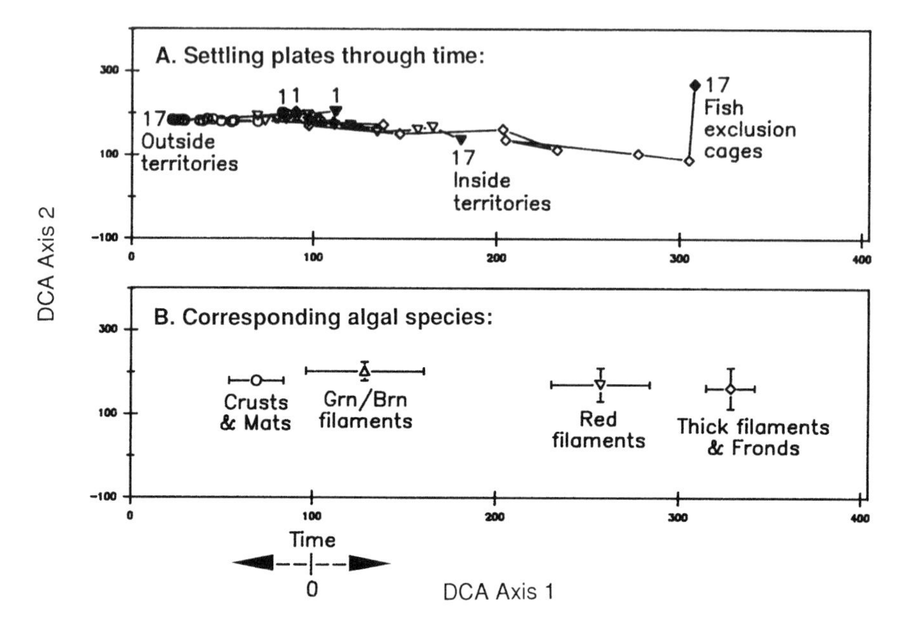

Figure 10–2. Effects of differential fish grazing on algal succession on a Hawaiian reef, illustrated by a multivariate ordination (detrended correspondence analysis, DCA) of algae growing on settling plates of different ages. (**A**) Each point represents the mean score values for a sample of about 21 settling plates; points closer together indicate more similar species compositions and relative abundances. There are 17 such samples for each of the three treatments indicated; sample 1 was after 7 days into the experiment (indicated as "time 0" along the x-axis), and sample 17 was at the end of one year. (**B**) Each point represents the mean (±SE) score values for groups of algal species corresponding to the samples in plot A. Thus, dominance during succession within cages proceeded from green and brown filaments to red filaments to thick filamentous and frondose species, whereas succession inside damselfish territories proceeded only to a mixture of red, green, and brown filaments. Succession outside territories proceeded from green and brown filaments directly to grazer-resistant crusts and mats (modified from Hixon and Brostoff, 1996).

10.2.2. Territorial Damselfishes

By defending small patches of dead coral, and grazing or "weeding" the algae in these patches in a particular way, territorial damselfishes often establish and maintain visually distinct mats of macroalgae on reefs (Vine, 1974; Brawley and Adey, 1977; Lassuy, 1980; Montgomery, 1980; Hixon and Brostoff, 1996). In general, these mats are sites of greater primary productivity than comparable areas outside territories (Montgomery, 1980; Russ, 1987; Klumpp et al., 1987). This production is an important food source for not only the resident damselfish, but also small invertebrate herbivores inhabiting the mat and larger intruding herbivores (Russ, 1987; Klumpp and Polunin, 1989).

Given that territory mats can cover well over 50% of shallow reef tracts (Sammarco and Williams, 1982; Klumpp et al., 1987), the local effects of damselfishes on the benthic community can be substantial. In particular, the defense, grazing, and weeding activities of these fish (possibly combined with localized fecal fertilization) strongly affect the local species diversity of reef algae. This effect has been demonstrated by three similar experiments in Guam (Lassuy, 1980), Hawaii (Hixon and Brostoff, 1983), and the Great Barrier Reef (Sammarco, 1983). Each experiment compared algal diversity on dead coral surfaces exposed to each of three different treatments: accessible to mostly damselfish grazing inside territories, accessible to intense grazing by other herbivores outside territories, and protected within fish-exclusion cages outside territories.

Although strict comparisons are precluded by differences in experimental design and laboratory analyses, some general patterns do emerge. For both damselfish species that Lassuy (1980) studied (*Stegastes lividus* and *Hemiglyphidodon plagiometopon*), he found caged surfaces exhibited the greatest algal diversity after 2 months. Hixon and Brostoff (1983) and Sammarco (1983) obtained the same result from samples taken after 2–6 months and 3 months, respectively. However, after a year, both the latter studies found that algal diversity was greatest inside damselfish territories. These data, combined with the fact that Sammarco studied one of the same species as Lassuy *(H. plagiometopon),* suggest that Lassuy's (1980) samples may have represented early successional stages.

In the Hawaii study, Hixon and Brostoff (1996) showed that moderate grazing by the damselfish *Stegastes fasciolatus* slowed and appeared to stop succession at a high-diversity middle stage dominated by red filaments (Fig. 10–2). Thus, rather than altering the successional trajectory like more intensive grazers (see above), damselfish appeared to simply decelerate algal succession. Territorial fish may maintain the midsuccessional algal community because these species provide a superior food source for the damselfish (Montgomery and Gerking, 1980) and/or a source of invertebrate prey and palatable epiphytes (Lobel, 1980).

Hixon and Brostoff (1983, 1996) further showed that grazing by damselfish inside their territories was of intermediate intensity relative to that within cages and outside territories. Correspondingly, the standing crop of algae was also at intermediate levels inside territories, while local species diversity was at its maximum. These results thus corroborated the intermediate-disturbance hypothesis (sensu Connell, 1978; Chapter 15). At low levels of grazing disturbance within cages, a few dominant competitors (coarsely branching species such as *Hypnea* and *Tolypiocladia*) were capable of locally excluding most other species. At high levels outside territories, only a few crustose species persisted. Inside damselfish territories, the coexistence of many algal species was maintained because their densities were apparently kept below levels where resources (presumably mediated by living space) became severely limiting (Fig. 10–3A).

Given that territorial damselfish can locally enhance species diversity, they can be considered a "keystone" species (sensu Paine, 1966; see also Williams,

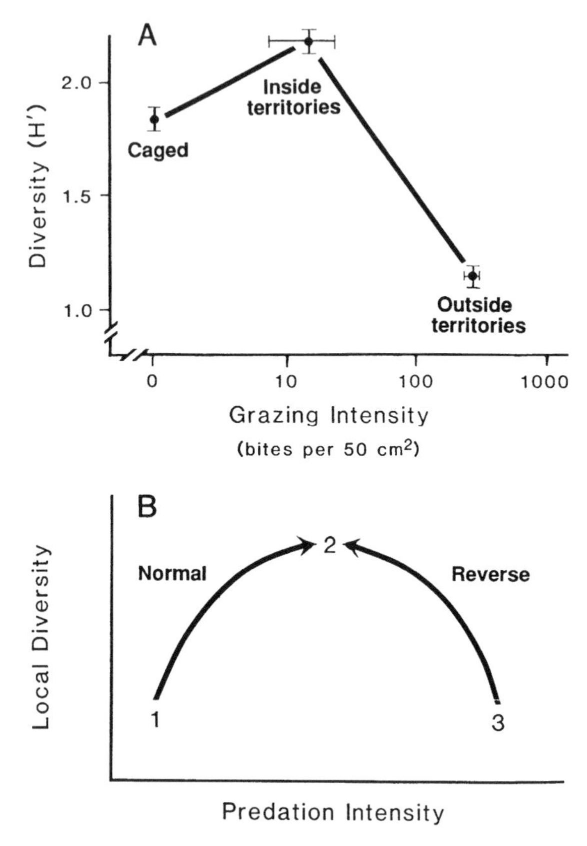

Figure 10–3. (**A**) Algal species diversity (measured by the Shannon-Wiener index, H') on settling plates exposed off Hawaii for one year to each of three grazing treatments: caged, exposed but defended inside damselfish territories, and exposed outside territories to parrotfishes and surgeonfishes. Each vertical bar represents 2 standard errors for mean diversity from 21 settling plates sampled at the end of one year; each horizontal bar represents the 95% confidence interval for mean grazing intensity, measured as the "standing crop" of fish bite marks per plate, from 63 settling plates sampled earlier in the year. (**B**) Graph of the intermediate-disturbance hypothesis, showing that a keystone species can enhance local species diversity either by increasing predation intensity from point 1 toward point 2 ("normal"), or by decreasing overall predation intensity by territorial defense from point 3 toward point 2 ("reverse"), as demonstrated by territorial damselfish (modified from Hixon and Brostoff, 1983).

1980). However, unlike the normal kinds of keystone species, which enhance diversity by increasing predation intensity on a prey assemblage, the territorial behavior of damselfish enhances diversity by decreasing predation overall (Fig. 10–3B). That is, in the absence of a normal keystone species, predation is low and diversity is low because a few prey species competitively exclude most others from the local system (e.g., Paine, 1966). However, in the absence of damselfish (keystone species in reverse), predation is high (due to schooling herbivores) and diversity is low because few prey are able to survive intense grazing.

There is evidence that this pattern documented in Hawaii is common. Assuming that grazing intensity was intermediate inside damselfish territories in Sammarco's (1983) study, *Hemiglyphidodon plagiometopon* is a keystone species where it is abundant at the Great Barrier Reef, and possibly Guam (Lassuy, 1980). More recently, Hinds and Ballantine (1987) found that the algal mats in territories of *Stegastes planifrons* off Puerto Rico decline in diversity when caged, also suggesting a keystone-species effect. Note, however, that not all damselfishes enhance local algal diversity; some species maintain near monocultures within their territories by intense nonselective grazing (Montgomery, 1980).

Regardless of whether damselfishes enhance local algal diversity, the greatly increased standing crop of erect algae inside their territories (compared to more heavily grazed surfaces outside) has important secondary effects on reef benthos. The algal mat serves as a refuge for invertebrate microfauna and/or various epiphytes (Lobel, 1980; Hixon and Brostoff, 1985; Zeller, 1988). Also, because accretion by crustose coralline algae adds to the reef framework, and such algae are overgrown by the algal mat, damselfish territories may be sites of weakened reef structure (Vine, 1974; Lobel, 1980).

Damselfish territories may also indirectly affect nitrogen fixation on reefs, although available data are somewhat contradictory. During the same study as Sammarco (1983) described above, Wilkinson and Sammarco (1983) found that nitrogen fixation by blue-green algae (cyanobacteria) was positively correlated with grazing intensity on the Great Barrier Reef, being lowest within cages, intermediate inside damselfish territories, and greatest outside territories. However, both Lobel (1980) and Hixon and Brostoff (1996) found considerably more blue-green algae inside than outside territories in Hawaii. Finally, Ruyter Van Steveninck (1984) found no differences in the abundance of filamentous blue-green algae inside and outside damselfish territories in the Florida Keys. These discrepancies suggest possible regional differences in local distributions of blue-green algae.

10.2.3. Conclusion

Herbivorous fishes strongly affect the distribution and abundance of reef macroalgae. Where there is ample shelter from predation and protection from strong turbulence, schooling herbivores can crop reef algae to very low-standing crops, leaving mostly grazer-resistant forms such as crusts, compact turfs, or chemically defended species. Such intense herbivory may be essential for reef-building corals to flourish. Indeed, Glynn (1990, p. 391) concluded that the "maintenance of modern coral reefs may be due largely to the activities of fish and invertebrate herbivores that prevent competitively superior algal populations from dominating open, sunlit substrates." In any case, it is important to realize that a myriad of factors are involved in these and other switches in dominance among algae and between algae and corals. For example, Littler and Littler (1984) see nutrient levels as pivotal in determining how herbivore activity will affect the dominant benthos on reefs (Fig. 10–4).

Besides the schooling herbivores, territorial damselfishes have particularly strong local effects on shallow-reef algae, effects that can cascade through the entire benthic community. The defensive and grazing activities of damselfishes and the resulting dense algal mats they defend can substantially affect reef

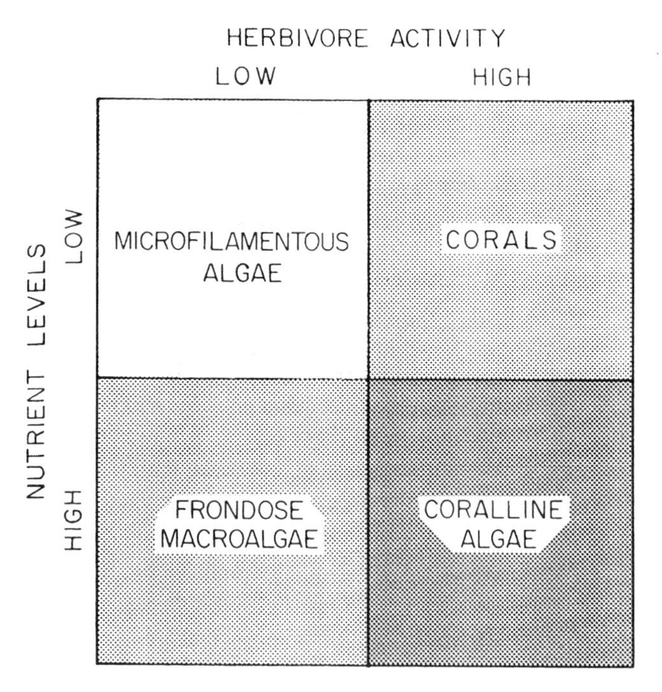

Figure 10–4. Predicted interaction of herbivore activity and long-term nutrient levels in determining the dominant benthos on shallow coral reefs. By consuming erect algae, herbivores shift the benthic community from filamentous or frondose algae (left) toward corals or grazer-resistant coralline algae (right). Secondarily, nutrient levels are predicted to shift dominance between different kinds of erect algae (left) or between corals and coralline algae (right) (from Littler and Littler, 1984).

accretion, nitrogen fixation, epiphytes and small invertebrates that inhabit the algae, and, as will be seen in the next section, corals. Clearly, herbivorous fishes are major players determining the character of shallow coral-reef communities, and territorial damselfishes in particular can act as a keystone species.

10.3. Fish Effects on Corals

10.3.1. Direct Consumption

Compared to the effects of herbivores on algae, surprisingly few studies have demonstrated that corallivorous fishes influence the distribution and abundance of reef-building corals. For example, although butterflyfishes are among the most obligatory of corallivores (Reese, 1977), these fishes appear to have a negligible effect on coral standing crops (Harmelin-Vivien and Bouchan-Navaro, 1981, 1983). At Aqaba in the Red Sea and Moorea in the South Pacific, butterflyfishes occurred at average densities of 69 and 51 fish per 1,000 m^2, yet consumed an average of only about 10 and 28 grams of coral polyps per 1,000 m^2 per day, respectively. It appears that corals often retract all their polyps in response to predation by butterflyfish, making polyps locally unavailable to predators for considerable periods (D. W. Meadows, personal communication). Such factors may preclude high densities of large-bodied obligate corallivores, perhaps necessitating the large feeding territories defended by butterflyfishes (see Tricas, 1989; Roberts and Ormond, 1992).

Nevertheless, the local distributions of several coral genera are strongly affected by coral-feeding fishes. Neudecker (1979) provided one of the first experimental demonstrations that fishes can potentially affect the depth zonation of corals. Off Guam, he transplanted small colonies of *Pocillopora damicornis* from a relatively fish-free lagoon (1–2 m depth) to reef slopes (15–30 m depth) where this coral was naturally absent and corallivorous fishes were common. Coral transplants survived well when caged, but exposed colonies were partially consumed by butterflyfishes and triggerfishes within one week.

The effects of fish-consuming corals can have ramifications for interactions among corals. Off Hawaii, Cox (1986) showed that the feeding preference of the butterflyfish *Chaetodon unimaculatus* for the coral *Montipora verrucosa* can reverse the competitive dominance of this coral over another species, *Porites compressa.* Inside fish-exclusion cages, *Montipora* overgrew *Porites,* yet outside cages, this dominance sometimes reversed due to differential grazing of *Montipora* by the butterflyfish.

Besides the strict corallivores, herbivorous fishes may also directly affect corals by occasionally consuming or otherwise killing them. Territorial damselfishes are known to remove polyps, thereby killing patches of coral on which the damselfish establish their algal mats. In the Caribbean, the damselfish *Stegastes planiforns* was observed killing *Montastrea annularis* and *Acropora cervicornis*

(Kaufman, 1977). Knowlton et al. (1990) suggested that such predation dramatically slowed the recovery of *A. cervicornis* off Jamaica following Hurricane Allen, inhibiting the usual dominance of this species. Similarly, off the Pacific coast of Panama, *Stegastes acapulcoensis* killed patches of *Pavona gigantea* (Wellington, 1982). Wellington's study demonstrated how this direct effect, combined with various indirect effects, strongly affected coral zonation (see below).

Outside damselfish territories, the reported direct effects of herbivorous fishes on corals are contradictory. On one hand, field observations have noted grazing fishes damaging juvenile corals (Randall, 1974; Bak and Engel, 1979). Littler et al. (1989) suggested that parrotfishes (*Scarus* spp. and *Sparisoma* spp.) substantially influence the local distribution of *Porites porites* off Belize by eliminating this delicately branching species from areas where these fish are abundant. They proposed that a combination of differential consumption of *P. porites* by parrotfishes and the relative availability of refuge holes for grazing fishes of different sizes among different microhabitats determined whether backreef bottoms were dominated by macroalgae, *P. porites,* or the relatively mound-shaped and grazer-resistant *P. astreoides* (Fig. 10–5). Similarly, recently recruited coral colonies survived intense parrotfish grazing in laboratory mesocosms in Hawaii only when structural refuges from grazing were provided (Brock, 1979).

On the other hand, there is evidence that herbivorous fishes avoid consuming living corals in the field, including recently recruited colonies (Birkeland, 1977), and only the largest species of parrotfish, the Indo-Pacific *Bolbometopon muricatum,* is reported to consume substantial amounts of live coral (Choat, 1991; see also Randall, 1974). Such differential grazing may moderate competition between algae and corals, preventing algae from excluding corals. Indeed, Lewis (1986) noted that macroalgae overgrew corals of the genus *Porites* when herbivorous fishes were excluded by fencing from a shallow reef off Belize for 10 weeks. However, given such contradictory evidence, whether nonterritorial herbivorous fishes have negative or positive effects on corals appears to depend on the particular system.

Finally, a poorly documented yet possibly substantial source of coral mortality is consumption of coral spawn by planktivorous reef fishes. At the Great Barrier Reef, Westneat and Resing (1988) noted that the guts of the planktivorous damselfishes *Abudefduf bengalensis* and *Acanthochromis polyacanthus* were packed with coral gametes during the annual mass spawning of corals.

10.3.2. Indirect Effects

Available experimental evidence suggests that indirect effects of territorial damselfishes influence the local distribution and abundance of corals more extensively than direct consumption by corallivores. By defending and maintaining their algal mats, damselfish produce patches in which juvenile corals are often smoth-

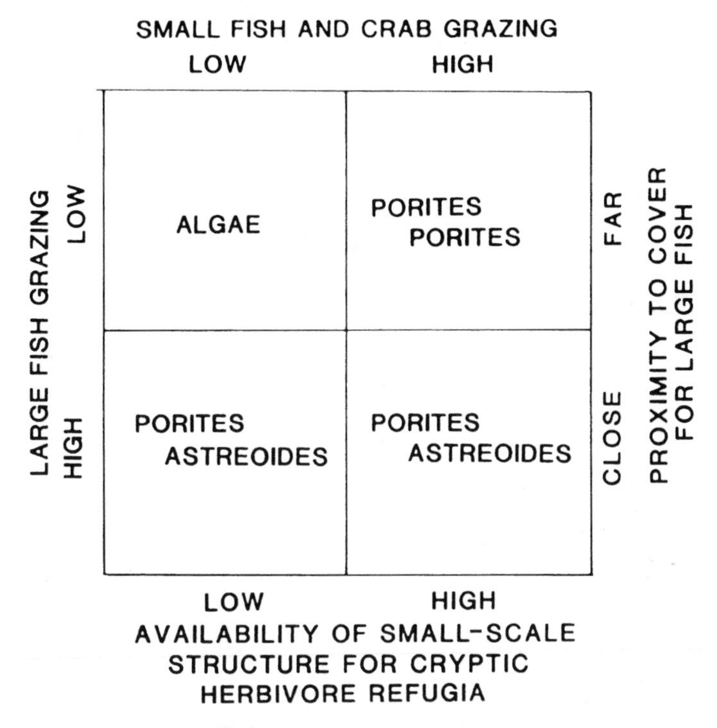

Figure 10–5. The direct influence of physical shelter from predation on the local abundance of grazing fishes, and subsequent indirect effects on the dominant backreef benthos in Belize. When large shelters are nearby, large fishes will be locally abundant and reduce both macroalgae and the delicately branching coral, *Porites porites,* leaving the relatively invulnerable mound-shaped *P. astreoides* to dominate (lower left and right). Where large shelters are rare and small shelters are abundant, small grazers will reduce algae, allowing the competitively subordinate *P. porites* to dominate (upper right). Only where all refuges for grazers are absent will algae dominate (upper left) (from Littler et al., 1989).

ered (Vine, 1974; Potts, 1977). Additionally, the algal mat provides microhabitats facilitating various boring organisms, which enhance bioerosion of the coral framework (reviews by Hutchings, 1986; Chapter 4). However, some coral species seem to recruit more successfully to damselfish territories than to adjacent undefended areas, suggesting that the territories may provide at least a temporary refuge from corallivores (Sammarco and Carleton, 1981; Sammarco and Williams, 1982; see below). If for any reason coral heads manage to reach a certain size, they may become invulnerable to algal overgrowth (Birkeland, 1977).

Given that damselfishes may have both positive and negative effects on corals, complex interactions can result. An example is provided by a study of coral zonation on the Pacific coast of Panama by Wellington (1982). In this system, branching *Pocillopora* corals dominated shallow areas (0–6 m depth), while the

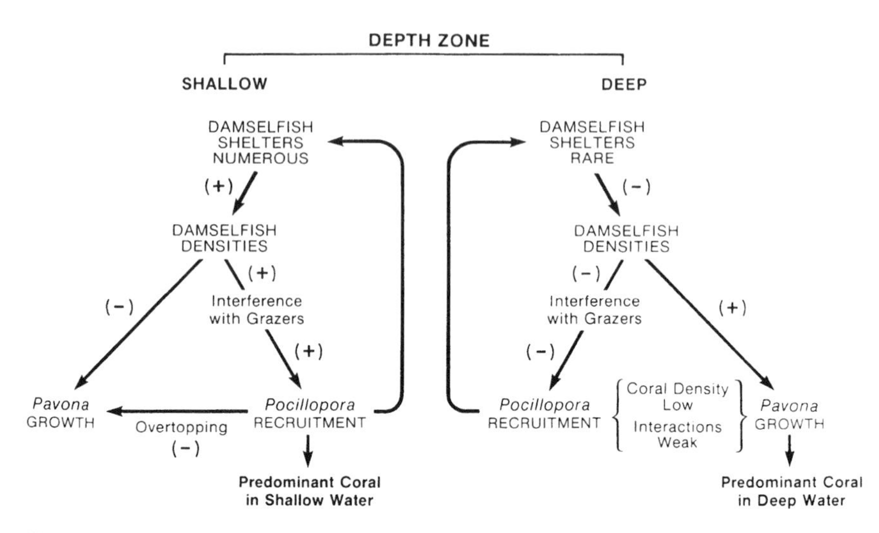

Figure 10–6. Interactive feedback loops influencing the depth zonation of *Pocillopora* and *Pavona* corals off the Pacific coast of Panama. Arrows indicate the direction of each effect; "+" and "–" indicate positive and negative effects, respectively. The direct negative effect of damselfish on *Pavona* in shallow water is due to polyp removal; the "direct" positive effect in deep water is actually an indirect effect mediated by the scarcity of damselfish shelters (from Wellington, 1982).

massive *Pavona gigantea* dominated deeper areas (6–10 m depth). Using a clever series of observations and experiments, Wellington discovered an interactive feedback loop whereby the damselfish *Stegastes acapulcoensis* may directly and indirectly cause this zonation (Fig. 10–6). When establishing territories in the shallow zone, damselfish differentially kill *Pavona* by polyp removal and maintain their algal mats on the exposed substrata; *Pocillopora* is apparently protected by its tightly branched morphology and rapid polyp regeneration. Moreover, *Pocillopora* colonies within the periphery of territories are protected from corallivores by the defensive behavior of the damselfish. These factors enhance the ability of *Pocillopora* to competitively dominate *Pavona* in shallow areas. The *Pocillopora* framework, in turn, provides the damselfish with shelter, a necessary requisite for a territory. In the deep zone, shelter sites and thus damselfish densities are low because overall coral cover (and thus intercoral competition) is low, apparently due to physical factors (attenuated light, reduced water motion, etc.). Here, transient fish corallivores (mostly puffers) differentially eat *Pocillopora,* whose branches they can ingest and masticate, leaving *Pavona* as the dominant coral.

10.3.3. Conclusion

Truly corallivorous fishes have been shown to affect the local distribution and abundance of some corals. However, the territorial activities of herbivorous damselfishes appear to have more substantial effects on corals in shallow water. This difference appears to be a function of the wide variety of direct and indirect effects manifested by damselfishes. In any case, interactions between fishes and corals seem to be largely indeterminate. It appears that, on exposed reef surfaces, coral recruits may initially experience enhanced survival where they are protected from extensive fish grazing, such as within damselfish territories, but will often be overgrown by algae eventually. Those coral larvae settling on surfaces exposed to grazing by herbivorous fishes outside territories may or may not initially suffer high mortality, depending on whether they are consumed along with other fish prey, but some colonies will eventually reach an invulnerable size where they are both immune to incidental predation and freed from competition with algae.

10.4. Fish Effects on Invertebrate Corallivores and Herbivores

Besides directly consuming corals or algae, reef fishes can also affect invertebrate corallivores and herbivores, causing subsequent indirect effects on the dominant reef benthos. Most obviously, some fishes consume these organisms, including the major invertebrate corallivore, the crown-of-thorns seastar *(Acanthaster planci),* and the major invertebrate herbivores, sea urchins (Chapters 9 and 11). At the Great Barrier Reef, Pearson and Endean (1969) noted planktivorous damselfish consuming early developmental stages of *Acanthaster.* In the Red Sea, Ormond et al. (1973) documented that triggerfishes and puffers killed 1,000 to 4,000 *Acanthaster* per hectare each year, a rate that accounted for an observed decline in the *Acanthaster* population.

Triggerfishes and puffers also consume sea urchins, as do large wrasses (Labridae) and porcupinefishes (Diodontidae; e.g., Randall, 1967). Field experiments have demonstrated that such predation can be intense (Glynn et al., 1979) and can force urchins to remain near shelter (Carpenter, 1984). Thus, the risk of predation by fishes limits the area over which urchins can overgraze algae and seagrass, resulting in discrete barren zones or "halos" around Caribbean reefs (Ogden et al., 1973). Hay (1984) suggested that overfishing of large wrasses and triggerfishes has resulted in unusually high urchin densities in populated regions of the Caribbean.

Besides the mechanism of direct consumption, fishes may negatively affect invertebrate corallivores and herbivores by competitive interactions. In defending their territories, several damselfish species in the South Pacific exclude *Acanthaster,* as first noted by Weber and Woodhead (1970). This exclusion apparently results in the preferred prey of the seastar (mostly acroporid corals) being more abundant and more diverse inside territories than outside (Glynn and Colgan, 1988). In contrast, the species diversity of new coral recruits on the Great Barrier Reef was smaller inside territories of the damselfish *Hemiglyphidodon plagiometopon,* although the density of coral spat (mostly acroporids) was greater there (Sammarco and Carleton, 1981).

In the Caribbean, the damselfish *Stegastes planiforns* can exclude the urchin *Diadema antillarum* from its territories (Williams, 1980, 1981), which may also serve as refuges for certain corals (Sammarco and Williams, 1982). Corals such as *Favia fragum* can apparently withstand competition with the macroalgae that dominate inside territories. Given that, first, damselfish can prevent urchins from overgrazing their territories, and second, the algae growing within the territories provide food for the damselfish, Eakin (1987) concluded that the relationship between damselfish and their algal mats is a case of mutualism.

Parrotfishes and surgeonfishes also compete with *Diadema* on Caribbean reefs, although the urchin appears to be the dominant competitor in this case (Carpenter, 1986a). In particular, increases in the local abundances of these fishes have been documented following experimental removals of, or natural declines in, populations of the urchin (Hay and Taylor, 1985; Carpenter, 1990; Robertson, 1991).

Finally, complex interactions between invertebrates and fishes can occur. Outbreaks of *Acanthaster* can kill large tracts of coral, presumably increasing the availability of substrata for macroalgal growth, which in turn may increase the local densities of herbivorous fishes and decrease those of corallivorous fishes. This sequence was documented for some fishes both at the Great Barrier Reef (Williams, 1986) and off Japan (Sano et al., 1987), although the response of herbivorous fishes was negligible. Clearly, there are many possible ecological linkages among algae, corals, invertebrate herbivores and corallivores, and reef fishes.

10.5. Ramifications for Reef Management

Even though the available evidence suggests that both corallivorous and (especially) herbivorous fishes can have strong local effects on the structure of benthic reef communities, the explicit utility of this knowledge for managing coral reefs seems limited. This is not to say that it is impossible to predict the consequences of some human activities. For example, the studies summarized here suggest that overharvesting herbivorous fishes and invertebrates can allow algae to outcompete corals, that removing fish predators of urchins can allow these herbivores to overgraze algae, and that altering the density of territorial damselfishes can drastically affect the local benthic community. However, more specific predictions may not be possible. The reasons for this less-than-optimistic view are basically twofold.

First, before predicting how harvesting fishes will secondarily affect the benthic community on a reef, we have to know what determines the local population sizes of fishes in the absence of harvesting, and subsequently, how those populations will respond to harvesting. Our knowledge of the population dynamics of unexploited reef-fish populations is rudimentary, so predicting even the direct effects of fishing is immensely difficult (see Russ, 1991; Chapters 16, 17, and 18). For example, there is evidence that territorial damselfishes inhabiting at least one site on the Great Barrier Reef are naturally recruitment limited (review by Doherty and Williams, 1988). Assertions that this pattern is typical of reef fishes in general has fueled a controversy that has remained unresolved for over a decade (see chapters by Doherty, Ebeling and Hixon, Hixon, Jones, Sale, and Williams, in Sale, 1991). Besides damselfishes, our knowledge of the population dynamics of other reef fishes is even more limited, although it is clear that both corallivores and herbivores are subject to overfishing (review by Russ, 1991).

Second, coral-reef communities are immensely complex, so that the demise or outbreak of a single species due to human activities may have unanticipated and severe ramifications for the remainder of the system (Chapter 11). As a keystone species, territorial damselfishes can manifest a very complex variety of direct and indirect effects on shallow reef systems (Fig. 10–7). The numerous

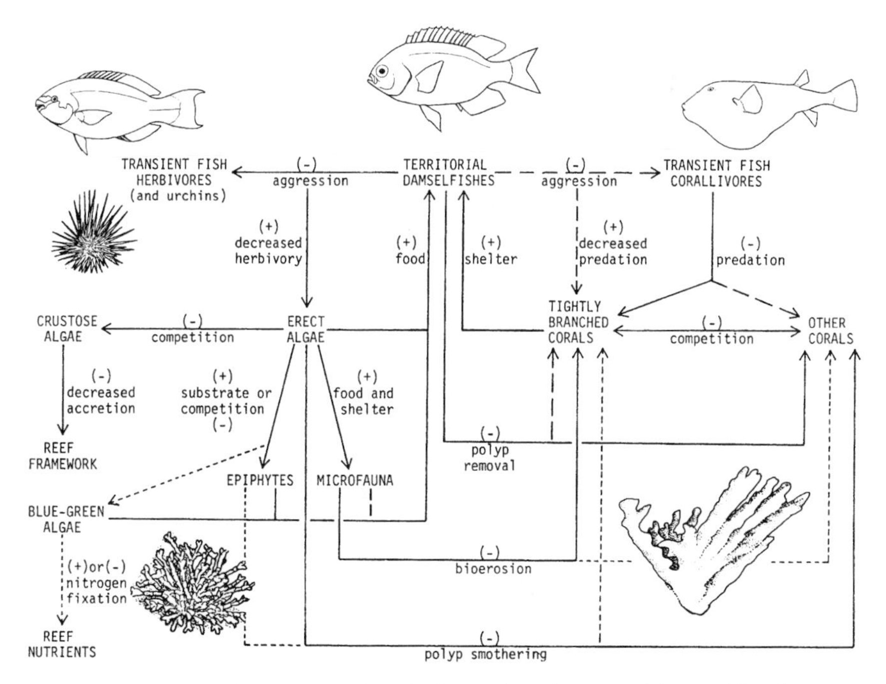

Figure 10–7. Flowchart synthesizing the interactions between fishes and benthos on a shallow coral reef where territorial damselfishes are abundant. Arrows indicate the direction of each effect; "+" and "–" indicate positive and negative effects, respectively. Dashed lines indicate relatively weak effects, and dotted lines indicate effects that are poorly documented and/or controversial. Where damselfishes are rare, some of these effects reverse. In particular, erect algae and their associates are often replaced by grazer-resistant crustose algae due to intense grazing by transient herbivores (modified from Hixon, 1983).

indirect interactions between such fishes and the reefs they inhabit may simultaneously be both positive and negative from a human perspective. For example, damselfish territories may be sites of high algal productivity and species diversity, but may also be sites of reduced coral growth and weakened reef framework. The multitude and complexity of these effects severely limit our ability to predict accurately the effects of harvesting any particular species (let alone multiple species).

Although reef systems may be too complicated to allow us to predict explicit outcomes of human activities, the patterns summarized in this and other chapters of this volume do provide an important lesson: managers should cast a skeptical and cautious eye on proposals to strongly alter the abundance of any coral-reef inhabitant. The secondary results of such alterations may be both unexpected and undesirable. More than any other natural system, coral reefs seem to verify John Muir's (1911) axiom: "When we try to pick out anything by itself, we find it hitched to everything else in the universe."

11

Indirect Interactions on Coral Reefs

Steven C. Pennings

When it comes to communities, environmentalists believe that "you cannot change just one thing." Species, they say, do not exist in isolation, but rather are embedded in a web of interactions, such that a change in one species can affect a surprising number of other species. Although this is an old idea, it has only recently received rigorous attention from ecologists. It is already clear, however, that in many cases reef managers must consider this connectedness of ecological systems in order to make effective decisions.

To describe the ways in which species are "connected" to each other, ecologists have coined the terms direct and indirect effects. For example, the crown-of-thorns starfish *Acanthaster planci* may have direct effects by dramatically reducing total cover and altering species composition of corals (Chapter 9; Colgan, 1987; Birkeland and Lucas, 1990; Glynn, 1990b). However, *Acanthaster* can also alter the densities of many species that it does not consume or otherwise directly interact with at all. Following outbreaks of *Acanthaster,* algal, alcyonacean, and sponge cover on the reef may increase, corallivorous fish may decrease, and urchins and herbivorous fish may increase (Birkeland and Lucas, 1990; Glynn, 1990b). These changes may then result in altered patterns of bioerosion and nitrogen fixation (Birkeland and Lucas, 1990). Such effects are all indirect effects, in that they originate from *Acanthaster* but are transmitted to other species through changes in coral cover.

There are numerous anecdotal examples of indirect effects in the older ecological literature. However, only recently have ecologists specifically designed experiments to study indirect effects. In the older literature, indirect effects were often regarded as irritating complications that one tried to avoid or at least explain in a post-hoc way. Now, ecologists see indirect effects as revealing in a more

Financial support during manuscript preparation was provided by NSF (OCE-9116307).

accurate way how communities are structured (Levine, 1976; Lawlor, 1979; Schaffer, 1981; Stone and Roberts, 1991). Consequently, experiments are now designed to tease apart direct and indirect effects, reveal their relative magnitudes, and discern their role in community structure and stability.

Indirect effects are common, perhaps ubiquitous, in ecological systems. Indirect effects produced "unexpected" results (e.g., the removal of a predator or competitor resulted in a decrease in a second species) in approximately 40% of studies on predation (Sih et al., 1985), 20% of studies on competition (Connell, 1983; Lawlor, 1979), and 10–33% of studies on plant competition (Goldberg and Barton, 1992). These studies illustrate that when the entire community is considered, the net outcome of indirect effects can alter the strength of, or even reverse, a direct interaction between two species (Lawlor, 1979; Stone and Roberts, 1991; Abrams, 1992). For example, the net effect of a predator on a prey species might actually be positive, if the predator concentrates its attacks upon the prey's major competitor. Because indirect interactions can commonly have such dramatic results, we must understand the degree to which coral-reef communities are structured by indirect interactions in order to manage reefs effectively.

We will begin with a more complete definition of indirect effects than is given above, together with examples of indirect effects from coral reefs. Then we will discuss whether or not indirect effects might be unusually important on coral reefs, and consider some practical problems of studying indirect effects on coral reefs. Finally, we will conclude with a section on future directions. A major theme of this chapter will be that indirect effects are potentially of major importance to reefs, but that many of our examples are incomplete, and therefore inadequate as guides, because most studies were conducted for other reasons and did not specifically address the importance of indirect effects.

11.1. Definition, Variety, and Examples of Indirect Effects

Miller and Kerfoot (1987) categorized indirect effects as trophic linkage, behavioral, and chemical response. The reader should beware of the early (and some of the current) literature, which uses a variety of terms, often inconsistently, for the same phenomena.

A trophic-linkage indirect effect occurs when one species alters the density of another, which in turn alters the density of a third. Because this effect operates through population dynamics, it occurs on a timescale related to the generation times of the species involved.

A behavioral indirect effect occurs when one species alters the behavior of a second, which in turn qualitatively alters the nature of the interaction of the second species with a third. Because this effect is mediated by the behavior of the second species, it can occur as quickly as the second species is able to gather and respond to information about the first species. Thus, behavioral indirect effects can occur on much shorter timescales than can trophic-indirect effects.

A chemical-response indirect effect occurs when one species causes chemical changes in a second, which in turn qualitatively alters the nature of the interaction of the second species with a third. Examples could include induced chemical defenses in plants that are grazed by multiple herbivores, or the sequestering of plant secondary metabolites by herbivores for their own defense. These chemical changes operate on a timescale that is usually much shorter than that of population dynamics, but may be longer than the behavioral changes discussed above.

Following Miller and Kerfoot (1987), we can represent these three categories of indirect effects with a simple equation modeling the rate of change of species C as some function of its own density (N_c) and that of species B (N_b),

$$dN_c/dt = f(N_b, N_c)$$

By definition, species A affects species C in a trophic-linkage indirect effect by altering N_b. In contrast, in a behavioral or chemical-response indirect effect, species A affects species C by altering *f*, the form of the function itself, so that the per-capita effect of B on C changes. Thus, behavioral and chemical-response indirect effects fall into the category of *higher-order effects* because the nature of the interaction between B and C changes if A is present. However, because the term higher-order effects has been the source of much confusion in the literature (Worthen and Moore, 1991), the nomenclature proposed by Miller and Kerfoot (1987) will be used.

Distinctions between trophic-linkage, behavioral, and chemical-response indirect effects are useful for understanding the variety of mechanisms that may operate in nature to produce indirect effects. They are also important considerations in designing appropriate experiments. Because of the different timescales involved in trophic-linkage versus behavioral and chemical-response indirect effects, the duration of the experiments we design to study them must also differ (section 11.4). Moreover, because different mechanisms are involved, the variables that we might choose to measure may also differ (Strauss, 1991).

There is, of course, no reason to expect these three types of indirect effects to be mutually exclusive in nature. In fact, interactions between a group of species might well involve two or three types operating simultaneously. If so, ecologists will face the challenging task of designing experiments to simultaneously explore very different mechanisms that operate on very different timescales.

11.1.1. Trophic-Linkage Indirect Effects

Historically, ecologists have paid the most attention to trophic-linkage effects. Because the number of potential indirect links between species is large for even very simple systems, a huge variety of trophic-linkage effects is possible. First, we will briefly review six simple trophic-linkage indirect effects that have received considerable attention. These will serve to illustrate the variety of indirect effects

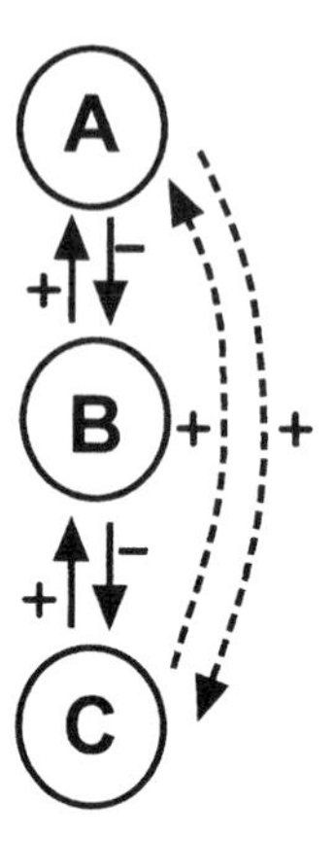

Figure 11–1. Trophic cascade. Direct effects are indicated by solid lines; indirect effects by dotted lines.

that are possible. Then, we will consider examples of trophic-linkage indirect effects from coral reefs.

Trophic cascades (Fig. 11–1) have commonly been documented in aquatic systems (Kerfoot and Sih, 1987, but see Vanni and Findlay, 1990). They occur across multiple trophic levels: species A consumes B, which consumes C. Species A and C may not directly interact, but A indirectly benefits C by consuming C's predator. If A consumes both B and C, A will have both a direct negative and an indirect positive effect on C, and the net effect of A on C may be positive, negative, or too small to measure, depending on the relative magnitude of the direct and indirect effects. Strong (1992) argues that trophic cascades will be important only in relatively low-diversity systems where a single species can dominate the dynamics of a single trophic level, and where omnivory is rare.

Apparent mutualisms, also called gratuitous mutualisms (Fig. 11–2), can occur in systems of competitors on the same trophic level. Species A and B compete, and species B and C compete, but A and C compete weakly or not at all. A and C indirectly benefit each other by affecting their mutual competitor, B. Apparent mutualisms should be common in systems of multiple competitors (Levine, 1976; Davidson, 1980; Bender et al., 1984; Stone and Roberts, 1991).

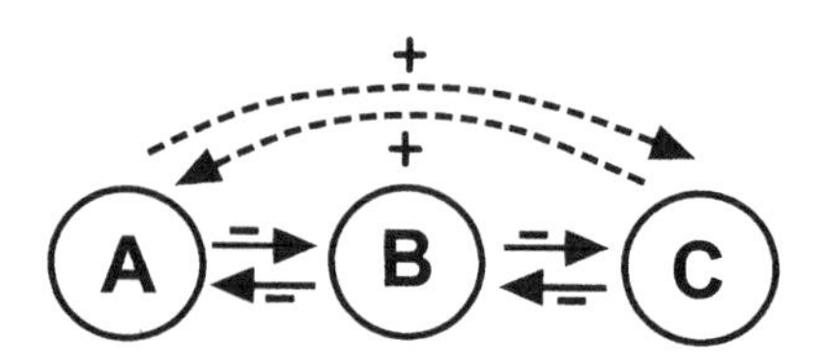

Figure 11–2. Apparent mutualism. Direct effects are indicated by solid lines; indirect effects by dotted lines.

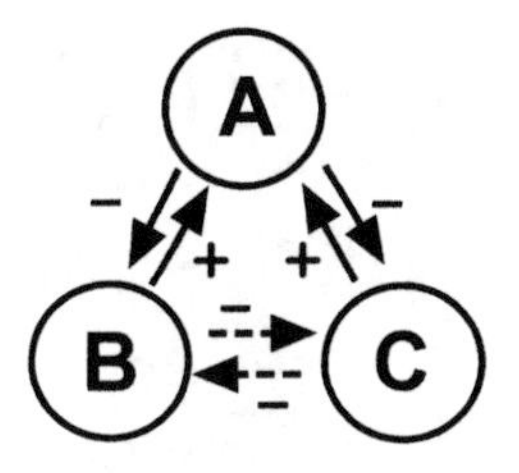

Figure 11–3. Apparent competition. Direct effects are indicated by solid lines; indirect effects by dotted lines.

Apparent competition can occur between two species that share a predator (Fig. 11–3). Species A consumes both B and C. B and C do not directly interact, but may have an indirect negative effect upon each other by affecting their mutual predator (Holt, 1977, 1984; Holt and Kotler, 1987; Schmitt, 1987). A variety of effects other than apparent competition is also possible between prey that share a predator (Abrams, 1984, 1987; Holt and Kotler, 1987).

Let us modify Figure 11–3 so that B and C compete (with B being the superior competitor), and so that A preferentially feeds upon B. In this case, A may have an indirect positive effect upon C, and is called a keystone predator (Brooks and Dodson, 1965; Paine, 1966; Lubchenco, 1978); this terminology also carries the connotation that the indirect positive effect of A on C is sufficient to cause a major change in the nature of the prey community.

Perhaps the best-studied indirect effect is exploitative competition (Fig. 11–4): species A and B both consume C, and thereby have a indirect negative interaction. Widespread recognition that exploitative competition was best described as an indirect interaction was probably hindered by the general practice of treating competition phenomenologically rather than mechanistically. Thus, both textbooks and simple mathematical equations commonly define competition only as a negative–negative interaction, obscuring potentially important differences between the variety of mechanisms that can lead to net negative–negative interactions (Schoener, 1983; Goldberg, 1990).

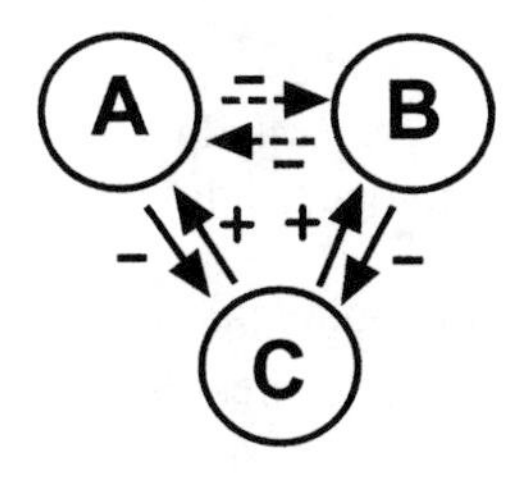

Figure 11–4. Exploitative competition. Direct effects are indicated by solid lines; indirect effects by dotted lines.

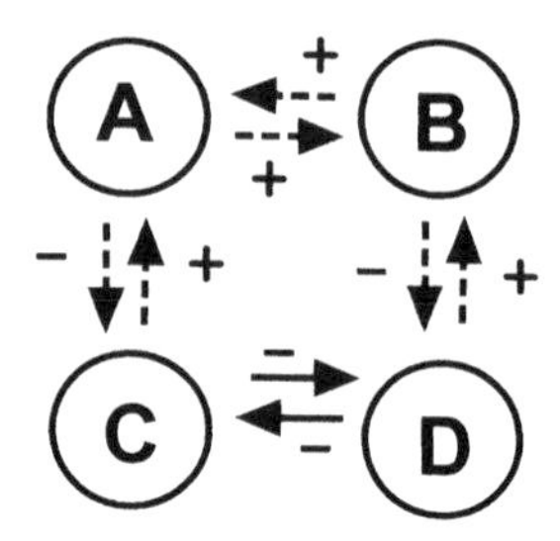

Figure 11–5. Indirect mutualism. Direct effects are indicated by solid lines; indirect effects by dotted lines.

Finally, indirect mutualisms can occur between predators of competing prey. Consider Figure 11–5. A and B are specialist predators on C and D, respectively; C and D compete. A and B will indirectly interact positively (Levine, 1976; Vandermeer, 1980). Each consumer, by eating its preferred prey, will free the preferred prey of the other consumer from competition (Dethier and Duggins, 1984).

We will now turn to actual examples of trophic-linkage effects from coral-reef systems.

Competition Between Fish and Urchins for Algae and Its Mediation by Predators

This topic is reviewed by Carpenter (Chapter 9) and Hixon (Chapter 10), so we will cover it very briefly here. As described above, exploitative competition is an indirect interaction mediated through a limiting resource. Both herbivorous fishes and urchins consume algae, and each has strong effects on algal standing stocks (Carpenter, 1986a; Lewis, 1986) (Fig. 11–6). Experimental removal of urchins is typically followed by an increase in fish foraging in urchin-removal areas (Hay and Taylor, 1985); however, it is not always clear whether this behavioral change in foraging patterns by fish leads to changes in fish population dynamics. The mass mortality of *Diadema* throughout the Caribbean and subsequent increase in algal cover was followed by an increase in recruitment and/or survival of herbivorous fishes in only some areas (Hughes et al., 1987; Robertson, 1991); severe overfishing in many areas may have prevented fish populations from responding. An alternate hypothesis is that local populations of tropical-reef fish and urchins are affected primarily by recruitment variability rather than by food availability, even though food availability may have strong effects upon individual size and growth.

Predators may also affect the population densities of both urchins and fish. Hay (1984b) argued that heavy fishing pressure on many Caribbean island reefs depleted numbers of both herbivorous fish and fish that prey on urchins. As a result, urchins on these reefs were freed from both competition and predation, and attained very high densities (until the mass mortality). The relative importance

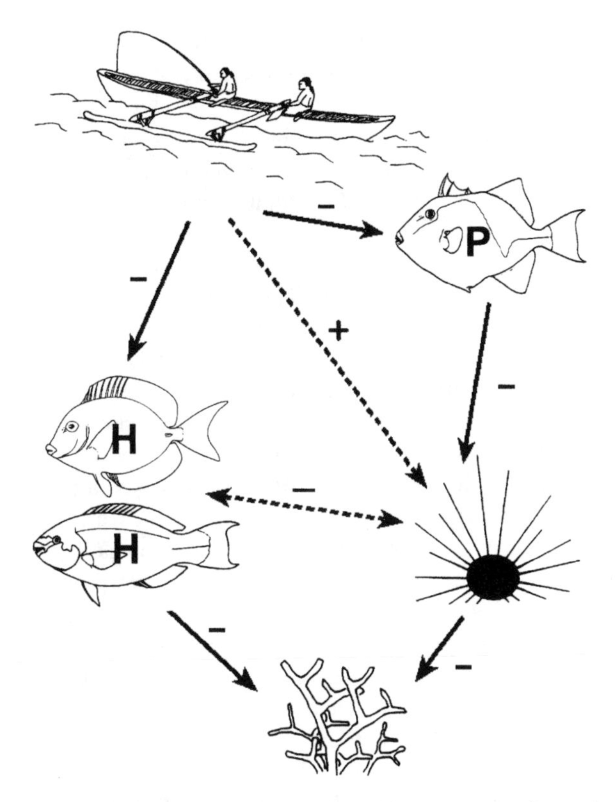

Figure 11–6. Both herbivorous fish and sea urchins consume algae; thus, they may compete exploitatively. Humans may have a positive indirect effect on sea urchins both by removing predatory fish (P), which eat urchins, and by removing herbivorous fish (H), which compete with urchins. Direct effects are indicated by solid lines; indirect effects by dotted lines.

of human removal of predatory and herbivorous fishes upon urchin numbers is not clear since fishermen usually take both predatory and herbivorous fish simultaneously. Human removal of fish that prey upon urchins has been implicated in causing unusually high urchin densities in Kenya (McClanahan and Muthiga, 1989; McClanahan and Shafir, 1990), although Kenyan fishermen also took herbivorous fish.

In sum, reef herbivores may be affected by a suite of indirect interactions mediated through their predators and their prey. Unfortunately, rigorous documentation of this is difficult. Relevant "natural experiments" (human fishing pressure, *Diadema* die-off) confound a variety of effects and therefore are difficult to interpret unambiguously, and experimental approaches to these interactions have been too brief to document population dynamics of any of the species involved except algae, and therefore offer only circumstantial evidence of the purported interactions. Because the life spans of both reef fishes and urchins can exceed a decade, rigorous experiments that could document, say, a chain of effects from humans to predatory fish, to urchins, to algae, to herbivorous fish, could require several decades, and are probably not feasible. However, this example has obvious management implications: should one wish to control urchin numbers (say, in a recreational area), it is unwise to allow exploitation of both urchin competitors and predators.

Herbivore Mediation of Algal-Coral Competition

This topic is covered extensively by Carpenter (Chapter 9) and Hixon (Chapter 10; see also Glynn, 1990b), so we will cover it very briefly here. Foliose algae are generally better competitors for space than are corals; however, algal standing stocks are usually kept very low by herbivores. When herbivores are absent, naturally in certain locations on the reef, experimentally within herbivore-exclusion cages, or as a result of overfishing (such as on some Caribbean reefs before and following the mass mortality of *Diadema*), algal biomass and cover increase rapidly, and corals decline. Thus, herbivores are indirectly responsible for the success of corals, and indeed the very existence of coral reefs (Fig. 11–7). On a very local scale, overgrowth of corals by algae can be mediated by the herbivorous crabs *Mithrax* spp., which shelter in selected coral heads (Coen, 1988a, b). With a few exceptions (notably *Acanthaster*), the direct effect of corallivores on corals is much smaller than the indirect effect of herbivores (Ogden and Lobel, 1978; Lessios, 1988; Glynn, 1990b).

Although the existence of this indirect link is well documented, key details remain to be addressed. For example, some herbivores exert a direct negative effect on corals by removing recently settled juveniles, and by eroding or browsing larger colonies (Hutchings, 1986; Glynn, 1990b; Chapter 4). This direct negative effect has not been experimentally isolated and compared with the indirect positive effect of algal removal. In some cases, the direct negative effect of herbivores (especially echinoids) can be similar to, or greater than, their indirect positive effect, and reef growth can be severely impeded by heavy bioerosion (Glynn et al., 1979; Glynn and Wellington, 1983; Hutchings, 1986; Glynn, 1990b; Chapter 4). The density and species composition of the herbivores at a particular site probably control which effect predominates. Because herbivores can so drastically affect coral success, a better understanding of these direct and indirect effects is needed. The short generation times of algae, and their ability to overgrow corals rapidly, should allow the link from herbivores to algae to corals to be fairly well documented within a reasonable time period. However, the negative effects of herbivores and algae on coral recruitment may not be manifested in the population structure and density of corals for a much longer time.

The management implications are clear: heavy harvesting of all herbivores is incompatible with vigorous coral growth.

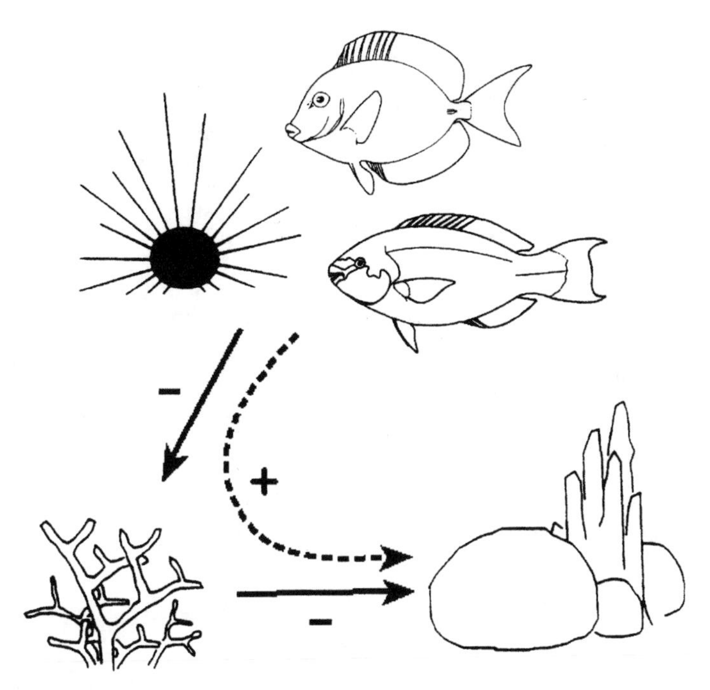

Figure 11–7. Herbivorous fish and urchins consume algae, which can overgrow coral. Consequently, these herbivores have an indirect positive effect on coral. Direct effects are indicated by solid lines; indirect effects by dotted lines.

Cascading Effects of Predation

In contrast to freshwater systems (Kerfoot and Sih, 1987), relatively little evidence exists for cascading effects of predation on reefs. Despite abundant circumstantial evidence that predatory fish affect populations of prey fish, experiments are few and inconclusive (Hixon, 1991). Moreover, although predatory fish are the most desired catch of fishermen on coral reefs, there is little evidence that removal of predatory fish results in increases in prey fish (Russ, 1991). However, studies of the effects of fishing on coral reefs suffer from a variety of unavoidable problems, including poor replication and controls, the fact that fishermen may take only a few of several ecologically similar predator species, and the fact that fishermen may simultaneously remove both predator and prey species (Russ, 1991). Both Hixon (1991) and Russ (1991) conclude that there is considerable scope for experimental approaches to this topic on reefs.

McClanahan and co-workers (McClanahan, 1989, 1990; McClanahan and Muthiga, 1989; McClanahan and Shafir, 1990) argue for the existence of cascading effects of human predation on fishes on Kenyan reefs (Fig. 11–8). These studies compared fauna on protected reefs to fauna on heavily fished reefs; this

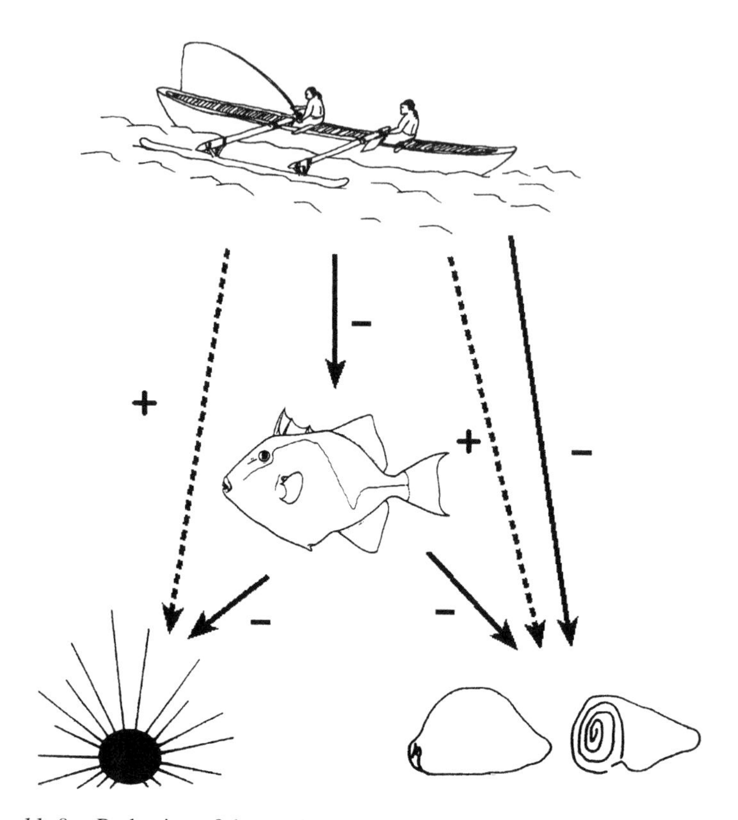

Figure 11–8. Both triggerfishes and wrasses feed on sea urchins and snails. Therefore, the taking of fish by humans has an indirect positive effect on both urchins and snails. Since humans also collect snails for the shell trade, the net effect of humans on gastropods depends on whether the direct negative or indirect positive effect is larger. Direct effects are indicated by solid lines; indirect effects by dotted lines.

approach suffers the flaw that reserves were certainly not designated randomly with respect to the biota present in each. Nevertheless, patterns suggestive of trophic cascades were observed. On fished reefs, triggerfishes and wrasses, which are predators of urchins and molluscs, were rarer, and urchins were much more abundant than on protected reefs. Moreover, although humans collected molluscs on fished reefs for the shell trade, densities of molluscs were actually higher on fished reefs than inside reserves, perhaps because of the absence of natural predators. Presumably the indirect positive effect of humans on molluscs, mediated through mollusc predators, was greater than their direct negative effect.

Cleaning Symbioses

A number of fishes and shrimp remove ectoparasites and diseased tissue from other organisms (Fig. 11–9). These "cleaners" may be fairly obligate, such as

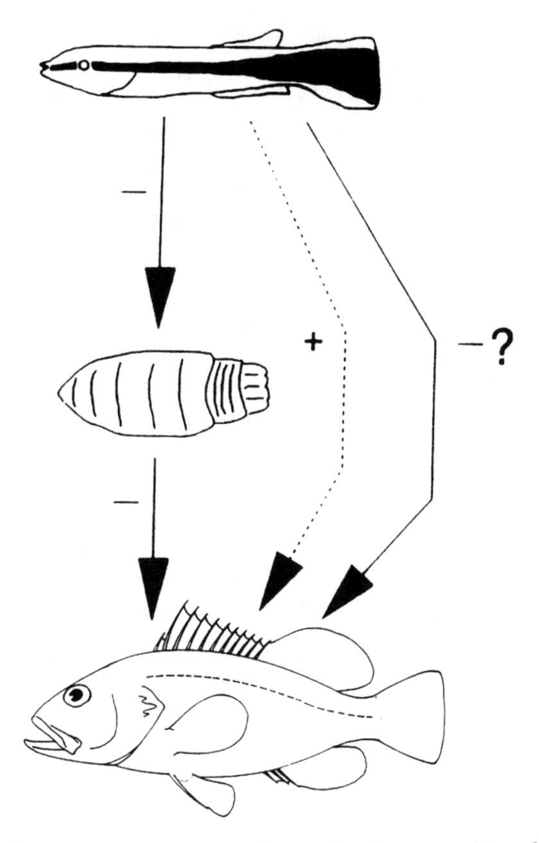

Figure 11–9. Cleaner wrasses remove ectoparasites from a variety of fishes and may thereby have an indirect positive effect on fish health or numbers. Cleaner wrasses may also have a direct negative effect on host fishes if they remove scales, mucus, or skin. Direct effects are indicated by solid lines; indirect effects by dotted lines.

wrasses of the genus *Labroides,* or facultative, such as juveniles of many reef fishes (Randall, 1958; Limbaugh, 1961; Feder, 1966; Itzkowitz, 1979; Moosleitner, 1980). If ectoparasites have a deleterious effect on fishes, and if cleaners are successful at removing ectoparasites, then cleaners might have a strong indirect positive effect on the numbers of reef fishes. The facts that reef fishes solicit cleaning via a variety of behaviors (Feder, 1966; Losey, 1971, 1974), that fishes may spend a considerable amount of time soliciting and receiving cleaning (Reinthal and Lewis, 1986), and that cleaning stations are locations of high fish diversity (Slobodkin and Fishelson, 1974) argue strongly that cleaners are beneficial to host fish.

However, little evidence exists that tropical cleaners actually reduce parasite loads on fishes (Gorlick et al., 1978), with the exception of a single study which found that fish on reefs that naturally lacked cleaners bore larger, but not more numerous, ectoparasites (Gorlick et al., 1987). Moreover, no published data support the contention that cleaners indirectly increase the numbers of reef fishes. An early study reported that, when cleaners were removed from a reef, fish numbers declined (probably through emigration) and the remaining fish suffered increased levels of disease and parasitism (Limbaugh, 1961). However, no data were presented, and the study was unreplicated and lacked controls. Two subsequent studies, both also unreplicated, failed to detect any effect of the removal of cleaners on other fishes (Youngbluth, 1968; Losey, 1972). Losey (1972) suggested that cleaners may have a beneficial effect on other fishes only when parasite levels are high. Finally, Gorlick et al. (1987) removed cleaners from two patch reefs but not from two control reefs. This experiment also failed to detect an effect of cleaners on numbers of other fishes.

Thus, although the importance of cleaners to reef fish communities has been widely accepted in the popular press, this contention is almost completely undocumented in the scientific literature. Limited evidence does suggest that cleaners may reduce the parasite load of reef fishes, but no evidence supports the hypothesized indirect positive link from cleaners to population sizes of the organisms that they clean. Moreover, fishes probably incur a direct negative cost to being cleaned through the concomitant loss of scales, mucus, and skin (Losey, 1979); this cost has not been quantified. A rigorous experimental approach to this issue is long overdue.

11.1.2. Behavioral Indirect Effects

Associational Defenses

An associational defense occurs when the presence of one organism protects another from predation (Hay, 1986). Only a handful of associational defenses have been documented from marine systems (Hay, 1986, 1992; Pfister and Hay, 1988; see also Dayton, 1985, for a possible example). These include several from coral-reef habitats.

Working in Belize, Littler et al. (1986, 1987) documented that a number of algal species were significantly more abundant close to the unpalatable alga *Stypopodium zonale,* the purple sea fan *Gorgonia ventalina,* and the fire coral *Millepora alcicornis* than at random points on the reef. Pieces of the palatable alga *Acanthophora spicifera* were grazed at lower rates when placed adjacent to these species than when placed farther away. Removal of *Stypopodium plants,* or of *Gorgonia* or *Millepora* colonies, led to significant decreases in algal abundance compared with unmanipulated controls. Plastic mimics of *Stypopodium* plants were only partially effective in defending associated algae, suggesting that part of the protective effect may have been due to secondary metabolites present in *Stypopodium.* Experiments were not done to determine if the protective effects of *Gorgonia* and *Millepora* were due to physical obstruction of grazing, the release of noxious chemicals, stinging by nematocysts, or some combination of these mechanisms. The net positive benefit of these three species on palatable algae probably represents the sum of an indirect positive effect (discouraging grazing) and a direct negative effect (shading, possible allelopathy [Sammarco et al., 1983; Nys et al., 1991]), with the indirect positive effect predominating (Hay, 1986; Pfister and Hay, 1988). However, the direct and indirect effects were not separated and individually quantified in these three coral-reef examples.

Glynn (1990b) describes some potential examples of associational defenses benefiting shrimp and fish. The well-known symbiosis between anemonefish and anemones could also probably be described as an associational defense, although it has not been rigorously studied from this context.

Mixed-Species Foraging Schools

Fish species can indirectly enhance each other's foraging effectiveness and thereby have indirect negative effects on prey species by foraging in mixed-species groups. This effect can benefit both herbivorous fish and predatory fish.

Coral-reef habitats are typically heavily grazed by herbivorous fish, and standing stocks of palatable algae are often highest inside the territories of herbivorous damselfish. Solitary invaders can be repulsed by damselfish, but schools of invaders can overwhelm the defensive capabilities of damselfish and feed on algal lawns inside their territories (Vine, 1974; Robertson et al., 1976; Foster, 1985). Because the success of individual invaders is positively related to group size (Foster, 1985), invading fish may form mixed-species schools (Reinthal and Lewis, 1986; Horn, 1989). Fish in these groups have indirect positive effects on each other by diluting attacks from damselfish, direct negative effects on algae through their own foraging, and indirect negative effects on algae through enhancing the effectiveness of foraging by other fish. Whether fish also compete for algae in damselfish territories, and thereby have indirect negative effects on each other, is unclear. Because of the difficulties involved in controlling the size of fish groups, these studies tend to be somewhat correlative and do not experimentally isolate individual effect pathways.

Furthermore, herbivorous fish (and other foragers; Glynn, 1976) may disturb invertebrates and small fish, which can then be easily caught by predators (Robertson et al., 1976; Ogden and Lobel, 1978; Dubin, 1982; Horn, 1989). Similarly, fish being pursued by eels may as a result be vulnerable to other predators (Karplus, 1978; Abrams et al., 1983). Thus, predatory fish may benefit from the activities of fish or other organisms that change prey behavior. If both species are predators, there may also be an indirect negative effect if they compete for prey. Most of these examples are little more than anecdotes, and it is unknown how much these effects actually contribute to the daily foraging success of predators.

Coral Zonation Patterns Indirectly Created by Damselfish

Territorial defense by herbivorous damselfishes creates a web of indirect effects. These are reviewed in more depth by Hixon (Chapter 10). In the Gulf of Panama, pocilloporid corals occupy 80–85% of the cover in shallow (0–6 m) water, whereas massive corals, especially *Pavona gigantea,* predominate at lower densities (ca. 18% cover) in deeper (6–10 m) waters. This zonation pattern is facilitated both directly and indirectly by the herbivorous damselfish *Eupomacentrus (= Stegastes) acapulcoensis. Eupomacentrus* directly damage *Pavona* in shallow water, thereby creating free space for their algal lawns.

Damselfish are less effective at removing tissue from highly branched pocilloporids, and therefore pocilloporid corals do better than *Pavona* inside fish territories. However, pocilloporids are more severely affected by pufferfish and parrotfish grazing than is *Pavona. Eupomacentrus* excludes these corallivores from its territories, thereby indirectly favoring the survival of pocilloporids in shallow water (Wellington, 1982). Presumably, the removal of *Eupomacentrus* would lead to a fairly rapid decrease in pocilloporid cover in shallow water, followed by a slow increase in *Pavona* cover; however, such an experiment has not been done.

Mediation of Acanthaster Predation on Corals by Coral Symbionts and Damselfishes

The crown-of-thorns starfish *Acanthaster,* when common, is the major predator on corals in most reef habitats. However, not all corals are equally likely to be eaten by *Acanthaster.* Corals that harbor invertebrate symbionts, or corals that occur in damselfish territories, gain indirect protection from predation via these associated fauna. The following accounts are based primarily on the reviews of Birkeland and Lucas (1990) and Glynn (1990b).

Crustacean obligate symbionts are the xanthid crabs *Tetralia* spp., which occur on acroporid corals; *Trapezia* spp., which occur primarily on pocilloporids; and the alpheid shrimp *Alpheus lottini,* which also occurs primarily on pocilloporids. Corals survive without symbionts in the absence of predators (but see Glynn, 1983), but these crustaceans cannot survive without the corals, from which they obtain shelter and food. If *Acanthaster* approach a defended coral, the crabs and shrimp detect the starfish via chemical cues (Glynn, 1980) and harass it so sufficiently that it may move away (Glynn, 1976). Effectiveness of the defense depends on the number of crustacea on the coral and how aggressively they attack the starfish. In two experiments, symbionts more than halved the occurrence of predation on defended colonies. The presence of crustacean symbionts presumably has a dramatic indirect effect on coral community structure following outbreaks of *Acanthaster* (Glynn, 1987), but this has not been experimentally addressed.

The presence of coral symbionts can lead to an associational defense between coral species if corals with symbionts happen to surround corals that lack symbionts, the latter deriving benefit from the defenses of their neighbors (Glynn, 1985, 1990b). Although this effect has been observed to occur at least occasionally, it has not been studied experimentally.

A few other invertebrates have been implicated in coral defense. Living massive *Porites* spp. are frequently colonized by boring bivalves and polychaetes; these organisms deter *Acanthaster* from feeding in their immediate vicinity, leading to remnant patches of coral that survive *Acanthaster* outbreaks (DeVantier et al., 1986; Birkeland and Lucas, 1990). These symbionts may also have negative effects on *Porites* by weakening its skeleton.

As discussed by Hixon (Chapter 10), the direct and indirect effects of territorial damselfishes are manifold. In some cases, territorial damselfishes may indirectly benefit corals by excluding *Acanthaster* from their territories (Glynn and Colgan, 1988; Birkeland and Lucas, 1990). Few studies have described this indirect interaction, most are observational or correlative rather than experimental, and it is not clear if, or under what conditions, the indirect positive effect of damselfishes on corals outweighs the direct negative effect of damselfishes killing corals to enlarge their algal lawns and the indirect negative effect of damselfish algal lawns limiting coral recruitment. These three effects probably operate on very different timescales, complicating an experimental approach to the issue.

Presence of Predators Affects Grazing Behavior of Herbivores

The intensity of herbivory is not uniform across the reef: in general, shallow-reef slope habitats suffer high rates of herbivory whereas deep sand plains, lagoons, and reef flats experience lower rates of herbivory (Hay, 1981b, 1984a; Horn, 1989). The border between high-relief reef habitat and low-relief seagrass beds is often characterized by a bare zone or "halo" of sand maintained by intense herbivory (Randall, 1965; Ogden et al., 1973; Ogden and Lobel, 1978; Glynn, 1990b). This spatial variation in herbivory is probably caused by predation. Herbivorous fishes and urchins do not stray far from high-relief habitats because these areas offer the best protection from predators. Similarly, herbivorous fishes may avoid deeper water because predators are more effective at lower light levels (Hobson, 1991; McFarland, 1991). As a consequence, plant abundance is higher and species composition is different in these "refuge" areas. Thus, this variation in plant community structure may be indirectly caused by the threat of predation on herbivores.

A variety of studies has documented the link between algal community structure and the intensity of herbivory in different reef habitats (reviewed in Horn, 1989), and *Diadema* homing behavior is known to correlate with predator abundance (Carpenter, 1984). However, little data exist on the actual rates of mortality of herbivores due to predation in these different habitats. In fact, Shulman (1985) demonstrated that rates of mortality of tethered juvenile fish were higher near to reefs than far away. Thus, the link between predators and herbivore behavior is still somewhat speculative (Hay, 1991), and the requisite experiments may now be impossible in many areas due to overfishing of predatory fishes. However, the general phenomenon of consumers staying close to shelter when predators are present is well documented in other aquatic systems (e.g., Werner et al., 1983; Werner and Hall, 1988; Power and Matthews, 1983; Power, 1984, 1987, 1990; Power et al., 1985; Sih et al., 1988; Huang and Sih, 1990; Hixon, 1991; Sih and Kats, 1991; Werner, 1992) and is the most plausible explanation for the patterns observed on the reef.

11.1.3. Chemical-Response Indirect Effects

Chemical ecology is a relatively new discipline, especially in marine systems, where the field has made major strides in recent years (Hay and Fenical, 1988; Paul, 1992). To date, we have only begun to identify the diversity of interactions in reef ecosystems that are mediated by secondary chemistry.

Sequestering of Dietary Secondary Metabolites

Many marine algae, angiosperms, and sessile invertebrates such as sponges produce high concentrations of secondary metabolites (so-called because they do not play a known role in basic metabolism and physiology). These secondary metabolites may serve a variety of functions; the one that has received the most experimental attention is that of deterring grazing (Hay and Fenical, 1988). However, a variety of opisthobranch gastropods such as nudibranchs, sacoglossans, and sea hares feed selectively upon chemically rich plants or animals and sequester secondary metabolites from their diets into their own tissues (Faulkner, 1992). If these sequestered compounds then defend the opisthobranchs from their own predators, this would represent a chemically mediated indirect negative effect of the plant or sessile invertebrate upon the predator(s) of the opisthobranch. What evidence supports this linkage?

Natural products chemists have isolated a large number of secondary metabolites from opisthobranchs and their foods. In many cases, identical compounds are found in both, suggesting a dietary origin for some opisthobranch metabolites (Faulkner, 1984a, b, 1986, 1987, 1988, 1992; Carefoot, 1987; Karuso, 1987), while other metabolites are thought to be produced by the opisthobranchs themselves. Conclusive evidence that some opisthobranchs sequester secondary metabolites from their diets has come from a variety of sources: radiolabeling of dietary metabolites (Stallard and Faulkner, 1974b), the demonstration that the particular metabolites found in opisthobranchs varied geographically and/or as a function of natural variation in diet (Yamamura and Hirata, 1963; Stallard and Faulkner, 1974a; Faulkner et al., 1990; Pennings, 1990; Paul and Pennings, 1991), and the demonstration that opisthobranchs could sequester a variety of secondary metabolites incorporated into artificial foods (Pennings and Paul, 1993).

Evidence that sequestered secondary metabolites affect the palatability of opisthobranchs to predators is rare. However, a number of studies have found that some isolated metabolites do deter feeding by potential predators (e.g., Cimino et al., 1982; Thompson et al., 1982; Paul and Van Alstyne, 1988; Pawlik et al., 1988; Hay et al., 1989, 1990a; Paul et al., 1990; Paul and Pennings, 1991; Faulkner, 1992), or that animals feeding upon chemically rich diets are less palatable to some predators than animals feeding upon chemically poor diets (Pennings, 1990).

This indirect linkage between the diet of opisthobranchs and their predators is especially interesting in the case of opisthobranchs that have a generalized diet or that feed upon hosts with variable secondary chemistry, because it presents the possibility that individual opisthobranchs might be differentially defended based on their diet. Many opisthobranchs naturally eat more than one host species; this affects their secondary chemistry (Yamamura and Hirata, 1963; Stallard and Faulkner, 1974a; Faulkner et al., 1990; Pennings, 1990), but the effect that this has upon their palatability has rarely been investigated (Pennings, 1990) and remains an interesting area for further research.

Use of Chemically Defended Plants as Domiciles

The herbivorous amphipod *Pseudamphithoides incurvaria* lives in a domicile that it constructs from the chemically defended alga *Dictyota barayresii.* Amphipods inside these domiciles were rejected by fish, but exposed amphipods or amphipods inside domiciles constructed from the palatable alga *Ulva* sp. were rapidly eaten by fish (Hay et al., 1990b).

Induced Defenses Affecting Multiple Grazers

Induced defenses occur when an organism increases its level of chemical or morphological defenses after detecting a potential grazer; similar facultative responses can also occur to competitors (Havel, 1987; Harvell, 1990). Induced defenses are common in terrestrial plants (Havel, 1987; Bryant et al., 1988; Myers, 1988) and in marine systems they have been documented in a byrozoan (Harvell, 1984, 1986), a barnacle (Lively, 1986), and two algae (Lewis, 1986; Lewis et al., 1987; Van Alstyne 1988, 1989). If these increased defenses reduce the ability or inclination of a second species of grazer to consume the prey, the two species of grazers will have an indirect negative interaction that is mediated by the induced defenses of their prey (Faeth, 1988). To my knowledge, there is no existing evidence for this sort of an indirect interaction in coral-reef systems, but such interactions are likely whenever prey are consumed by multiple grazers and can facultatively alter their commitment to defense (Havel, 1987; Faeth, 1988).

11.2. Relative Importance on Coral Reefs

Are indirect interactions more or less important on coral reefs than elsewhere? Coral-reef communities are composed of many species, most of which are rare (Endean and Cameron, 1990). Must we consider not only the direct but also the multitude of indirect effects between these species in any ecological study on the reef? If the answer is yes, the impossibility of doing all the requisite manipulations (section 11.4) could prohibit comprehensive experimentation and severely impair our ability to explain ecological phenomena. As Schoener (1993) has previously stated, ecologists have much at stake in this issue.

Two intuitive arguments with opposite conclusions could be advanced concerning the importance of indirect interactions in high-diversity systems. On one hand, we could argue that indirect interactions are very important, that the high diversity of coral-reef communities results in every species being affected by a complex web of direct and especially indirect interactions (Levine, 1976). Although both direct and indirect interactions might be weak because most species occur at low densities, the huge number of indirect effects could result in their summed effects being considerable.

On the other hand, we could argue that indirect interactions are not particularly important on coral reefs. The magnitude of any single indirect effect should quickly diminish along a chain if it has many branches. For example, if a predator eats five prey species, each of which in turn eats five prey species, and so on, then a species three links away indirectly contributes to only $\frac{1}{125}$ of the diet of the first predator. In a perfectly deterministic world, a multitude of weak effects could have a considerable summed effect.

However, the real world is often stochastic, and it seems likely that weaker effects would be swamped by stochastic "noise" in the system (Schoener, 1989; Mills et al., 1987). For example, recruitment limitation can preclude even very strong effects from manifesting themselves. Although *Diadema* and *Echinometra* were found to compete in Jamaica (Williams, 1981), mass mortality of *Diadema* at the same site did not result in any increase of *Echinometra,* presumably because of the absence of recruits (Hughes et al., 1987). Moreover, although *Diadema* clearly competed intraspecifically for food (survivors of the mass mortality increased in size dramatically), the density of *Diadema* has not increased since the mass mortality, due to an almost complete lack of recruitment (Hughes et al., 1987; Hughes 1994; de Ruyter van Steveninck and Bak, 1986; Lessios, 1988; Levitan, 1988; Karlson and Levitan, 1990). Low and/or variable recruitment is an important factor affecting the population dynamics of many coral-reef species (Hughes, 1985; Wellington and Victor, 1985, 1988; Warner and Hughes, 1988; Doherty, 1991). Recruitment variation, together with other sources of stochastic variability, may often prevent indirect effects from operating along long chains (Mills et al., 1987; Schoener, 1989, 1993; Birkeland and Lucas, 1990).

More generally, Paine (1980, 1988, 1992) argues that many species interact weakly, and that these weak links can usually be ignored in community theory (however, Endean and Cameron [1990] take exactly the opposite view for reefs). If weak links can safely be ignored, it may be that the theory required to understand coral-reef communities will be surprisingly simple considering the diversity of the communities. If many coral-reef species interact only weakly with others, we may be able to focus only on direct and indirect interactions of species of unusually high influence and thereby understand the primary biotic forces structuring the community (assuming that postrecruitment effects structure reef communities at all—a much debated issue [Sale, 1991]). Unfortunately, determining which species interact strongly and which weakly usually requires a great deal of experimental work—it cannot necessarily be deduced simply from food web structure (Paine, 1980, 1992).

Finally, many extended chains of indirect effects may be of opposite sign and effectively cancel each other. From a practical point of view, then, it may be possible to ignore them in certain situations. This is a potentially dangerous route, however, because if all the indirect effects are not well studied, it is unclear how to decide which can and cannot be safely ignored.

The preceding arguments suffer the limitations of all verbal models. Unfortunately, theoreticians have only just begun to address these issues more rigorously, and there is debate over the appropriate theoretical approaches (Wiegert and Kozlowski, 1984; Paine, 1988; Loehle, 1990; Patten, 1990; Kenny and Loehle, 1991; Polis, 1991). It is certainly premature to draw any firm conclusion from the preliminary results to date. The interested reader should consult Briand (1983), Yodzis (1988), and Schoener (1993) for more information.

Similarly, our understanding of indirect interactions in natural communities is in its early stages, and is probably insufficient to resolve this issue. For example, the best documented examples of indirect interactions to date involve strongly interacting species in relatively simple systems (e.g., Dungan, 1986, 1987; Wootton, 1992, 1993). Is this because indirect interactions are most important in these systems, or merely that simple systems are most amenable to experimentation? Sih et al. (1985) found in a survey of the literature that indirect effects were more commonly observed in experiments on predation in temperate systems than in tropical systems, but they felt that this conclusion was premature given the relatively small number of experiments performed in tropical systems. Many purported examples of indirect interactions on coral reefs involve species of unusual individual effects on the reef (e.g., *Acanthaster, Diadema*). Does this indicate that weakly interacting species do not have important indirect effects, or only that strongly interacting species have been best studied?

Further purported examples of indirect effects on reefs involve groups of ecologically similar species that have been lumped (e.g., all herbivorous fishes, all predatory fishes, all corals, all algae). This could indicate that similar species are important when taken as a group, but have negligible influence alone, or it could simply reflect that little experimental attention has been paid to the interactions of individual species. For example, acanthurids and scarids are often lumped in experiments as "herbivorous fishes," yet they almost certainly differ strongly in their feeding preferences and effects on algae (Lewis, 1985; Choat, 1991; Schupp and Paul, 1994). Paine (1988) comments on the dangers of lumping species together while constructing theory.

In summary, although we know enough about indirect effects to consider (and perhaps worry about) their importance on the reef, we do not know enough about them to resolve some of the troubling issues that this consideration raises. To make the situation worse, considerable practical difficulties constrain our ability to experimentally study indirect effects on reefs. This is the topic of the next section.

11.3. Some Practical Considerations

We saw earlier that most examples of indirect effects on coral reefs lack complete documentation. To understand why this is so, we must consider the practical problems of studying indirect effects, and of doing fieldwork on coral reefs in general.

The study of indirect effects, especially in complex communities, is severely constrained by the requirements of sound experimental design. This is best illustrated by a hypothetical example (Fig. 11–10). Consider a simple experiment consisting of two treatments: the removal of species A and a control. Suppose that we knew that species A often ate species B, but rarely ate species C, and that we had specifically designed the experiment to determine if A significantly reduced the density of B. Furthermore, suppose that densities of species B increased and species C decreased in plots where A was removed, compared with

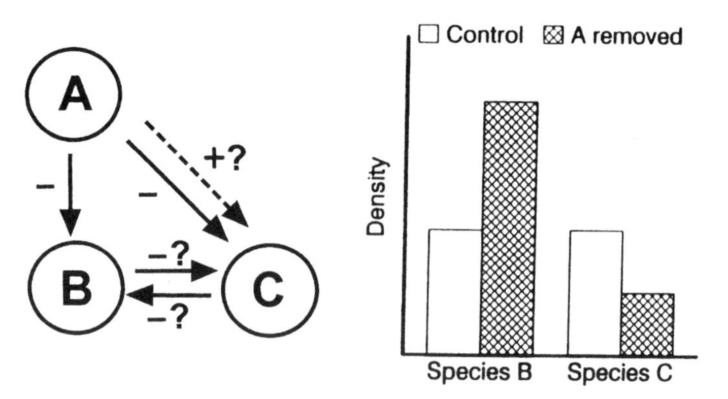

Figure 11–10. Hypothetical food web and experimental results. The predator, A, feeds primarily upon prey species B. When A is removed, the density of B increases, and the density of C decreases. We might conclude that B and C were competitors, and that A had an indirect positive effect on C.

control plots. We would probably conclude that this experiment demonstrated, as expected, a direct negative effect of A on B. The decrease of C when A was removed would probably be explained in a post-hoc way as resulting from competition between B and C, and thus representing a positive indirect effect of A on C. The more we knew about the natural history and ecology of all three species involved, the more confidently we would assign this indirect link from A to C.

The reader will recognize that studies resembling this hypothetical scenario are frequently encountered in the literature. In most cases, the explanation offered by the author(s) for observed indirect effects is probably correct. However, the experimental design is not sufficient for a complete understanding of the system, because the relative contributions of the direct negative effect of A on C (caused by occasional predation) and the indirect positive effect of A on C (caused by A suppressing B's density) were not teased apart and individually quantified. All we know is their combined effect. In a more disconcerting example, if the direct and indirect effects of A on C were similar in magnitude, they would cancel, and we would probably wrongly conclude that A interacts with C weakly or not at all.

To explicitly measure the direct and indirect effects of A on C, we must perform two more experiments with at least two treatments each: (1) manipulate the density of A while holding B constant, and (2) manipulate the density of B while holding A constant. If we were also interested in the direct and indirect effects of C on A, of C on B, or of B on C, even more treatments would be required. When we consider that we will often wish to include more than two levels of each manipulation, that many more than three species may be involved, and that each treatment must be adequately replicated, the demands of the experimental design can easily become very severe (so much so that Loehle [1990] suggests that the attempt to study such things in complex communities is futile). For example, even to replicate the minimal six-treatment design five times would require 30 experimental units. If manipulating the densities of species A or B required large cages or isolated patch reefs, this design might prove to be logistically impossible.

A second concern is that to detect trophic-linkage indirect effects, long-term experiments may be called for (Bender et al., 1984; Yodzis, 1988). Trophic-linkage effects, by definition, operate through changes in the densities of the species involved. For long-lived, sporadically recruiting, or slow-growing species, such as many reef fishes and corals (Hughes, 1985; Warner and Hughes, 1988; Doherty, 1991), the required dynamics might occur on a scale of decades or even centuries. For example, a perturbation that precluded successful recruitment of corals might not manifest itself as reduced coral cover until the adult colonies gradually died off over the course of decades. The more species involved in the trophic-linkage chain of interest, the longer it will take for a change in one species to manifest itself as a change in the target species.

As the chain of species dynamics caused by a given manipulation unfolds, a variety of "transient dynamics" may occur (Bender et al., 1984). For example, a species with a short generation time might first increase following the removal of its generalist predator, then decrease as a superior competitor with a longer generation time, also freed from predation by the same manipulation, slowly increased in numbers. Thus, short-term experiments might completely mislead us about the long-term consequences of a given perturbation.

Yodzis (1988) suggests that experiments should be run twice as long as the sum of the generation times of all the species in the chain of interest. To run experiments on indirect interactions involving most species of corals or reef fish using this criterion would require herculean efforts. Consequently, the experiments necessary to detect some trophic-linkage indirect effects may be impractical for most experimenters, and alternatives such as natural experiments may have to be used, perhaps in combination with a variety of other approaches such as long-term monitoring, short-term experiments, and modeling.

In many cases, researchers studying trophic-linkage indirect effects have observed changes in densities caused by movement of individuals into or out of experimental areas (e.g., Williams, 1981; Hay and Taylor, 1985). The link between such redistributions of individuals and changes in overall population densities through altered recruitment or death rates is assumed, but rarely documented. Although some long-term experiments have been performed, the overwhelming majority of ecological field studies are less than 3 years in duration (Schoener, 1983; Goldberg and Barton, 1992). In contrast, behavioral and chemical-response indirect effects may occur much more rapidly, and thus be more amenable to experimentation. For further discussion on appropriate approaches to the study of indirect effects, the interested reader should begin with Bender et al. (1984), Yodzis (1988), Fairweather (1990), Huang and Sih (1990), and Strauss (1991).

The above issues are compounded by the practical difficulties of doing field experiments of any sort in coral-reef habitats. Many corals grow slowly, so disruptive experimental manipulations may be avoided for ethical reasons, or may be legally prohibited. Many of the species likely to be involved, such as fish, *Acanthaster,* or *Diadema,* are sufficiently mobile to make it difficult to constrain them to isolated patch reefs; however, cages alter light and current regimes, and quickly foul with a variety of algae and invertebrates, introducing a suite of potential cage artifacts. Moreover, time spent underwater is limited by the physical rigors involved and the risk of decompression sickness.

These constraints are formidable. However, similar obstacles have been overcome in other systems. For example, replicated long-term experiments involving very large cages have revealed a variety of indirect effects operating in desert granivore systems (e.g., Davidson et al., 1984; Brown et al., 1986; Thompson et al., 1991). Similarly, replicated experiments using stream pools have provided insight into indirect effects operating between birds, fishes, and algae (Power and Matthews, 1983; Power, 1984, 1987, 1990; Power et al., 1985). Finally, the use of surface-supplied air and heated water allowed Schmitt (1987) sufficient bottom time to demonstrate indirect effects operating in a temperate marine system of molluscs and their predators. Thus, the logistical problems facing coral-reef researchers are not unique, and, with the possible exception of detecting trophic-linkage effects among long-lived organisms, can probably be overcome.

11.4. Future Directions

Indirect interactions are probably very important in structuring reef communities; however, our understanding of indirect interactions on the reef is incomplete. Consequently, there is much that remains to be done, and the area should be a productive one for future research. In the context of this book, three areas stand out as especially needing further attention.

First, most examples of indirect interactions from coral reefs are incompletely documented. Often, indirect effects are known to exist, and perhaps to override direct effects of an opposite sign, but the two have not been separated and individually quantified. Without information on each effect alone, it is difficult to predict how the summed effect would change in a slightly different ecological context or after a novel perturbation (such as an anthropogenic disturbance). Future work must overcome the difficulties inherent in performing rigorous field research on coral reefs, and must individually quantify direct and indirect effects. There is a real need for long-term experiments to document trophic-linkage indirect effects; however, these experiments may be difficult for most researchers to sustain.

Second, it is currently unclear whether indirect interactions have more or less influence on coral reefs than elsewhere. Verbal arguments can be made on both sides, but rigorous theory is only beginning to develop. This issue can be approached on two fronts. The further development of mathematical theory should substantively clarify the issues involved and suggest patterns that might occur in the field. At the same time, further studies on indirect effects in coral reefs can be compared with similar studies in other systems, perhaps leading to the emergence of an empirical generalization regarding the importance of indirect effects on reef communities.

Third, one is struck, upon reading the coral-reef literature, by the paucity of mathematical modeling compared with the literature from other habitats (but see Hughes, 1984; Done, 1987; Warner and Hughes, 1988; Paulay and McEdward, 1990). Attempts at modeling coral-reef communities might clarify the current state of our knowledge, indicate critical experiments that need to be done, and suggest possible responses of reefs to anthropogenic disturbances. Because of the difficulties in running long-term experiments, mathematical models could be critical in guiding management decisions.

11.5. Conclusion

Because a species can affect another with which it does not directly interact, or can affect another through both direct and indirect pathways, it can be difficult to predict the effect on reef communities of a given perturbation. Some of the indirect interactions discussed in this chapter, such as those involving herbivores, algae, and corals, play fundamental roles in structuring the reef community. However, in most cases, we poorly understand the exact nature and relative importance of direct and indirect links. Knowledge about these links is vital to an understanding of coral-reef communities, especially today when reefs are experiencing a variety of novel perturbations.

Human beings are affecting reef communities in a variety of ways including fishing, collecting for the aquarium and shell trades, pollution, and sedimentation. Although we sometimes understand these direct effects moderately well, we usually have little insight into important indirect effects that they may generate. Moreover, typical experiments may not last long enough to detect important trophic-linkage indirect effects. Without a better understanding of coral-reef community structure, some management decisions (or lack thereof) may lead to unpleasant surprises (Loehle, 1990). Managers must bear in mind that any given management policy may affect not only the species that are explicitly the focus of the policy, but also a host of other species.

12

Geographic Differences in Ecological Processes on Coral Reefs

Charles Birkeland

On the western coasts of continents, some coral communities receive concentrated nutrient input by upwelling of deeper waters from below the photic zone. Along eastern coasts of continents, coral reefs receive sediment and nutrients from higher ground by terrestrial runoff (Fig. 12–1). The coral reefs of atolls in central oceanic regions receive nutrients at lower rates and lower concentrations than do coral reefs along continental coasts. These large-scale differences in nutrient input among geographic regions (Fig. 12–2) create conditions that favor

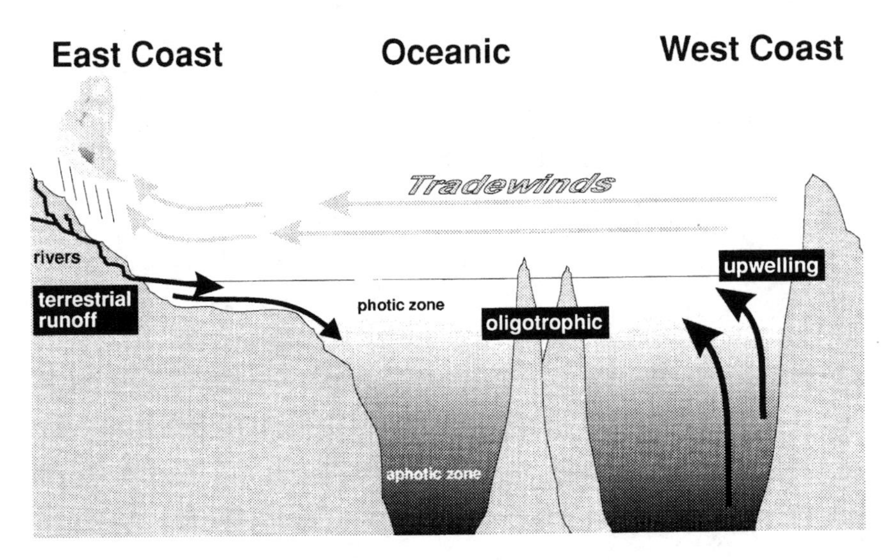

Figure 12–1. The trade winds bring both upwelling of nutrients to the eastern tropical oceans and rains with terrestrial runoff of nutrients to the western tropical oceans. The oceanic regions tend to have lower concentrations of nutrients in the surface waters.

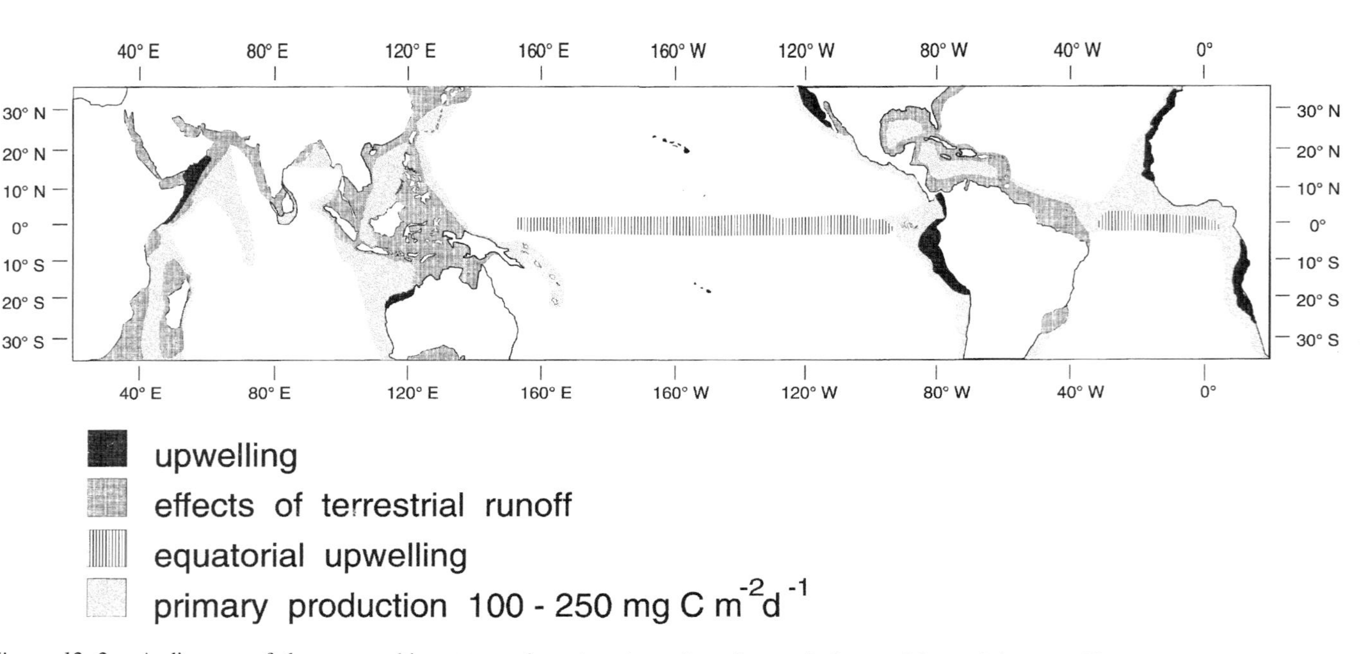

Figure 12–2. A diagram of the geographic pattern of nutrient input into the tropical seas. Most of the upwelling occurs on the western coasts of continents and terrestrial runoff is greatest on eastern coasts. The unshaded central portions of the oceans have average rates of primary production of less than 100 mg C m^{-2} d^{-1}. Compare with Figure 12–1.

the dominance of different kinds of organisms among regions (Fig. 12–3), with more abundant recruitment in areas of regular pulses of nutrient input. These regional differences are important to take into account when developing coral-reef management policies (Chapter 18).

Large-scale oceanographic processes bring about profound differences in patterns of recruitment and growth (Birkeland, 1987; Sherman, 1994). Ryther (1969) calculated that upwelling areas take up only about 0.1% of the ocean surface, but provide 50% of the world fishery catch, a 500-fold difference in yield per unit area in comparison with the world ocean. Abundant recruitment of fishes and other organisms with planktotrophic larvae occurs each year in regions of upwelling unless nutrient input is inhibited by an El Niño–Southern Oscillation (ENSO) event. The effects of a failure of upwelling and nutrient input are manifested at all levels in the food web. During the 1982–1983 ENSO (Glynn, 1990a), there was a decrease in primary productivity and density of phytoplankton, a temporary closure of much of the fishery industries, and large-scale mortalities of sea birds (85% in Peru) and marine mammals (30% for fur seals in the Galápagos). But on most years, the nutrient input from upwelling is associated with abundant recruitment and rich fisheries.

About 82% of all the sediment that is carried into the oceans of the world drains off the landmasses in the western tropical seas; only 2% drains off the coasts in the eastern tropical seas (Milliman and Meade, 1983; section 1.2 in Chapter 1; Fig. 12–1). Half the sediment being deposited into the oceans of the

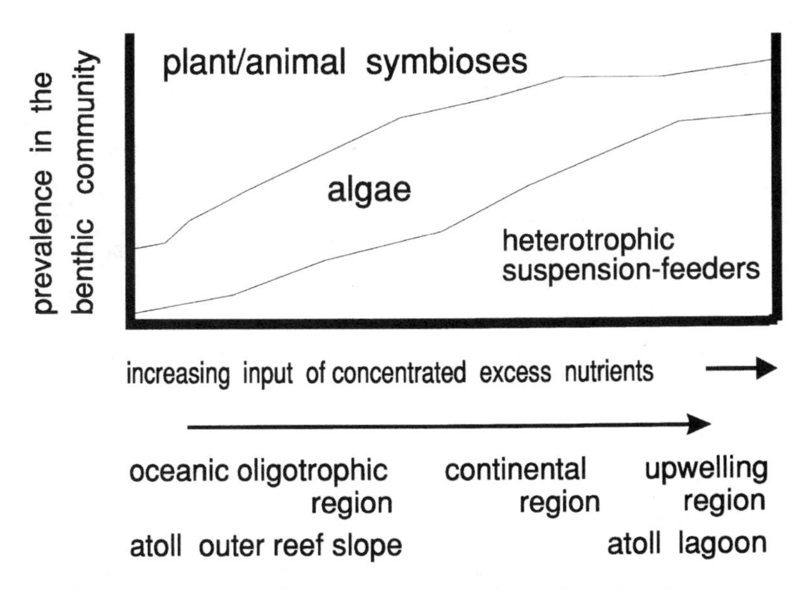

Figure 12–3. Change in dominant trophic characteristics of coral-reef communities as a function of nutrient input both geographically and locally.

world is coming off the high islands of the southwestern tropical seas, the center of coral-reef diversity (Milliman, 1992; Fig. 12–2). The smaller rivers running off the steeper slopes of the high islands have greater sediment yields (load per unit drainage basin area) than do larger rivers off the continents because the larger rivers such as the Amazon tend to deposit sediment in their own beds and deltas.

Plant-animal symbioses dominate areas of low nutrient input, such as oceanic atolls, where efficient recycling of nutrients is favored (Fig. 12–3). At intermediate levels of nutrient input, such as along eastern coasts of continents and in lagoons of oceanic high islands, algae are favored because small plants respond and grow more rapidly than do plant-animal symbioses (e.g., corals and giant clams) to pulses of concentrated nutrients from terrestrial runoff. Corals are usually able to persist because grazing of algae by herbivores allows space for recruitment of juvenile corals in regions of intermediate nutrient input. At very high levels of nutrient input, phytoplankton preempts light and nutrients and restricts to shallow water the **compensation depth** (the depth at which there is just enough light that photosynthesis equals respiration) of benthic algae. The phytoplankton-based food web in which space is dominated by heterotrophic suspension feeders replaces the benthic alga-based food web in regions of concentrated nutrient input (Fig. 12–3).

This chapter reviews the physical processes that bring about geographic differences in sources and rates of nutrient input into coral communities, and how these differences in nutrient input affect community structure and the relative strengths of ecological processes. Rate and concentration of nutrient input change the abundance and reliability of recruitment, the growth rates of organisms, the intensity of competition and predation, and the dominant trophic structure of the coral-reef ecosystem. The relationships between nutrient input and biodiversity, and the effects of these factors on community function, will be described. This chapter will conclude with considerations of how these differences in community structure and ecological processes in coral communities in different geographic regions determine that the resources should be managed differently in these different regions.

12.1. Physical Processes That Bring About Geographic Differences

12.1.1. Longitudinal Differences

The fundamental force behind both longitudinal and latitudinal differences in ecological processes on coral reefs is the sunlight falling at a high angle on equatorial regions throughout the year. Since the angle is more direct near the equator, the heat energy is concentrated over a smaller surface area and so the air is warmer on the average near the equator (Fig. 12–4A). Furthermore, although the Earth receives an average of 12 hours per day of sunlight everywhere, the

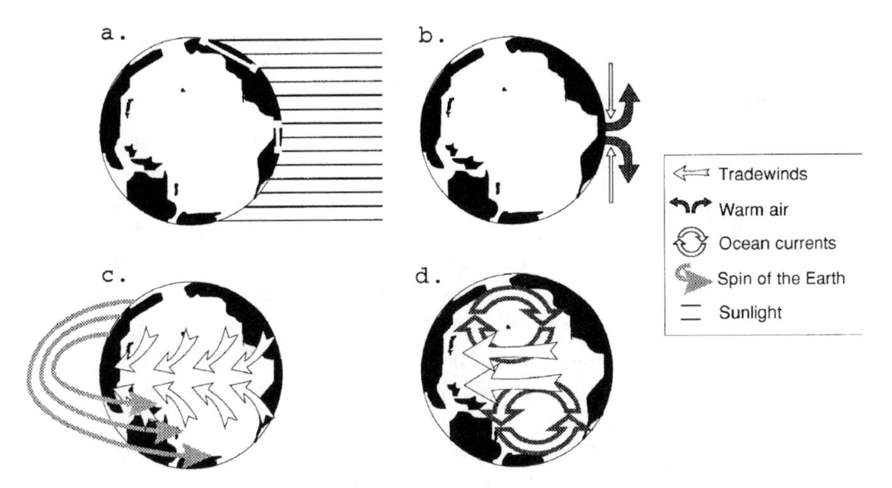

Figure 12–4. The physical forces that produce the geographic variation in ecological process of coral reefs. **(A)** Energy from the Sun is more concentrated throughout the year in the tropics. **(B)** Excess heat in the tropics produces low pressure, rising air, with replacement air coming in from higher latitudes. **(C)** The spin of the Earth causes the replacement air to move westward relative to the surface of the Earth, creating the trade winds. **(D)** The trade winds and the spin of the Earth cause the Coriolis effect, bringing warm seawater and rainfall to the western tropical seas and upwelling of cool nutrient-laden seawater to the eastern tropical seas.

sunlight is distributed more evenly through the seasons in the tropics and so there is no annual cooling.

The climatic and oceanographic patterns of the world are driven by the excess heat in the tropics and the deficit at higher latitudes. As the equatorial air warms, it expands and rises, and is replaced by cooler air from the north and the south (Fig. 12–4B). Since Earth spins on its axis from west to east, the trade winds (replacement air) effectively (from the perspective of creatures on Earth's surface) move westward rather than directly north or south (Fig. 12–4C). The trade winds blow the main equatorial currents westward. The equatorial currents turn away from the equator in the Atlantic and the Pacific when they are blocked by the continents, moving clockwise in the Northern Hemisphere and counterclockwise in the Southern Hemisphere (Fig. 12–4D). This pattern is called the Coriolis effect.

From the geographic perspective of coral-reef management, the Coriolis effect produces the most important factor, the pattern of nutrient input. The nutrients come up from the depths on the western coasts of continents and come down from land with sediments on the eastern coasts (Fig. 12–1). On the western margins of continents, the trade winds push ocean currents away from the coast. The westward-moving waters are replaced in part by the nutrient-rich upwelling waters from deep below the photic zone. Sites of upwelling do exist east of continents, for example, on the coasts of Somalia, Yemen, Indonesia, and

Venezuela; but the major sites of upwelling are prevalent on the western coasts of continents, and this determines important geographic differences in the nature of coral communities (Fig. 12–3).

By the time the waters in the photic zone reach the oceanic central Pacific, the nutrients have been taken up by phytoplankton and the oceanic waters are generally nutrient poor (Figs. 12–1 and 12–2). In these oligotrophic waters, a relatively high proportion of nutrients is bound into the biomass in the shallow waters that encompass coral-reef communities. On coral reefs of atolls in these central oceanic regions, there is considerable recycling of nutrients in the coral-reef system and comparatively little input from outside sources (Chapter 7).

Trade winds blowing across the tropical ocean waters onto the eastern coasts of continents are laden with moisture (Fig. 12–1). Water has a higher latent heat capacity than does soil, and so the continental landmasses heat more rapidly during the summer than do the oceans. Therefore, the warmer air rises over the continents and the moist air is drawn off the tropical oceans. As these moisture-laden monsoonal winds rise over the coastal hills, the moisture precipitates. Therefore, the major rivers that carry sediment into tropical coastal areas are generally on the eastern coasts of continents (Figs. 12–1 and 12–2; section 1.2 in Chapter 1).

The influence of rivers on marine life can be observed where human activities affecting rivers have had international consequences on biota. For example, immediately following the construction of the Aswan Dam on the Nile, the multinational fisheries catch for the eastern Mediterranean Sea decreased to 3.7% of its former levels (Aleem, 1972). Age-class strength of economically important marine invertebrates and fishes are correlated with river discharge strength (Sutcliffe, 1972, 1973; Wolff et al., 1987). The Amazon River creates a large gap in coral distribution and separates the corals of the western tropical Atlantic into two biogeographic provinces. Likewise, major breaks in the distribution of corals in the Bay of Bengal and Andaman Sea (Indian Ocean) are formed by river outflow.

A typical influence of nutrients on geographic scale can be witnessed in one day by a visitor to Panama. The Pacific and Atlantic coasts are only about 70 km apart at the Isthmus of Panama, yet the Pacific coast supports a large fisheries industry while the Atlantic coast of Panama supports only subsistence fishing and small-scale fishing enterprises. Vast schools of fishes and hundreds of sea birds can be seen on an outing to coral communities in the regions of upwelling along the Pacific coast, while a diverse assemblage of fishes, with each species swimming alone, in pairs, or in small schools, is the usual case on coral reefs in regions with less concentrated nutrient input.

Geographic and climatic patterns are usually more dependable than are smaller-scale local weather trends, but conversely, when large-scale climatic patterns are disrupted, the effects can be more dramatic. While the Coriolis effect and upwelling in the east are relatively dependable, they can be disrupted by El Niño–

Southern Oscillation (ENSO). ENSO can cause major changes in the marine ecosystems, mass mortalities throughout the food web, and long-term deterioration of the coral reefs, especially in the eastern Pacific where the ecosystem normally receives abundant nutrient input (Glynn, 1990a).

The Coriolis effect also forces typhoons or tropical cyclones to generally move from east to west because Earth spins eastward below the atmosphere in which cyclones are suspended (Fig. 12–4). Typhoons or hurricanes (the names for major cyclones in the Pacific and Atlantic, respectively) are created by rising warm air and rapid replacement by air in "feeder bands" moving in laterally. Since rising warm air creates and provides the energy for cyclones, tropical cyclones will probably increase in frequency and intensity with global warming. Typhoons usually increase in intensity as they move westward. The increases will probably be greater in the western tropical oceans (e.g., Guam) than farther east (e.g., Samoa). This is an example of how differences of some variables among geographic regions, for example, storm frequency and strength, will probably become even greater than they are now.

The Coriolis effect determines the direction of major current patterns away from the equator in the western sides of tropical oceans and toward the equator on the eastern sides (Fig. 12–4D). Thus, coral-reef communities are more easily replenished with recruits from tropical waters after local extinctions in the western tropical oceans, especially at higher latitudes. This is a factor leading to greater species richness in coral-reef communities in the western sides of oceans (Chapter 14). Coral-reef communities have a greater latitudinal range in the western sides of oceans. Aggregations of individuals of species carried by currents to environments with ecological conditions beyond their limits of successful reproduction ("pseudopopulations") are also more frequent at high latitudes in western sides of oceans.

The greater longitudinal extent of the Pacific produces a greater decrease in species and habitat diversity from west to east across the Pacific, counter to the main currents (Chapter 14). A lesser degree of trophic and population interconnections between coastal habitats in the Pacific than in the Atlantic may result partially from less reliable availability of noncoral habitats in Oceania (Birkeland and Amesbury, 1988). The occasional ENSO changes direction of prevalent currents and periodically allows for species introductions and replenishment (Richmond, 1990).

12.1.2. Latitudinal Differences

Throughout much of their range, coral reefs generally exhibit rates of metabolism and calcification independent of latitude (Smith and Kinsey, 1976; Crossland, 1988). However, when the latitudinal limits of reef growth are approached, a relatively steep decrease of reef growth with latitude becomes apparent. Gross carbonate production of individual corals and coral-reef accretion by corals decline in a linear fashion as a function of latitude in the Hawaiian Archipelago (Grigg, 1982). The point at which the rate of reef accretion becomes less than the rates of reef subsidence and/or erosion, beyond which the reef begins to "drown," is termed the Darwin Point (Fig. 3–6 in Chapter 3). The Darwin Point is the threshold for oceanic atoll formation, and is potentially at higher latitudes on the western sides of tropical oceans than on the eastern sides.

As a rule of thumb, hermatypic corals are restricted to regions where the average seawater temperatures are 18–30°C. Because of the Coriolis effect on major current patterns, warm waters and coral reefs extend further from the equator on the western sides of oceans. However, the temperature range tolerated by corals differs among geographic regions, with the lethal limits of particular species being higher where the mean seawater temperature is higher (Fig. 15–3 in Chapter 15).

Light penetrates the ocean to greater depths when and where the sun is directly overhead. The proportion of the year in which the Sun is directly overhead decreases with latitude, and the Sun is never directly overhead north of the Tropic of Cancer or south of the Tropic of Capricorn. Therefore, we might have predicted that the maximum depth of reef formation would decrease with latitude. However, this may not be the case (Veron, 1995).

12.2. Geographic Differences in Characteristics of Dominant Species

In regions with rich nutrient input from external sources, small organisms can preempt resources from larger organisms. Whether planktonic or benthic, small plants take nutrients from the water column more rapidly, and grow and reproduce more rapidly, than do larger plants with more complex structural attributes (Odum et al., 1958; Geider et al., 1986). An experimental comparison of two macroalgae of the Ulvophyceae showed that *Ulva curvata* is able to respond to brief inputs of ammonium with more effective uptake and more rapid growth than the more structurally complex *Codium fragile* (Ramus and Venable, 1987).

Phytoplankton in the water column can preempt light and reduce the compensation depth of benthic algae under conditions of high nutrient input (Twilley et al., 1985). If the compensation depth of benthic algae were at 100 m in water without phytoplankton, then at a density of only 2 g m^{-3}, phytoplankton would raise the compensation depth of the benthic algae to 3.5 m, restricting benthic algae to only the upper 3.5% of their potential range (Ryther, 1963). The single-celled phytoplankters take up nutrients and reproduce rapidly, decreasing both the nutrients and light available to the benthic algae (Geider et al., 1986). In some areas with especially rich nutrient input, phytoplankton virtually exclude the entire benthic algal food web (Kamura and Choonhabandit, 1986; Birkeland, 1987).

The same trends can be seen in coral-reef animals. The growth rates of small suspension-feeding animals, such as mussels and barnacles, are strongly associated with the amount of food in the water column (Page and Hubbard, 1987). The growth rates of large, colonial, plant-animal symbionts such as hermatypic corals do not respond as rapidly or to as great a degree to concentrated nutrient input (Kinsey and Davies, 1979). The growth of small corals has been found to be much slower than previously assumed (Wallace, 1985b; van Moorsel, 1988).

There may be some physiological constraints and rate limits associated with the physiological complexities of plant-animal symbioses (Chapter 5). When primary production of phytoplankton increases, zooplankters can increase rates of consumption and the secondary production of the zooplankton community can increase several-fold (12–16 mg C m^{-2} day^{-1} in tropical oceanic waters to about 75 mg C m^{-2} day^{-1} in upwelling regions [Barnes and Hughes, 1988]). In contrast, zooxanthellae and corals are in a complex physiological balance (Chapter 5). Increased nutrient input can sometimes cause a rapid increase in zooxanthellae. But this overcrowding can disrupt the physiological conditions of the coral, which may react by expelling the excess zooxanthellae. This constrains a response to nutrient fertilization with more rapid growth.

Community structure on shallow-water tropical hard substrata is determined to a large extent during recruitment. Competition for space among adult colonies is easier to see, but the events during recruitment are of greater effect on the eventual nature of the benthic community. In nutrient-rich upwelling regions in the tropics, the rapid takeover of space and accumulation of biomass by small, fast-growing suspension feeders such as barnacles, bivalves, sponges, ascidians, and bryozoans often preempt space from coral recruits (Birkeland, 1977). Although it is not unusual to find adult corals surviving and growing well in regions of upwelling, in nutrient-rich lagoons of high islands, near the mouths of rivers, and near sewer outfalls, these examples are most often individual colonies that have reached a refuge in size. These exceptional cases suggest that the physical environment itself is favorable for growth and survival of corals, but excessive concentrations of nutrients have an indirect effect of differentially favoring the algae and suspension-feeding animals over coral recruits in the competition for space.

Greater rates of nutrient input intensify competition for space by two mechanisms: faster growth of algae and suspension feeders and a greater density of recruitment (Barnes, 1956; Sutcliffe, 1972, 1973; Birkeland, 1977, 1982). In contrast, space is available for a longer period of time in the oligotrophic waters of the western Pacific, and the supply of larvae in the water column, rather than availability of space, is a major determinant of the size of recruitment (Birkeland et al., 1982).

Plant-animal symbioses, in contrast, are better adapted to low-nutrient environments (Muscatine and Porter, 1977). Coral communities are able to grow to a large size and live for years in clear, nutrient-poor oceanic habitats where active suspension feeders are unable to maintain positive energy budgets (Page and

Hubbard, 1987). The plant-animal symbioses inherit space in low-nutrient environments by default, not by outcompeting the suspension feeders. Likewise, the benthic algae do not outcompete the phytoplankton, but exist to the extent allowed by the factors that limit phytoplankton standing stock.

Oceanic coral reefs in oligotrophic waters are dominated by plant-animal symbionts (Fig. 12–3). As we move along a gradient of increasing nutrient input, for example, from an atoll to a high island or approaching a sewage outfall, benthic algae become more prevalent. In regions of even stronger nutrient input, the phytoplankton-based food web dominates and the substratum becomes occupied predominantly by heterotrophic suspension feeders (Smith et al., 1981; Rose and Risk, 1985).

The qualitative nature (life-history characteristics, physiology, morphology) of the dominant biota may influence ecosystem-level processes. The "turn-on/turn-off" point of reefs has been defined as the point at which the dominant processes change from accretion to bioerosion or back (Buddemeier and Hopley, 1988). Reefs are generally growing (accretion is dominant) when scleractinian corals and crustose coralline algae are prevalent. Bioerosion becomes more prevalent in areas of high nutrient input in which boring bivalves and/or sponges become abundant within the reef framework (Chapter 4; Highsmith, 1980), and sponges and algae become more prevalent than corals on the exposed surfaces.

Hatcher describes in Chapter 7 how energy and materials from filamentous algae tend to enter the herbivore food web, and energy and materials from more rigidly structured macroalgae tend to enter the detrital food web. Small plants of relatively simple morphology generally have higher net productivity per unit biomass and a potential for more rapid growth than do larger, more complex algae (Littler and Littler, 1980). If smaller organisms of relatively simple morphology have a higher net productivity per unit biomass (Geider et al., 1986), then in view of the prevalence of filamentous algae and small, rapidly growing suspension feeders (mussels, oysters, and barnacles) such as in regions of upwelling, we predict that the system yield may be greater in upwelling regions than in oligotrophic regions where scleractinian corals and crustose coralline algae are relatively prevalent. If the geographic pattern of nutrient input affects the life-history characteristics of the dominant organisms, then these matters should be taken into account when developing resource management programs (Chapter 18).

12.3. Regional Differences in Ecological Processes

12.3.1. Grazing and Predation Pressure

Areas with nutrient input from upwelling produce large stocks of fishes (Ryther, 1969; Gulland, 1976) and the age-class strengths of fishes and invertebrates are often correlated with annual river-discharge strength (Aleem, 1972; Sutcliffe, 1972, 1973; Wolff et al., 1987), presumably because of increased survival of planktotrophic larvae with a concentrated supply of food. Dense populations of predators and grazers that result from strong recruitment exert intense predation and grazing pressure. For example, the grazing pressure on benthic communities of the Pacific coast of Panama where upwelling occurs is greater than on the Caribbean coast (Glynn, 1972; Earle, 1972; Vermeij, 1978).

Standardized experimental comparisons of the intensity of grazing on sponges by fishes at three geographically separated locations in both the Caribbean and in the eastern tropical Pacific showed the grazing pressure in the eastern Pacific to average 25 times greater than the grazing pressure in the Caribbean (Birkeland, 1987). The eastern Pacific reefs were more densely populated with sponge-grazing angelfishes (*Holacanthus* and *Pomacanthus*) than were Caribbean reefs, and the sponges were smaller and more crytic in the eastern Pacific even though they grew rapidly when protected from grazing (Birkeland, 1987).

Two behavioral changes of predators can occur as population density increases, and these amplify the intensity of grazing and predation pressure. First, the *Holacanthus passer* in the eastern Pacific tend to forage in schools, occasionally as large as 4 or 5 dozen, while members of this genus in the Caribbean tend to forage alone or in pairs. This compounds the effects of large populations because each fish forages more efficiently when it is part of a school than when it is alone (Pitcher et al., 1982; Pitcher and Magurran, 1983; Pitcher, 1986; Wolf, 1987). Concentrated grazing by an aggregation of fishes could also have more intensive effects than the grazing by the same number of individual fishes that are more randomly or evenly distributed.

Second, when population densities of predators and grazers become large and prey become scarce, the consumers must broaden their diet in order to get enough to eat (Ivlev, 1961; Werner and Hall, 1974). Intense grazing by generalists can set back succession, increasing the proportion of early successional biota with a greater ratio of net to gross productivity (Birkeland et al., 1985). Refuge in patchy distribution and irregular abundance is not as effective for small prey against a variety of common generalist grazers or predators that forage in schools (Pitcher et al., 1982; Wolf, 1987), because generalists can forage widely, being maintained on alternative foods between patches of the prey in question. When foraging by generalists is intense, turnover in occupation of substratum increases and many small individuals of a variety of prey species may be grazed indiscriminately.

Chemical and morphological defenses are more characteristic of organisms in nutrient-poor environments (Coley et al., 1985). Morphological and chemical defenses may be ineffective if the prey individual is too small to be recognized by the grazer or predator. If grazing is intense enough, the prey may be bitten off incidentally before they reach a recognizable size. This may apply to coral recruits as well (Sammarco, 1980, 1985). Under these circumstances, chemical and morphological defenses may be less effective for survival to reproduction than is rapid growth and early reproduction. As grazing intensity increases along a gradient of increasing concentration of nutrients, the smaller, faster-growing species of benthic organisms are favored both in competition for space and by reproducing before being eaten.

Both the tendency of the predators toward more generalized diets and the responses of the prey toward investing in rapid growth and reproduction rather than defense (Coley et al., 1985)—conditions brought about by concentrated pulses of excess nutrient input—tend to increase net production, increase rates of population turnover, and shorten the food web.

12.3.2. Food Webs

The number of substantial trophic levels in food webs between the primary producers and humans decreases as nutrient input increases (Ryther, 1969). Nutrient input tends to shorten food webs by several distinct mechanisms (Hallock, 1987). Concentrations of rapidly growing and rapidly reproducing large-celled or chain-forming diatoms are supported in regions with nutrient input from upwelling and terrestrial runoff. These dense concentrations of large-celled phytoplankters support dense concentrations of relatively large herbivores such as anchovy and krill, which are then directly fed upon by tuna, sea birds, and marine mammals. In contrast, in nutrient-poor waters such as the tropical gyres, tiny coccolithophorids and other nanoplankton and picoplankton are the predominant primary producers. The primary consumers of the tiny phytoplankters are usually small, leading to added levels and interconnections in the food web before harvest by the largest animals in the sea.

Although coral reefs exist in tropical regions of upwelling, they are more prevalent in environments with low concentrations of nutrient input. Plant-animal symbioses are favored in competition for space in regions of dilute nutrient input (Muscatine and Porter, 1977; Lewin et al., 1983; Wilkinson, 1986; Birkeland, 1987). Large colonial animals such as corals add topographic complexity to the habitat. The development of topographically complex biological substrata facilitates the accommodation of more species in the habitat. The diversity of species on coral reefs and the advantages of recycling of nutrients both tend to lead to more complex food webs. On coral reefs, even fish feces have been observed to be fed upon (recycled) by corals and fishes. In fact, Robertson (1982) deduced that some fish fecal material may be recycled through five fishes before it reaches the coral or other benthic substrata.

Hallock (1987) further elaborates on several ways in which a low level of nutrient input leads to habitat diversity and species diversity, and therefore to a more complex food web. Habitat diversity is increased in low-nutrient environments because gradients extend across wider areas and are more stable. Oligotrophic waters are relatively clear and so the gradient in light attenuation is extended over a longer depth gradient. Low-nutrient waters are generally more stable in depth stratification because vertical mixing is characteristic of regions of nutrient input by upwelling and internal waves. A stable pattern of environmental heterogeneity could increase the potential for specialization and species diversity. Regions with strong pulses of nutrient input are less stable not only because of the fluctuations in magnitude of nutrient input, but because of the physical factors that accompany upwelling (water temperature change) and river discharge (salinity changes, turbidity, and sedimentation).

In regions of short food webs, the ratio of fisheries yield to gross primary production is higher than in complex communities with more trophic levels, such as coral reefs, because energy is lost at each step as matter is passed up the food web. Therefore, in regions of nutrient input such as upwelling areas, the fisheries yield per unit gross productivity would be substantially higher than in oligotrophic areas most favorable to coral reefs (Fig. 18–1). Ryther (1969) assigns one or two steps between phytoplankton and humans in upwelling systems, acknowledging that nutrient-rich areas of the world ocean have the fewest trophic levels in their food webs. Grigg et al. (1984) assigned six trophic levels to coral-reef systems. The implications of these levels for management will be developed in Chapter 18.

12.3.3. Diversity and Ecosystem Function

The geographic differences among coral reefs that first come to mind concern species richness. The diversity of species, genera, and families of corals, fishes, and most other taxonomic groups of coral-reef animals are very much greater in the western Pacific compared to the Atlantic or eastern tropical Pacific (Table 14–1), and most of the causal explanations for these differences involve geographic patterns and processes such as isolation and current direction (Chapter 14).

But does the diversity of corals affect the growth of the reef? Do reefs on Guam with 267 species of coral grow any faster than those in the Caribbean with less than a quarter the number of species or in French Polynesia with less than half?

Notwithstanding the spectacular differences among geographic regions in size, age, and species richness (Chapter 14), these characteristics do not appear to affect the overall growth or physiology of coral reefs (Kinsey, 1982; Crossland, 1988). About 400 species of hermatypic corals occur in Japanese waters (Veron, 1993), while a total of only about 155 species occur in French Polynesia (Society Islands, Tuamotus, Gambiers, Australs, Marquesas; see Chevalier, 1982). Nevertheless, the reefs and atolls of French Polynesia often appear to be accreting or growing at the same rate or better than those in Japanese waters. Likewise, Caribbean and other western Atlantic reefs have only about 65 species of hermatypic corals, about one-sixth the species richness of Japanese waters, yet the reefs in the Atlantic seem to to be accreting at rates comparable to those in the Pacific (Kinsey, 1982). The characteristics of the physical environment overshadow the influence of differences in biodiversity on the functioning of coral reefs, as long as representatives of each of the performers of key ecological roles are present.

Primary productivity and fisheries yield are not obviously greater or lesser on central Pacific reefs than on more diverse reefs in the western Pacific.

Despite the concept of convergent evolution, the richer species pool in the western Pacific provides kinds of predators with no counterparts in the western Atlantic; the crown-of-thorns seastar, *Hymenocera* (shrimp that prey on large seastars), giant clams, anemone fishes, and sea snakes are examples. Some of the individual species such as the crown-of-thorns starfish *Acanthaster planci* (Indo-West Pacific), *Diadema antillarum* (western Atlantic), and *Eucidaris thouarsii* (Galápagos Islands) cause large-scale phenomena that are unique to the particular geographic region (Chapter 9). Mutualistic associations are also more diverse in the western Pacific, with some groups of associations, such as giant clam-zooxanthellae and anemone-anemonefishes, being characteristic of the Indo-West Pacific (Vermeij, 1978), but not the Atlantic or eastern tropical Pacific.

Some entire classes, orders, and families of organisms are absent from coral reefs in the central Pacific. There are at least 91 species of crinoids in the shallow waters of Indonesia, at least 55 in the Philippines, 21 in Palau, 6 in Guam and the Marshall Islands, and none at all on coral reefs in Hawaii, French Polynesia, the Line Islands, or the eastern tropical Pacific (Birkeland, 1989). Although individual species of echinoderms have overwhelming influences on the functioning of the coral-reef ecosystems in their respective regions, some other species in their classes do not, and the overall species diversity of their classes does not seem to matter (Birkeland, 1989).

Nevertheless, coral-reef communities appear susceptible to diseases when large monocultures exist. Diseases of commercial sponges have spread widely in the tropical western Atlantic about six times since the mid-nineteenth century. A disease killed about a million mother-of-pearl–producing oysters in the Gambier and Tuamotu archipelagoes in the late 1980s. A massive mortality of *Diadema antillarum* occurred in the tropical western Atlantic in the early 1980s. The density of *A. antillarum* prior to the mortality may have been unnaturally high as a result of fishing pressure on predators of urchins (Hay, 1984b; Hughes, 1994). Evenness in species abundance is an aspect of diversity. Regardless of species richness, the thin dispersal of species may facilitate stability of the system by reducing the spread of communicable disease and making the overexploitation of certain species uneconomical.

12.4. Management Considerations

The tropical western Atlantic is only about a tenth the area of the tropical western Pacific. The interconnectedness of the entire tropical western Atlantic, as evidenced by the spread of diseases of urchins and sponges throughout the region within a year on each occasion (Chapters 6 and 9), and by the relatively uniform faunal distribution compared to the tropical Pacific (Chapter 14), indicates that management of coral-reef resources requires international cooperation. Sea turtles from nests in one country may never return because they have been harvested by hunters in other countries. As exemplified by the effects of the Aswan Dam on the fisheries of the eastern Mediterranean, a project undertaken hundreds of kilometers inland in one country can have major effects on the marine fishery resources of several other countries (Aleem, 1972). Unfortunately, there are presently no international legal mechanisms analogous to domestic court cases that can be applied to prevent international downstream damages. Each nation is an independent legal unit in an ecological continuum.

The coral reefs of the world cover about one-sixth of the world's coastlines, have greater gross productivity, and have a far greater standing stock of fishes than the combined stock of all the regions of upwelling. Yet upwelling favors abundant recruitment that supports industrial fisheries, while most commercial export fisheries on coral reefs have been especially sensitive to overexploitation. Because of the spectacular standing stock of coral-reef fish biomass, government agencies often consider it their mandate to develop export fishing industries to help improve the economies and standards of living of the indigenous people. Although well intentioned, these government agencies most often bring principles of fisheries science from temperate regions to coral reefs, and this often undermines more effective and appropriate management systems already in existence, having been developed over centuries by the indigenous people (Chapter 16).

Why are principles of fisheries science from the temperate regions often inappropriate for coral-reef fisheries? The large numbers of species and topographic complexity of coral reefs make wholesale harvest and processing relatively uneconomical. Although the biodiversity, the large number of steps in the food web, and the life-history characteristics of species adapted to low-nutrient environments all contribute to the maintenance of high-standing stocks of many species per unit of gross production, a relatively small portion of this gross production is transformed into secondary production that is meaningful for human consumption (Chapters 7 and 18). Furthermore, the life-history traits of target species of coral reefs (residential postlarval stages, slow growth, long life, dependence on multiple reproduction) make coral-reef animals particularly vulnerable to overexploitation. In Chapter 18, a new paradigm for resource management is presented for coral reefs that takes into account the unique aspects of coral-reef ecosystems.

13

Ecosystem Interactions in the Tropical Coastal Seascape

John C. Ogden

A major component of the high biological diversity of the tropics is the complex mosaic of interacting ecosystems, or seascape, of the coastal marine zone. This narrow band of shallow water, formed by recent sea-level rise, in intimate contact with watersheds on land and open to the sea, is the source of much of the organic production and diversity of species in the tropical marine environment. It provides many important resources, and channels, dilutes, disperses, and metabolizes the effluents of human activities.

A tropical seascape will often contain some combination of coral reefs, seagrasses, and mangroves. Their development and complexity of interaction largely depend on the size of the adjacent landmass and the amount of terrestrial runoff. Of course, these ecosystems can thrive in isolation. Atolls, for example, have extensive coral-reef development usually in the absence of either seagrasses or mangroves. Despite the fact that the biogeographical realms of the Atlantic and the Indo-Pacific have distinct differences in species diversity and composition, the dominant biostructural components of each ecosystem—mangrove trees, seagrasses, and reef corals—have similar physiological and ecological roles (Birkeland, 1987).

Although the component ecosystems of the tropical seascape, particularly coral reefs, have been studied at well-known marine laboratories and field stations for many years, relatively few studies have concentrated on their interactions (Ogden and Gladfelter, 1983). This chapter will consider biological, biochemical, and physical interactions between tropical coastal ecosystems and their importance to coral-reef structure and function and to the stability of the seascape (Ogden, 1988).

13.1. Biological Interactions in the Seascape

13.1.1. Edge Effects

The boundary, or edge, between coral reefs, seagrasses, and mangroves is an important component of each of the adjoining habitats, exhibiting abruptly altered

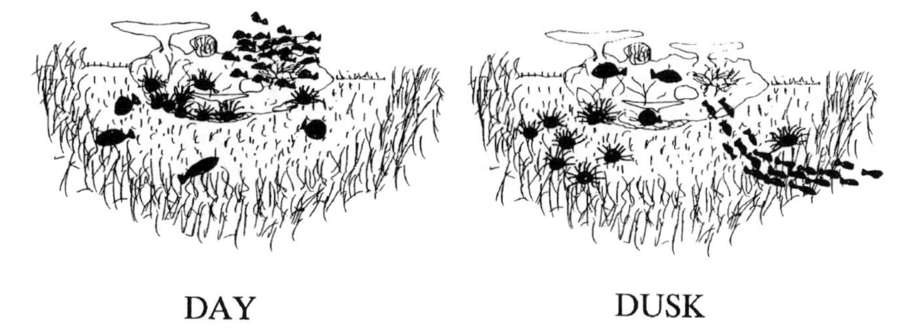

Figure 13–1. Sketch of the edge of a Caribbean patch reef during the day and at night showing dense seagrasses, grazed halo, and the position of sea urchins *(Diadema)*, herbivorous fishes (large), and grunts (small).

community composition and physical regime, and providing sites for foraging and spawning (Johannes, 1978b). For example, diurnal and nocturnal foraging movements of herbivorous fishes and sea urchins from coral reefs into seagrass beds (Fig. 13–1) are known to create conspicuous grazed "halos" around Caribbean and some Indo-Pacific coral reefs (Randall, 1965; Ogden et al., 1973; Birkeland, 1985). Migration at dusk of large schools of benthic-feeding grunts (Haemulidae) and snappers (Lutjanidae) into seagrass beds from coral reefs and return at dawn are a characteristic of the Caribbean (Ogden and Ehrlich, 1977). This ritualized behavior on fixed migration pathways may allow the fish to partition large seagrass foraging areas (Ogden and Zieman, 1977; Ogden and Quinn, 1984). Similar migrations are known in Indo-Pacific reef fishes, but have not been implicated as vectors of organic and inorganic material to reefs as they have in the Caribbean (Birkeland, 1985; Meyer et al., 1983).

13.1.2. Nurseries

In the Caribbean, seagrass beds and mangroves function as nurseries for a variety of fishes and invertebrates that spend their adult life on coral reefs. The salient features of the nursery are (1) location away from the heavy predation characteristic of reefs; (2) protection afforded to small organisms by the structural complexity of masses of leaves and roots; and (3) a rich food supply based on plant detritus and associated microorganisms and small invertebrates (Adams et al., 1973; Young and Kirkman, 1975; Ogden and Gladfelter 1983).

The life cycle of the French grunt *(Haemulon flavolineatum)* in the Caribbean provides a good example of the function of nurseries. French grunts spawn throughout the year. After about 2 weeks as planktonic larvae, postlarvae settle by night into seagrass beds. Schools of postlarvae gradually move toward the reef, arriving as small juveniles in adult livery (Fig. 13–2) after about one month (McFarland et al., 1985). Postsettlement mortality of French grunts is very high,

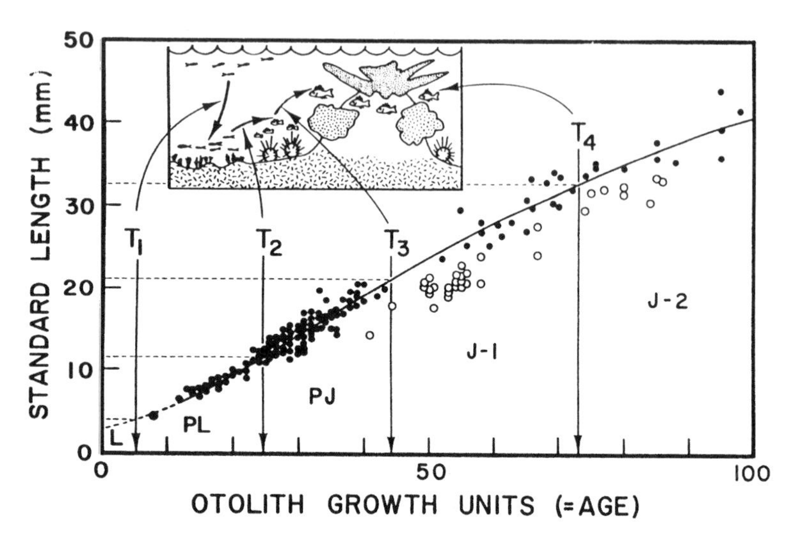

Figure 13–2. The age-growth curve for young French grunts. Each otolith growth unit represents one day. T_1, T_2, T_3, and T_4 show size and age at transition points as the grunts move from the seagrass nursery to juvenile populations on coral reefs (after McFarland, 1979).

about 90% in the first month and nearly 99% in the first year. However, the nursery may provide the critical small-percentage increase in the probability of survival of postlarvae that will assure a healthy adult population (Shulman and Ogden 1987).

Studies of the importance of nurseries in the Pacific are equivocal. Birkeland and Amesbury (1988) studied populations of fishes in mangroves, seagrasses, and reefs in Papua New Guinea and found that while reefs in the vicinity of seagrasses and mangroves showed increased recruitment, this did not result in increases in the adult population. Birkeland and Amesbury (1988) suggest that removal of adults by fishing may be more sustainable on reefs with neighboring habitats as there is an increased chance of recruitment.

13.2. Fluxes of Nutrients and Organic Material in the Seascape

Surface waters in tropical oceans are low in concentrations of dissolved nutrients (Morris et al., 1981). The high productivity of coastal ecosystems is often regarded as a paradox, the resolution of which comes from hypothetical nutrient subsidies. What evidence do we have that nutrients are limiting to the productivity of coastal systems and that they are dependent on subsidies from adjacent systems?

While nutrients may limit productivity of components of the coral-reef ecosystem, particularly benthic algal turfs (Hatcher and Larkum, 1983; Williams and

Carpenter, 1988), there is little evidence that the coral-reef ecosystem is nutrient limited (D'Elia and Wiebe, 1990; Wiebe, 1987). Crossland and Barnes (1983) found no flux of nitrogen, phosphorus, silicate, organic nitrogen, and organic carbon across the reef system at Lizard Island, Australia. Smith (1984) found no nutrient limitation where hydrodynamic fluxes were large and Grigg et al. (1984) suggested that nutrient concentrations did not limit the primary productivity of French frigate shoals in Hawaii.

Increased nutrient concentrations may harm coral reefs by suppression of calcification of corals and shifts in dominance to fleshy benthic algae (Kinsey and Davies, 1979; Littler and Littler, 1988). Hallock and Schlager (1986) suggest that the drowned reefs of the Caribbean were unable to pace Holocene sea-level rise due to increased nutrient concentrations and sediments in the water. The *Halimeda* algal bioherms of the western Caribbean and elsewhere in the Pacific may be similarly related to the suppression of coral-reef accretion in favor of calcareous algae by elevated nutrient concentrations (Hine et al., 1988).

Productive mangrove forests are found along all tropical coastlines, in basins and coastal floodplains, and in other protected areas with abundant runoff of sediments and nutrients from land. Mangroves also grow in scrub form on desert coasts and low islands with minimal runoff, and their productivity is low (Lugo et al., 1973). Boto and Wellington (1984) found nitrogen limitation in a northern Australian mangrove forest, but the nitrogen cycle in mangroves has not been well studied (Nedwell, 1975; Wiebe, 1987).

The available data do not generally support the assertion that mangroves enhance the productivity of downstream systems through export of inorganic and organic material (Wiebe, 1987). Boto and Bunt (1981) estimated export of only 1 g carbon m^{-2} day^{-1} in a *Rhizophora* mangrove forest in Queensland, and concluded that nitrogen and phosphorus are conserved within the mangrove system. Golley et al. (1962) reported a gross productivity to respiration ratio of 0.9 for a *R. mangle* forest in Puerto Rico, evidence for little or no export from the system. Nixon et al. (1984) found no evidence for outwelling from several Malaysian mangrove forests and concluded that they were sinks for organic and inorganic material.

Studies of the nitrogen cycle in seagrasses have been concentrated in the Caribbean on *Thalassia testudinum.* McRoy (1983) found a positive relationship between the depth of the sediment-organic layer and the successional state of the seagrass ecosystem. *Thalassia,* the "climax" species, required the deepest sediments and presumably has the greatest nutrient requirements. Nitrogen fixation has been implicated in supplying much of the nitrogen requirements of the system (Capone et al., 1979) or very little (McRoy et al., 1973). *Thalassia* beds appear to recycle and retain ammonium (Wiebe, 1987), but studies of denitrification are few (McRoy and Lloyd, 1982). Williams (1987) showed that experimental fertilization of the sediments increased biomass of *Syringodium filiforme* and *Thalassia.*

Estimates of export from Caribbean seagrass beds range from 1% to 10% of leaf productivity for *Thalassia* and over 50% for *Syringodium* (Greenway, 1976; Zieman et al., 1979). The decomposition of detrital seagrasses can cause local pulses in nutrients and organic material in bays (Zieman, 1982). Suchanek et al. (1985) showed by comparison of stable carbon isotope ratios that seagrass detritus in the deep sea was an important carbon source for sea urchins and deposit-feeding holothurians. In contrast, Zieman et al. (1984) found that the stable carbon isotope ratios of consumers in seagrass beds and adjacent mangroves reflected the dominant plant of each system and there was little evidence of carbon subsidies by either system.

The shallows of Florida Bay at the southern tip of the Florida peninsula support a large percentage of the seagrasses of the Gulf of Mexico region. Beginning in 1987, seagrasses (mostly *Thalassia testudinum*) began to die (Robblee et al., 1991). The die-off spread relentlessly and by early 1994 affected an area of more than 40,000 hectares, almost 20% of the area of the bay. The cause of the dieback has not been established, but its effects are widespread. Catches of the commercially important pink shrimp that use the bay as a nursery have declined. The disappearance of the seagrasses has allowed waves to suspend formerly trapped sediment nutrients into the water column, contributing to huge algal blooms that have killed sponges over large areas and have threatened water quality downstream over the coral-reef tract. While there are signs that seagrasses are recovering, the dieback is a continuing, complex natural experiment whose full resolution awaits further research (Boesch et al., 1993).

13.3. Physical Interactions in the Seascape

13.3.1. Lagoons

Vertical coral-reef growth depends on sea-level rise over the past 6,000 years. If the coastal topography includes a broad platform, prograding sea level may create a shallow lagoon shoreward of a platform-edge coral reef. As this expanse of shallow water equilibrates rapidly with the atmosphere, lagoon water, either too hot or too cool for coral reefs, may cause the eventual demise of the seaward reefs. Neumann and Macintyre (1985) call this phenomenon "reefs shot in the back by their own lagoons." In the Florida Keys, the seaward movement of water inimical to coral-reef growth from shallow Florida Bay through passes between the Keys, caused the demise of reefs in the central coral-reef tract about 3,000 years ago. Reefs to the north and south, protected from the influence of bay water, continued to thrive and are presently the best developed in the region (Ginsburg and Shinn, 1993).

Periodically, Florida Bay continues to exert a controlling influence on coral-reef development in southern Florida. The severe winter of 1976–1977 chilled Florida Bay water as low as 13°C. Intrusion of this water over the reefs of the

Dry Tortugas killed 96% of the corals surveyed by Porter et al. (1982). Decades of water management in the Everglades, draining the wetlands for development and agriculture, have severely decreased the amount of freshwater flowing to Florida Bay (McIvor et al., 1994). During hot, calm summer periods, salinities over twice that of normal seawater (70‰) have been observed in the bay. In the Summer, hot, high-salinity water (over the suspected coral bleaching threshold of 32°C) has been observed over the reef tract (Ogden et al., 1994). Incursions over the Florida reef tract of low-salinity water from summer floods in the upper Mississippi River resulted in anomalous data in physiological experiments with in-situ corals and may be a long-term influence on coral-reef growth (Ogden et al., 1994).

13.3.2. Buffers and Sinks

Coral reefs require clear water for vigorous growth. They are strongly influenced by the sediments in terrestrial runoff (Fig. 13–3). Riverine discharge containing a heavy sediment load can destroy or severely restrict coral-reef community development to only the most sediment-tolerant species. The clearing of watersheds for agriculture, industry, and tourism, and the destruction of coastal estuaries, seagrass beds, and mangrove forests, which act as sediment traps, are among the most damaging influences on coral reefs around the world (Kühlmann, 1988; Ogden and Gladfelter, 1986; Pannier, 1979). The coral reef of Cahuita National Park, on the Atlantic coast of Costa Rica, has been virtually destroyed by decades of siltation delivered by rivers originating in highlands destabilized by agriculture. Petersen et al. (1987) report on a study of two rivers, one flowing from a watershed dominated by sugarcane fields and one from the naturally vegetated Black River

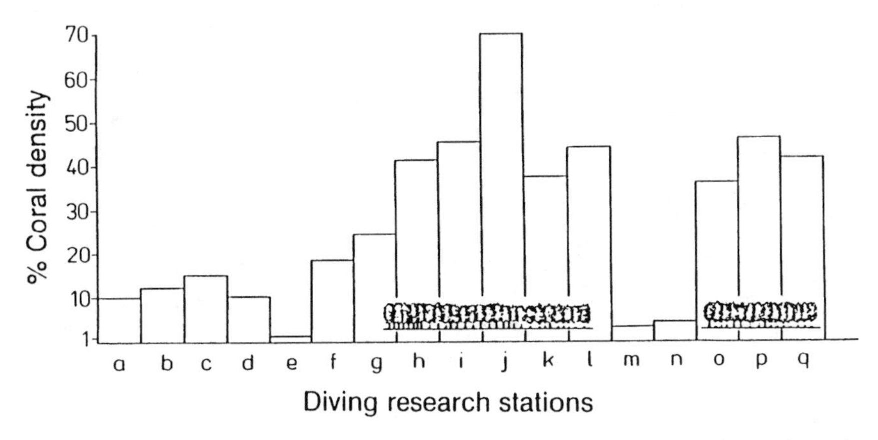

Figure 13–3. The average coral cover around Ishigaki Island, Ryukyus, Japan, depends on shoreline forests to absorb damaging pollutants (from Kühlmann, 1988).

Table 13-1. Comparison of Chemical and Physical Parameters of Two Rivers in Jamaica (The exogenous river arises in a sugar-cane field and the endogenous river in the naturally vegetated Black River morass.)

Parameter	Units	Exogenous	Endogenous
PO_4–P	$\mu g\ P \times L^{-1}$	9–11	3–5
NO_3–N	$mg\ N \times L^{-1}$	0.48–0.87	0.21–0.36
Particulate organic carbon	$mg\ C \times L^{-1}$	1.4–3.8	0.25–0.50
Total solids	mg dry mass $\times L^{-1}$	5.4–21.5	0.7–0.9
Organic leaching of detritus	$g \times g^{-1}$	0.16 ± 0.23	0.08 ± 0.04
Inorganic sedimentation	g ash $\times g^{-1}$ organic	0.24 ± 0.06	0.03 ± 0.02
Leaf decomposition[a]	$g \times g^{-1} \times yr^{-1}$	8.6–13.4	5.2–6.7
Microbial respiration[a]	Mg $O_2 \times g^{-1}$ organic $\times h^{-1}$	0.47 ± 0.05	0.34 ± 0.03

Source: From Petersen et al., 1987.

[a]Data for sawgrass, *Cladium jamaicensis* only.

morass in Jamaica. Table 13–1 dramatically shows the differences between the two rivers. All of the factors elevated in the river originating from the sugar-cane field are inimical to the growth of coral reefs.

At the shelf edge, coral reefs stabilize the seascape by dissipating the impact of ocean waves, creating over geologic time lagoons and sedimentary environments that favor the growth of seagrasses and mangroves (Fig. 13–4). The long-

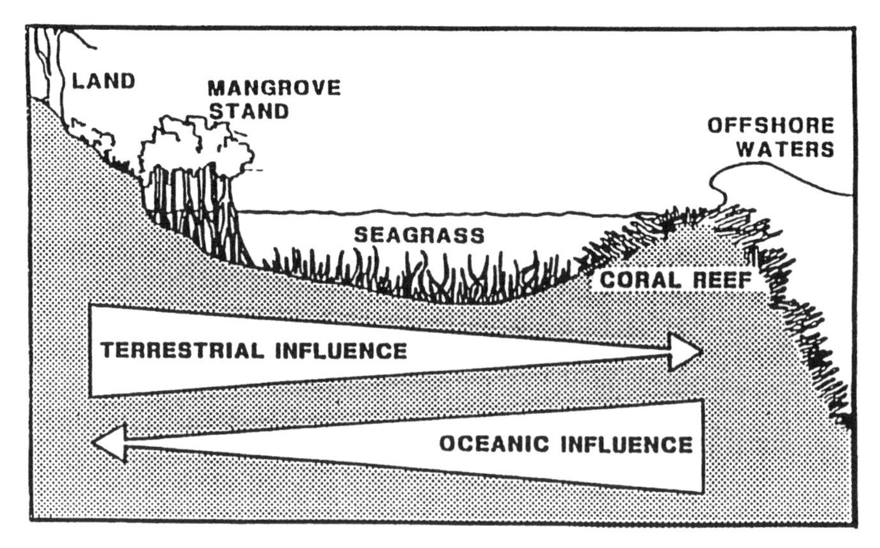

Figure 13–4. Schematic diagram of the tropical coastal seascape. The opposing arrows show the buffering of land influence by shoreward ecosystems and the buffering of ocean influence by the coral reef (from Ogden, 1987).

term dynamics of the physical relationship between ocean waves, coral reefs, and lagoon circulation are virtually unknown. At the land-sea boundary, coastal forests, mangroves, and seagrasses interrupt freshwater discharge, stabilizing the salinity of the coastal zone; trap and bind sediments, reducing sediment loads in the water column; and absorb nutrients, promoting the growth of coral reefs offshore (Wiebe, 1987).

Periodic events such as hurricanes and heavy rains flush great quantities of accumulated material from mangrove and seagrass sinks. These events "reset" the sinks and have a potentially long-term effect upon the downstream ecosystems. In August 1992, Hurricane Andrew swept over the Florida peninsula, defoliating and eventually killing approximately 150 km of mangrove trees (Ogden, 1992). Suspension of organic-rich sediments within the mangroves caused large fish kills from anoxia, and the long-term nutrient release associated with submerged mangrove detritus may stimulate persistent phytoplankton blooms. Phytoplankton and benthic algal blooms are also associated with runoff water and groundwater seepage (Marsh, 1977; Johannes, 1980). Outbreaks of the crown-of-thorns starfish *(Acanthaster planci)* have been correlated with storms on high Pacific islands that deliver pulses of nutrient-rich runoff water to the coastal zone and may enhance the survival and growth of the starfish larvae (Birkeland, 1982). The discovery of fluorescent bands of fulvic acids from long coral cores in Australia that correlate with river discharge records will be a useful technique to infer recent past events that may have influenced the structure of coral reefs (Isdale, 1984a, b; Smith et al., 1989).

The buffering capacity of coastal ecosystems is threatened by the projected rate of sea-level rise under scenarios of global warming. Given a conservative estimate of 15 ± 3 mm yr increase in sea level, the vertical carbonate accretion rates of protected coral-reef flats may be insufficient to keep up. These zones will become inundated and subjected to erosion by progressively larger waves (Buddemeier and Smith, 1988). Seagrass and mangrove communities will be eroded and will become less effective buffers, releasing nutrients, turbidity, and sediments, further slowing coral-reef growth rate.

13.4. Reef Management, Global Change, and Comparative Research

It is obvious that coral-reef ecosystems cannot be understood or managed in isolation from the complex mosaic of interacting ecosystems in the tropical seascape including watersheds. Critical adjacent habitats functioning as nurseries, foraging areas, and physical and chemical buffer zones are often difficult to identify and may be overlooked in management strategies that focus only on coral reefs. Nevertheless, they are potential bottlenecks, causing major changes and even collapse of reef populations if they are damaged or disturbed.

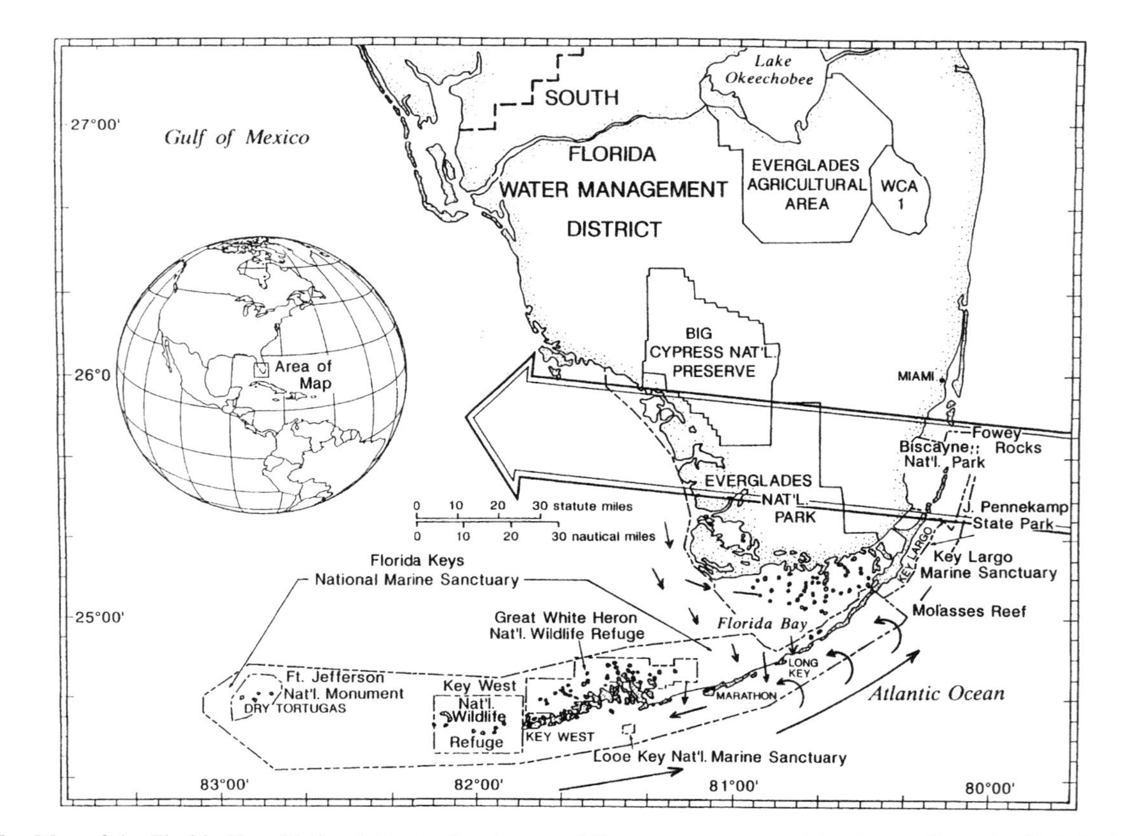

Figure 13–5. Map of the Florida Keys National Marine Sanctuary and the ocean currents and freshwater flows that link the interdependent ecosystems of the seascape. The large arrow is the path of Hurricane Andrew, August 1992 (from Ogden, 1992).

In response to the rapid decline of Florida coral reefs, the 2,500 n.mi.^{2} Florida Keys National Marine Sanctuary was established by Congress in 1990 (Fig. 13–5). It is contiguous with Biscayne National Park, Everglades National Park, and Big Cypress Reserve, totaling over 5,000 sq. mi. Since 1991, the National Oceanic and Atmospheric Administration (NOAA), as the lead agency, and the Environmental Protection Agency (EPA), with responsibility for water quality, along with numerous state and local agencies, have been engaged in drafting the sanctuary management plan, one of the most complex coastal resources management programs ever attempted. The plan, while focused on the sanctuary, erases the artificial boundaries of agency jurisdiction and treats southern Florida as a mosaic of interdependent, land-margin ecosystems.

At a larger geographic scale, sea-level rise and other regional and global events affecting the potential long-term survival of coral reefs (Buddemeier and Smith, 1988), including mass mortalities (Lessios et al., 1984), coral diseases (Gladfelter, 1982), and phenomena such as coral-reef bleaching (Brown and Ogden, 1992; Glynn, 1985; Williams et al., 1987; Ogden and Wicklund, 1988), argue strongly for the establishment of regional, long-term, coordinated study sites in the tropical western Atlantic and the Indo-Pacific. Within these large, ecologically coherent regions, coral-reef ecosystems should be comparatively studied along environmental gradients that represent the full range of development of ecosystem interactions in the seascape (Ogden, 1987).

14

Diversity and Distribution of Reef Organisms

Gustav Paulay

The most species-rich marine communities probably occur on coral reefs, a habitat in which many groups of organisms reach their greatest diversity. The diversity of reef organisms, spatial patterns in their distribution and diversity, and ecological and historical explanations for these patterns are presented in this chapter.

The chapter begins with definitions of diversity and a discussion of how measures of diversity depend on the taxonomic and spatial scales being considered. The taxonomic richness of coral reefs, how many species occur on reefs, how commonly are taxa restricted to reef habitats, and possible explanations for this high diversity are considered in section 14.2. The influence of diversity in the functioning of reef ecosystems is examined in section 14.3, focusing especially on photosymbiotic, reef-constructing and reef-eroding organisms. The basic question being addressed here is, How important is biodiversity to ecosystem function? Section 14.4 is a brief overview of the process and pitfalls of taxonomic identification of reef organisms, with a guide to some of the more important taxonomic literature presented in the second appendix.

Sections 14.5, 14.6, and 14.7 comprise the bulk of the chapter and consider diversity from a geographical standpoint. Present latitudinal and longitudinal patterns of distribution (section 14.5) and the historical origins of these patterns (section 14.6) are given as background with which to approach the question (section 14.7), How did speciation among and within biogeographical regions lead to the present diversity and distribution of reef organisms? The chapter concludes with a look at the future of biodiversity on coral reefs: the impact of humans on reefs and likely consequences for the diversity and distribution of reef organisms (section 14.8).

14.1. Definitions and Scales

14.1.1. What Is Diversity?

Biological diversity can be considered from a variety of perspectives: taxonomic, genetic, ecologic, or functional (Steele, 1991). The first of these is the focus of this chapter. Taxonomic diversity has come to encompass different meanings in different disciplines. In ecological studies, diversity frequently refers not only to the number of taxa (taxon "richness"), but also to a measure of equitability or evenness in the abundance of taxa. By this definition, *N* species that are equally common are more diverse than *N* species that vary in abundance. Several mathematical diversity indices are available that incorporate a measure of equitability of abundance (Peet, 1974). However, including such a measure is not only impractical at large spatial and temporal scales, but also ignores the fact that rare species are often as important as common ones for biogeographic and evolutionary considerations (Brown, 1988; Schluter and Ricklefs, 1993). In these latter fields, the term diversity usually refers only to the number of taxa ("richness"), and it is used in this sense in this chapter. Taxonomic diversity can also be considered at a variety of spatial (e.g., within habitat, between habitat, regional) and phylogenetic (e.g., species, families, phyla) levels.

14.1.2. Taxonomic Diversity: Spatial Scales

Within the continuum of spatial scales, three levels are often discussed with respect to taxonomic diversity: within habitat (alpha), between habitat (beta), and regional (gamma) diversity (Whittaker, 1972; Brown, 1988; Clarke, 1992). Ecological studies usually focus on finer spatial scales (e.g., Jackson, 1991), in part because they address the immediate factors regulating the local coexistence of species in communities. In contrast, studies addressing the long-term historical controls of diversity usually emphasize regional-scale processes of speciation, extinction, and changes in species distributions.

Regional diversity is the product of between-habitat as well as within-habitat species richness (Whittaker, 1972). Thus, it depends in part on the diversity of available habitats and the degree to which species are ecologically restricted to individual habitats, and on the geographic fragmentation of the region and the degree to which species are geographically restricted to areas within the region. The greater the ecological or geographical restriction (endemicity) of species, or in other words, the less overlap there is in species composition among different habitats and among different areas, the higher the regional diversity. High regional diversity can thus be a result of high within-habitat species richness or of great ecological or geographical heterogeneity accompanied by ecological or geographical restriction of species.

Abele (1974) found that while within-habitat diversity in decapod crustaceans does not increase from high to low latitudes, regional diversity increases considerably, due to greater differences in species composition among habitats in the tropics than at higher latitudes. A striking example of differences in geographic restriction can be found among fishes. While within-habitat diversity of fish species is typically lower in fresh waters than in comparable marine habitats (especially on coral reefs), on a global scale, freshwater habitats account for 39% of the fish diversity in 1.5% of the available habitat area. This is largely the result of the much greater degree of geographic differentiation of freshwater than saltwater and the correspondingly much higher geographic restriction of freshwater than marine fish species (Nelson, 1984; Brown, 1988). On reefs, within-habitat diversity of fishes in some habitats is comparable between the western Atlantic and Indo-West Pacific, although regional diversity is much greater in the latter (Thresher, 1991; see below).

14.1.3. Taxonomic Diversity: Phylogenetic Scales

Very different patterns of taxon richness can result when different levels of the taxonomic hierarchy are considered. While rain forests are much more diverse in species than are coral reefs, they are considerably poorer in phyla. Similar differences are evident between marine and terrestrial systems in general (May, 1994). Such differences in species richness reflect both ecological and evolutionary differences (May, 1994). Thus, the greater vertical habitat complexity of rain forests compared to coral reefs may facilitate the coexistence of more animal species, and the greater opportunities for geographic isolation in terrestrial than in marine systems contribute to higher potential rates of diversification in the former (Kay and Palumbi, 1987; Paulay, 1994; Rapaport, 1994). The greater phylum diversity of reefs compared to forests reflects general differences between sea and land, resulting from the marine origin and early diversification of life. Thus, 17 of the approximately 34 recognized animal phyla are restricted to marine habitats; 16 occur in both marine and nonmarine environments (most, if not all, of these had a marine origin); and only one phylum, the Onychophora, is restricted to nonmarine (in this case, terrestrial) habitats (although it appears to have had a marine origin) (Table 14–1).

Trends in the temporal and spatial distribution of organisms can be different or become less clearly marked as one moves up the taxonomic hierarchy (Valentine, 1985). Higher taxa have a longer history and a greater diversity than the lower taxa they encompass, and are thus less likely to exhibit crisp patterns in distribution and diversity. Differences in regional diversity, as well as patterns in ecological and geographic restriction, are much greater and thus more evident (and accessible to analysis) at the species than at higher taxonomic levels. Nevertheless, patterns in diversity and ecological or geographical restriction of higher taxa are of interest as they may be indicative of more fundamental or older patterns.

Table 14-1. Animal Phyla on Coral Reefs

Known from Coral Reefs		Not Known from Coral Reefs
Porifera	Placozoa	Onychophora (nonmarine)
Cnidaria	Ctenophora	Orthonectida (3 genera, 2 monospecific)
Dicyemida	Platyhelminthes	Pogonophora (largely deep sea)
Gnathostomulida	Gastrotricha	Cycliophora (monospecific)
Nematoda	Nematomorpha	**Total: 4**
Kinorhyncha	Priapula	
Loricifera	Acanthocephala	
Rotifera	Entoprocta	
Tardigrada	Nemertea	
Echiura	Sipuncula	
Mollusca	Annelida	
Arthropoda	Phoronida	
Bryozoa	Brachiopoda	
Chaetognatha	Echinodermata	
Hemichordata	Chordata	
Total: 30		

14.2. Biota of Reefs

Before examining biodiversity on reefs, the restriction of taxa to reef habitats, and potential reasons for the particularly high diversity encountered on coral reefs, it is important to consider what is encompassed by reefs as habitats. Reefs can be defined either narrowly, to include only the hard substrata built by reef-constructing organisms, or widely, to include expanses of soft sediment that are largely derived from reefs and interspersed among the hard substrata. In the widest sense, entire atolls can be considered as reef systems, even though they are commonly dominated by soft-bottom habitats by virtue of their large lagoons (e.g., Jokiel and Maragos, 1978). We will follow the wider definition of reefs, because (1) at least smaller expanses of soft bottoms are intimately connected to the hard bottoms by migration, different stages in life histories, and trophic links (Chapter 13), and (2) available data on reef diversity are rarely restricted to the narrower definition (but see Choat and Bellwood, 1991).

14.2.1. Species Richness on Reefs

Coral reefs, especially in the central Indo-Pacific and the Caribbean, are usually thought to hold the greatest diversity of marine life, at least on a per-unit-area basis (the deep sea also holds great diversity, partly because of its large area; Grassle and Maciolek, 1992; May, 1994). Diversity on reefs, however, is strongly influenced by environmental conditions and geographic location, so that remote or high-latitude reefs often have relatively low species richness (see below). There are little comparative data presently available with which to evaluate the overall species richness of reefs relative to other habitats. Estimates of the numbers of reef species in many groups may be very inaccurate, due to the abundance of poorly understood sibling species complexes. Knowlton (1993) suggests that when all sibling species are recognized, the number of marine species will be found to be an order of magnitude greater than currently estimated.

Reaka-Kudla (1994) estimates that 35,000–60,000 species of reef-dwelling animals and plants have been described, but proposes that this may be a fraction of the total. Reasonably comprehensive diversity figures are not yet available for any individual coral-reef area. Preliminary species checklists are available for a few tropical locations, but each includes only a small portion of the local biota; at any given location, few taxa have been sampled and studied in detail. Over 7,000 marine species have been recorded from the Hawaiian Islands (Fig. 14–1), 3,800 from French Polynesia (mostly shallow depths; Richard, 1985), and 3,400 from Guam (mostly shallow depths). About 2,000 species of marine invertebrates are recorded from the reef-poor Galápagos Islands (Peck, 1993).

The inadequacy of these regional diversity estimates becomes readily apparent when the few well-studied taxa are considered (Table 14–2). The Australian sponge fauna alone comprises about 5,000 species (this includes all habitats, deep and shallow; estimate by Hooper and Lévi, 1994). About 6,500 species of marine molluscs occur in New Caledonia (estimate by Bouchet, 1979). Over 3,400 opisthobranch molluscs are known in the Indo-West Pacific, with almost

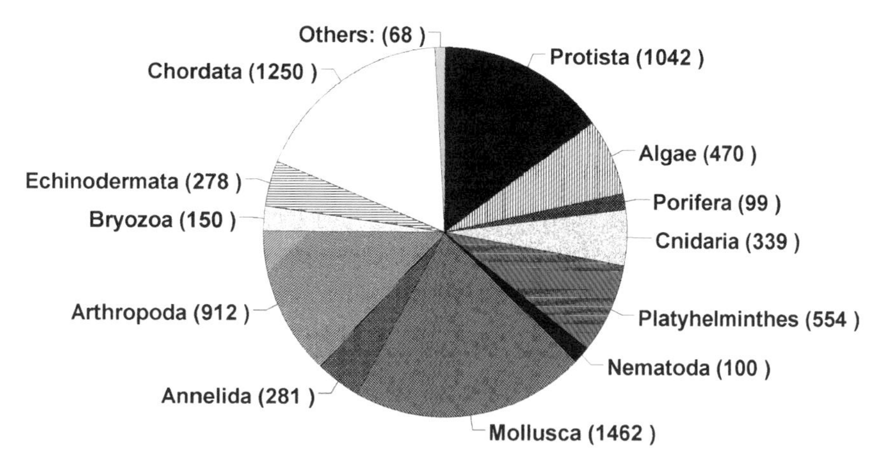

Figure 14–1. Known diversity of marine organisms in the Hawaiian Islands, from all depths, based on Eldredge and Miller (1995), with modifications from L. Eldredge (pers. comm.), Moretszohn and Kay (1995), Gosliner and Draheim (1996), and Kelly-Borges and Valentine (1995). Note that differences in diversity reflect both actual differences as well as variation in extent of taxonomic knowledge, ranging from very good (e.g., Chordata), to poor (e.g., Nematoda).

Table 14-2. Tropical Regional Diversity of Selected Taxa

	IWP	EP	WA	EA
Magnoliophyta: Tropical seagrasses[1]	14	4	6	1
Granuloreticulosa: photosymbiotic foraminifera[2]	36	4	23	
Porifera[3]	5000+		561	500+
Reef poriferan genera[4]	244		117	
Cnidaria				
Reef scleractinian genera[5]	90	9	25	7–8
Reef scleractinian species	700[6]	30[7]	65[7]	14[7]
Alcyoniina	690+[8]	0[9]	6[9]	
Mollusca				
Shelled Gastropoda[10]		2475	2643	
Cypraeidae[11]	178	24[12]	6	9
Naticidae: Naticinae and Sininae[13]	85	21	20	20
Conus[14]	316	30	57	22
Bivalvia	2000[15]	564[9]	378[9]	427[16]
Arthropoda				
Stomatopoda[17]	249	50	77	30
Caridean genera[18]	91	28	41	
Echinodermata	1200[19]	208[20]	148[20]	
Chordata				
Reef fish[21]	3000	300	750	
Shore fish	4000[21]	650[21]	1400[22]	450[21]
Chaetodontidae[23]	98	4	7	5
Pomacentridae[24]	268	22	19	12

Selected data on the regional diversity of tropical marine taxa. Numbers are of number of species except where otherwise indicated. Data from: (1) Phillips and Meñez, 1988; (2) Hallock, 1988; (3) Hooper and Lévi, 1994: IWP estimate based on their 5000 estimated shallow and deep Australian sponges; EA estimate based on 538 species recorded from temperate EA; (4) Soest, 1990; (5) Appendix 14-1; (6) Veron, 1995; (7) personal compilation, numbers approximate; (8) Gawel, 1977; (9) Bayer et al., 1970; (10) Allmon et al., 1993; (11) Burgess, 1985; (12) Emerson and Chaney, 1995; (13) Kabat, 1996; (14) Röckel et al., in press; A.J. Kohn, pers. comm.; (15) personal estimate; (16) von Cosel, 1991; (17) Reaka and Manning, 1987; (18) Bruce, 1976: WA and EA not separated; (19) Clark and Rowe, 1971, with an estimated 150 new species described since then; (20) Maluf, 1988; (21) Lieske and Myers, 1994; (22) Robins, 1991; (23) Allen, 1979; (24) Allen, 1991.

a third of these being undescribed; new species are still found regularly, indicating that the fauna remains incompletely known (Gosliner and Darheim, 1996).

At least 4,000 species, or almost a third of the world's marine fish species, occur on coral reefs (Choat and Bellwood, 1991; Lieske and Myers, 1994). In the tropical Indo-West Pacific, three-quarters of the 4,000 shore species are reef associated (Lieske and Myers, 1994). Indonesia, the Philippines, and Papua New

Guinea each boast about 2,500 species of reef fishes (Lieske and Myers, 1994; in comparison, British Columbia has a total of only 325 shore fish species; Tunnicliffe, 1992).

While the Scleractinia is best known for the striking reef corals, it also includes a large number of solitary and colonial, ahermatypic (not reef-forming) species, which range into high latitudes and deep waters. About half of the world's currently known scleractinian species and genera (ca. 800 of 1,400 species and ca. 108 of 230 genera) are reef associated. This proportion may decrease with future work because many more ahermatypic than hermatypic corals remain undescribed (S. Cairns, pers. comm.). The Philippines, Palau, and Japan each have over 400 species of reef scleractinian corals (Veron, 1992a; Maragos and Meier, 1993) (in contrast, British Columbia has fewer than 10 scleractinian species; Kozloff, 1987).

Altogether, the diversity estimates of these better-known taxa, several of which include many thousands of species in single archipelagoes of the central Indo-West Pacific, indicate that the true diversity of reef organisms on such moderate-sized reef systems is at least in the tens of thousands of species. In comparison, the well-studied biota of cold temperate British Columbia comprises about 5,000 species of organisms from shore to deep sea (Tunnicliffe, 1992).

Most regional diversity estimates do not distinguish between species that occur on coral reefs and those that occupy other habitats (Table 14–2). A few groups, such as algae (Silva, 1992) and amphipods (Barnard, 1991), are known to be more diverse in regions and habitats other than tropical coral reefs. However, many large groups, including anthozoans, gastropods, stomatopods, holothuroids, and fish, clearly reach their greatest diversities on reefs.

14.2.2. Ecological Restriction to Reefs

The great biological diversity found on reefs reflects both obligate reef associates (often in extensive symbioses that beget further diversity) and taxa that also occur in nonreef environments (see below). To what degree are various taxa restricted to reef environments?

Reefs are home to most, if not all, marine animal phyla and all algal divisions. The only marine phyla that I have not seen recorded from coral-reef habitats are (1) the Pogonophora, whose members rely on symbiotic, chemoautotrophic bacteria for their nourishment and typically live in deep water (including hydrothermal vents); (2) the Orthonectida, a poorly known phylum of mesozoan-grade, minute, parasitic organisms classified in three known genera and a handful of species; and (3) the Cycliophora, a recently discovered group based on a single species living symbiotically on the Norwegian lobster. In harboring such diversity, reefs are not unusual; no phyla or divisions are restricted to coral reefs, and most have almost cosmopolitan distributions among shallow-marine habitats. For example, the only phyla not known from the cold temperate waters of Washington

State are the monotypic, warm-water Placozoa, the Nematomorpha (known from <10 marine species), and the newly discovered Loricifera and Cycliophora (Kozloff, 1987; pers. obs.).

At lower taxonomic levels, restriction in latitudinal range (see below) and habitat is more common, so that some families and many species are restricted to reef habitats. Restriction to reef habitats, especially of families and genera, is most prevalent among reef-forming taxa and their obligate associates (Jackson et al., 1985). Among scleractinian corals, the major constructors of reefs, 11 families are comprised strictly of zooxanthellate, reef-dwelling species, 7 are strictly azooxanthellate, and 5 include species with both strategies. A number of other, mostly species-poor, photosymbiotic families are also reef specialists, such as the octocoral families Tubiporidae, Xeniidae, and Helioporidae, and the bivalve family Tridacnidae (giant clams).

In addition to the few families that are entirely restricted to coral reefs, many of the more conspicuous reef inhabitants belong to families that are largely restricted to, or reach their greatest diversities on, reef habitats. These include some very diverse gastropod families (e.g., Mitridae, Cypraeidae, Conidae), several large fish families (e.g., Scaridae, Chaetodontidae, Pomacanthidae, Acanthuridae; Choat and Bellwood, 1991), crustaceans (e.g., Gonodactylidae), and anthozoans (e.g., Alcyoniidae, Nephtheidae). Similarly, many families of coral symbionts reach their greatest diversity on coral reefs, although they also occur, with their hosts, in deep water and at high latitudes. Examples include the coral-associated crab families Cryptochiridae and Trapeziidae (R. Kropp, pers. comm.), and copepod families Corallovexiidae and Xarifiidae (A. G. Humes, pers. comm.).

Several reef-associated families originated relatively recently, coincident with the latest, Cenozoic expansion of reefs (Chapter 2), including cone shells (Fig. 14–2: Conidae; Kohn, 1990) and giant clams (Tridacnidae; Rosewater, 1965) in the Eocene, and parrotfish (Scaridae; Choat and Bellwood, 1991) in the Miocene. Although such close temporal and ecological association with reefs suggests that some of these families originated in reef habitats, the association of others is clearly secondary. Thus, while most species of Conidae currently reside on coral reefs, the family originated in deeper water on soft bottoms (Kohn, 1990).

In contrast to the youth of many of the above reef-associated families, caves and interstices of the reef framework harbor an unusual fauna where many groups are of much greater age. Here, an archaic assemblage of brachiopods, sphinctozoans, sclerosponges, arborescent and encrusting foraminifera, and primitive bivalves and gastropods predominate (Fig. 14–3; Jackson et al., 1971; Basile et al., 1984; Kase and Hayami, 1992; Hayami and Kase, 1993). Some of these cryptic cave sponges appear to be descendants of the stromatoporoids, sphinctozoans, and chaetetid tabulate "corals" that dominated many exposed reef communities during the Paleozoic and Mesozoic (Hartman and Goreau, 1975; Hartman, 1982; Vacelet, 1985). Their restriction to cryptic habitats today is thought to be the result of their inability to cope with the biological pressures of modern reef

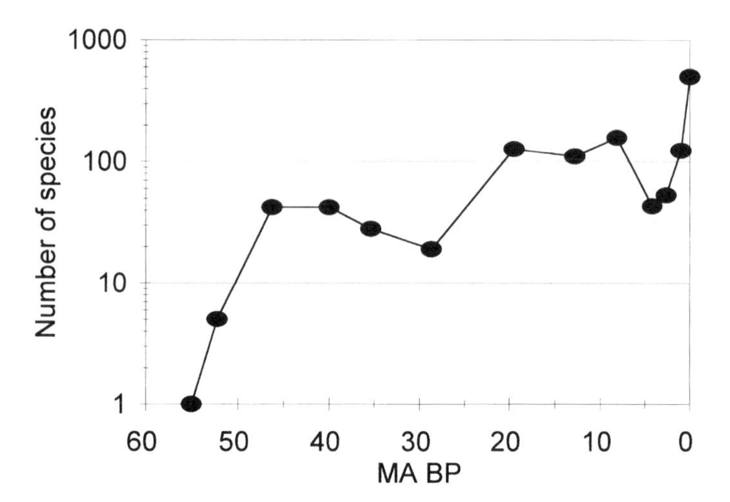

Figure 14–2. Diversification of the gastropod family Conidae. Although cones appear to have originated in soft-bottom habitats, their greatest diversity and much of their diversification is tied to reefs, and is indicative of the Cenozoic expansion of coral reefs. Graph depicts raw diversity data of number of species known from each epoch subdivision, from the group's lower Eocene origin to present, uncorrected for the duration of the epoch subdivisions (after Kohn, 1990).

environments (Jackson et al., 1971; Vermeij, 1987a). They also represent some of the oldest lineages that appear to have remained largely reef associated throughout their history.

At the species level, restriction to reefs is also variable, and comparisons again are limited by paucity of data. While many species are strictly reef associated, many others occur in a variety of nonreef habitats as well. Choat and Bellwood (1991) divide fishes into "reef fishes" (comprised of three groups of related families) that are restricted to hard-bottom reef habitats and "reef-associated fishes" that occur in other habitats as well. The authors conclude that the difference between these two groups is the result of different factors attracting them to reefs. While reef fishes have an "obligate association with the coral reef biota," reef-associated fishes appear to be attracted largely by the reefs as structures. Recent work is showing that many reef-associated organisms occupy niches that are narrower than had been previously assumed (Knowlton and Jackson, 1994), suggesting that more species may be restricted to reef habitats than had been thought.

14.2.3. Why Are Coral-Reef Environments so Diverse?

Explanations for the high diversity of organisms must involve both the evolutionary origin and ecological maintenance of the diversity (Ricklefs and Schluter,

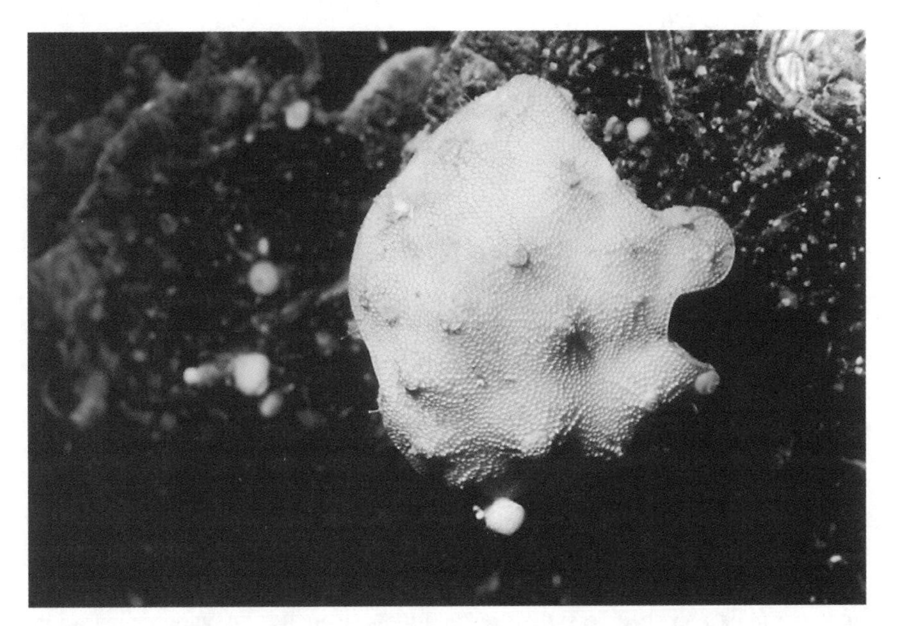

Figure 14–3. *Acanthochaetetes wellsi,* a sclerosponge whose genus originated in the Jurassic. Together with a diversity of other coralline sponges, sphinctozoans, brachiopods, primitive bivalves, and other unusual organisms, it lives in an anachronistic community in reef caves of the western Pacific.

1993). The latter has been the subject of much debate and discussion with regard to coral reefs (e.g., Connell, 1978; Huston, 1985; Jackson, 1991) and is only briefly considered here; the former is discussed below under diversification. Our understanding of the environmental factors that control species diversity is still rudimentary, partly because so many mutually complementary processes appear to be at work, at a variety of spatial and temporal scales. Brown (1988) provides an insightful review of the various factors that are thought to affect species diversity: environmental favorability, productivity, habitat heterogeneity and niche diversity, niche breadth, interspecific interactions, and the evolutionary processes of historical stability, species origination, and extinction.

Although a general understanding of the relative importance of these factors is lacking, coral reefs are favorable for high diversity in a number of regards. Reefs occur in moderately warm waters with little or moderate seasonal variation and generally moderate fluctuations in abiotic parameters. They occur in systems of effectively intermediate primary productivity; in most ecosystems, there are more species at intermediate productivities than at especially high and low productivities (Rosenzweig and Abramsky, 1993). In general, marine systems that have especially high productivities, such as those with substantial upwelling, are characterized by relatively few species and simple, short food webs (Hallock, 1988b).

At the other extreme, in areas with very low productivity, species numbers are often low because of limitations on biomass (Brown, 1988). Although reefs occur in oligotrophic (low-productivity) waters, their endogenous productivity is substantial due to efficient nutrient recycling (Chapter 7) through complex food webs; they consequently can have high biomass and high diversity (Hallock, 1988b).

Niche diversity on reefs is great due to the three-dimensional complexity of the reef framework. Coral reefs probably have the highest level of habitat heterogeneity in the oceans. Niche diversity is further increased by the abundance and importance of symbiotic interactions (Chapters 5 and 11), through which, as in rain forests, diversity begets diversity (May, 1994). Choat and Bellwood (1991) draw attention to how these two factors increase niche diversity for reef fishes. While planktivores and other reef-associated species appear to be largely attracted by the complex surface heterogeneity of reefs, reef fishes in the strict sense rely on the diverse reef biota with which they are obligately associated. Finally, reef communities are renowned for the high degree of niche specialization of component species (e.g., Kohn, 1959; Knowlton and Jackson, 1994; but see Connell, 1978). This high degree of specialization may reflect the abundance of often complex and specific symbioses and the seasonally less variable nature of tropical rather than temperate environments (cf. Stevens, 1989; see below).

14.3. Role of Diversity in Reef Ecosystems

Diversity in species is naturally associated with diversity in modes of life. The variety of niches exploited by coral-reef organisms is spectacular. Diversity is also associated with the adoption of similar basic ecological roles by a variety of different taxa, and thus to potential differences among areas in ecosystem structure and function, depending on which groups fill a given role in different areas. There are two important aspects of ecological diversity on reefs: those of photosymbiotic associations and of reef builders and destroyers. They both lead to the question of how diversity may influence ecosystem function.

14.3.1. Diversity in Photosymbiotic Relationships

Perhaps the most striking ecological attribute of coral reefs is the great efficiency of nutrient recycling, which allows reefs with great biomass and productivity to flourish in oligotrophic oceans (Chapter 7). Much of this recycling is the result of symbiotic associations between microscopic plants and their animal hosts (Chapter 5). These associations characterize modern and many ancient reefs, and are much more pervasive on reefs than in any other habitat. Photosymbiosis is phylogenetically diverse, with a wide range of both host animals and symbiotic algae involved. Multitudes of combinations occur on reefs (Fig. 14–4), although in many cases the metabolic association between host and alga is poorly understood.

Among reef algal symbionts are several genera of dinoflagellates (zooxanthellae). *Symbiodinium,* which itself exhibits considerable ultrastructural and genetic diversity (Blank and Trench, 1985; Rowan and Powers, 1991a, b), is the most ubiquitous of these. Other algae occurring symbiotically in reef organisms include a diversity of cyanobacteria (including the green prokaryote *Prochloron*), chlorophytes (zoochlorellae), diatoms, rhodophytes, chrysophytes, and even free, ephemeral chloroplasts from chlorophytes and rhodophytes (Zann, 1980; Rützler, 1990; Lee and Anderson, 1991). Algal symbionts have been found in a variety of hosts, including other algae, ciliates, foraminifera, sponges, cnidarians, flatworms, molluscs, echinoderms, and ascidians (Zann, 1980). The varied physiological requirements of different algal groups allow host species (e.g., milioline foraminifera) harboring a diversity of algal symbionts to thrive under especially varied environmental conditions (Hallock, 1988b).

The most conspicuous photosymbioses on reefs are those between zooxanthellae and a great diversity of cnidarians: some scyphozoan medusae, zoanthids, coralliomorpharians, actinians, scleractinians, octocorals, and all milleporans. Also important are zooxanthellae in some cardiid and all tridacnid bivalves, a variety of algal symbionts in benthic foraminifera, cyanobacterial symbionts in sponges, and *Prochloron* and other cyanobacteria in didemnid ascidians (Zann, 1980; Rützler, 1990; Monniot et al., 1991; Lee and Anderson, 1991). The paucity of scleractinian families (5 of 23; see above) and genera (7; see Appendix 14–1) that have both zooxanthellate and strictly heterotrophic members speaks for the relatively long-term evolutionary stability of photosymbiotic associations.

Each new generation of host must acquire the symbionts, either on its own or from its parents. The progeny of giant clams, many broadcast-spawning scleractinians, and broadcast-spawning or surface-brooding octocorals obtain their zooxanthellae from free-living algal populations only after larval settlement. Other hosts, such as some broadcast-spawning and brooding scleractinians, internally brooding soft corals, and didemnid ascidians, transfer their symbiotic algae maternally by putting algal cells into or onto the eggs or brooded embryos (Beckvar, 1981; Richmond, 1990; Monniot et al., 1991; Benayahu, 1994).

14.3.2. Diversity Among Reef Constructors and Eroders

Major constructors of reef framework include a variety of corals, especially colonial scleractinians, helioporans, and milleporans, as well as coralline algae. Minor contributors to the framework, especially in crevices, include solitary scleractinians, stylasterine and tubiporan corals, sclerosponges, sessile molluscs (vermetid gastropods, chamid bivalves, oysters, etc.), bryozoans, foraminifera, and algae (Kühlmann, 1985). Many of the latter three groups have expansive sheetlike growth forms and are important reef binders, as are some poorly skele-

Figure 14–4. Diversity of photosymbiotic organisms. Clockwise from upper left: the ascidian *Didemnum molle* harboring the green prokaryote *Prochloron;* the encrusting "killer" sponge *Terpios hoshinota* with photosymbiotic cyanobacteria and attempting to overgrow the scleractinian *Porites rus;* the foram *Marginopora vertebralis* harboring symbiotic dinoflagellates; the scleractinians *Stylophora mordax* and *Acropora danai,* the hydrocoral *Millepora platyphylla,* and the soft coral *Astrospicularia randalli,* all harboring symbiotic dinoflagellates.

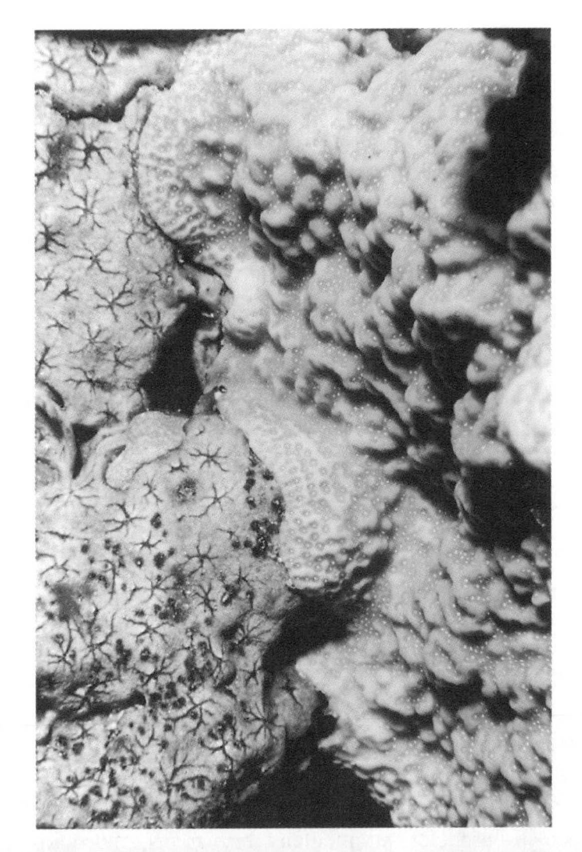

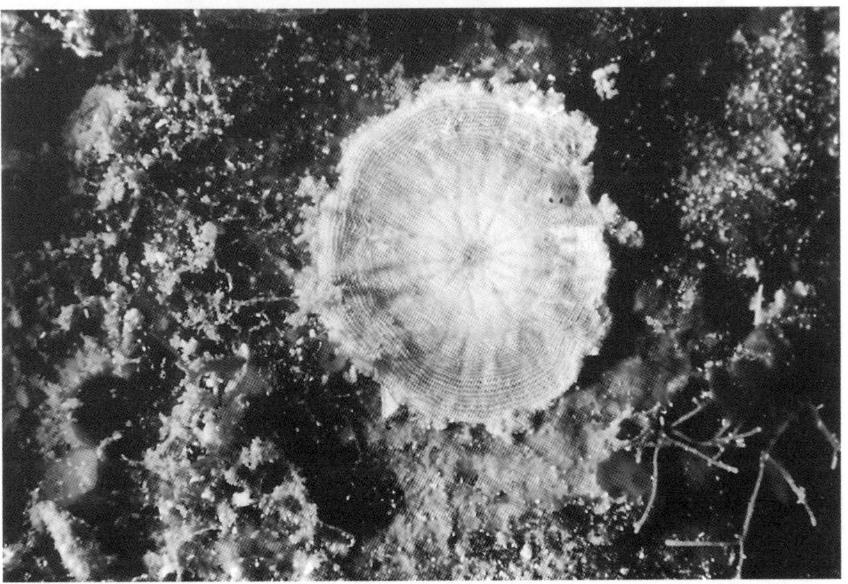

Figure 14–4. Continued

tonized taxa, like demosponges (Wulff, 1984). Indigenous reef sediments are comprised of the spicules and mineralized skeletal fragments of numerous algae (especially *Halimeda,* but also many other genera), sponges, octocorals, arthropods, echinoderms, and ascidians, together with the shells of molluscs and brachiopods, and fragments of framework and binding organisms (Kühlmann, 1985).

Of the two common mineral forms of calcium carbonate, aragonitic skeletons are the most important in modern reef systems. Aragonite is the major or sole mineral in scleractinians, milleporans, helioporans, stylasterines, most molluscs, some sponges and bryozoans, many algae (including *Halimeda*), and most ascidians (Carter, 1990). Nevertheless high-magnesium calcite can be locally dominant, especially where there is an abundance of coralline algae whose skeletons are composed of this mineral (Johnson, 1971; Borowitzka, 1977). Other calcitic reef biota include foraminifera, some sponges and bryozoans, alcyonacean spicules, echinoderms, brachiopods, and certain molluscs (Carter, 1990). Silicious skeletal elements are the least common, and include demosponge and hexactinellid spicules and some protist tests (notably diatoms). Reefs have not always been dominated by aragonitic framework builders, however: rugose and tabulate corals that prevaled on many Paleozoic reefs apparently had calcitic skeletons (Johnson, 1971).

A diversity of organisms contributes to reef bioerosion (see Chapter 4). These include microorganisms: bacteria, cyanobacteria, fungi, an assemblage of endolithic green and red algae, and possibly some forams. Also contributing to erosion are macroborers, including numerous sponges, polychaetes, bivalves, acrothoracican and thoracican barnacles, sipunculans, bryozoans, phoronids, coralliophilid gastropods, and alphaeid and callianassid shrimp (Carriker et al., 1969, Table 5.1 in Vermeij, 1987a; Chapter 4). A number of grazers attack reefs from the outside, most notably echinoids and fish (especially scarids), but also gastropods, chitons, and decapods (Fig. 4–1 in Chapter 4; Hutchings, 1986; Table 5.2 in Vermeij, 1987a).

14.3.3. Diversity and Ecosystem Function

As can be seen from the previous discussion, a variety of organisms partake in nutrient recycling through photosymbiosis and contribute to reef growth and erosion. The same can be said of most other ecological processes on reefs. This high diversity within many ecological guilds is one of the major characteristics of coral reefs (cf. Choat and Bellwood, 1991). How does such diversity influence ecosystem function? Are differences in species diversity reflected in differences in reef structure and functioning?

The high within-guild species diversity of most reefs adds considerable redundancy to reef ecosystems. As a result, large differences in species diversity may lead only to minor differences in ecosystem functioning, as long as crucial guilds remain (Birkeland, in press; see Baskin, 1994, for terrestrial analog). Nevertheless, the absence/presence of entire guilds, or ecologically important keystone species, may lead to important, regional-scale differences among reef ecosystems (see longitudinal comparison discussion later in this chapter; see also Birkeland, 1996). Keystone species are organisms whose presence/absence or abundance strongly influences the structure of the community they inhabit (Paine, 1966).

The species diversity of most groups decreases markedly across the Pacific, from high diversity toward the western continental margins, to much lesser diversity on central Pacific oceanic islands (see below). This includes not only a west to east decrease in species diversity, but also the disappearance of entire classes and orders of organisms (Kay, 1979b). Thus, crinoids, dendrochirote holothuroids, and sepioid and nautiloid cephalopods (all taxa with non-planktotrophic development) drop out of reef faunas east of western Polynesia. Nevertheless, eastern Polynesian reefs are not strikingly different from reefs nearer the center of diversity.

The large drop in diversity from more than 400 to less than 100 reef coral species from the western Pacific Ocean to eastern Polynesia (see below; cf. Fig. 14–6) is not accompanied by obvious correlated changes in reef structure. Large and robust atolls exist and have kept up with sea level for many millions of years, even in the eastern Tuamotus (Labeyrie et al., 1969). Most striking in this regard is the almost-atoll of Clipperton in the eastern Pacific, where the atoll reefs were built by the vigorous growth of a total zooxanthellate coral fauna of only eight or nine species (Fig. 14–5; Wellington et al., 1995; Glynn et al. 1996).

However, in some cases, differences in coral diversity are correlated with differences in reef structure. Some of the striking differences in physiography between Caribbean and western Pacific reefs may partly reflect the greater number of species and growth forms of scleractinian corals in the western Pacific. Since at least the mid-Pleistocene, Caribbean shallow-reef fronts have been dominated by largely monospecific zones of the coarsely branching *Acropora palmata* (near the reef crest) and the finely branching *Acropora cervicornis* (in slightly deeper water; Jackson, 1992). West Pacific reef fronts have a more varied and diverse assemblage of encrusting, massive, and branching coral species. Kojis and Quinn (1993) argue that the greater resilience of Pacific than Caribbean reefs to disturbance may be a reflection of the lower coral diversity in the latter, particularly in the shallowest reef zones dominated by these monospecific stands.

Differences in reef form among regions, however, often reflect differences in factors other than species composition and diversity. For example, the perceived lack of algal crests on reefs in the Caribbean and their great development in the Pacific were historically viewed as the result of region-scale differences in coralline algal species composition and diversity. However, subsequent work showed that crest development varies greatly within both regions and is probably under oceanographic control, being especially dependent on the degree of wave exposure (Adey and Burke, 1976; Chapter 3). The lack of reefs in the eastern Atlantic and their poor development in the eastern Pacific appear to be largely

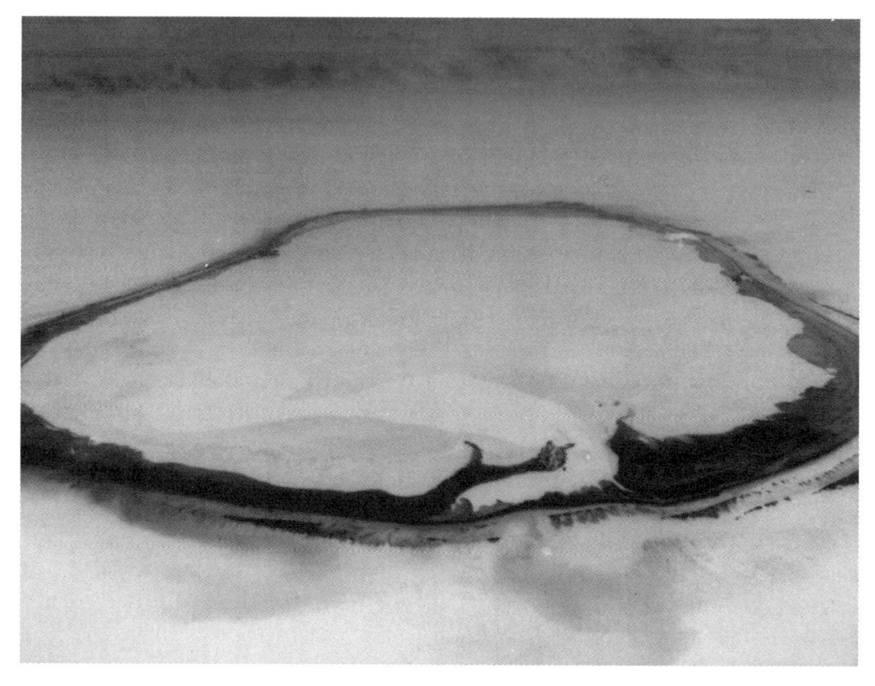

Figure 14–5. Clipperton Island. This well-developed eastern Pacific almost-atoll was built by the prolific growth of only eight or nine zooxanthellate corals (photo courtesy of John D. Jackson).

the result of unfavorable oceanographic conditions on eastern continental shores (Chapter 12), although the low diversity of available reef corals may also play a part. The robust reefs of Clipperton, outside the eastern Pacific upwelling zone but with very low coral diversity, demonstrate the importance of physical rather than diversity controls in the eastern Pacific.

14.4. Practical Biodiversity: Identification of Reef Organisms

A brief guide to some of the taxonomic identification literature is provided in Appendix 14–2, including publications with a geographic as well as taxonomic focus. The former cover a variety of taxa in a circumscribed area; they tend to be less accurate overall, but provide useful characters that differentiate locally recognizable species. The latter are more accurate, but taxonomically more limited.

Except for a few well-studied groups (mostly scleractinian corals, gastropods, crustaceans, echinoderms, and fish), tropical organisms are relatively poorly known taxonomically. Although many species have been named over the past two and a third centuries, few groups have been critically revised. Without taxonomic revisions, identification is difficult because (1) the literature is scattered; (2) most older species descriptions are inadequate for identification, generally do not describe or evaluate the limits of intraspecific variation, and often do not compare and distinguish the species described from related forms; and (3) many species have been described several times, so that different names may be used by different authors, or in different areas within the species' range.

Species identification in many groups can remain difficult even after suitable revisions have appeared, both because intraspecific variation can be great and because related species may have few (or no) discernible morphological differences. Although several regional revisions are now available for reef corals (see Appendix 14–2), identifications remain problematic: species display considerable ecophenotypic, genetic, geographic, and geological variability (Veron, 1981), and it is not uncommon for previously well-characterized species to turn out under closer scrutiny to be species complexes (Knowlton, 1993; Knowlton and Jackson, 1994). Identification keys, which work well for groups with relatively well-delineated species like insects, are often difficult to construct for groups as variable as corals (Wallace and Dallwitz, 1981).

Field guides relying on color pictures are often inadequate for species identification, which in many groups relies on microscopic characters not evident in illustrations of entire, live animals. The converse is also true: in many groups (e.g., opisthobranchs, crustaceans, echinoderms), differences in field appearance, especially in color patterns among related species, can be striking. Thus, despite similar morphologies, identification in the field may be much more straightforward than from preserved specimens. A limitation to easy field identification, however, is that most of the taxonomic literature to date is based largely on pickled specimens. Taxonomic treatises rarely consider color patterns or field appearance. Traditional taxonomy in many groups yet awaits translation into, and improvement from, such useful field characters.

Problems of species distinction are typically greatest for the most diverse groups, because the larger the number of species, the greater the occupation of available morphospace. Furthermore, very diverse genera are often those undergoing rapid evolutionary radiation, resulting in numerous, recently diverged, closely related species (e.g., Palumbi and Metz, 1991; Wallace and Willis, 1994). Difficulties in species identification are exacerbated by the abundance of sibling species complexes, as marine organisms often use cues that are not so obvious to land-based species such as humans (Knowlton, 1993). Within diverse genera, like the corals *Acropora* and *Porites,* the soft coral *Sinularia,* and the gastropods *Cypraea* and *Conus,* are many examples of such complexes of closely related species and accompanying taxonomic problems (e.g., Verseveldt, 1980; Burgess, 1985; Potts and Garthwaite, 1991; Potts et al., 1994; Wallace and Willis, 1994; Röckel et al., 1995).

Because the identification of reef organisms can be so problematic, and because morphologically similar species may differ substantially in their ecology, geological history, and so forth (Knowlton and Jackson, 1994), it is crucial that research on all but well-documented species be supported by appropriate voucher specimens. These should be deposited in a permanent and readily accessible collection (preferably at a National Museum), such that future workers may verify the identity of the species involved.

14.5. Distribution of Diversity: Large-Scale Patterns

There is great variation across the tropics both in the relative dominance of shallow-water habitats by reefs and in the diversity of their biotas. I will now examine the geographic patterns of distribution and diversity of reef organisms before considering how a variety of processes may have contributed to the observed patterns evident today.

14.5.1. Variation in Diversity with Latitude

Perhaps the most obvious worldwide biogeographic pattern is the increase in regional species richness with decreasing latitude (e.g., Pianka, 1966; Stevens, 1989; Rex et al., 1993). This general latitudinal diversity gradient is striking, and parallel gradients occur within many, if not most, major taxa (Schluter and Ricklefs, 1993). Several potential and not mutually exclusive hypotheses have been proposed to explain this trend (see Pianka, 1966, 1983; Stevens, 1989). These include, among others, hypotheses about the time available for diversification, the rate of diversification and extinction, and the maintenance of diversity. Thus, the potentially greater age of tropical than cold polar biotas would have allowed more time for diversification in the former. Rates of diversification in the tropics may be higher than at high latitudes, as has been found at the ordinal level in marine invertebrates (Jablonski, 1993). The potential for niche specialization may be greater in the tropics than at higher latitudes, due to the smaller range of environmental variation experienced by organisms within their lifetimes in the former (Stevens, 1989). The amount of habitat area within isothermal belts is an order of magnitude greater in the tropics than in the temperate zone (Terborgh, 1973).

However, the trend of increasing diversity toward the poles is by no means universal, with some taxa showing little variation in diversity with latitude, and others reaching their greatest richness in temperate to polar regions, especially in the Southern Hemisphere (e.g., Barnard, 1991; Clarke, 1992; Silva, 1992).

Reef corals and their associated biota are limited to tropical and subtropical seas, and thus at least on a coarse scale conform to the general latitudinal pattern. Relatively diverse assemblages of reef corals extend to 28–35° latitude in the North and South Pacific, North and South Indian Ocean, and North Atlantic, reaching the greatest extremes in Japan (35°N), Lord Howe Island (32°S), and

Bermuda (32°N). The latitudinal limits of reef development and reef coral growth are much narrower along eastern than western shores of the Pacific and Atlantic, the former being areas poorly suited (see below) for reef growth and possessing low coral diversity (Fig. 14–6; Veron, 1974, 1992a, 1993, UNEP/IUCN, 1988a; Chapter 12). Worldwide, the extreme latitudinal limits for strictly zooxanthellate corals are approximately 38°N (Azores; Veron, 1993) and 38°S (Victoria, Australia; Cairns and Parker, 1992). A few facultatively zooxanthellate scleractinians are known from slightly higher latitudes (e.g., the Mediterranean; Zibrowius, 1976).

The major proximal causes for the present-day latitudinal limits of coral growth appear to be temperature and competition with macroalgae (Veron, 1995). Low temperatures are lethal to reef corals, and reef-coral diversity falls in a marked and tight regression with decreasing water temperatures along north–south coastlines, as along Japan or Australia (Veron and Minchin, 1992; Veron, 1995). Reef corals extend to much higher latitudes along the western Pacific where boundary currents carry warm waters toward the poles than in areas lacking boundary currents. Reefs extended to considerably higher latitudes during several time periods in the Phanerozoic, and this has been used as evidence that subtropical waters reached further poleward at those times (Adams et al., 1990; Copper 1994). Strikingly species-rich high-latitude reefs are also known in the Holocene of Japan, coincident with higher than present temperatures (Veron, 1992b).

Although authors have also suggested that light is a potential limiting factor in the polar distribution of corals, the existence of deep-water coral communities near the high latitudinal limits of coral growth indicates that light per se may be much less important than temperature in setting limitations (Veron, 1995).

The fact that reef corals do not do well at high latitudes leads to the question of why they should be so limited. Neither scleractinian corals nor photosymbiotic anthozoans are limited to the tropics, so why haven't zooxanthellate corals with lower temperature tolerances evolved? Vermeij (1978) suggests that the latitudinal limits to zooxanthellate corals are ultimately set by competition with macroalgae, facilitated by differences in productivity and rates of herbivory between low and high latitudes. Thus, in the low-productivity environments typical of reefs, the efficient recycling capabilities of animal-plant symbioses (Chapter 5) allow these to outcompete macroalgae, while the latter take over where productivity is higher (Birkeland, 1988b).

The impact of productivity is evident within the tropics also, with reefs being poorly developed in areas of high productivity, where competition from macroalgae is a serious threat, especially to coral recruits (Birkeland, 1977, 1988b; Chapter 12). Furthermore, a great diversity of herbivorous fish crop algae in the tropics and subtropics, so high algal biomass is typically limited to areas that have only limited access to herbivores (Hay, 1991). In contrast, specialized fish herbivores are rare or absent at higher latitudes (Vermeij, 1978). As a consequence, algal biomass is much greater per unit area at high than at low latitudes (Ebeling and Hixon, 1991). Finally, Silva's (1992) observation that temperate

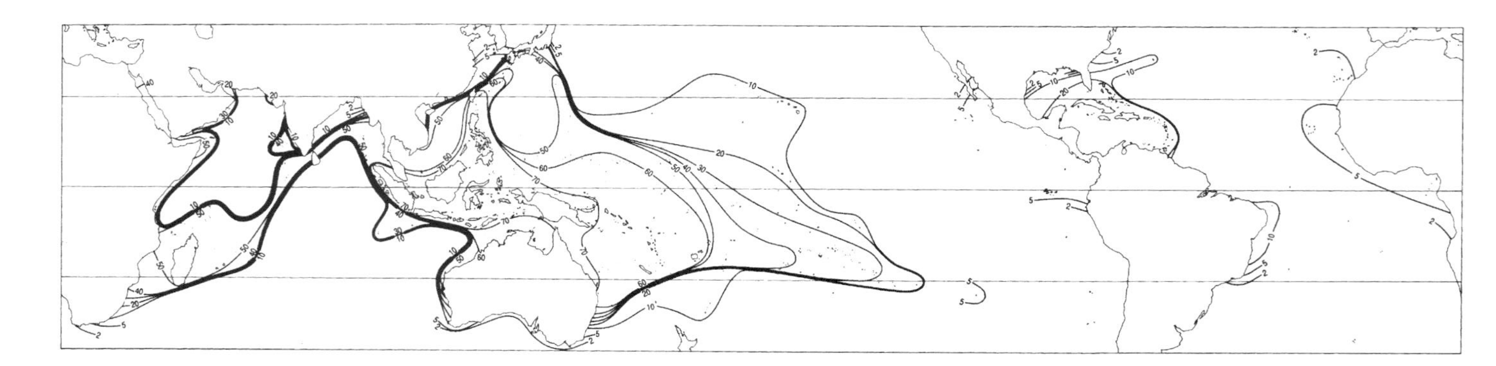

Figure 14–6. Coral diversity at the species level. Contours were produced by combining distribution ranges of genera and thus indicate maximum generic diversity over large regions (from Veron, 1993; reprinted by permission of the Australian Institute of Marine Science).

algal diversity is as high as, if not higher than, tropical algal diversity fits these hypotheses and contrasts markedly with the latitudinal diversity gradient seen in corals.

Although photosymbiotic relationships are less common at high latitudes, a variety of such associations occur, some involving zooxanthellae living in actinian anemones (Rosen, 1984). At least along the eastern Pacific, such photosymbiotic anemones live in physically harsh areas where algal cover is low, and this may in part explain their occurrence. Thus, the photosymbiotic anemone *Anthopleura elegantissima,* which ranges north to Alaska (Morris et al., 1980), inhabits tide pools in the middle intertidal, above the denser algal zones. Its relative, *Anthopleura xanthogrammica,* prefers the bottoms of heavily scoured surge channels, where algal growth is again limited.

Fautin and Buddemeier (1992) suggest an alternate explanation for the restriction of zooxanthellate corals to low latitudes: that this may be due in part to limitations to calcification at high latitudes, as well as to metabolic constraints on the symbiotic association (Chapter 5). Phylogenetically, anemones may be corals that have lost their skeletons (Hand, 1966; Fautin and Lowenstein, 1994). If so, zooxanthellate (and zoochlorellate) "corals" actually do extend to high latitudes.

Although both temperature and solar irradiance increase gradually from pole to equator, generic diversity patterns of reef corals in a given region show a step-and-plateau pattern: within about 10° of latitude from their latitudinal maxima, the number of genera increases from zero to maximum diversity levels, remaining at uniformly high levels within this wide, equatorial band (Rosen, 1984; see also Fig. 3–5 in Chapter 3). A similar, striking, steplike increase in diversity around 40°N latitude in predatory gastropods was observed in the northeastern Atlantic, and explained by latitudinal differences in food-web stability attending the change from tropical aseasonal to high latitude seasonal climates (Taylor and Taylor, 1977). In comparison, bivalves exhibit a more gradual increase in diversity with latitude from pole to equator (Stehli et al., 1967).

At the species level, corals that are not otherwise greatly restricted geographically (as judged from the extent of their longitudinal range) tend to range widely latitudinally within the limits of reef growth (Veron, 1986, 1995). Some, for example, the blue coral *Heliopora coerula,* are limited to a narrow range of tropical latitudes around the equator. The restriction of *Heliopora* to strictly tropical waters appears to be a result of its narrow temperature tolerance, especially with regard to reproduction (Zann and Bolton, 1985; Babcock, 1990).

Latitudinal restriction to the subtropical fringes of reef growth, with absence from lower latitudes, also occurs. A distinctive, subtropical faunule is especially obvious among the numerous islands of the southern Pacific, extending between 24°S and 33°S: from Easter Island and Sala y Gomez through the southernmost outliers of Polynesia, the Kermadecs, Lord Howe, Norfolk Islands, and the subtropical Australian mainland. This faunule consists of latitudinally restricted but longitudinally often widespread species, and has been documented in molluscs, barnacles, echinoderms, and fish. In some areas, it constitutes a large portion (up to 42.5% in Easter Island molluscs) of the local fauna (Randall, 1976; Rehder, 1980; Paulay, 1989). A few of the species from this faunule are also known from northern subtropical latitudes (especially Hawaii), thus having an "amphitropical" distribution. Some also occur at comparable latitudes in the southern Indian Ocean (Houbrick, 1978; Briggs, 1987; Newman and Foster, 1987).

Such southern endemic and amphitropical distributions may reflect physical or biotic limitations. If these species are so distributed because of temperature or light limitations, their restriction to such a narrow latitudinal band indicates considerable stenotopy with regard to those physical factors. There is more support for the hypothesis that they are limited by biotic factors, because they are restricted to marginal reef zones where diversity is relatively low. Fewer species may mean fewer intense biotic interactions, which may allow the survival of species that are otherwise excluded at more diverse tropical localities (Briggs, 1987, Newman and Foster, 1987; Paulay, 1989).

14.5.2. Variation in Diversity with Longitude

Diversity and species composition vary considerably with longitude, both among and within ocean basins. Longitudinal patterns are more poorly understood than latitudinal patterns and biogeographic explanations for them are more varied (Rosen, 1988). Much of the remainder of this chapter is concerned with examining the patterns of longitudinal variation in diversity and species composition and the processes that may be responsible for these patterns. Variation among the major regions will be examined first, while variation within these regions is discussed under patterns of speciation below.

The tropical oceans of the world can be divided into four major biogeographic regions: the Indo-West Pacific (IWP), East Pacific (EP), West Atlantic (WA) and East Atlantic (EA) (Fig. 14–7). Each is characterized by having largely endemic biotas (Briggs, 1974). These areas differ greatly not only in their biotas, but also in geographic extent, oceanographic conditions (Chapter 12), diversity of available habitats, and relative abundance of different habitats. Whereas the IWP and WA have diverse and abundant reef communities, the EP and EA exhibit only limited reef development.

The IWP extends from the Red Sea and eastern African coast through the Indian Ocean to eastern Polynesia (Hawaii, Line Islands Marquesas, and Easter Islands), between roughly 30°N and 30°S. It is by far the largest and most diverse biogeographic region on Earth. Reefs are common in or dominate much of this area, although in some localities, such as along the southern Asian coastline from Burma to the Persian Gulf, they are rare (UNEP/IUCN, 1988b). Practically all tropical, oceanic islands in the IWP are encircled, and many were built, by reefs.

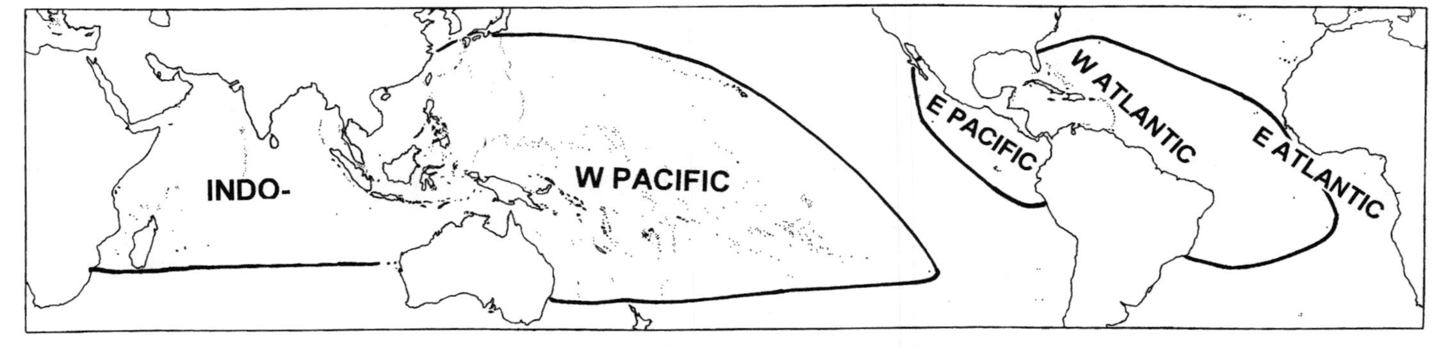

Figure 14–7. The four tropical regions: Indo-West Pacific, East Pacific, West Atlantic, and East Atlantic. The limits of these four regions, as well as interregional speciation, are well demonstrated by the distribution (marked by the lines) of the circumtropical goatfish complex of *Mulloidichthys vanicolensis* (IWP), *M. dentatus* (EP), and *M. martinicus* (WA and EA insular areas). Distributional data from various sources.

The vast IWP is recognized as a single biogeographic entity because of its considerable biotic homogeneity. Kay (1984) summed up the distributions in the IWP of a variety of animal groups that have been taxonomically revised. She found that about 37% of the species ranged widely in the IWP, while the remainder were confined to one of three subregions: the Indian Ocean, Western Pacific, or Pacific Ocean. Lieske and Myers (1994) estimated that about 25% of the Indo-Pacific shore fish species range from the western Indian Ocean to the central Pacific, while Forest and Guinot (1961) noted that about 60% of Polynesian crabs range to the western Indian Ocean. Such wide ranges are not characteristic of all groups, however. Sponges provide a striking example of much greater levels of geographic restriction, with only about 5% of regional species being widespread IWP forms (Hooper and Lévi, 1994).

The EP comprises the tropical western American coastline as well as nearby offshore islands, including the Revillagigedo, Clipperton, Cocos, and Galápagos Islands, all of which harbor a predominantly EP biota (Briggs, 1974). It is separated from the IWP by the East Pacific barrier, a stretch of about 6,500 km of open ocean between the Line Islands and the above-mentioned island groups adjacent to the American mainland. Although the region is biologically very diverse, little of this diversity is associated with reef communities. With physical conditions being generally unfavorable for reef growth, reefs in the EP are uncommon and harbor relatively few obligate reef bionts (Dana, 1975; Glynn and Wellington, 1983; Chapter 12).

The tropical WA comprises the tropical eastern American coast, neighboring islands in the Caribbean and off Brazil, and Bermuda (Briggs, 1974). It is the second largest reef region, with reefs dominating a large portion of shallow-marine habitats, especially around islands. The diversity of reef-associated organisms in the WA is second only to that in the IWP (Table 14–2 and Fig. 14–8), and a large portion of the biodiversity of the region is reef associated.

The EA comprises tropical coastal western Africa and neighboring island groups, including the Cape Verde Islands and islands in the Gulf of Guinea (Briggs, 1974). It is the smallest and least diverse tropical region, and reef development is practically absent. While coral communities do exist, most of the biota of the region is not reef associated.

For marine groups in general, the IWP is the most species-rich and the EA is the most species-poor region, while the EP and WA fall in between (Fig. 14–8; Briggs, 1974). Some taxa, for example, fiddler crabs, are more diverse in the EP than in the WA, while others, including most reef-associated groups, have greater representation in the WA (Fig. 14–8; Table 14–2). For a given group, regional species richness is typically several times greater in the IWP than in other regions, with smaller differences in diversity among the WA, EP, and EA. Reef coral species richness is an order of magnitude greater in the IWP than in the WA, which itself has more than twice the diversity of the EP and more than four times the diversity of the EA. The very high diversity of the IWP relative to the

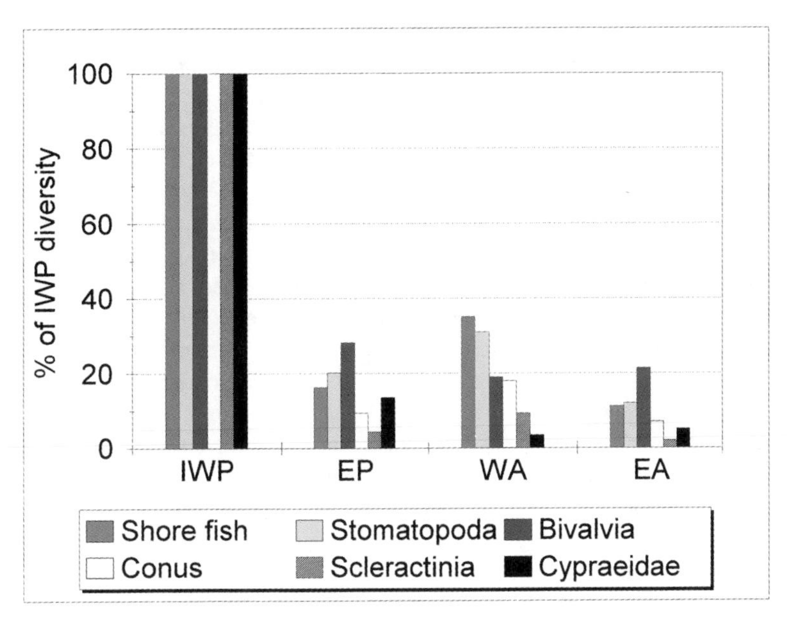

Figure 14–8. Diversity of selected taxa in the four tropical regions (sources: as in Table 14–2).

other regions is most striking among reef-associated taxa. Thus, while there are about 10 times as many reef fishes in the IWP as in the EP, there are only 6 times as many shore fishes (Table 14–2).

In addition to such differences in species richness, the biotas of these regions differ substantially in taxonomic composition. The proportion of a region's biota that is restricted to it varies among regions and depends on the taxa and habitats being considered. For marine habitats in general, the IWP has the most distinct biota: the vast majority of its species, many of its genera, and even several of its families (e.g., tridacnid clams, xeniid, tubiporid and helioporid octocorals, pectiniid scleractinians, caesionid, kuhliid, and plesiopid fish) are geographically restricted (endemic) to the region. However, a number of predominantly IWP species range into the EP, having dispersed across the eastern Pacific barrier (see below). The EA and WA also share a number of species, potentially a result of trans-Atlantic dispersal (Scheltema, 1971; Briggs, 1974; Vermeij and Rosenberg, 1993). The EP and WA share numerous genera, attesting to their common history prior to the Pliocene uplift of the Isthmus of Panama (see below). However, genetic divergence and local extinction have left few species in common between the two regions (Briggs, 1974; Jackson et al., 1993; Knowlton et al., 1993).

The effect of habitat on regional endemism and biotic affinity is strikingly evident when reef-associated taxa are compared with other shelf inhabitants. For example, the general biota of the EP is highly diverse, endemic to the region at the species level, sharing affinities with the WA. However, the reef biota is much less diverse, has low endemism, and shares many species with the IWP. A similar but less striking disparity is evident among reef and nonreef biotas in the EA. Both these examples illustrate the different histories of reef versus nonreef habitats in these eastern regions, where conditions are poor for reef development (see below).

Regional differences in the abundance and diversity of important reef taxa—including large groups as well as single, especially influential, keystone species—have resulted in some striking differences in ecology among the different regions. For example, on IWP reefs, soft corals (Alcyoniina) are diverse (over 700 species in the IWP; Gawel, 1977), ubiquitous, dominant space occupiers (Fig. 14–9) that partake in complex, competitive interactions with scleractinians and other reef organisms (e.g., Sammarco et al., 1985; Dai, 1990; Alino et al., 1992). A large proportion of soft corals are zooxanthellate, some forming colonies many meters across, blanketing the reef. In contrast, soft corals are virtually absent from reef communities in other regions, and the handful of species present live at greater depths or in nonreef habitats (e.g., Bayer, 1961; Humann, 1993).

Gorgonians exhibit a contrasting pattern: they are strikingly diverse, abundant and dominant in the Caribbean (Fig. 14–9), but uncommon at best on shallow, oceanic reefs in the central Pacific and Indian oceans. In deeper waters, however, and in the central, Indo-Malayan area of the IWP, gorgonians are more species rich than in the WA, albeit they are generally less important as space occupiers (Bayer, 1961; pers. obs.).

Sponges are also more conspicuous in the tropical WA (Fig. 14–9) than in the oceanic Pacific, a reflection of differences not in species richness, but in body size and total biomass of species (Wilkinson, 1987a). However, sponges are conspicuous and dominant space occupiers in the productive waters of the continental western Pacific. These differences in the abundance of sponges among and within regions appear to be largely the result of contrasting productivity patterns (Birkeland, 1987). The importance of productivity in structuring sponge communities is supported by the fact that photosymbiotic sponges, adapted by virtue of their symbioses to oligotrophic waters, are virtually absent from the Caribbean. Photosymbiotic sponges also increase in diversity from inshore to offshore on the Great Barrier Reef, reflecting decreasing nutrient input along this gradient (Wilkinson, 1987a, b). Differences in productivity among regions or "subregions" are likely responsible for a variety of other ecological differences among regional biotas (see, e.g., Thresher, 1991, on reef fish communities).

Even single species can make major ecological differences to a region. Most obvious in this regard are keystone species that affect the abundance and community structure of major space occupiers. The large-scale mortality of *Diadema antillarum* in the WA demonstrated the crucial role this sea urchin plays in keeping algal biomass down and thus facilitating coral growth and settlement (e.g., Carpenter, 1990a, 1990b; Chapter 9), albeit the importance of this species

Figure 14–9. IWP vs. WA reef scene. Note dominance of soft corals (Alcyoniina) in addition to corals in the western Pacific, compared with lack of soft corals but abundance of gorgonians and sponges in WA (photos by C. Birkeland).

itself may be the result of human predation on fishes and other invertebrate herbivores (Hay, 1984b; Jackson, 1994). Several groups of coral predators are important in structuring IWP coral communities, but are absent from the Atlantic. These include the asteroids *Acanthaster planci* and two species of *Culcita,* species of the muricid gastropod genus *Drupella,* and coral-feeding parrotfishes (Goldman and Talbot, 1976; Moyer et al., 1982; Birkeland, 1989a). The potential invasion of the WA by *Acanthaster* (and other especially "undesirable" taxa, like sea snakes), is one of the important concerns regarding the construction of a sea-level canal across Panama (Bayer et al., 1970).

14.6. Distribution of Diversity: History of Regionalization

Differences in taxonomic composition and diversity among regions can be explained in part by present-day conditions. Thus, habitat area clearly plays a role, with about 85% of the world area of reefs lying in the Indo-Pacific, compared with only 15% in the Atlantic (Smith, 1978). Present oceanographic conditions are also important in constraining the extent, structure, and composition of reef communities. The cool, nutrient-rich, upwelling waters that dominate the eastern shores of ocean basins present the most obvious environmental limitation to reef biotas on a regional scale (Chapter 12). These upwelling waters greatly limit the development of reefs in the EP and EA, by limiting reef corals directly and by enhancing the abundance of algal competitors and bioeroding organisms that attack the reef framework (Dana, 1975; Birkeland, 1977, 1988b; Glynn and Wellington, 1983; Glynn, 1988a).

Fluctuations in oceanographic conditions, as during El Niño–Southern Oscillation (ENSO) events, can lead to massive mortalities and even extinctions of reef taxa, especially in upwelling regions (Glynn and de Weerdt, 1991; Glynn and Colgan, 1992; Glynn and Feingold, 1992). On a geological timescale, upwelling waters and ENSO-like oceanographic fluctuations may explain the low diversity of coral faunas and meager reef development on the eastern shores of ocean basins (Dana, 1975; see below): EP reefs are rarely more than a few meters thick (Glynn, 1988a; Glynn and Colgan, 1992) and EA coral communities do not even form "true" reefs (UNEP/IUCN, 1988a). In contrast, reef sequences over a kilometer thick are not uncommon in the IWP (Ladd, 1961).

Present-day conditions alone, however, cannot account for the observed patterns in taxonomic composition, richness, and uniqueness of regional biotas. Essential to any analysis of these patterns is an understanding of the development of the biota in historical terms, how changing geographic and environmental conditions led to the observed patterns of species distributions and diversity. How the four major tropical reef provinces arose and became increasingly isolated during the Cenozoic, and how this led to their biotic differentiation, through species originations, extinctions, and range modifications, is explained below.

14.6.1. Tectonic Provincialization

Considerably less biogeographic regionalization existed in the marine tropics in the Cretaceous than today. During the Cretaceous, the circumequatorial Tethyan seaway spread from the New World east to the proto-Pacific Ocean (e.g., Dhondt, 1992, but see Rosen and Smith, 1988). In the early Tertiary, the EP and WA were not yet differentiated, as the two were connected via tropical waters between North and South America. The young Atlantic was still relatively narrow, such that the shallow, tropical waters of the New and Old Worlds were in much greater proximity than they are today, and the eastern shores of the Atlantic were continuous with the Indian Ocean (Fig. 2–5 in Chapter 2).

Thus, in the Cretaceous and early Tertiary, the circumtropical band of shallow-marine waters had only one major apparent barrier: the vast proto-Pacific ocean, extending between Asia and the Americas (Vermeij, 1978; Coudray and Montaggioni, 1982; Grigg and Hey, 1992). However, other types of barriers, for example, in the form of wide areas unfavorable to reefs, may have subdivided this vast tropical ocean (Rosen and Smith, 1988). Whether the eastern or western (if any) part of the proto-Pacific served as a major dispersal barrier depends on the placement of Cretaceous mid-Pacific reefs, not yet well resolved (Schlanger and Premoli-Silva, 1981; Kay, 1984; Rosen and Smith, 1988; Grigg and Hey, 1992).

Through the late Cretaceous and Tertiary, the relative physical continuity of this vast tropical area gradually gave way to the isolation of the regions we recognize today (e.g., Dhondt, 1992). Through the Cenozoic, seafloor spreading gradually widened the Atlantic. Between the late Oligocene and mid-Miocene, the Indian Ocean separated from the Mediterranean–EA. Falling sea levels in the late Miocene temporarily severed even the connection between the Mediterranean and EA (Adams et al., 1983; Rosen, 1988; Aharon et al., 1993; McCall et al., 1994). Finally, shoaling and eventual emergence of the Isthmus of Panama obliterated the connection between the EP and WA by the middle Pliocene (Coates et al., 1992; Waller, 1993).

14.6.2. Dynamics of Species Biogeography During the Late Cenozoic

The physical separation of the four biogeographic regions was accompanied by the divergence of regional biotas (e.g., Kay, 1990; Dhondt, 1992). While reef organisms clearly underwent range contractions and expansions throughout their Cenozoic existence, the increased geographic isolation of regions toward the late Cenozoic made interregional range expansions more difficult (WA-EA) or virtually impossible (EP-WA; EA-IWP). The extinction of regional populations of previously widespread taxa (relictual endemism), together with postisolation origination within regions (neoendemism), created the endemic biotas observed today. Examples of both relictual endemism and neoendemism can be found among families that are presently endemic to the IWP. Thus, giant clams (Tridacnidae) and blue corals (Helioporidae) once had circumtropical distributions, being recorded from the Tertiary of both Europe and America; they are restricted to the IWP today (Rosewater, 1965, Colgan, 1984). In contrast, the coral families Fungiidae and Pectiniidae have always been largely or entirely restricted to the Indo-Pacific, where they diversified in the Neogene (Wells, 1966; Appendix 14–1). Regional extinctions appear to have been facilitated by the changing oceanographic conditions and increased climatic and sea-level fluctuations that characterized the late Cenozoic (e.g., Stanley, 1986; Jackson et al., 1993; see below).

Western Atlantic

The present coral communities of the two most diverse reef regions, the IWP and WA, differ greatly, sharing only 8 of their combined 107 zooxanthellate genera. The IWP and WA share no zooxanthellate scleractinian species, with the possible (though unlikely) exception of *Siderastrea radians* (Veron, 1986; A. F. Budd, pers. comm.). The extent of this differentiation is largely due to (1) the Cenozoic extinction of 21 genera in the WA that survive in the IWP, and (2) the appearance of numerous new endemic genera in both reef provinces during the Neogene (Appendix 14–1; Budd et al., 1994).

Major local extinctions in the WA occurred during the late Oligocene and early Miocene and again in the Plio-Pleistocene (Budd et al., 1994). The proportion of genera that were shared with the IWP declined from about half to one-third of the WA fauna during the Plio-Pleistocene extinction, an event that eliminated a third of the WA coral fauna (Budd et al., 1994). Although the generic diversity of WA corals has been declining throughout the late Cenozoic, species diversity has remained relatively stable, as a result of high rates of species origination (Budd et al., 1994; Johnson et al., 1995).

Similarly high late Cenozoic extinction and origination rates occurred among tropical WA molluscs. Although about 70% of Pliocene WA molluscs are now extinct, this extinction was balanced by high rates of origination; thus, while species turnover was great, overall diversity changed little (Allmon et al., 1993; Jackson et al., 1993; Waller, 1993).

The large-scale extinctions in the WA were likely the result of changing oceanographic conditions associated with the Pliocene closing of the Isthmus of Panama, and may have also been enhanced by the increasing global refrigeration that began in the middle Pliocene (Stanley, 1986; Coates et al., 1992; Budd et al., 1994).

Eastern Pacific

The closure of the Isthmus of Panama precipitated even greater turnover in the reef biota of the EP than in the WA. Although the coral fauna bordering the present EP (e.g., California) was very similar to that of the WA (e.g., Florida) through much of the Cenozoic, this changed drastically when the two oceans became isolated in the Pliocene (Dana, 1975; Budd, 1989). After the emergence of the isthmus, species that went extinct in the generally unfavorable environments of the EP (cf. Glynn and de Weerdt, 1991; Glynn and Colgan, 1992) could no longer recolonize from the east. While the number of American species so declined, colonization from the central Pacific remained possible, albeit across a formidable open-ocean barrier (Dana, 1975; Glynn and Wellington, 1983; Rosenblatt and Waples, 1986; Budd, 1989). As a result, all of the modern EP reef coral genera as well as most (>90%) of the species are shared with the IWP today, while only a third of the genera and none of the species are shared with the WA (Appendix 14–1; Veron, 1995; personal compilation).

The extinction of the American coral fauna in the EP was not complete: some EP corals are survivors from the older amphi-American fauna. Evidence includes (1) the EP fossil record, (2) the observation that some of the differences between the present WA and EP are due to the extinction of WA populations and not necessarily to novel colonizations to the EP, and (3) the occurrence of at least one sister species pair. Several EP species have a fossil record extending back to the Plio-Pleistocene, indicating that they may have had a continuous presence in the region since before the emergence of the isthmus (Reyes Bonilla, 1992). The majority of the genera and at least one of the species that are today confined in the Americas to the EP also lived in the Caribbean prior to their Plio-Pleistocene extinctions there (Budd et al., 1994). Finally, the recent discovery of a species of *Siderastrea* in the EP, a genus not otherwise known in the Pacific but common in the WA, provides direct evidence for the survival of at least one American species in the EP (Budd and Guzmán, 1994).

In any case, in the EP, the large number of clearly IWP species that lack a Caribbean fossil record provides strong support for the dispersal hypothesis for the majority of EP reef corals. Although such a dispersal connection between the EP and IWP has been argued to be novel (due to the Pliocene migration of the Line Islands into the path of the equatorial countercurrent; Dana, 1975), the similarity of Miocene and earlier American and IWP faunas (Budd et al., 1992, 1994) may be evidence that a connection has been in existence for a long time.

For most shallow-water taxa other than corals, very few species are shared between the EP and IWP. The vast majority of these pan-Pacific species live on coral reefs, and, like reef corals, represent a large portion of the strictly reef-associated fauna of the EP today (Fig. 14–10). This reflects the reef-dominated nature of central Pacific islands, the nearest source region, and the extinction-prone history of EP reef habitats since the middle Pliocene.

Among an estimated 2,475 species of EP shelled gastropods (Allmon et al., 1993), only 65 (2.6%) occur also in the IWP (Emerson, 1991; Emerson and Chaney, 1995), and a few others appear to be geminate (i.e., sister) species of IWP taxa. Almost all 65 pan-Pacific gastropods are reef associated, and they

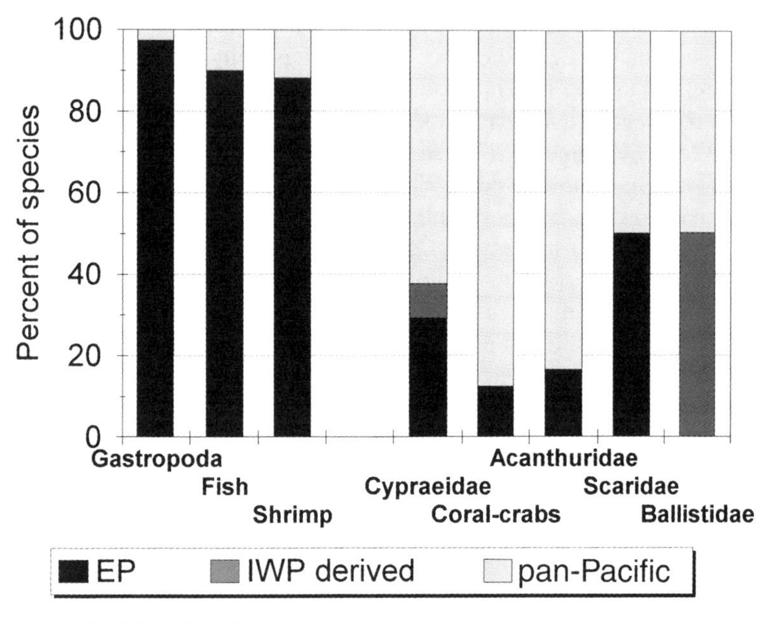

Figure 14–10. Diversity of EP, IWP-derived, and pan-Pacific elements in the EP fauna. The three major groups on the left, inhabiting a diverse array of habitats, show high endemicity, indicating a predominantly in-situ origin. In contrast, the five reef-associated taxa on the right show that the reef biota has a predominantly IWP origin. EP species are those whose range does not extend west into the IWP. IWP-derived species have a distribution like EP species, but their sister species are known and of IWP distribution, indicating that they probably originated through relatively recent allopatric speciation from a pan-Pacific ancestor. Pan-Pacific species are widespread in the IWP with their range extending to the EP.

comprise a large portion of the EP's reefal gastropod fauna. Among 227 EP penaeoid and caridean shrimp, 27 (12%) range into the IWP (Wicksten and Hendrickx, 1992). Among cowries, a family in large part confined to reefs, 15 of the 24 EP species are pan-Pacific and 2 additional species are geminate to IWP taxa (Emerson and Chaney, 1995).

Similarly, seven of eight scleractinian-associated crabs in the EP occur in the IWP (Garth, 1974). Among EP reef fishes, five of the six acanthurids, three of the six scarids, and three of the six balistids also occur in the IWP, and the remaining three balistids are sister species of IWP taxa (Berry and Baldwin, 1966; Rosenblatt et al., 1972; Thomson et al., 1979). While overall, less than 10% of the region's fish fauna are shared with the IWP, 24% of the fish species found in coral-reef communities in the Gulf of Chiriqui have trans-Pacific distributions (Rosenblatt et al., 1972; Fig. 14–10; Table 14–2). That species with pan-Pacific distribution indeed arise through long-distance dispersal was demonstrated genetically by Rosenblatt and Waples (1986) for 12 fish species. Because of the potential enrichment of EP reef biotas by successful colonists from the very diverse IWP region, some groups of reef-associated organisms with sufficient powers of dispersal (e.g., cypraeids; Table 14–2) are more diverse in the otherwise species-poor reefs of the EP than in the WA.

In contrast to reef-associated organisms, the general, shallow-water marine biota of the EP is endemic and distinctive, and shows much closer relationship with the WA than with the IWP (Ekman, 1953; Bayer et al., 1970; Vermeij, 1978; Emerson, 1991). This reflects the common history of the two regions prior to their separation in the Pliocene, a history that is further evidenced by the abundance of geminate species: closely related pairs of species on either side of the isthmus, hypothesized to be the products of divergence after separation (see, e.g., Bayer et al., 1970; Vermeij, 1978; Voight, 1988; Knowlton et al., 1993). About 45% of brachyuran and 12.4% of the mollusc species in the Americas appear to be geminate (Bayer et al., 1970).

Eastern Atlantic

In the more poorly known EA region, reef corals apparently underwent a process similar to that in the EP: loss of most of the original Tethyan fauna following the Oligocene/Miocene closure of the seaway, due presumably to unfavorable eastern oceanographic conditions, and subsequent invasion of species from the west (Boekschoten and Best, 1988). Of the 49 or 50 genera of extant reef corals known in the European Tethys and eastern Atlantic, only 7 survive in the region today, and the local extinction of all 42 or 43 extant genera coincides with the Oligocene–Miocene closure of the Tethyan seaway (Appendix 14–1; Tethyan age data from Veron, 1995). Three of the locally extinct genera survive in both the WA and IWP, 13 in the WA, and 27 in the IWP.

The diversity of zooxanthellate corals in the EA is low today, with 7 genera and approximately 14 species (Laborel, 1974; Cairns, 1991b). The majority (71%) of the species are shared with the WA, and appear to be the result of long-distance colonization (as has been demonstrated for a number of invertebrates by Scheltema, 1971); the remainder are endemic. The eurytopic genus *Siderastrea* may represent a Tethyan survivor, as evidenced by its persistent occurrence in the EA fossil record (Boekschoten and Best, 1988).

As in the EP, most species in taxa other than corals are largely endemic to the region. Thus, only about 6% of the tropical EA marine molluscs have amphi-Atlantic distributions (123 of 2,200 species; Vermeij and Rosenberg, 1993). In the EA as in the EP, offshore islands have a greater representation of western species than does the continent (Vermeij and Rosenberg, 1993).

As the separation of EA and IWP regions preceded the separation of EP and WA regions by several million years, the biotic affinities between the former two are much less than between the latter two. The majority of EA species appear to be related to WA forms in most habitats (Briggs, 1974; Woodroffe and Gindrod, 1991). However, ties with the IWP have also remained; for example, the arcoid bivalves of West Africa show much greater affinity with the IWP than with the WA fauna (Oliver and von Cosel, 1992).

Indo-West Pacific

In contrast to the high rates of species turnover observed in the EP, WA, and EA, the limited available data on fossil faunas in the vast IWP indicate that rates of extinction in molluscs and reef corals were no greater than background levels, at least since the Pliocene (Stanley and Campbell, 1981; Vermeij, 1987b; Veron and Kelley, 1988; Paulay, 1991, 1996a). While all other regions show high levels of regional extinction among corals that are globally extant today, no extant coral genus is known to have undergone regional extinction in the IWP region (Appendix 14–1). Some of the 13 coral genera that are now confined to the Americas but are known from the European Tethys may have ranged into the IWP; if so, they represent regional extinction in the IWP. However, seven or eight of these genera arose during, or subsequent to, the late Oligocene to middle Miocene (Rosen, 1988; McCall et al., 1994) isolation of the Mediterranean and Indian oceans (Appendix 14–1). Further, Rosen and Smith (1988) argue that the European Tethys and the Indian Ocean were partially isolated biotically even prior to the severance of a marine connection between them. Vermeij (1987b) similarly notes that no subgenus-level mollusc taxon is known to have gone extinct in the western Pacific during the late Cenozoic.

14.7. Distribution of Diversity: Evolutionary Diversification

The high diversity of species on reefs is ultimately due to speciation, and diversification in the sea is nowhere more apparent than in the great adaptive radiations that characterize numerous reef taxa. Examples include the several hundred species of the predominantly reef-associated gastropod genus *Conus,* and the more than 100 species each of the prosobranch gastropod *Cypraea,* nudibranchs *Chromodoris* and *Cuthona,* hard coral *Acropora,* soft coral *Sinularia,* and snapping shrimp *Alpheus.* Many of the common genera encountered on reefs include dozens of species, especially in the IWP.

What is the geography and tempo of these radiations? Speciation is generally thought to require the isolation of populations (Futuyma, 1986). This can result either from the division of a relatively widespread species by external events (vicariance), or by the creation of an isolated population by founder events: the rare, chance dispersal of propagules across a wide barrier (Valentine and Jablonski, 1982). In this section we examine how these processes may have contributed to speciation both among and within regions. It should be noted that some authors have suggested that simple geographic isolation is not the only process responsible for speciation, as evidenced by recent diversifications within the confined area of the Caribbean (see below) and the largely sympatric distributions of many IWP species groups (Veron, 1995). However few data are available at present to evaluate such alternative propositions (Paulay, 1996b).

14.7.1. Interregional Speciation

The gradual physical differentiation of the four tropical regions during the Cenozoic (see above) provided numerous opportunities for isolation by vicariance. This is most clearly evident in the relatively recent evolutionary divergence of species across the Panama land bridge, giving rise to numerous pairs of geminate taxa in the WA and EP (e.g., Bayer et al., 1970; Knowlton et al., 1993; see above).

The widening of the Atlantic also created opportunities for speciation by vicariance. However, cross-Atlantic geminate species pairs may have arisen alternatively via founder events involving long-distance dispersal across the Atlantic. Certain groups almost certainly diverged by vicariance: those with no or poor opportunities for dispersal and showing a long history of divergence. Examples include a variety of apparently ancient crustacean lineages that inhabit marine caves on opposite shores of the Atlantic (Iliffe et al., 1984). In contrast, closely related species pairs, often in genera whose origins postdate the opening and initial widening of the Atlantic, clearly diverged from populations established via dispersal across the basin. This latter mode of divergence has been demonstrated in some sea-turtle populations (Bowen et al., 1989) and scallops (Waller, 1993). It also likely applies to most strictly reef-associated groups, as most have planktonic larvae capable of long-distance dispersal (cf. Scheltema, 1971), the majority are conspecific with WA populations (indicating recency of origin), and the marginal reef environment of the EA is not conducive to their long-term persistence (see above).

Speciation between the tropical IWP and EP regions can occur only after successful colonization by long-distance dispersal across the eastern Pacific barrier. Historical changes in patterns of ocean circulation or island positions may have at times facilitated or impeded dispersal opportunities and thus contributed a vicariant component (Dana, 1975; Rosen, 1988; Grigg and Hey, 1992). That dispersal across this tremendous expanse of open ocean occurs today is demonstrated by the occurrence of larval stages of a diversity of benthic invertebrates and fish in the water column of the eastern Pacific barrier (Johnson, 1974; Leis, 1983; Scheltema, 1988), as well as by evidence of gene flow between central and eastern Pacific populations (Rosenblatt and Waples, 1986).

Nevertheless, the barrier presents a major impediment to dispersal, and successful colonization across this filter must be a very rare event for most species (Scheltema, 1988). In addition to dispersal limitations, the establishment of IWP species in the EP is limited by the substantial differences between the ecology of these regions. Several IWP species are known from the EP on the basis of single or few specimens, evidently representing failed or limited colonization attempts (e.g., Emerson, 1991; Emerson and Chaney, 1995). Successful founders thus have the potential to diverge into novel species, relatively unimpeded by significant gene flow across the barrier (cf. speciation among islands; Carson, 1983). Several sister-species pairs, for example, the cowries *Cypraea isabella* and *C. isabellamexicana* (Burgess, 1985) and the stomatopods *Lysiosquilla monodi* and *L. sulcirostris* (Reaka and Manning, 1987), between the IWP and EP attest to speciation across the eastern Pacific barrier.

Interregional speciation events can lead to circumtropical genera with different sister species occupying each region, as with the goatfish *Mulloidichthys vanicolensis* (IWP), *M. dentatus* (EP), and *M. martinicus* (WA and EA) (Fig. 14–7; Stepien et al., 1994), and the stomatopods *Acanthosquilla biminiensis* (EP and WA), *A. septemspinosa* (EA), and *A. acanthocarpus* (IWP) (Reaka and Manning, 1987). Although interregional diversification is clearly important, especially for taxa with great powers of dispersal, most of the diversity of the two great reef regions appears to have been generated in situ, by intraregional mechanisms.

14.7.2. Speciation within the Indo-West Pacific

Diversity within the IWP region peaks in the Indo-Malayan area, is relatively even across the Indian Ocean, and falls gradually eastward across the western and central Pacific (Fig. 14–6). This diversity pattern is found in the majority of species-rich and widespread taxa, and has led to much discussion about the source of the pattern as well as the diversity (Kay, 1979b).

For many years, with the dominance of the center-of-origin concept in biogeography, high-diversity areas such as the Indo-Malayan triangle were looked upon as centers of diversification (e.g., Ekman, 1953; Briggs, 1974, 1992). From these centers, new taxa were considered to have expanded their range into areas with lesser diversity, such as the insular Pacific. A decrease in the average generic age of corals and barnacles from the central Pacific toward Indo-Malaya may support this view (Stehli and Wells, 1971; Newman, 1986); alternatively this age pattern may be a reflection of (1) the fact that widespread genera are more likely to occur at peripheral localities and to persist through time (being less susceptible to extinction) than are genera with smaller ranges (cf. Hansen, 1980; Jablonski, 1986), or (2) the fact that older genera have more species and that the combined ranges of species in a larger genus are likely to be greater than in a smaller genus by chance alone (Jokiel and Martinelli, 1992).

The center-of-origin paradigm views peripheral endemic species as either (1) relicts of previously widespread species, which lost out in the "evolutionary arms race" (cf. Vermeij, 1987a) in the high-diversity center where they originated, or (2) species that originated in situ, but are destined to remain so localized until their eventual extinction. The permanent geographic restriction of such neoendemics could be the result of their difficulty to expand their range "upstream" against a gradient of increasing diversity that may also represent increasing intensity of biotic interactions (cf. Terborgh, 1973). Such a general inability to disperse toward areas of increasing diversity, apparently reflecting competition, has been demonstrated by Diamond and Marshall (1976) for Pacific island birds.

The opposite view, that the Indo-Malayan area owes its high diversity to being a sink for species originating on the remote archipelagoes of the oceanic Pacific, was proposed by Ladd (1960) (based on a suggestion by Fenner Chace). Ladd argued that the predominantly west-flowing currents, winds, and storm tracks in the oceanic Pacific facilitate dispersal to, rather than from, Indo-Malaya. He noted that the long and diverse history of marine life in the central Pacific, attested to by the geological record, provided diversity from which to draw. In Ladd's scenario, the diversity peak in the Indo-Malayan area would be a result of species accumulation together with the extinction of peripheral populations of previously widespread species. Jokiel and Martinelli (1992) proposed a similar scenario for reef corals.

Concepts of the Indo-Malayan area as a cradle of diversification or as a museum for species that originated elsewhere are not mutually exclusive propositions, nor are they the only ones (Rosen, 1988). Both the geographic patterns of species originations and the nature of subsequent distributional changes are likely to vary among groups. One way to evaluate the importance of alternative modes of diversification is to consider how geographic isolation may arise and lead to speciation within the vast IWP.

Whether and how a species can undergo allopatric speciation in the IWP depends in large part on (1) the dispersal ability of the organism, which limits both the species' potential geographic range and gene flow among its populations; (2) historical changes in the distribution and connectivity of suitable shallow-marine habitats, caused by changes in tectonics, climate, sea level, and ocean circulation; and (3) the rate of evolution, or how rapidly isolated populations can diverge. The last question becomes especially relevant when considering the duration of episodic vicariant events, such as those driven by Milankovitch cycles (Bennett, 1990), like sea-level fluctuations (e.g., McManus, 1985; Springer and Williams, 1990; see below). The 10^4–10^5-year time intervals over which isolating barriers arise and break down during late Cenozoic sea-level cycles may be of sufficient duration to lead to species-level divergence in some cases, but are certainly much shorter than the average 10^6–10^7-year life span of typical marine species (cf. Stanley, 1979).

Paradoxically, many taxa that have undergone large radiations in the IWP consist predominantly of highly dispersive, widespread, sympatric species (Kohn, 1971). Recent phylogenetic studies have shown that many sister species in the IWP are widespread and have broadly sympatric ranges (e.g., Palumbi and Metz, 1991; Wallace et al., 1991; Pandolfi, 1992). Indeed, dispersal subsequent to species originations appears to have masked presumed patterns of allopatry that lead to speciation in most cases (Gosliner and Darheim, 1996; Kabat, 1996).

How do such readily dispersing organisms become sufficiently isolated to allow for allopatric speciation? Although genetic divergence of populations with good dispersal abilities may be facilitated by strong selection or homing behavior, large-scale geographic isolation is also needed. Such geographic isolation may result from the separation of populations either by geological/oceanographic vicariant processes, or by sheer distance following a founder event (Palumbi, 1992).

Speciation among islands by founder event may be common among marine species (Knowlton, 1993). For widespread species (e.g., *Mulloidichthys vanicolensis;* Fig. 14–7), isolation sufficient for genetic divergence is likely to be limited to sites near the edge of their range. The potential for allopatric divergence in such peripheral populations is evident in genetic studies of widespread IWP species; these studies frequently show very limited genetic variability in remote insular populations, indicative of their origin through rare founder events (Palumbi and Metz, 1991; Benzie, 1992; Palumbi, 1994).

The importance of insular central Pacific endemics in contributing to diversity in the IWP was underscored by Kay (1984), who showed that such endemics are numerically abundant, comprising about 40% of the insular marine species of the Pacific tectonic plate. Although the restriction of a species' range to the central Pacific can also be the result of secondary reduction of a formerly wider range (relictual endemism) (e.g., Newman and Foster, 1987), many species with such ranges are clearly the result of peripheral divergence from widespread, allopatric sister species (Fig. 14–11A; e.g., Kay, 1967; Rehder, 1980; Blum, 1989; Pandolfi, 1992; Randall, 1995).

Peripheral endemics in the insular Pacific had the powers of dispersal to colonize their remote home (Newman, 1986), and may or may not retain these powers following divergence. For example, endemic Hawaiian angelfish have some of the longest larval life spans known for the family (Thresher and Brothers, 1985), presumably attesting to their ancestor's ability to arrive there. Many, if not most, endemics on remote islands probably represent evolutionary dead ends, which will go extinct without issue in their isolated homes (cf. Kohn, 1980). However, those insular endemics that manage to later expand their ranges can complete the cycle of speciation for widespread species (Jokiel and Martinelli, 1992), albeit direct evidence for such range expansion is lacking.

Isolation leading to the development of peripheral endemics appears to be possible via climatic factors (perhaps through vicariance) as well as by distance. The subtropical–warm temperate fringes of the IWP (South Africa, Japan, southeastern and southwestern Australia) harbor a diversity of endemics, many of which have sister taxa at tropical latitudes (e.g., Pandolfi, 1992; Gosliner and Darheim, 1996).

While groups with good powers of dispersal are unlikely to find sufficient geographic opportunities for isolation in Indo-Malaya (cf. Jokiel and Martinelli, 1992), groups with poor dispersal capabilities may find ample opportunities. The

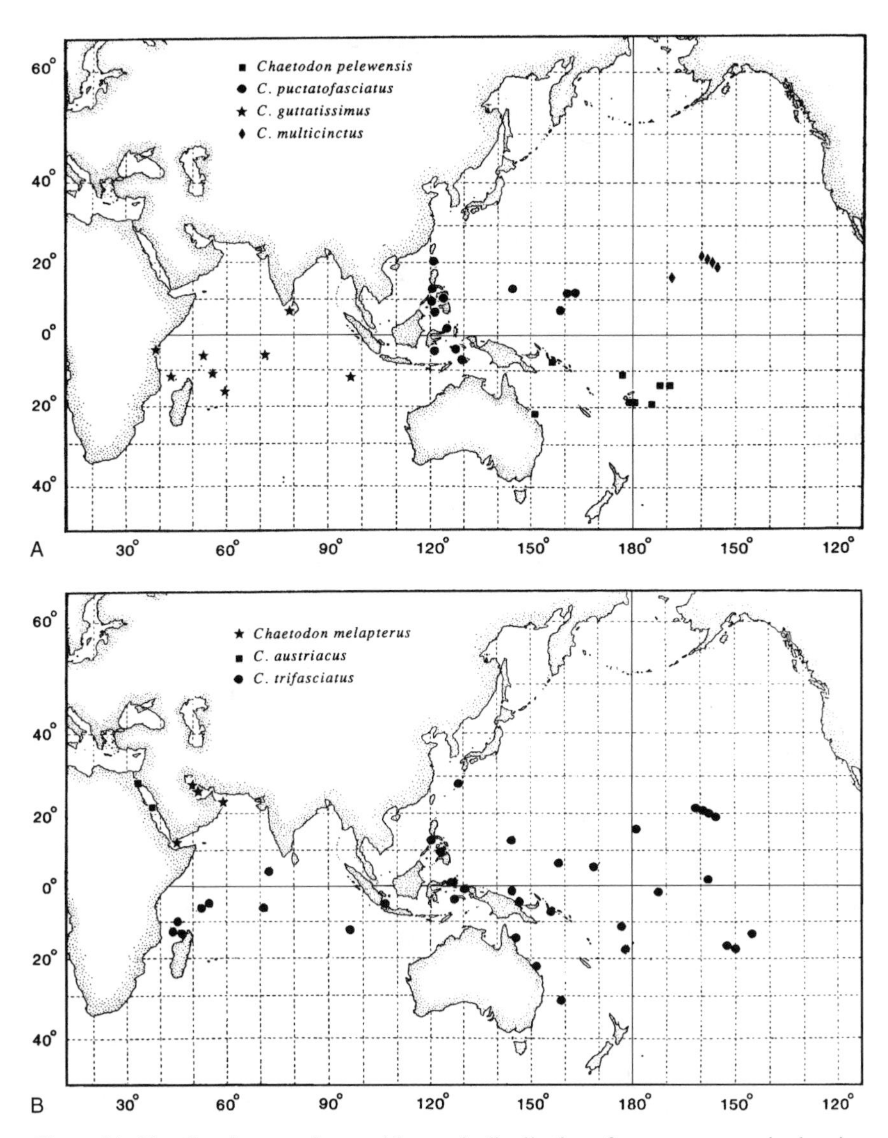

Figure 14–11. Species complexes with mosaic distribution of component species bearing evidence to patterns of speciation in the IWP. **(A)** Species in the *Chaetodon punctatofasciatus* complex show divergence in the remote oceanic Pacific (Hawaii: *C. multicinctus*) and apparent speciation across the Indonesian barrier *(C. guttatissimus–C. punctatofasciatus)*. **(B)** Species of the subgenus *Corallochaetodon* show peripheral speciation in the Red Sea and in the Arabian Gulf region. (From Blum, 1989; reprinted by permission of Kluwer Academic Publishers, Dordrecht.)

importance of island size and dispersal ability in controlling opportunities for speciation has been clearly shown by Diamond (1977). He noted that intraisland speciation for vagile groups like birds and mammals is possible only on some of the largest islands (Madagascar and Luzon being the minimum size, respectively), while for land snails and flightless insects, islands a few squared kilometers in area suffice for local diversification (cf. Paulay, 1994). An analogous situation clearly exists in the sea.

Groups that lack a planktonic period in their development and do not raft well, such as volutid gastropods, sepioid cephalopods, and numerous species of the gastropod genus *Conus,* are largely or entirely restricted to the continental shores of Indo-Malaya, being absent from the vast oceanic Pacific (Weaver and duPont, 1970; Kohn and Perron, 1994; Bouchet and Poppe, 1988). Despite their relatively limited continental distributions, such groups may be diverse. For example, over 130 species of large-bodied volutid gastropods are known from the warm temperate to tropical continental shores of Australasia (Weaver and duPont, 1970). For these groups, Indo-Malaya is clearly an important site of species origination: extremely limited dispersal abilities have allowed extensive allopatric speciation within the region, with resultant closely related species often having very restricted and largely allopatric distributions (e.g., Weaver and duPont, 1970; Fleminger, 1986; Poppe and Goto, 1992). This is clearly shown in several volutid genera, comprised of geographically greatly restricted, allopatric to narrowly sympatric species in Southeast Asia–Australia (Fig. 14–12). Some groups, most notably sponges, although widely distributed as a group across the IWP, show highly restricted species distributions, creating a complex mosaic patchwork of species ranges across the entire region (Hooper and Lévi, 1994).

Thus, Indo-Malaya is likely both a center of diversification, especially for groups lacking planktonic dispersal stages, and accumulation, especially for groups with great powers of dispersal. Other possibilities for vicariant isolation, as would be provided by sea-level fluctuations, further increase the modes of diversification available within the IWP (Newman, 1986).

Our present-day perspective of the world is biased in that we live at a time of maximum sea level with respect to the glacio-eustatic cycles that characterize the Quaternary. Reefs and shallow continental shelves are today generally covered by water (cf. Chapter 2); the Red Sea is open to the Indian Ocean, and the Indian and Pacific oceans communicate through a wide, island-strewn, shallow seaway. At glacial low stands, however, the Red Sea becomes isolated (Sheppard and Sheppard, 1991). Communication between the Indian and Pacific oceans is greatly restricted by the emergence of the shallow continental shelves that dominate the Indo-Malayan region, by changes in circulation patterns, and by increased upwelling in the remaining channel between the two great oceans (Fleminger, 1986; Blum, 1989). Furthermore, numerous semienclosed basins are created in the Indo-Malayan region, providing opportunities for isolation, especially for poorly dispersing taxa (McManus, 1985).

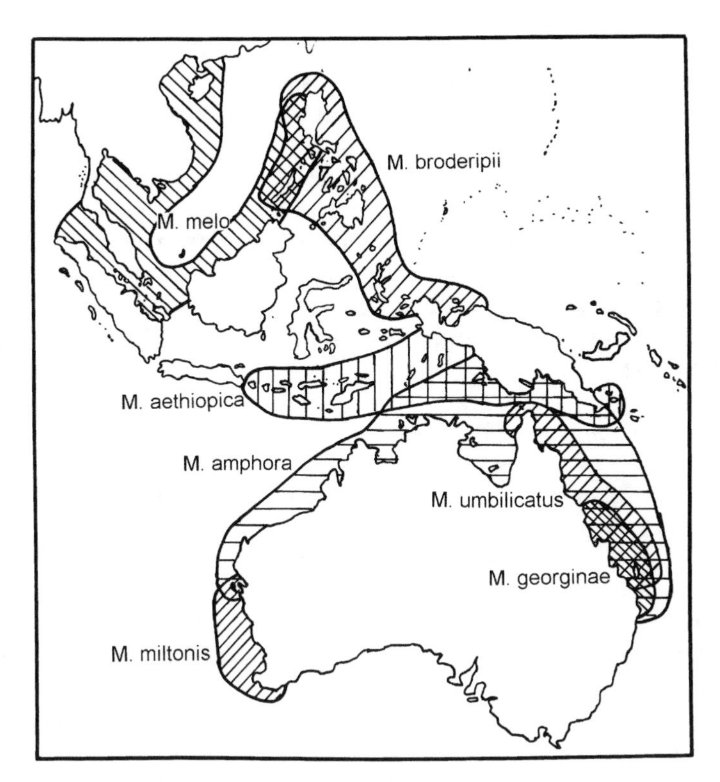

Figure 14–12. Distribution of species of the volutid gastropod genus *Melo.* Volutids lack pelagic larvae and illustrate how groups with poor dispersal abilities can have restricted, allopatric to narrowly sympatric ranges and diversify within continental Southeast Asia–Australia. An eighth, poorly known species, *M. nautica,* not included on the figure, is known only from the southern Philippines. Distributional data from Poppe and Goto (1992).

These vicariant changes are clearly mirrored in species distributions: numerous species are endemic to the Red Sea, and many of these have allopatric sister species that inhabit the Indian Ocean (Fig. 14–11B; Blum, 1989). Species pairs that straddle the Indonesian barrier are also common, and include not only benthic invertebrates and fish, but even pelagic organisms associated with upwelling regions (Fig. 14–11A; McManus, 1985; Fleminger, 1986; Rudman, 1987; Blum, 1989; Springer and Williams, 1990).

The previous examples demonstrate some of the less ambiguous modes of speciation operating within the IWP. The narrowly allopatric, abutting ranges of sister species pairs in a variety of other areas of the IWP point to additional, less obvious modes of speciation within the region (Blum, 1989). Some coincident species range boundaries, such as those of many species at the reef-poor, central stretch of the Indian Ocean, may reflect effective dispersal barriers, allowing speciation by founder events; others may point to past vicariant events. In many cases, allopatric sister species have closely abutting ranges that meet in areas lacking obvious dispersal or vicariant barriers; here, the mechanism of isolation is unclear, attesting to the complexity and diversity of modes of speciation operating within the IWP (Blum, 1989). Understanding the importance and relevance of different modes of speciation in the IWP is one of the most intriguing problems in marine biogeography and a fruitful field for future research. Phylogenetic studies of closely related species with allopatric or narrowly sympatric ranges are an excellent, but currently rarely used, method for elucidating patterns of diversification in the IWP (e.g., Gosliner, 1987a; Blum, 1989; Myers, 1991; Pandolfi, 1992).

14.7.3. Speciation within Other Regions

Evolutionary radiations of marine organisms are not uncommon within the other three tropical regions, although they are generally less spectacular than in the IWP (e.g., Reaka and Manning, 1987). Diversifications of reef-associated organisms are limited on the eastern shores of ocean basins, where the poverty and instability of coral reefs have allowed little time or opportunity for speciation.

In the WA, however, numerous examples of in-situ radiation of reef organisms are evident. The WA endemic coral genera *Mycetophyllia* and *Agaricia* have at least five and seven species, respectively (Budd et al., 1994), and Atlantic *Montastrea* (Knowlton et al., 1992) and *Porites* (Potts and Garthwaite, 1991; Potts et al., 1994) have undergone both obvious and cryptic speciation. Budd et al. (1994) and Johnson et al. (1995) showed that, despite the Plio-Pleistocene extinction of numerous coral-reef species and genera in the WA, species numbers were maintained because of concomitant high species origination rates in surviving lineages. The great Plio-Pleistocene extinction of marine molluscs was likewise accompanied by high rates of origination, such that net diversity remained virtually unchanged (Allmon et al., 1993; Jackson et al., 1993). Waller (1993) documents the rapid radiation of five species of *Caribachlamys,* a scallop genus that originated after the closure of the Isthmus of Panama.

The WA basin is limited in size and has numerous closely set islands; it lacks major dispersal barriers. Thus, genetic isolation may be more difficult to attain in the WA than in the IWP, especially for groups with good powers of dispersal. Accordingly, many of the endemic radiations documented are of relatively poorly dispersing organisms. Reaka and Manning (1987) tabulate several WA radiations of stomatopods, noting that such radiations are especially prevalent among taxa whose larvae settle and metamorphose at small sizes, thus spending shorter periods of time in the plankton. The rapid diversification of the five *Caribachlamys* species mentioned above was certainly facilitated by the unusual, nonplanktotrophic development characterizing the genus (Waller, 1993). Volutid gastropods, a group that lacks pelagic larvae (Bouchet and Poppe, 1988), also show localized radiations (e.g., five species of *Falsilyria* off Honduras) that may have originated through habitat separation caused by Pleistocene sea-level fluctuations (Petuch, 1981a).

In some of these groups, as in *Falsilyria,* the species resulting from recent speciation have retained localized, allopatric distributions (see also Bert, 1986). However, the vast majority of WA species, including the products of many relatively recent endemic radiations, are widespread throughout the basin (Liddell and Ohlhorst, 1988; Robins, 1991). We are thus left with no obvious longitudinal gradients in diversity and no geographical clues to the origin of most species within the region.

Localized endemics in the WA are rare and fall largely into two categories: species with poor dispersal abilities, whose distributions may still reflect the geography of recent speciation events, and species with formerly widespread distribution that are now restricted to small areas of suitable habitat. Among reef corals, subregional endemics occur only on the isolated reefs of eastern Brazil, to which eight species are restricted (Budd et al., 1994). These species appear to include both recently diverged as well as relictual endemics. Thus, although the majority lack a Caribbean fossil record and are allopatric with putative sister species in the Caribbean, the Brazilian endemic genus *Mussismilia* (including one of its three species) and *Meandrina braziliensis* are known from the Caribbean in the Neogene (Budd et al., 1994).

Relictual endemics in the WA have been especially well documented among molluscs, and are best represented in upwelling regions of Colombia and Venezuela (Petuch, 1981b, 1982). The survival of numerous Neogene species and genera in these areas of upwelling implies that their extinction elsewhere was likely the result of changing temperature and nutrient conditions resulting from the closure of the Isthmus of Panama (Keigwin, 1982b; Vermeij and Petuch, 1986; Jackson et al., 1993).

14.8. Effects of Human Activities

14.8.1. Extinction

Because of the generally great dispersal abilities and resultant wide ranges of marine organisms, human-caused species extinctions are less likely in the sea than on land. The only presently documented case of extinction among marine invertebrates during historical times is that of *Lottia alveus,* a limpet species that was restricted to eelgrass in the northwestern Atlantic (Carlton et al., 1991). Extinctions among endothermic marine vertebrates, however, have been common (Vermeij, 1989). The logistic difficulties of sampling underwater, however, pose a methodological problem for demonstrating extinction. This is because many marine species are known from single collections and many others remain undiscovered, so data with which to judge the status of species can be very limited.

The resilience of species to total extinction depends in large part on the extent of their distributions; while widespread marine species are unlikely to become endangered in the immediate future, localized endemics can be vulnerable. An important unknown in this regard is the actual range of many species. Numerous species that were previously considered to be widespread have been found, under closer scrutiny, to represent complexes of geographically restricted sibling species; thus, marine species may be more vulnerable to extinction than previously thought (Knowlton and Jackson, 1994).

In the IWP, endemics are most common in remote island groups, in peripheral continental areas such as the Red Sea, South Africa, southern Japan, and southeastern and southwestern Australia, and in the Indo-Malayan center of diversity. The Indo-Malayan area contains by far the largest number of species with narrow distributions (cf. Fig. 14–12), including species with poor dispersal abilities that have diversified in the area as well as relictual taxa that have gone extinct elsewhere. This area also suffers some of the greatest exploitations and degradations of coral reefs and other coastal environments, due to large and rapidly increasing human populations and economies (Chapter 15; Wilkinson, 1994).

Wilkinson (1994) notes that all Indo-Malayan reefs, except for those in Australia, are in critical or threatened condition—categories applied to reefs that are likely to collapse within the next 40 years, if management and conservation measures are not implemented. This combination of limited species distributions and rapid environmental degradation has raised the very real possibility of marine species extinctions in the area (Wilkinson et al., 1993). Veron (1992a) has discussed similar concerns for southern Japan, an area with a number of coral-reef endemics, some of which are known from very limited populations. This region is also being rapidly degraded.

Because of the highly enclosed nature of the WA, oceanographic and biological events can be region-wide there. This is evidenced by the rapid spread of the *Diadema* plague, as well as the numerous recent, region-wide, bleaching events (Lessios et al., 1984; Williams and Bunkley-Williams, 1990). Both disturbances underscore the vulnerability of this enclosed basin and its biota, further illustrated by the high rates of extinction suffered in the last couple of million years since the Caribbean has become isolated from the Pacific (Jackson, 1994; see above).

Despite the rarity of known marine species extinctions, localized areas can experience regional diversity loss. A species may thus go extinct over part of its range, without likely reestablishment of populations in the foreseeable future. Extinction of local populations is most obvious for species with large body size, and such species are also the most vulnerable to overexploitation by fisheries. Fisheries have already led to the complete extinction of several marine mammals and birds (Vermeij, 1989). Among marine invertebrates, giant clams appear to be especially vulnerable; several species underwent large-scale range contractions even prior to human exploitation, perhaps as a result of habitat loss caused by sea-level fluctuations (Fig. 14–13; Taylor, 1978; Paulay, 1996a). The recent

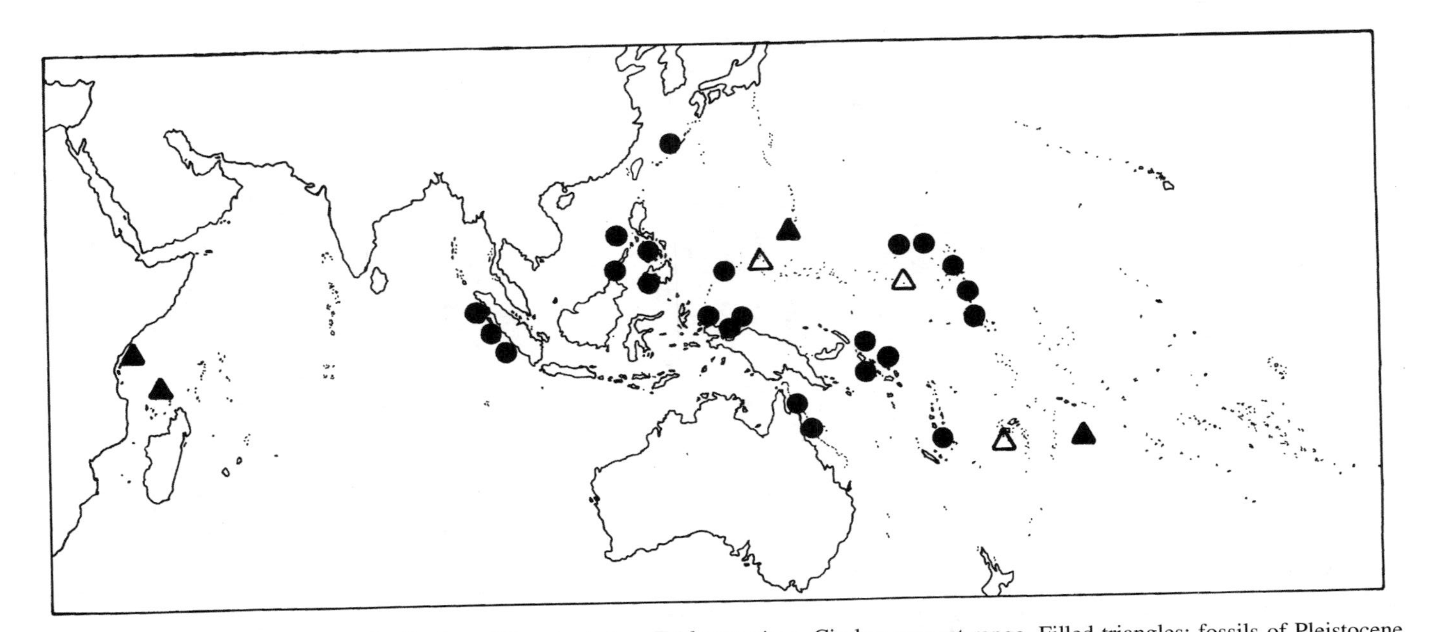

Figure 14–13. Recent and past distribution of the giant clam *Tridacna gigas.* Circles: present range. Filled triangles: fossils of Pleistocene or older age; empty triangles: Holocene records, representing potential/likely extirpation by humans (after Paulay, 1996a).

extinction of the giant clams *Hippopus hippopus* in Fiji, the Mariana Islands, and possibly Tonga, and of *Tridacna gigas* in Fiji and the eastern Carolines, were likely the result of human agencies (Fig. 14–13; Paulay, 1996a), and demonstrate the vulnerability of regional populations and thus endemic taxa. Giant clam stocks in numerous IWP nations are now severely depleted as a result of local and high-seas fisheries (Copland and Lucas, 1988).

14.8.2. Introduced Species

A problem of opposite nature to extinction, but also serious in its consequences, is the human-facilitated dispersion of species. Accidental and deliberate introductions of species into areas outside their range have accompanied the development of long-distance ocean voyaging, and have been rapidly increasing with improved transportation technologies and increasing volume of transport (Carlton, 1985).

Deliberate marine introductions are limited to a relatively few fish and shellfish species of commercial interest (e.g., Fitzgerald and Nelson, 1979; Randall, 1987; Eldredge, 1988). The majority of marine introductions are the result of accidental transport by oceangoing vessels between ports, either on (or in) the hulls of vessels, or in ballast water taken in near one port and discharged at another (Carlton, 1985, 1987; Morton, 1987; Carlton and Geller, 1993). Species so introduced usually first establish in harbor areas, including many large reef lagoons. As a result, harbor biotas are becoming increasingly cosmopolitan, with numerous exotic species being established at all major ports. Although some introduced species remain confined to harbor habitats, others actively expand their distribution (e.g., Monniot et al., 1985; Carlton, 1985, 1987; Morton, 1987).

Introduced species can have substantial impact at several levels. They can lead to striking changes in recipient ecosystems; they can cause large-scale mortality in, or even extinction of, indigenous populations; they can lead to millions of dollars of economic damage in fisheries, aquaculture, and industry; and they can have negative impacts on human health. For example, the recent establishment and enormous success of the corbulid clam *Potamocorbula amurensis* in the San Francisco Bay estuary has caused large-scale shifts in benthic mollusc communities, eliminated spring phytoplankton blooms, and greatly altered zooplankton communities (Carlton et al., 1990; Hedgpeth, 1993; Kimmerer et al., 1994). Although extirpations of native populations by introduced organisms are well documented on land (e.g., Murray et al., 1988; Howarth, 1991; Paulay, 1994), they appear to be uncommon in marine settings (Vermeij, 1989). Introduced species can, however, lead to rapid evolutionary changes in endemic species (e.g., Vermeij, 1982); among closely related species, replacement of endemic forms by introduced taxa may be possible (McDonald and Koehn, 1988; Meehan et al., 1989).

Economic damage to fisheries and aquaculture includes the transport of a variety of shellfish parasites and predators along with broodstock (e.g., Ledua and Adams, 1988; Humphrey, 1988). For example, giant clam parasites and pathogens have been carried around with clam stocks. Parasitic pyramidellid gastropods have been spread with clam stocks to Guam, Hawaii, the Marshall Islands, and Florida and have led to a high mortality of juvenile clams, at least in hatchery settings (Cumming, 1988; pers. obs.; B. Smith, pers. comm.; J. Wise, pers. comm.).

14.9. Conclusion

Coral reefs are probably the most diverse habitats in the oceans, with tens of thousands of species occurring on the richest reefs. Our documentation of the extant of this diversity is still woefully incomplete. Most groups still await taxonomic revisions and the proportion of undescribed species is high in all but the best-known taxa. Identification of species in all but the best-known groups is difficult. A guide to the taxonomic identification literature is given in Appendix 14–2. Nevertheless, general patterns are evident in the extent, nature, role, distribution, and origin of this diversity, mostly based on work in some better-known groups, and a variety of processes has been invoked to explain the origin of observed patterns of distribution and diversity.

While many species and genera are endemic to reef habitats, few higher taxa are so restricted, and most of these are either reef builders or their obligate associates. The high diversity of organisms on reefs creates both a considerable diversity of guilds as well as high within-guild diversity. Diversity in three ecologically important guilds: photosymbiotic associations, reef constructors, and reef eroders are great, and briefly reviewed above. The great among- and within-guild diversity on reefs creates considerable redundancy in ecosystem function. Well-developed reefs can exist not only at centrally located, high-diversity areas, but also in remote, low-diversity areas, as long as environmental conditions are favorable.

Coral-reef organisms, as well as most but not all major taxa of marine organisms, exhibit an increase in diversity from pole to equator. Reef corals and reefs are restricted to within 30–35° latitude, but exhibit an evenly high diversity across much of the equatorial belt. The restriction of zooxanthellate corals to low latitudes appears to be due proximally to lack of tolerance of low temperatures as well as to competition with macroalgae. In an evolutionary context the latitudinal restriction of reef corals may be due largely to competition with macroalgae. Algae do better at the high-nutrient concentrations and low herbivory found at high latitudes, whereas photosymbiotic organisms like corals thrive at the low-nutrient concentrations and high herbivory characteristic of much of the tropics.

The marine tropics can be readily divided into four major biogeographic regions: the Indo-West Pacific, eastern Pacific, western Atlantic, and eastern Atlantic. Reefs dominate large areas of the IWP and WA, but are limited in their development and diversity in the EP and EA. The poverty of reefs in the latter two regions is due in part to oceanographic conditions inimical to reef development along eastern shores, and in part to the isolation of these regions through the Cenozoic. Tectonic forces were largely responsible for the origin of the four major regions, and also had major influences on the diversification and extinction of their biota. The connection between the EP and WA was severed during the Pliocene, and was followed by the extinction of most of the original reef biota of the EP. As a result, the present reef fauna of the EP is depauparate and largely the result of colonization from the IWP across the vast eastern Pacific barrier. A similar scenario, of extinction following isolation by the Oligocene–Miocene closure of the Tethys, and subsequent recolonization across the Atlantic, appears to best explain the EA "reef" biota. The IWP and EP, and the WA and EA, are separated by wide-open ocean barriers, which do allow for some interregional dispersal. These vicariant separations of some, and chance dispersal between other, regions have led to interregional isolation of, and speciation in, many taxa.

The biota of the WA has diverged markedly from the other regions through the Cenozoic, through tectonic isolation (above), several bouts of mass extinction of the biota, and local diversification of surviving taxa. Evidence of similar large-scale extinctions during the late Cenozoic in the IWP is lacking.

IWP is the most diverse and extensive marine biogeographic region on Earth. It is recognizable from the considerable homogeneity of its biota, with many species distributed from eastern Africa to the central Pacific. Nevertheless, subregional endemism is also prevalent in many taxa. Diversity in the region peaks between Southeast Asia and Australia, is relatively high across the Indian Ocean, but decreases in a striking diversity cline eastward across the central Pacific. Patterns of species distribution and biotic diversity appear to be the result of a variety of modes of speciation and extinction, and do not fit single explanations. Diversification occurs in a variety of patterns: both in peripheral areas as well as in the diversity center (and elsewhere), through both vicariance and founder events. Diversification has been striking in several reef taxa, with several genera in the IWP, including well over 100 species. Our understanding of the origin of diversity and distributional patterns in the IWP is still rudimentary.

The WA exhibits considerably greater biotic homogeneity than the IWP, with most species distributed through most of the region. As a result, geographical clues to speciation are rarely evident. Nevertheless, moderate endemic radiations are known, several in taxa with relatively poor dispersal abilities.

Human impacts on reefs have been increasing greatly in recent decades and will likely escalate further in the future. Marine organisms typically have wider ranges and live in less accessible habitats than terrestrial species, and consequently, documented marine extinctions are rare. Nevertheless, local extinctions are well documented for some conspicuous species in several island groups. The destruction of reef habitats portends potential species extinctions in areas where both endemism and human populations are high, as in the Indo-Malayan diversity center or in southern Japan. Anthropogenic transport of species is also changing the composition of marine communities worldwide.

Appendix 14-1. Distribution and Stratigraphic Range of Living Zooxanthellate, Scleractinian Genera

	IWP	EP	WA	EA	W Tethys
Astrocoeniidae					
Stylocoeniella Eoc	X		F		F
Stephanocoenia ?Cret, Eoc			X		F
***Pocilloporidae**					
**Madracis* ?UCret,[1] Eoc	X	A[2]	X	X	F
Palauastrea Rec	X				
Pocillopora Eoc	X	X	F		F
Seriatopora Mio	X				
Stylophora Pal	X		F		F
Acroporidae					
Acropora Eoc	X	X	X	?	F
Anacropora Rec	X				
Astreopora UCret[3]	X		F		F
Montipora ?Eoc, Olig	X				F
Poritidae					
Alveopora Eoc	X		F		F
Goniopora MCret[3]	X		F		F
Porites ?Cret, Eoc	X	X	X	X	F
Stylaraea Plio-Plei	X				
Siderastreidae					
Anomastrea Rec	X				
Coscinarea Eoc	X		F		
Horastrea Rec	X				
Psammocora Mio	X	X	F		
Pseudosiderastrea Plio	X				
Siderastrea Cret	X	X	X	X	F
Agariciidae					
Agaricia Olig[4]			X	?	F
Coeloseris ?Mio	X				
Gardineroseris Mio	X	X	F		
Leptoseris Olig	X	X	X		F
Pachyseris ?Eoc, Mio	X				F
Pavona ?Eoc, Olig	X	X	F		F

	IWP	EP	WA	EA	W Tethys
Fungiidae					
Cantharellus ?Mio	X				
Ctenactis Mio	X				
Cycloseris ?Cret,[5] Pal[6]	X	X			
Fungia Mio	X				
Halomitra Mio	X				?F[5]
Heliofungia ?Mio	X				
Herpolitha Plio	X				
Lithophyllon ?Olig,[5] Mio[6]	X				
Podabacia Plio	X				
Polyphyllia Rec	X				
Sandalolitha Plio	X				
Zoopilus Rec	X				
***Oculinidae**					
Acrhelia Rec	X				
Galaxea Mio	X		F[4]		
**Oculina* ?Cret, Olig	(A)[7]	A[8]	X	I(A)[9]	F
Parasimplastrea Mio	X				
Schizoculina Rec				X	
Simplastrea Rec	X				
Pectiniidae					
Echinophyllia Mio	X				
Mycedium Mio	X				F
Oxypora Plio	X				
Pectinia ?Plio,[5] Plei[3]	X				
Physophyllia Plei	X				
Mussidae					
Acanthastrea ?Mio	X				
Australomussa Plio	X				
Blastomussa Plei	X				
Cynarina ?Olig, Plio	X				
Indophyllia Olig	X				
Lobophyllia Mio	X				F
Scolymia Olig	X		X		F
Symphyllia Plio	X				
Mussa Mio			X		
Mussismilia Mio			X		F
Isophyllia Mio			X		
Isophyllastrea Mio			X		F
Mycetophyllia Olig			X		F

	IWP	EP	WA	EA	W Tethys
Meandrinidae					
Ctenella Rec	X				
Meandrina ?Eoc, Olig			X		F
Dichocoenia ?UCret, Eoc			X	?	F
Dendrogyra Mio			X		F
Merulinidae					
Boninastrea Rec	X				
Hydnophora ?Cret, Olig	X		F		F
Merulina Plio	X				
Paraclavarina Rec	X				
Scapophyllia Mio	X				F
***Faviidae**					
Astreosmilia Rec	X				
Australogyra Rec	X				
Barabattoia Rec	X				
Caulastrea Eoc	X		F		F
**Cladocora* ?UCret,[1] Eoc		A[2]	X	X	F
Colpophyllia Eoc			X		F
Cyphastrea ?Olig, Mio	X				F
Diploastrea ?LCret,[1] Eoc	X		F		F
Diploria UCret,[1] Eoc			X		F
Echinopora Mio	X				
Erythrastrea Rec	X				
Favia ?Cret, Eoc	X		X	X	F
Favites Eoc	X		F		F
Goniastrea Eoc	X		F		F
Leptastrea Olig	X				F
Leptoria ?Cret, Eoc	X		F		F
Manicina Olig			X		
Montastrea ?UJur;[1] Eoc.	X		X	X	F
Moseleya Rec	X				
Oulastrea Rec	X				
Oulophyllia ?Olig	X				F
Platygyra ?Eoc, Olig	X		F		F
Plesiastrea Mio	X				F
Solenastrea Olig			X		F
Trachyphylliidae					
Trachyphyllia Eoc	X		F		F
***Caryophylliidae**					
Catalaphyllia Rec	X				
Euphyllia ?Eoc, Olig	X		F		F

	IWP	EP	WA	EA	W Tethys
Gyrosmilia Rec	X				
Montigyra Rec	X				
Physogyra Rec	X				
Plerogyra Rec	X				
Eusmilia ?Olig, Mio			X		F
**Heterocyathus* Plio	X[10]	A/?X[11]			
***Dendrophyllidae**					
Duncanopsammia ?Mio,[5] Plio[3]	X				
**Heteropsammia* ?UCret,[10] Olig	X		F[10]		F
Turbinaria Olig	X		F		F

Notes

Only genera with some zooxanthellate members are listed; genera and families marked with * also contain azooxanthellate representatives (based on Wells, 1956; Veron, 1986; and various sources). In addition to the genera listed, zooxanthellate specimens are also known in *Balanophyllia* (*B. europaea*; Zibrowius, 1976; Dendrophyllidae) and *Astrangia* (*A. poculata*; Boschma, 1925; Rhizangiidae).

Regional records are marked as: A = extant, azooxanthellate only; I = introduced; F = fossil occurrence only; X = extant, at least some zooxanthellate. Regional zooxanthellate occurrences are based primarily on Veron (1993), with the following modifications. *Diaseris* is subsumed in *Cycloseris*, as it was shown to be a polyphyletic derivative (Hoeksema, 1989); *Goreaugyra* is subsumed in *Meandrina* (Cairns, pers. comm), *Heterocyathus* is included (Hoeksema and Best, 1991) and the following dubious regional records are omitted: EA records of *Acropora* (see Laborel, 1974; Laborel in UNEP/IUCN, 1988a), *Agaricia*, and *Dichocoenia* (not recorded in recent EA literature); EP records of *Montipora* (see Durham and Barnard, 1952). The stratigraphic range and fossil occurrences of genera are based on Veron (1995) and other references noted by superscripts. Regional fossil and azooxanthellate occurrences and human introductions are based on references identified by superscripts. As the EP and WA were contiguous until the Pliocene, Tertiary American fossil occurrences are all listed only under WA. The fossil occurrence of genera in the European and East Atlantic Tertiary is noted separately from the recent EA, as at the time of their occurrence the European Tethys was contiguous with the proto-IWP region. Note that the first stratigraphic appearance of many genera is based on old literature, with potentially inaccurate dates. These dates have only been systematically reassessed in the American Cenozoic (Budd et al., 1992, 1994), but not elsewhere.

EA = East Atlantic; EP = East Pacific; IWP = Indo-West Pacific; WA = West Atlantic; W Tethys = Fossils from Europe, W and N Africa and E Atlantic islands. Cret = Cretaceous; Eoc = Eocene; Jur = Jurassic; LCret = Lower Cretaceous;

MCret = mid Cretaceous; Mio = Miocene; Olig = Oligocene; Pal = Paleocene; Plei = Pleistocene; Plio = Pliocene; Rec = Recent; UCret = Upper Cretaceous; UJur = Upper Jurassic. References: (1) Wells, 1956; (2) Wells, 1983; (3) Veron and Kelley, 1988; (4) Budd et al., 1994; (5) Veron, 1995 (source of all fossil data emphasized here for records with mixed data); (6) Hoeksema, 1989; (7) Squires, 1958 (record from New Zealand); (8) Cairns, 1991b; (9) Zibrowius, 1974; (10) Hoeksema and Best, 1991; (11) Durham and Barnard, 1952.

Appendix 14-2. Taxonomic References for Identification of Reef Biota

"The time is far in the future when a tropical marine ecologist can identify the components of an ecosystem or a food chain from his handy pocket guide" (Chace, 1969, in Abele, 1972).

The purpose of this compilation is to give an introduction to some of the more generalized taxonomic reviews and identification guides available on reef biota. General references to identification are listed first, followed by a list of taxonomically broad but regionally localized guides, with publications focusing on major taxa given last, arranged by the tropical marine region to which they apply (IWP = Indo-West Pacific, EP = East Pacific, WA = West Atlantic, EA = East Atlantic). The taxonomic literature for reef organisms is vast, but for most groups no general reviews are available and thus accurate identification generally requires specialist knowledge. The small selection presented here is meant to give an introduction to some of the more general and reasonably accurate large-scale reviews available, and is certainly incomplete. References below concern mostly the macrobiota. Higgins and Thiele (1988) give a useful entry into the meiofaunal literature, and Lee et al. (1985) to the Protista, with Loeblich and Tappan (1988) giving a detailed review of Foraminifera, the most conspicuous protist group on reefs.

Most of the reviews below provide desciptions, keys, or illustrations of the species or genera included; however, a few are biotic inventories or bibliographies; these are marked by *. Some of these are excellent introductions to the taxonomic literature; others are included mostly for groups/areas poorly covered by more general publications.

General References

*Sims (1980) provides a useful general overview of the taxonomic identification literature for marine animals, providing references for identification for all marine phyla, with both systematic and geographic breakdown in most. Parker's (1982) treatise is an excellent entry into the family-level literature of all living organisms. Moore et al. (1953–) provides reviews to the genus, mostly for those invertebrates with some fossil record.

Regional Guides

The following is a selection of major reference books and field guides that cover the biota of specific areas where reefs are important. Although an attempt was made to select the more accurate publications, field guides covering a wide range of organisms are prone to have misidentifications in some groups, and those relying largely on color pictures can be problematic to use, as gross apperance is, in many groups, insufficient for accurate identification. The *UNEP/IUCN (1988a, b, c) review of coral reefs is also a useful entry into some of the site-specific taxonomic literature, especially for corals. The *Zoological Catalogue of Australia* (*Houston, 1983–), now complete for sponges and echinoderms among marine taxa, is becoming an excellent source of taxonomic information for the IWP. It is planned to cover the entire Australian fauna, listing all known species and providing a vast introduction to the literature; however, these volumes do not in themselves serve as identification guides.

Tropical seas: Ehrhardt and Moosleitner (1995).

IWP: Colin and Arneson (1995), Allen and Steene (1994), Gosliner et al. (1996), Siboga-Expeditie Monographie (1901–1961: 39 large volumes covering all marine taxa; the most comprehensive published marine study of an IWP fauna). Hawaii: *Reef and Shore Fauna of Hawaii* (Devaney and Eldredge, 1977, 1987, in prep.; Kay, 1979a); Australia: Mather and Bennett (1993); Red Sea: Vine (1986); South Africa: Branch and Branch (1981); Japan: numerous publications available in Japanese.

EP: Brusca (1980), Kerstitch (1989), *James (1991).

WA: Humann (1989–1993), Colin (1988), Sterrer (1986), Rützler and Macintyre (1982).

Taxonomic Reviews

Algae: IWP: Dawson (1954, 1956, 1957), Trono and Ganzon-Fortes (1988); Hawaii: MacGruder and Hunt (1979); WA: Littler et al. (1989), Taylor (1960).

Seagrasses: Phillips and Meñez (1988).

Porifera: IWP: deLaubenfels (1954), Bergquist (1965), *Hooper and Wiedenmayer (1994), *Kelly-Borges and Valentine (1995); WA: Wiedenmayer (1977), Zea (1987).

Hydrozoa: Kramp (1961), Totton and Bargmann (1965); IWP: Millard (1975). Stylasterina: Cairns (1983, 1991).

Corals (esp. Scleractinia): Wood (1983); IWP: Veron et al. (1976–1984); Veron (1986), Sheppard and Sheppard (1991), Randall and Myers (1983); EP: Durham and Barnard (1952), Wells (1983); WA: Humann (1993), Cairns (1982), Smith (1971); EA: Chevalier (1966).

Alcyonaria: Bayer (1981); IWP: Williams (1993); major alcyonacean generic revisions by Verseveldt (1980, 1982, 1983), Verseveldt and Bayer (1988); WA: Bayer (1961), Cairns (1977).

"Anemones": Calgren (1949), Fautin and Allen (1992).

Platyhelminthes: Cannon (1986); Polycladida: Prudhoe (1985).

Nemertea: *Gibson (1995); IWP: Gibson (1979–1983); WA: Corrêa (1961).

Sipuncula and Echiura: Stephen and Edmonds (1972).

Polychaeta: Fauchald (1977), Day (1967).

Mollusca: Abbott and Dance (1982—see end for useful guide to molluscan taxonomic literature); IWP: Kira (1962), Habe (1964), Cernohorsky (1972, 1978), Kay (1979a), Wilson (1993); Bivalvia: Oliver (1992), Lamprell and Whitehead (1992); Opisthobranchia: Willan and Coleman (1984), Bertsch and Johnson (1981), Gosliner (1987b), Wells and Bryce (1993); EP: Keen (1971); WA: Warmke and Abbott (1961), Humfrey (1975); EA: Gofas et al. (1981), Bernard (1984).

Crustacea: IWP: *Siboga-Expeditie Monographie* (1901–1961: volumes 17–27 cover the Crustacea); Decapoda: Burukovskii (1982); IWP: *Yaldwyn (1973), Dai and Yang (1991), Miyake (1983—includes stomatopods, in Japanese), Sakai (1976), Chace (1983–1988), Chace and Bruce (1993); WA: Williams (1984); EA: Manning and Holthuis (1981); Peracarida: Amphipoda: Barnard and Karaman (1991); Isopoda: WA: Kensley and Schotte (1989); Cirripedia: IWP: Foster (1980); Acrothoracica: Tomlinson (1969).

Bryozoa: *Soule and Soule (1976) provide a guide to the major taxonomic monographs by region.

Echinodermata: IWP: Clark and Rowe (1971), Clark and Courtman-Stock (1976), Guille et al. (1986), *Rowe and Gates (1995); WA: Hendler et al. (1995), Clark and Downey (1992).

Ascidiacea: IWP: Monniot et al. (1991), Kott (1985, 1990a, b), Monniot and Monniot (1996); WA and EP: Van Name (1945).

Fish: Lieske and Myers (1994); IWP: Randall et al. (1990), Myers (1989), Smith and Heemstra (1986); EP: Allen and Robertson (1994), Thomson et al. (1979); WA: Bohlke and Chaplin (1993).

15

Disturbances to Reefs in Recent Times

Barbara E. Brown

The view that coral reefs evolved under stable, benign conditions, where fluctuations in physical and chemical variables were limited, was challenged in the early 1970s when reefs were described as a temporal mosaic of communities in different stages of recovery from various sources of disturbance. Subsequently, ecological theories and models have suggested that the high diversity that characterizes many coral reefs is actually maintained by intermediate disturbance (Connell, 1978).

Coral reefs have always been subject to some form of disturbance operating at different levels. The community response, it is now recognized, may be complex, depending not only on the timing and intensity of the disturbance, but also on the history of events that have already taken place prior to the latest disturbance (Hughes, 1989).

The last decade has offered a number of opportunities for reef scientists to study the effects of disturbance on coral reefs with increasing numbers of reports of coral mortality as a result of natural perturbations (Fig. 15–1). Threats to reefs from human disturbance have also escalated during this period, leading to concern about the general balance of life and death on coral reefs worldwide (Wilkinson, 1993) and the potential not only for corals to adapt to a changed climate, but also their capability of coping with human influences.

In considering the fate of reefs in future years there has been much recent controversy over whether coral reefs are robust (having withstood environmental rigors over geological time) or fragile, since there are many examples where reefs appear particularly susceptible to human-induced disturbances. There is now general agreement that neither the view that coral reefs are robust nor

This chapter benefited from support from the Overseas Development Administration's Natural Resources and Environment Department.

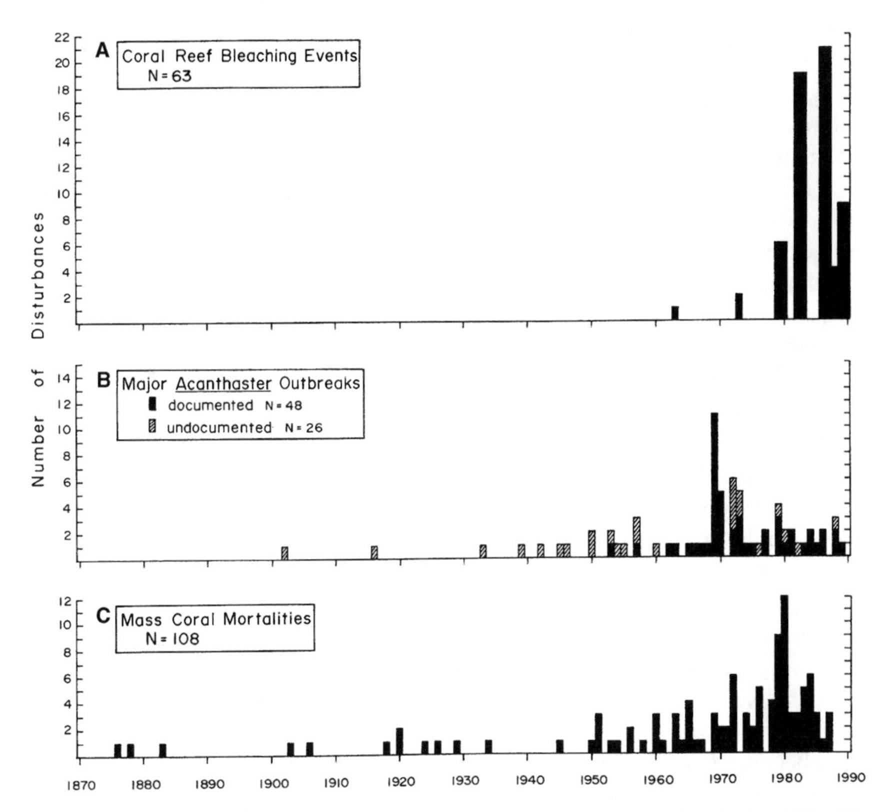

Figure 15–1. Reported natural disturbances to coral reefs worldwide (1870–1990). **(A)** Coral bleaching events. **(B)** Major crown-of-thorns, *Acanthaster planci* outbreaks. **(C)** Mass mortalities of corals resulting from physical stressors such as violent storms, heavy rains, and disease outbreaks (after Glynn, 1993).

inherently fragile is true over all timescales. The coral-reef ecosystem is clearly robust with respect to variability over periods of millions of years; reef fragility is, however, reflected on timescales that are relevant for human society.

In this chapter we shall be looking at some of the known tolerances of reef corals to selected environmental parameters; the susceptibility of corals to natural and human disturbances and the timescale for recovery following damage; natural versus human damage and possible synergistic interactions between the two; and finally, a prediction of how coral reefs may fare in their response to future environmental changes.

15.1. Conceptual Framework

From the outset it is important to establish a framework for discussion of recent disturbances on coral reefs. We will focus on the scleractinian or stony corals.

While information on the effects of disturbance on other important reef organisms such as fish exists in the literature, it is much more fragmentary than that available for corals, which constitute the major reef builders.

First, it is important to distinguish between natural disturbances and human disturbances. This distinction may not be clear in all cases. For example, there have been recent claims that coral bleaching (loss of algae and/or their pigmentation, discussed in Chapter 6) may represent an early signal of global warming. Similarly, it has been suggested that crown-of-thorns outbreaks may be stimulated by human activities. I have chosen to categorize these influences on coral reefs as natural disturbances in the absence of substantive scientific evidence proving otherwise.

Second, the danger of overgeneralization must be emphasized. A key conclusion emerging from long-term ecological studies reported in recent years has been the site- or location-specific nature of the results, whether this be the use of modeling tools to predict recovery, the actual recovery process itself, or the act of defining the susceptibility of particular coral species or the reef to human or natural disturbance. While broad generalizations are valuable, particularly for management purposes, it is important to recognize the limitations of such statements.

Third, it is important to define adaptation as used in this chapter. Adaptation to given environmental conditions is made up of a genetic and a nongenetic component. Genetic adaptation is the basis for evolution and it is likely, though not proven, that the latitudinal and perhaps between-habitat tolerances of corals have become fixed in the population over a large number of generations and therefore have a genetic basis. Nongenetic adaptation, commonly known as acclimatization, involves changes in the tolerance of a colony during its lifetime and may be responsible for seasonally adjusted tolerances observed in reef corals. Such adaptations may very well account for significant within- and between-reef variation in responses to disturbance that limit the usefulness of our generalizations.

Fourth, the importance of scale in defining responses of reef organisms to disturbance cannot be overstated. Considering timescales, then, ecological processes are slow and changes cannot always be detected in the short term. Many corals are long-lived and therefore studies must be scaled to their lifetime, which may be several hundred years (it is perhaps worthwhile reflecting that the longest existing reef-monitoring program has spanned only 20 years in the period up to 1993!). Spatial scales are also important with geographic, regional, and local processes, each playing increasingly important roles in defining responses of reef communities to disturbance and their recovery patterns (Chapter 12).

Finally, the definition of that point in time when recovery might be considered as "complete" varies from study to study (see Done, 1992, for review). In many accounts, recovery constitutes restoration of predisturbance levels of coral cover, a definition that ignores previous levels of diversity, sizes of colonies, mix of

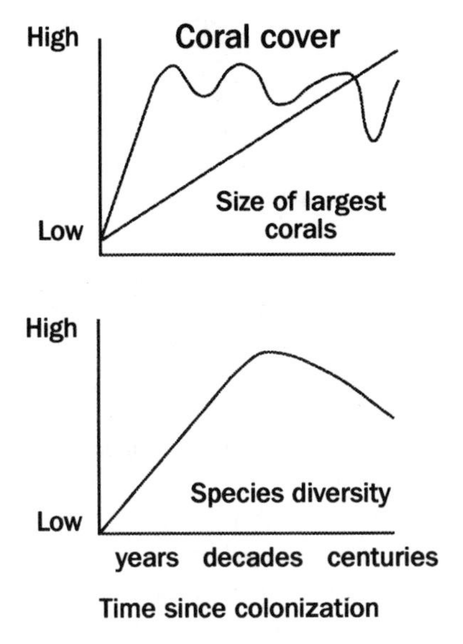

Figure 15–2. Changes in some coral community attributes plotted against increasing time since colonization and without major disturbance (after Done, 1992).

growth forms, and structural complexity of the reef community and framework. Recovery that incorporates all these criteria may take considerably longer to effect than recovery that simply restores predisturbance percent coral coverage (Fig. 15–2). In discussion of recovery in this chapter, distinctions will be made on the criteria used.

15.2. Tolerances of Reef Corals

It is recognized that corals cannot be viewed in isolation when considering the demise, recovery, and scope for survival of reefs faced with disturbance. Ecological interactions of corals (Chapter 11) and other organisms (borers [Chapter 4], herbivores, predators, macrophytic algae [Chapters 9, 10, and 11]) will play important roles in succession and reef function, factors that in turn will dictate the ultimate fate of the reef. Unfortunately, our understanding of such ecological interactions under disturbed and nondisturbed conditions is poor.

As a first step in understanding the ability of corals to adapt to disturbance, particularly physical changes in temperature and ultraviolet light levels, some estimate of their physiological tolerance levels is required. Some of the most fundamental characteristics of a living reef, namely the distribution, abundance, and diversity of reef corals, are governed in part by the physiological tolerance limits of corals. For example, the temperature control of reproduction in corals has been cited as the principal factor limiting corals to tropical and subtropical localities (Rosen, 1981). Similarly the symbiotic relationship between algae and coral, discussed in Chapter 5, appears to be governed by temperature, thus limiting the geographical range of reef-building corals (Rosen, 1981). Variation in the tolerance limits of corals to environmental factors ultimately results in a broad variation in responses to perturbation and scope for subsequent recovery not only across latitudinal and longitudinal gradients, but also within a reef and sometimes even within a colony.

It is surprising that so little is known about the physiological tolerances of corals at any of these levels. The situation is complicated because in considering the coral we must take account of both plant and animal components within a single colony, and it is likely that we are dealing with an overall tolerance to extreme environmental conditions—for example, high seawater temperature, ultraviolet radiation—that is not the sum of the tolerances of the separate components but actually some lower threshold. Results of experiments on temperature-stressed symbiotic anemones show that they are more stressed by elevated seawater temperatures than aposymbiotic (lacking symbiotic algae) members of the same population (Suharsono et al., 1993), suggesting that under extreme conditions there may be "costs" in maintenance of the symbiotic relationship. Additionally, consider the fact that the plant component in the coral symbiosis may potentially comprise one or two algal species or strains, sometimes within a single coral host (Rowan and Powers, 1991), and the complexity of understanding coral responses at the organism level to environmental change becomes apparent.

Nevertheless, from the limited work carried out on the tolerances of reef corals, there are some interesting conclusions. Most concern natural environmental factors such as temperature, light, and salinity.

15.2.1. Geographic Variations in Environmental Tolerances

Field and laboratory experiments suggest that tropical corals from Enewetak have an upper lethal limit of 34°C while subtropical corals from Hawaii can survive only up to 32°C. These thresholds correspond to the 2°C difference in the normal maximum seasonal temperature between the two areas (Fig. 15–3).

In some parts of the world this upper lethal limit may be elevated by at least 4°C as a result of extreme local conditions. For example, reef corals in the Western Arabian Gulf are exposed to the most rigorous temperature and salinity regimes in the world. The hardiest species survive exposure to maximum temperatures of 36–38°C and minimum temperatures of 11.4°C, while salinities of 39–46‰ have been recorded on inshore reefs. At least 24 species of corals have adapted, probably genetically, to such extremes, which are beyond the tolerances of corals from most other regions of the world (Sheppard, 1988). Table 15–1 describes the temperature tolerances of some of these species.

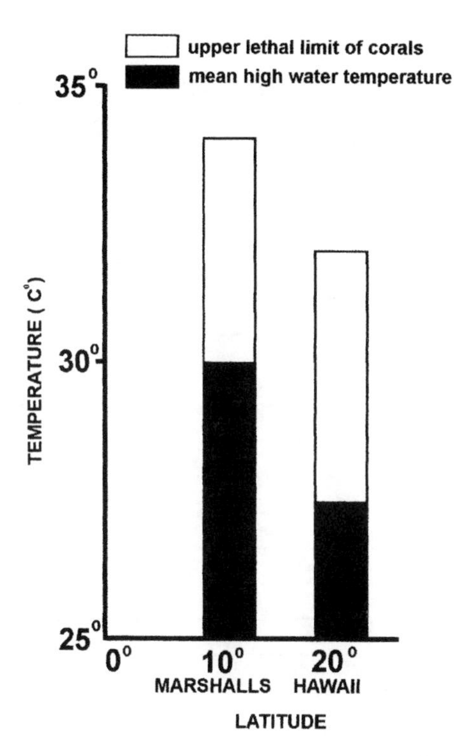

Figure 15–3. Latitudinal variation in physiological limits of corals (after Jokiel and Coles, 1990).

15.2.2. Within-Site Variation in Environmental Tolerances

Within a site, corals from different habitats may show significant variation in environmental tolerances. Good examples of such differences are the bleaching responses of corals to elevated seawater temperatures. During a bleaching episode in Bermuda in 1988, offshore corals showed a greater bleaching susceptibility than inshore populations (Cook et al., 1990); similar effects were observed in Thailand in 1991 when submerged reef-slope corals suffered more pronounced bleaching than intertidal corals belonging to the same species (Brown et al., 1993). In these cases it is likely that corals subject to wide temperature ranges (i.e., lagoonal inshore reefs in Bermuda and intertidal reefs in Thailand) are "adapted" to local conditions, which in turn render them less sensitive to temperature variations.

15.2.3. Between-Species Variations in Environmental Tolerances

Considerable variation in environmental tolerances exists between species, with the most studied factors being sedimentation and temperature. Coral bleaching

Table 15-1. Coral Species That Survive Temperature Fluctuations of the Range Indicated

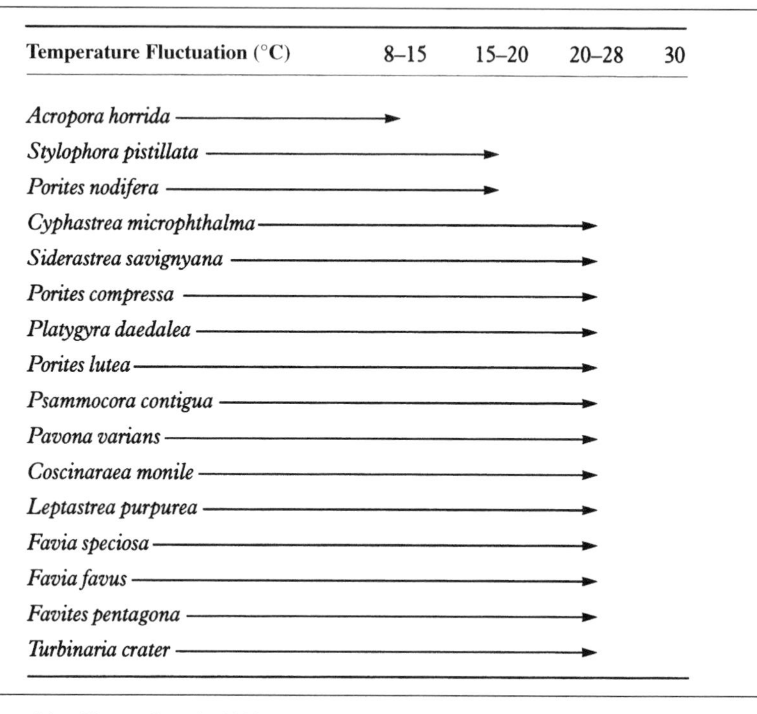

Temperature Fluctuation (°C)	8–15	15–20	20–28	30
Acropora horrida	→			
Stylophora pistillata	→	→		
Porites nodifera	→	→		
Cyphastrea microphthalma	→	→	→	
Siderastrea savignyana	→	→	→	
Porites compressa	→	→	→	
Platygyra daedalea	→	→	→	
Porites lutea	→	→	→	
Psammocora contigua	→	→	→	
Pavona varians	→	→	→	
Coscinaraea monile	→	→	→	
Leptastrea purpurea	→	→	→	
Favia speciosa	→	→	→	
Favia favus	→	→	→	
Favites pentagona	→	→	→	
Turbinaria crater	→	→	→	

Source: After Sheppard et al., 1992.

due to elevated seawater temperature (and possible increased ultraviolet radiation) has been shown to affect certain coral species more than others, though significant variability in response exists even within species. General patterns of species susceptibility to increased seawater temperature are now emerging, however. Studies in the Indo-Pacific suggest that the branching corals *Acropora* and *Pocillopora* are more susceptible to bleaching than the massive species (Fig. 15–4), while in the Caribbean the hydrozoan coral *Millepora* is particularly sensitive in its bleaching response (Williams and Bunkley-Williams, 1990).

15.2.4. Within-Species Variations in Environmental Tolerances

The best documented within-species variations are displayed in photoadaptive responses of corals to irradiance. Photoadaptive changes effected by corals may take minutes (polyp expansion/contraction), hours (changes in photosynthetic pigments in a zooxanthella), days (changes in the density of zooxanthellae), or years (changes in growth form and size and possible genetic selection).

Corals of the same species showing photoadaptation may be adapted to either

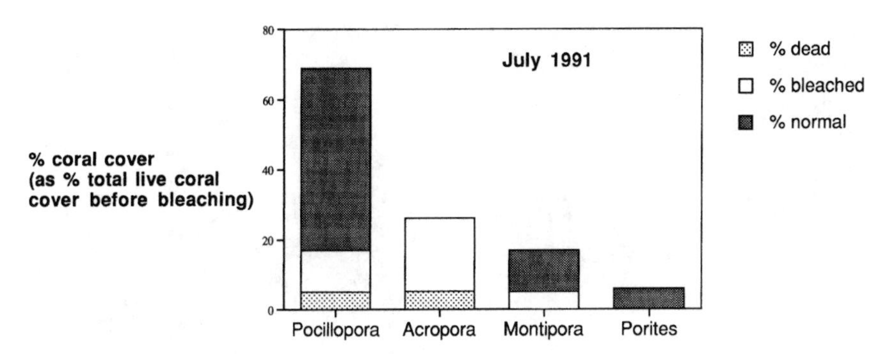

Figure 15–4. Variation in coral species susceptibility to bleaching and consequent mortality, Moorea, Society Islands (after Salvat, 1992b).

high or low light levels depending on their habitat. When the photosynthetic parameters of the branching coral *Stylophora pistillata* from high- and low-light habitats were compared at different light regimes, results showed that colonies living in low light had a much greater photosynthetic capacity at low-light levels than high-light counterparts (Porter et al., 1984). Interestingly, the low-light colonies were equally as capable in high-light levels as the high-light colonies.

The adaptation of corals to different light levels may include changes in pigmentation, zooxanthellae density, polyp density, and gross morphology. In many cases, the most obvious manifestation of photoadaptation to low-light levels has been an increase not in the density of zooxanthellae but in the average pigment content of each zooxanthella cell.

As well as observing adaptive responses of coral species within different habitats, acclimatization by reef corals in localities subject to considerable seasonality has also been recognized. Reef corals in these localities show marked seasonal photoadaptation and temperature acclimatization (Al-Sofyani and Spencer Davies, 1993). Similar photoadaptive effects have been observed in symbiotic anemones subject to a seasonal regime, with chlorophyll content of the symbiotic algae varying inversely with mean solar radiation (Dykens and Shick, 1984). Such responses could potentially result in corals showing greater or lesser susceptibility to environmental rigors at different times of the year.

15.2.5. Within-Colony Variations in Environmental Tolerances

It is quite likely that there are major differences in environmental tolerances even within the colony of a single species. Bleaching responses of corals in the field frequently involve only parts of the colony, for example, upper surface of massive species (Fig. 15–5). It has been suggested that upper surfaces may be subject to the interaction of different stressors (ultraviolet and increased seawater temperature) while other parts of the colony may only be exposed to increased seawater

Figure 15–5. Within-colony susceptibility to bleaching in *Goniastrea pectinata,* Phuket, Thailand (photo by B. E. Brown).

temperature, thus accounting for observed responses. Alternatively, physiological gradients in the coral could equally account for the bleaching pattern observed. The existence of calcification gradients and translocation pathways for the rapid transfer of resources in coral colonies has been demonstrated; also, specific areas of the colony may be reproductive while others are not, parts of the colony may be specialized for aggression with neighboring species, and genetic diversity within a single colony is possible. Such a spectrum of physiological and possibly genetic variation could give rise to extreme differences in within-colony responses to environmental stress.

15.3. Natural Disturbances on Coral Reefs

While extensive coral mortality on reefs may be attributed to a single or several natural factors that include low tides, volcanic eruptions, low temperature, and red tides, it is clear that during the last decade the attention of reef scientists has been focused on five major causes of natural disturbance: storms and hurricanes, coral bleaching, diseases of reef organisms (Chapter 6), outbreaks of coral predators such as *Acanthaster planci,* and mass mortalities of reef herbivores such as *Diadema antillarum* (Chapter 9). Some of these problems (e.g., *Acanthaster planci* outbreaks) are restricted to specific locations; others have higher incidence in certain parts of the world than others (e.g., coral and urchin diseases in the

Caribbean), while the remainder (storm damage and coral bleaching) have occurred with increasing frequency on a global scale since the mid-1980s. In this section we shall restrict our attention to four of these major factors.

15.3.1. Effects of Storms, Cyclones, and Hurricanes

The effects of storms and hurricanes on coral reefs are determined by a number of factors that include the directional approach of the storm, its intensity and resultant wave height and energy, the vertical relief of the site and its protection from direct influences, the reef community type and its susceptibility to high-energy conditions, together with past history of disturbance at the site. Factors causing damage to corals may include physical destruction of reef organisms by wave action and subsequent movement of coral rubble, increased sedimentation and turbidity, increased runoff after heavy rain, and release of nutrients from breakdown of moribund tissues following the storm.

The effect of vertical relief of reef slopes on the type and degree of damage inflicted from hurricanes was shown very clearly at sites in French Polynesia (Fig. 15–6). Between December 1982 and April 1983, six hurricanes ravaged reefs in the area. Hurricane intensities and tracks were similar to those reported in 1903–1905 that caused catastrophic damage to the Polynesian Islands. On steep outer-reef slopes (angle >45°) of the atoll at Tikehau, coral destruction

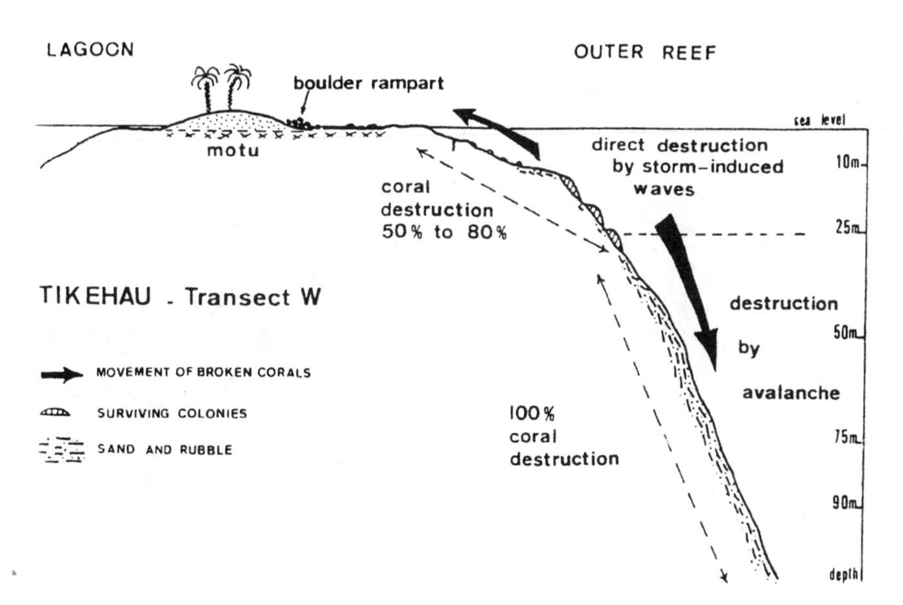

Figure 15–6. Cross-section of the west coast of Tikehau, French Polynesia, showing destruction of coral at different depths as a result of storm-induced waves (after Harmelin-Vivien and Laboute, 1986).

varied from 50% to 100% and was a function of depth. Between 12 and 30 m, coral mortality ranged from 60% to 80%, whereas below 35 m, 100% mortality was found with fragile platelike corals completely destroyed. On low-angle slopes, such as the northeastern coast of Moorea, damage occurred only between 0 and 20 m. As a general pattern, most storm damage by massive waves occurs at depths of 0–20 m, but coral blocks falling down steep slopes can cause damage down to greater depths of 35 m.

The Caribbean, in particular, appears to have suffered from a series of hurricanes since the 1980s, though there are few reports of reefs having been repeatedly disturbed. In 1979, the effects of Hurricanes David and Frederick were described in St. Croix. In 1980, Hurricane Allen caused reported damage in Barbados and Jamaica. In 1988, Hurricane Gilbert struck in Jamaica. In 1989, reefs in St. Croix were affected by Hurricane Hugo. In 1991, Hurricane Andrew caused very localized damage to reefs in the Florida Keys.

In terms of susceptibility of corals to storm damage in both the Indo-Pacific and Caribbean provinces, branching corals appear to suffer most from hurricane damage (Fig. 15–7). Results of studies in the Caribbean suggest that the branching staghorn and elkhorn corals *Acropora palmata* and *Acropora cervicornis* are most prone to storm damage, followed by the branching finger coral *Porites porites*. Such corals tend to "fragment" after physical damage, and regeneration is possible providing that tissue damage is not excessive and further disturbance

Figure 15–7. Coral rubble (mainly *Acropora cervicornis*) at 8 m depth on the reef slope at Discovery Bay, Jamaica, in 1983, showing effects of Hurricane Allen in 1980 (photo by T. P. Scoffin).

to the reef is minimal. The massive Caribbean coral *Montastrea annularis* appears to be much more resilient to storm effects.

Different coral species show varying susceptibilities to storm and cyclone effects with age. As corals grow, their vulnerability to breakage and dislodgement increases. Stands of branching corals that have rapid growth rates become vulnerable to cyclones within a few years while slow-growing massive corals become vulnerable to cyclones only after several decades.

Recovery times for reefs, subject to storm damage, are becoming increasingly more difficult to estimate, particularly in the Caribbean where the incidence of other natural factors (disease, coral bleaching, competition from algae and human influences) complicates the pattern of reef regeneration. In the Pacific, recovery periods for reefs in the Society Islands (which were almost totally devastated by the 1982–1983 hurricanes) were cited as being in the order of at least 50 years (based on restoration of predisturbance coral-cover values), a figure that matches estimated times of recovery for totally destroyed reefs in the Caribbean and Pacific. Since 1986, however, reefs in the Society Islands have been subject to widespread bleaching as a result of increased seawater temperatures. Should such disturbances continue over the next decade, the projected times for recovery may have to be extended.

15.3.2. Coral Bleaching

Bleaching responses in corals and other symbiotic reef organisms have been widely reported in the 1980s, both in the Indo-Pacific and the Caribbean (see Brown and Ogden, 1993, for review). While bleaching may not lead to coral mortality (under less severe circumstances, corals may recover their complement of symbiotic algae and, as a result, their pigmentation), many of the extensive bleaching events of the 1980s did result in considerable mortality of corals. In 1979, coral bleaching and mortality were reported in four areas of the Pacific Ocean and in two areas of the Caribbean. In 1982–1983, during a remarkably strong El Niño–Southern Oscillation event (ENSO), severe bleaching resulted in mass mortalities of corals around Costa Rica, Panama, Colombia, and Ecuador. Coral bleaching was also reported at 12 sites in the Indo-Pacific and 5 locations in the Caribbean at this time. During 1986–1987, coral bleaching was recorded at 12 new locations, including reefs in the Red Sea and Caribbean. In 1989, coral bleaching appeared to be restricted to the Caribbean; in 1991, extensive bleaching, leading to coral mortality, was reported in the Andaman Sea, Thailand, and French Polynesia.

Bleaching is a generalized response shown by corals to stress, since corals bleach upon exposure to a wide variety of pollutants, as well as to extremes of temperature, salinity, and light irradiance. Many of the bleaching events since the mid-1980s have been associated with elevated seawater temperatures, though in some instances exposure to harmful ultraviolet and human-induced disturbance have been cited as possible causative factors. The connection of worldwide bleaching events with anomalous seawater temperatures is a contentious issue because of the lack of long-term, high-quality temperature data to support ecological observations.

Nevertheless, there is strong evidence, particularly from the central and eastern Pacific, that widespread bleaching is associated with unusual increases in seawater temperature, with laboratory experiments demonstrating that high temperatures can induce bleaching in a manner consistent with field observations (Glynn and D'Croz, 1990). The bleaching response can be induced by short-term exposure of corals (about 2 days) at temperatures of 3–4°C above the seasonal maximum or by long-term exposure (i.e., several weeks) at elevations of 1–2°C. It appears that bleaching is not induced by thermal shock of rapidly fluctuating temperatures, but rather is a response to prevailing mean temperature (Jokiel and Coles, 1990).

The potential physiological and ecological effects of bleaching are wide-reaching. At the organism level, effects of bleaching include declines in protein, lipid, and carbohydrate; skeletal growth; reproductive output; and tissue necrosis in the coral host (Glynn, 1993). Ecological changes following bleaching include invasion of dead coral framework by benthic algae, a phenomenon reported following bleaching in Costa Rica, Panama, Galápagos, and Indonesia (Glynn, 1993). Dead corals also provide shelter and grazing surfaces for infaunal molluscs and sponges and grazing sea urchins and fishes. On eastern Pacific reefs that suffered high coral mortality in 1982–1983, bioerosion now exceeds carbonate production, threatening to convert the reef structure into sediment. A preliminary calcium carbonate budget for a Panamanian reef shows that the reef is currently eroding at a rate of 5.9 tonnes of $CaCO_3$ per year, whereas prior to the 1982–1983 ENSO, the reefs showed a net deposition rate of 24.7 tonnes $CaCO_3$ per year (Eakin, 1993).

Preliminary results on coral community recovery following mass bleaching and mortality events reveal highly variable rates both in the short and long term. In the 1991 bleaching events in Thailand and Tahiti, many bleached corals recovered their zooxanthellae within months of the onset of bleaching. These corals survived the event with either no coral mortality or limited mortality. In all field examples of bleaching studied so far, branching corals appeared to be the most susceptible, with 25% of all *Acropora* showing mortality in Tahiti. In longer-term studies, coral cover on an Indonesian reef attained 50% of its former level after 5 years, though in the Galápagos Islands virtually no recruitment to affected coral reefs has been observed 7 years after the major bleaching event (Glynn, 1991).

15.3.3. Outbreaks of the Crown-of-Thorns Starfish Acanthaster planci

A predator that produces devastating effects on coral reefs is the crown-of-thorns starfish *Acanthaster planci,* whose juveniles and adults feed directly upon hard corals (Fig. 15–8). Normal densities of starfish on coral reefs range from 6 to 20 km^{-2}, whereas outbreaks of *Acanthaster* may result in numbers in excess of 500 km^{-2}. Outbreaks were first recorded in the late 1950s in Japan and early 1960s on the Great Barrier Reef, Australia. Since then, major areas of outbreaks have been the Great Barrier Reef, Micronesia, Japan, Samoa, Fiji, Society Islands, Malaysia, Thailand, and the Maldives.

The Great Barrier Reef outbreaks are among the best studied (see Moran, 1986, for review) and have involved two apparent cycles of starfish invasion—one begun in the late 1960s and another in the 1980s. Although the data for the first peak of activity (1966–1975) are relatively limited, the data for the most recent peak of activity (1981–1989) support the hypothesis of southward-moving waves of outbreaks. Physical oceanographic studies illustrate that *A. planci* larvae would be carried southward during the summer spawning months where they could be entrained on reefs (Dight et al., 1990b).

Much of the Great Barrier Reef has been surveyed during the second phase of outbreaks. Reefs surveyed in the central third of the Great Barrier Reef have been affected to varying degrees over the last 8–9 years. Approximately 10% of the reefs had extensive, high coral mortality, with midshelf reefs being more affected than outer-shelf reefs (Doherty and Davidson, 1988). Reductions in coral

Figure 15–8. *Acanthaster planci* aggregation on the Great Barrier Reef in 1972.

cover were considerable, with coral cover declining from 78% to 2% in 6 months at some sites.

The starfish generally favors feeding on faster-growing, branching coral species such as *Acropora* and *Montipora* spp., with others such as the branching *Pocillopora* spp. protected by commensal crustaceans. Some massive species such as *Porites* are not favored, but will be eaten if supplies of preferred corals are exhausted.

Recovery from crown-of-thorns outbreaks depend at least in part on recruitment of juvenile corals, and this may take place within several months of an outbreak. It has been estimated that it probably takes 12–15 years for a reef to recover from an outbreak, that is, if the reef is dominated by branching *Acropora* species. The definition of *recovery* here implies return to predisturbance coral cover levels, species composition, and relative abundances. Mathematical simulations indicate recovery times for reefs dominated by slow-growing massive coral species to be in excess of 50 years assuming no further disturbance (Done, 1988).

There is no overall consensus on the causes of outbreaks of crown-of-thorns. Hypotheses include the role of nutrients in terrestrial runoff as a food supply for larval starfish, particularly during high rainfall periods when nutrients are flushed into the ocean. Another theory involves a reduction in predation pressure on adult starfish. Most scientists believe that it is unlikely that one factor alone accounts for *A. planci* outbreaks (Birkeland and Lucas, 1990).

15.3.4. Mass Mortality of the Sea Urchin Diadema antillarum in the Caribbean and Consequences on Coral Communities

The sea urchin *Diadema antillarum* is a major herbivore on coral reefs in the Caribbean (Chapter 9). Mass mortality of the sea urchin in 1983 as a result of a waterborne pathogen (Chapter 6), first noted in Panama and then subsequently throughout the Caribbean, has had dramatic effects upon many reefs in the area. On Jamaican reefs, the mortality of sea urchins together with the effects of two hurricanes have resulted in a precipitous decline in coral cover over the period 1970–1990 (Fig. 15–9) and a *phase shift* (i.e., shift from a coral-dominated community to one dominated by algae, as discussed by Done [1992]) of the community for a period of years.

Coral species diversity, however, showed little change from 1983 to 1987. The reasons are complex and indicate how the community response depends not only on the intensity and frequency of disturbance, but also on the history of events that have already taken place on the reef prior to the latest disturbance (Hughes, 1993). In this example, the first hurricane studied, Hurricane Allen, resulted in the death of the abundant branching corals *Acropora palmata* and *Acropora cervicornis,* leaving smaller encrusting and platelike species dominant, such as *Agaricia agaricites.* Subsequently, the algal bloom, which was a consequence of reduced grazing pressure resulting from the death of sea urchins,

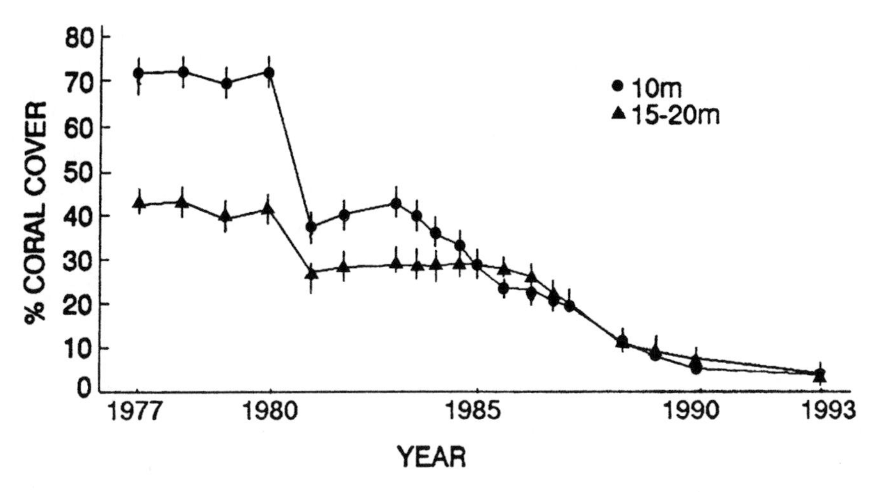

Figure 15–9. Percentage cover (mean ± SE) of corals at 10 m and 15–20 m depths at Rio Bueno, Jamaica, from 1977 to 1993 (after Hughes, 1994).

caused considerable mortality and failed recruitment of the encrusting corals, which were smothered by the algae. Recruitment of the surviving branching *Acropora,* which reproduce by fragmentation, was less affected by the algal bloom. As a result, the abundance of surviving corals became more equitable as abundant species (i.e., encrusting corals) declined faster than rare species (branching corals), with the species diversity remaining constant over this period.

The combination of the sea-urchin mortality and the two hurricanes have led to dramatic coverage and biomass of algae on Jamaican reefs, with little prospect of coral-reef recovery in the near future. Even severe storms, such as Hurricane Gilbert in 1988, fail to redress the balance, with algal areas cleared by the storm quickly becoming reinvaded by algae before corals can successfully settle. Hughes (1993) argues that herbivory is the most likely process that can reduce algal abundance, but since the density of sea urchins remains low and Jamaican reefs are chronically overfished, it is unlikely that coral assemblages in Jamaica will achieve total recovery to pre-1970 status for several decades.

Data from other reefs in the Caribbean, such as Curaçao and St. Croix, confirm the massive increase in algal coverage on the reefs since the sea-urchin dieback and the probable lack of any change in reef status until herbivores are available in sufficient numbers to redress the balance. Dramatic changes in coral cover in Curaçao and St. Croix were not as evident as in Jamaica, where the effects of two major hurricanes negatively interacted with sea-urchin mortality.

It is interesting to contrast the effects of sea-urchin mortality in the Caribbean and the Pacific (Chapter 12). The disease affecting the sea-urchin population has tended to spread widely in the Caribbean because of the gyre-like current patterns and the relatively small and enclosed nature of the sea, resulting in algal overgrowth on many reefs. Prevailing current patterns in the Pacific are mainly easterly; the main current flows north of the equator and misses most reef areas while the second current is weaker and in most years is a poor vehicle for larval transport, apart from those in which an El Niño occurs. As a result of these current patterns in the Pacific and the vast distance between islands, the effects of mass mortalities of sea urchins have been much more self-contained and restricted to certain geographic regions, for example, the Hawaiian islands (Birkeland, 1989a).

15.4. Human Disturbances to Coral Reefs

The current world population is estimated at 5.3 billion, of whom 3 billion live in coastal areas. Much of this population is in tropical countries with fringing and offshore coral reefs. By the year 2055, these population numbers are set to more than double, with the greatest increases in population occurring in tropical developing countries. Such population pressures bring additional stresses to coral reefs in the form of terrestrial runoff from poor land practices, sewage pollution, tourism, and industrial developments, the effects of which are discussed below.

15.4.1. Eutrophication

Eutrophication and its effects upon coral reefs is perceived as an increasingly major problem for coral-reef managers, with claims of damage to coral-reef ecosystems by sewage and agricultural runoff on both the Great Barrier Reef and Florida Keys in recent years. While these claims remain to be scientifically substantiated, there are a number of case histories from both the Pacific and the Caribbean illustrating the potential effects of eutrophication on corals (Pastorak and Bilyard, 1985).

Unlike other human disturbances, such as oil pollution and thermal effluents, the effects of increased nutrients are not always straightforward, partly because of the interplay of natural environmental factors in the field. Two examples of effects of eutrophication on coral reefs have been described in Barbados and in Kaneohe Bay, Hawaii. In Barbados, the stress is considered to be a combined function of nutrient enrichment, increased sedimentation, groundwater discharge, and introduction of toxins (Tomascik and Sander, 1987a). In Kaneohe Bay, the stresses include sewage discharge, sedimentation, and agricultural runoff, compounded by occasional freshwater inundation following storms (Kinsey, 1988).

In Barbados, nutrient concentrations along the coast increased substantially from 1972 to 1987. Effects of eutrophication were studied along a pollution gradient and were described as causing substantial changes in community structure on the reefs. Such changes included reduced coral species diversity and high abundances of macrophytic and filamentous algae at polluted sites, a reduction in skeletal growth in the massive coral *Montastrea annularis,* and a reduction of reproductive effort in the finger coral *Porites porites* (Tomascik and Sander, 1985, 1987a, b).

In Kaneohe Bay, secondary sewage was discharged into the bay from three outfalls from 1963 to 1977, with a total peak flow rate of 1.9×10^4 m^3 day $^{-1}$. During peak sewage discharge, the corals were subjected to both a soluble nutrient input and an increased organic particulate load in the form of phytoplankton and associated zooplankton. As in Barbados, considerable changes occurred in the reef community structure. In 1974, the green bubble alga *Dictyosphaeria cavernosa* became a major component of the community of the southern bay, smothering the reef corals and associated fauna. By 1977, particle feeders were dominant, the principal members being zoanthids, sponges, and barnacles. In 1979, the sewage was directed offshore, away from the bay to a deep ocean outfall.

After this date, the reef-community response to the reduced sewage loading was substantial. In the northern bay, where the only major stress was from sediment, no changes were evident. In the central bay, where community responses to earlier stresses were not particularly marked, little change was reported. In the southern bay, by 1982, the previously heterotrophic reef flats had lost most of their filter feeders and the dead reef substratum had become totally covered with red macrophytic algae. By 1985, algal populations had declined and coral recruits were observed over all the reef flats, with a very high percentage cover in some areas. It was clear by 1985 that recovery of reef flats in southern Kaneohe Bay was the result of diversion of sewage from the bay. Recovery was well in place by 1988 when a freshwater inundation from a severe storm caused a significant reduction in salinity in waters overlying the reef and considerable mortality of reef corals occurred. The rate of recovery in the years following the flood has been rapid, in contrast to the lack of recovery following an earlier freshwater inundation in 1965, when the reef was already suffering from the chronic effects of sewage discharge. Such results illustrate the important additive effects of natural and human disturbances that are considered later in this chapter.

Examples of coral-reef exposure to increased nutrients alone, isolated from other factors, are limited to manipulative field experiments such as that carried out by Don Kinsey at One Tree Island reef on the Great Barrier Reef in 1971–1972. In this experiment, a small lagoonal patch reef 25 m in diameter was subjected to a concentration of 20 μm nitrogen over 8 months. Effects noted were a 25% increase in primary production and a 50–60% decrease in reef calcification. Effects on community production continued for one month after the cessation of nutrient addition. Currently, further manipulative nutrient experiments are under way at One Tree Reef in a research program entitled ENCORE (**E**levated **N**utrients on **CO**ral **RE**efs). Basically, the study will involve fertilization of a number of reefs, followed by observation of recovery. The experiment will attempt to partition the effects of nitrogen and phosphorus, separately and combined, using as sample units patch reefs and microatolls of similar size and nature as those used in the original Kinsey project.

15.4.2. Sedimentation

Sedimentation is one of the most ubiquitous human disturbances on coral reefs (Hatcher et al., 1989). It is also a disturbance that has regularly affected coral reefs in recent years, particularly in developing countries, as a result of dredging, land erosion, and coastal engineering projects. Rogers (1990) cites sediment rates and suspended sedimentation concentrations for reefs not subjected to human activities as <1 to about 10 mg cm^{-2} day^{-1} and 10 mg l^{-1} respectively. Sediment rates on affected reefs suffering from moderate to severe sedimentation stress have been cited as 10–50 mg cm^{-2} day^{-1} while those described as severe to catastrophically affected by sediment received in excess of 50 mg cm^{-2} day^{-1} (Pastorak and Bilyard, 1985).

Sediments may smother reef organisms and reduce light available for photosynthesis, but there appears to be no reliable way in which the effects of sedimentation can accurately be predicted for any reef site. Some of the reasons for such variability have been alluded to in earlier discussion of variation in interspecies coral tolerance levels to environmental factors. The reef response to sedimentation will also very much depend on the reef setting, the hydrographic regime, the nature of the sediment, and the severity of its loading.

To illustrate the variability in responses of reefs to sedimentation, it is worth looking at a number of case histories. Dredging activities in Castle Harbor, Bermuda, approximately 30 years ago, caused a catastrophic mortality to corals in areas of confined water circulation (Dodge and Vaisnys, 1977) while dredging activities in the summer months only, near Miami Beach, Florida, caused relatively little mortality to scleractinian corals, though sublethal symptoms of stress (loss of symbiotic algae, production of mucus, tissue swelling) were obvious in affected corals (Marszalek, 1981).

At Ko Phuket, Thailand, dredging for a deep-water port over an 8-month period resulted in significant reduction in coral cover on intertidal reefs adjacent to the activity (Brown et al., 1990). Here, the sheltered reefs were dominated by massive species such as *Porites* and faviid corals known to be tolerant to sedimentation. One year after the dredging, the reefs showed a rapid recovery (in terms of coral cover and coral diversity), with coral tissue regenerating over areas that had shown partial mortality as a result of sediment loading.

A final example illustrates an almost complete lack of effect of sedimentation (Dollar and Grigg, 1981). In 1980, a Greek freighter carrying 2,200 tons of kaolin ran aground on a reef in the French Frigate Shoals, northwestern Hawaii. The ship was refloated after the cargo was thrown overboard. Field investigations conducted 14 days after the dumping of kaolin revealed a highly localized and very minor environmental effect. Coral damage was restricted to a small area where the ship's hull carved a channel through the reef and a zone less than 50 m from the affected channel where thick clay deposits buried coral colonies. Beyond this distance no corals appeared to be affected by the turbidity plumes, apparently because of the rapid dispersal and nontoxic nature of the kaolin.

In the mid-1980s another form of sediment stress on coral reefs caused concern. The potential problem was drilling fluids used to remove drilling cuttings and lubricate drill bits in the oil exploration industry. Experiments were carried out in the laboratory to evaluate effects of the drilling fluids on corals (Dodge and Szmant-Froelich, 1985). In many of the experiments, a drilling fluid concentration of 100 ppm was required to produce a harmful effect. Such concentrations, however, would rarely be encountered in the field, a fact that highlights the difficulties of extrapolating the effects of such pollutants from laboratory studies to the real world.

15.4.3. Oil Pollution

Our understanding of the effects of oil pollution on corals has most rapidly advanced as a result of documented cases of either chronic or acute pollution in the field. Many laboratory experiments have been carried out since the mid-1970s, but extension of these (often contradictory) results to the field has always proved difficult.

Early observations on the effects of oil on coral reefs suggested that oil had little damaging effect on corals unless it came into direct contact with coral surfaces. In many cases of acute and chronic oil pollution, no damage to reef communities was observed. Again it was almost impossible to make comparisons between studies because of the number of variables involved—these would include the type and volume of oil spilled, the use (or otherwise) of detergents to clean up the oil, the degree of shelter or exposure of the site and water movement in the area, the tidal range, and nature of the reef sites affected.

Some of the earliest demonstrated negative effects of chronic oil pollution were described on reef flats in the northern Gulf of Eclat, Red Sea, from 1974 to 1975 (Loya and Rinkevich, 1980). These reef flats were in close proximity to the oil terminal at Eclat. Chronically oil-polluted areas of the reef showed higher mortality rates of the dominant coral *Stylophora pistillata,* smaller numbers of breeding colonies, a decrease in the average number of ovaria per polyp, smaller numbers of planulae produced per coral head, and lower settlement rates of planulae on artificial objects when compared with a control reef nearby.

In 1986, our appreciation of the effects of oil pollution on intertidal and subtidal reef communities was substantially extended as a result of a major oil spill in Panama at a site a kilometer east of the Caribbean entrance to the Panama Canal (Jackson et al., 1989). After the spill, which involved 8 million liters of crude oil, most corals on reefs to depths within the area showed signs of recent stress, including bleaching, swelling of tissues, or conspicuous production of mucus. Three months after the spill, total coral cover decreased by 76% at depths of 0.5–3 m and by 56% at depths of 3–6 m on heavily oiled reefs. The decrease in coral cover on moderately oiled reefs was somewhat less.

The spill affected various species of corals very differently, the branching coral *Acropora palmata* suffering more at oiled reefs than massive species. Numbers of corals, total coral cover, and species diversity decreased substantially with increased amounts of oiling. Frequency and size of recent injuries on massive corals increased with level of oiling while growth of three massive species was less at oiled reefs in the year of the oil spill than during the previous 9 years. Estimated minimum times for recovery of the reef at this site were 10–20 years on the assumption that no other events would further depress coral populations. Scientists at Panama now expect a slow shift in the reef community toward a greater relative abundance of corals with a brooding reproductive strategy, since brooded larvae settle more rapidly than larvae produced by broadcasting species, the latter being more susceptible to oil slicks formed during heavy rains in the area by flushing of oil still bound to mangrove sediments. The original conclusion that oil had little harmful effect on corals now has little scientific basis, with predictions from laboratory experiments generally not appearing to be easily scaled up to the effects observed in a major oil spill.

15.4.4. Coral Mining

Coral mining activities have caused extensive degradation of reefs in a number of countries (Maldives, Indonesia, Sri Lanka, Tanzania, and Philippines). Corals have traditionally been used as building materials in the Maldives and other Indian Ocean islands since no rock or stone is available. In the Maldives, corals are removed by hand from the shallow reeftops (2–3 m depth) in lagoonal settings for use in the construction industry. Favored corals include the slow-growing massive species such as *Porites, Goniastrea, Favia, Favites,* and the branching coral *Acropora humilis.*

A rapidly growing population, together with massive expansion of the tourism industry, resulted in great demand for construction material in the mid- to late 1980s. It has been calculated (Brown and Dunne, 1988) that a minimum of 93,450 m^3 of coral has been extracted from 1972 to 1985 in North Malé Atoll alone.

The effects on the coral and fish communities have been profound, resulting in significant declines in coral cover, diversity, and associated fish assemblages. Apart from such localized damage, there is also concern about the rate of recovery of such communities. Reefs that were mined before the mid-1970s have shown little recovery (Brown and Dunne, 1988) since the physical environment of the shallow reeftops has become greatly altered as a result of mining activity, with increased wave effects and sedimentation posing difficult conditions for larval recruitment.

Another cause for concern, particularly for the Maldives as a low island state, is the need to conserve reef habitats on the outer edges of the atoll. Here the shallow reefs have an important protective role against the erosive forces of storm waves and potentially higher sea levels. In the Maldives, as the supply of corals on shallow reeftops in the lagoon is exhausted, coral miners turn to shallow-reef areas on the outer sides of the atoll to satisfy construction demands—a development that will only increase island vulnerability to storm influences.

Alternatives to the use of coral rock are available in the Maldives; they include the greater use of properly constructed concrete blocks manufactured from coral sand and the quarrying of dead coral material from a single location that could satisfy aggregate demands for North Malé Atoll until the year 2050. Both options have been investigated and have been shown to be economically and environmentally viable.

15.5. Human Versus Natural Influences on Coral Reefs

As the database on effects of disturbances on coral reefs has built up over the last decade, scientists have tended to categorize disturbances as either human or natural. Although the causes of disturbances may fall into these two broad categories, often the reef responses are very similar, particularly if the disturbance is severe. At this point it is useful to look at the attempts that have been made to compare and contrast the responses of reefs to the two types of disturbances and to assess the value that this approach might have both in predicting reef responses and recovery potential.

Some have described natural disturbances such as storms, crown-of-thorns starfish, mass bleachings, and El Niño as acute stresses, whereas human disturbances such as eutrophication and regular sedimentation have been considered as chronic influences with inputs being discharged over a long period of time (Kinsey, 1988). Such a division between chronic (human) and acute (natural) stresses cannot, however, be supported by a close inspection of the literature. Human disturbances, for example, may also be acute; a ship grounding or temporarily increased sedimentation from a dredging event or coastal engineering project may affect reefs over a very short timescale (days or months), while a natural disturbance, for example, the water from Florida Bay that moves across the open-shelf areas of the Florida Reef Tract, may affect reefs for thousands of years (Ginsburg and Shinn, 1993). Kinsey also suggested that a reef subjected to a single acute stress would likely recover relatively quickly (10–50 years) and certainly this seems to be the case for most natural stresses acting alone. The position is not so clear for acute human influences. Case histories, such as the dredging in Castle Harbor Bay, Bermuda, the dredging in Thailand, and kaolin spills in Hawaii, show a spectrum of reef response from complete degradation to full recovery to no effect.

When comparing the outcome of natural and human disturbances, the responses of reefs to single natural stresses seem much more predictable and also more amenable to broad generalizations than human disturbances. Following a severe storm, a crown-of-thorns starfish outbreak, or mass bleaching event, changes to the reef generally follow a similar pattern involving a phase shift from a coral-dominated community to one dominated by macroalgae. The persistence of the algal community may be brief, or it may last for 10 years or more. In the latter case, examples of protracted algal coverage appear to be restricted to case histories where other natural factors—for example, hurricanes followed by mass mortalities of sea urchins—have synergistically interacted in delaying the succession process (Hughes, 1994). Predication of the outcomes of human disturbances, even as single events, are more problematical, relating more to the local setting and circumstances, for example, type of sediment, type of oil and the location of the event in sheltered embayment, exposed headland, and so forth.

What is becoming abundantly clear is that recovery from stress (human or natural) is affected by intervention of other stresses. Most case histories of combined stresses suggest that reefs that are chronically polluted by human influences, and then affected by a natural stress, show very poor recovery following the second stress. Remove the chronic stress, as in the case of diversion of sewage from Kaneohe Bay, Hawaii, and the recovery from any ensuing natural stress, such as freshwater inundation, is much more rapid. Furthermore, the effects of multiple stresses on the reef may be more than additive—that is, synergistic—causing a rapid decline in reef health.

Single stresses are rarely found in the real world. Furthermore, in a world subject to resource exploitation and potential global warming, multiple stresses are more likely to be the norm during the next two decades. Such stresses may act simultaneously or consecutively. Our present database on the effects of multiple stresses on coral reefs is too small to predict what the ecological effects of interacting human and natural stresses may be, although improved information on the demography of dominant corals should provide a basis for predictive modeling of reef ecosystems. Simulation models could prove to be a very powerful tool in the prediction of stress effects on reefs.

So far, the use of simulation models to predict the effects of repeated disturbances on coral reefs has been limited, being restricted to the effects of natural disturbances. Early models were developed to predict the length of time a reef might take to recover to predisturbance status given that either the fastest- or slowest-growing coral was dominant. Subsequently, Done (1987, 1988) has developed a simulation model to predict the effects of various frequencies of crown-of-thorns starfish outbreaks on massive *Porites* populations. Although the model cannot be used to provide generalizations for a total reef system such as the Great Barrier Reef, it does provide a predictive tool for evaluating effects on *Porites* on a reef-by-reef basis. Given access to data on key measurable parameters for different coral species—that is, age/size structure, damage characteristics, recruitment rates, and growth rates—it should ultimately be possible to extend models to address questions such as, Are repeated bleaching-related coral mortalities ecologically sustainable?

15.6. Prediction of Responses of Reefs to Climate Change

The key factors that might be expected to affect coral reefs during a period of climate change are sea-level rise, increasing seawater temperature, altered carbonate mineral saturation, increases in ultraviolet radiation, and possibly a strengthening of currents and storm activity (Smith and Buddemeier, 1992), though present climate models give no clear indication if storms would increase in frequency or intensity should climate change.

15.6.1. Sea-Level Rise

An average rate of global mean sea-level rise of about 6 cm per decade over the next century (with an uncertainty of 3–10 cm per decade) has been predicted. Although there will be significant regional variation in sea-level rise, it does appear that within the uncertainties of the estimates of both past sea-level rise and calcium carbonate production rates, reef ecosystems could apparently keep pace with a 6 mm yr^{-1} sea-level rise. Estimations of reef accretion rates range from 1 to 10 mm per year, with a rate of 10 mm per year being accepted as the maximum sustained-reef vertical accretion rate.

Sea-level rise will affect some reef communities more profoundly than others. Reef flats constrained by present sea levels in protected waters might be expected to show increased diversity and productivity with progressively higher sea levels. However, in the short term (over the next decade), such changes will probably be swamped by natural variations in mean sea level. Many reef flats already show seasonal variations in mean sea level of 20 cm, with interannual variations of up to 30 cm between some years, so changes in the order of 6 mm per year would likely have relatively little effect, at least for the next 10 years. Although rapid responses of fauna and flora in such habitats might be expected to occur with progressively higher mean sea level, should sea-level rise be intermittent and the level remain static over some years, then shallow-water communities may show considerable mortality. This would result from accelerated growth by corals to keep up with sea level, which could ultimately render them more susceptible to subaerial exposure if sea levels remain steady for a number of years.

For other much deeper reefs there is the possibility that they may drown and not be able to keep up with sea-level rise, whereas others may be subject to greater physical wave stress. This topic has already been discussed in Chapter 3, where Figure 3–3 highlights the fate of reefs with different accretion patterns over time.

15.6.2. Effects of Increased Seawater Temperature Increase

The effects of increased temperature on reef corals with respect to bleaching have already been discussed (section 15.3.2). It has been argued that the potential effects of future temperature increases will depend on the scales of such changes. An increase in sea-surface temperature and/or its variability could produce an increasing frequency of bleaching, resulting in sublethal responses that might include reduced growth or reproductive potential of corals and increased partial mortality. Alternatively, more extreme temperature stress could lead to rapid changes in reef diversity and community structure. Whether corals have the scope to adapt to temperature changes in the order of 0.3°C per year (the predicted global mean temperature, given no substantial changes in greenhouse gas emissions) is unknown and will most likely vary depending on habitat. Corals are already living close to their lethal upper temperature, and although hardy reef-flat and shallow-water corals may show considerable scope for nongenetic adaptations (section 15.2) to increased seawater temperatures (e.g., in the production of stress proteins that protect both plant and animal cells in adverse conditions), this capability may be much reduced for species living subtidally in less variable surroundings. The extent of the ability of reef corals and their zooxanthellae to genetically adapt to higher temperatures, and the meaningful timescales involved, are not known.

15.6.3. Carbonate Mineral Saturation State

Another factor expected to alter with continued atmospheric CO_2 emissions is a change in carbonate mineral saturation states in seawater. It has been calculated that should the partial pressure of CO_2 increase, the carbonate saturation state would decrease. If it is assumed that the calcification of many reef organisms would be proportional to carbonate mineral saturation state, then it might be postulated that as calcification decreases, the ability of reefs to keep up with rising sea level might decline. Smith and Buddemeier (1992) also suggest that there may be an increase in biological and physical erosion of skeletons as a result of reduced calcification, which might similarly reduce the capability of reefs to keep pace with sea level in the long term.

15.6.4 Ultraviolet Radiation

An increase in ultraviolet radiation would not be expected as a result of climate change, interpreted in its narrowest sense, but should destruction of the stratospheric ozone layer by chlorofluorocarbons be significant, then an elevation in ultraviolet-B (UVB) radiation levels could also act as an additional stress on coral reefs.

It is known that many corals and reef-dwelling organisms possess UV-blocking pigments (Jokiel and York, 1982, 1984) and that projected increases in UVB exposure in the tropics are expected to be small. Average low-latitude UVB surface exposures are expected to increase by 1–10% (Smith and Buddemeier, 1992), though these figures could be confounded by climate-induced changes in cloud cover. However, it may well be that coral-reef organisms are living close not only to their lethal upper temperature but also to their UV-tolerance thresholds. A lack of knowledge of wavelength-specific effects of UV on corals inhibits our accurate prediction of their responses, though many scientists have attributed bleaching to a combination of temperature and UV stresses. Smith and Buddemeier (1992) consider that changes in UVB exposure are likely to be secondary factors in determining the response of coral-reef systems to climate change.

In all our predictions of effect of climate change on coral reefs we must bear in mind the potential interaction of the stresses outlined previously—that is, increased seawater temperatures and increased UV levels may lead to selective mortality of fast-growing branching corals, in turn resulting in less rapid $CaCO_3$ accretion and perhaps greater susceptibility of corals to bioerosion, factors that could reduce the ability of a reef to keep up with sea-level rise. In addition, negative influences that might accompany climate change, such as increased storms and changes in precipitation, leading to greater sediment discharge in coastal areas, all highlight the potential extra pressures on coral reefs in the next 40 years. Given the possible rapid decline in reef health that may result from synergistic interaction of such factors, it becomes apparent that much work needs to be done to minimize human influence on coral reefs worldwide.

One question that has often been asked is whether it is possible to predict from responses of coral reefs to natural disturbances their response to (1) human change and (2) climate change. The answer to the first part of the question is probably no, because the response to each human disturbance seems to be governed by local circumstances. The answer to the second part is probably a restricted yes, since responses to increased seawater temperatures and exposure to UV radiation give us some insight into the tolerances of reef corals to these parameters, as do responses of reefs to environmental changes throughout geological time (Chapter 2). However, we have relatively little information on the outcome of interactions between stresses on modern reefs and so our prediction capability regarding overall climate change and coral reefs is very limited.

16

Traditional Coral-Reef Fisheries Management

Robert E. Johannes

Coral reefs have provided coastal populations in the tropics with a vital source of animal protein for many centuries. However, biologists and governmental marine-resource managers face formidable difficulties in trying to understand reef and lagoon fisheries well enough to be able to manage them. Because of the complex physical structure of coral reefs, industrial-scale fishing methods are generally not feasible. Purse seines or trawls, for example, would quickly snag on the reefs and be ruined.

Reef fishes are thus still caught mainly by fishermen operating out of canoes or small boats, using methods that have changed little from those used centuries ago. An exceptionally wide range of fishing methods is employed, including many different kinds of nets and traps, along with spears (thrown from above the water or used by divers), poisons, trolling, and dropline fishing. Gleaning small fishes and invertebrates on foot at low tide on the reef flat is also an important method in many areas. Each of these methods selects different species and different size ranges within species.

More kinds of fishes and invertebrates are harvested from coral reefs than from any other kind of fishery because of the exceptional species diversity. There are also more fishermen per unit of catch than in most commercial fisheries, because of the type of methods used. Typically, catches are landed in small quantities at many different landing sites at all times of day and night, and the catch is spread through many distribution channels.

These characteristics combine to pose the most complex challenges to fisheries management. Unfortunately, the budgets for fisheries research and management are typically very limited in the less-developed countries where most coral reefs are found. Obtaining the information on catch, effort, and populations of fishes that is necessary for conventional scientific fisheries management is therefore nearly always prohibitively expensive. Under these conditions, the marine resource manager cannot realistically aim for some quantitative ideal, such as the maximum sustainable biological yield, maximum economic yield, or optimum yield so often featured in fisheries textbooks.

16.1. Value of Traditional Knowledge

Information that can be used to reduce ecological and economic stresses, reduce conflicts among fishermen, and attain a more equitable distribution of benefits can be obtained relatively cheaply. The knowledge of reef fishermen offers a shortcut to some of the basic natural history needed for a better understanding of reef resources. Reef fishermen are vastly more numerous than reef biologists, and have been plying their trade and passing on their accumulated knowledge for many more generations than the latter. It is hardly surprising that they know much about their local marine environments that is unknown to science. What is surprising is how long it has taken for us to become aware of the value of this knowledge. Only recently have biologists begun to make serious efforts to record and verify it.

One subject concerning which traditional fishermen often possess important information is the timing and location of reef-fish movements. For example, year after year, many reef fish migrate to specific locations on the reef for several days, usually on a particular moon phase, in order to spawn (Johannes, 1978). Some species travel in large schools, using consistent pathways, to such spawning sites. Local fishermen often know the precise timing and location of these migrations and/or subsequent aggregations. The fishermen of Palau, for example, provided information on the lunar periodic spawning aggregations of more than twice as many species of reef fishes for which similar information could be found in the scientific literature for the entire world (Johannes, 1981). Such information is valuable for fisheries management. Populations of most species of coral-reef fishes are normally scattered over large areas. Under these conditions, it is almost impossible to make a useful estimate of their population sizes. The difficulties are greatly reduced if the biologist knows where and when these fishes aggregate to spawn. Just as salmon are far easier to count on their spawning runs than at other times, so are many reef fish easiest to census when they are concentrated during their spawning periods.

Spawning migrations and aggregations also provide useful foci for the regulation of fishing pressure, because exceptionally large catches are often made from them. Johannes (1981) gives a detailed discussion of the subject. Some spawning populations are known to have been wiped out by overfishing before marine resource managers became fully aware of the need to regulate their exploitation, for example, grouper aggregations in Puerto Rico (Olsen and La Place, 1979) and bonefish in Kiribati (see below).

After the obliteration of a grouper-spawning aggregation by an ill-advised export fishery in Palau, the government of Palau recognized the need for locating and monitoring remaining spawning aggregations of these and several other reef fishes. At the request of local fishermen, who do their own monitoring of the spawning aggregations on which they fish, the Palau Division of Fisheries therefore instituted a ban on fishing for a number of species during their spawning periods (Marine Protection Act of 1994).

Coastal engineering projects can also create serious problems if planners ignore local information on spawning migrations. For example, some of the spawning migration routes of the bonefish, *Albula* sp., have been destroyed on Tarawa Atoll, Gilbert Islands, by solid causeways built to link islands. This, plus gillnet fishing to exhaustion of other spawning runs on this atoll, have put this fish—once the most important shallow-water species in local catches—in danger of imminent local extinction.

Fishermen could see this happening. But they were helpless to do anything about it. First, their traditional system of marine tenure had been destroyed by the British colonial government, leaving village leaders unable to control exploitation on their fishing grounds by traditional methods (see below). Second, they were unable to draw the attention of the postcolonial government of Kiribati to the problems of either overfishing or causeway building, and to persuade them to take the necessary preventive measures. A better example of the folly of ignoring the customs and knowledge of local fishermen can hardly be imagined.

Many more reef-fish-spawning aggregations throughout the tropics have undoubtedly been destroyed without any record. Biologists need to catalog as many as possible of those remaining in order to help ensure their protection. The task can be greatly simplified by tapping the knowledge of local reef fishermen.

Traditional knowledge can also play an important role in the siting and managing of coastal protected areas. Such knowledge is often superior in important respects to information gained by conventional resource surveys performed by foreign consultants constrained by insufficient time and money.

Local fishermen's knowledge of the timing and location of significant biological events is not restricted to spawning of reef fish. Certain otherwise unremarkable beaches may serve as rookeries for nesting sea turtles, or come alive with spawning land crabs during certain lunar periods and seasons (Johannes, 1981). What may look like an insignificant and relatively barren islet to a reserve planner during a site inventory made in one season may be thronged with breeding sea birds, sea turtles or, in rarer cases, sea snakes.

Traditional environmental knowledge can also play an important role in environmental impact assessment on reef and lagoon resources (Johannes, 1993). Fundamental to such assessment is recording the spatial distribution of living and nonliving resources and amenities by means of mapping. With this in mind, the U.S. Army Corps of Engineers sought and used the environmental knowledge of local fishermen extensively in producing marine resource atlases for a number of islands in Micronesia (Maragos and Elliott, 1985).

16.2. Putting Management Back at the Village Level

Traditional fishermen are not only knowledgeable about their local marine resources; in some parts of the world they have also developed an awareness of the need to protect these resources from overfishing. Because they lack the wide, productive continental shelves that typically border continental coastlines, Pacific island marine resources have been restricted largely to the narrow fringes of reef and lagoon. Because terrestrial sources of animal protein in most of these islands were scarce, these marine resources were vital.

As human populations and fishing pressure increased, the limits of these resources became obvious to the islanders centuries ago. In many of these cultures, in consequence, there evolved a marine conservation ethic, that is, an awareness of the finite nature of their marine resources and a commitment to prevent their overexploitation. Almost all of the basic fisheries conservation measures employed in the West today had long been in operation in Oceania when the Westerners first discovered the region (Johannes, 1978). These practices include closed seasons, closed areas, and size limits. Most importantly, they also include customary marine tenure systems (often referred to as traditional fishing rights), which embody the right to exclude outsiders from the fishing grounds.

If a group of fishermen possesses the right to control access to its fishing grounds, it is in their best interest to harvest in moderation. The benefits of doing so—sustained future yields—will go directly and entirely to them. When such rights are absent and everyone has access to the fishing grounds, local fishermen try to catch as much as they can, because what they do not catch is liable to be caught by outsiders rather than to form the basis for continuing good yields. Fish stocks inevitably dwindle under these circumstances.

16.3. Applying Traditional Knowledge in Modern Settings

Traditional fishing rights are quite at odds with Western legal concepts of the freedom of the seas, which are based on a naive belief in the inexhaustibility of sea fisheries. Many systems of customary marine tenure eroded as a consequence of the actions of Western colonial governments, and some have been completely destroyed. In the past few decades, ironically, Western fishery biologists have come to recognize that some form of limited access to marine resources is the cornerstone of sound management.

There is growing interest among government fishery managers in giving fishing peoples greater responsibility for managing their own fisheries. It is a notion that has been gaining currency among tropical fishery advisers around the world. Panayatou (1982, p. 48) has said, "The revival and rejuvenation of traditional customary systems . . . with limited but crucial government involvement, is one of the most promising policy options for upgrading and managing artisanal fisheries."

Public acceptance of resource management schemes is especially critical in the less-developed countries, where government enforcement is very expensive and usually inadequate. Enforcement must therefore come largely through informal social pressure and the authority of local village leaders. It is more likely to succeed if management measures are familiar and their purposes understood. Modern resource management programs that make use of traditional knowledge and integrate local traditional management practices are more likely to be understood and respected by local fishermen. Recognition of these facts has led to increasing efforts in the tropics to strengthen or resurrect customary marine tenure and associated management measures, record traditional knowledge about marine resources, integrate Western and local approaches to management, and restore greater localized management authority.

Recent experimental studies have demonstrated that the traditional custom in Pacific Island fishing cultures of setting aside a refuge for fishes can pay substantial dividends. Alcala and Russ (1990) found that both the total catch for the area and the catch per unit effort for individual fishermen could be increased by such reserves. When fishing was prohibited on 25% of the reefs along an island's coastline, the yield from the entire island was 54% greater despite the fact that only 75% of the island was fished (Fig. 18–3 in Chapter 18).

The difficulties of recording and evaluating traditional knowledge should not be underestimated. This knowledge not only must be verified (for, as with scientific knowledge, not all of it is valid), but it must also be blended with technical biological research—population dynamics, population genetics, physiology, and so forth—before it can be put to best use. This is no small matter.

It would be a mistake to infer from what has been said above that local environmental knowledge and management systems will solve all the environmental problems of coral-reef communities. Not all traditional controls over fishing function as conservation measures. Taboos on the catching of certain species, for example, may direct pressure away from abundant species toward other less abundant ones. It seems unlikely that the traditional local, nearshore controls over the harvesting of wide-ranging pelagic species such as some tuna and flying fishes would have any significant conservation effect.

Some traditional management systems may have worked only because of relatively small human populations. Such systems cannot be expected to solve today's resource management problems in areas where demands on resources greatly exceed their productive limits. No known form of resource management will recreate the miracle of the loaves and the fishes.

In some areas marine resource management was never practiced traditionally because marine resources were always greatly in excess of local needs. Today, when population pressures, new export markets, or other contemporary changes bring about an unanticipated decline in fish stocks in such areas, there is no local conservation ethic or traditional management system to provide a localized framework for responding to the problem. Here both the ethic and the response must be imported and introduced to local fishing communities with the aid of culture-sensitive education (Johannes and MacFarlane, 1991).

Village authority systems that have been the backbone of traditional marine resource management are eroding in many areas (Ruddle and Johannes, 1990). But cultures are not static; tradition has always been a dynamic force, changing in response to new circumstances. Japan offers a useful example in connection with fisheries management. There, inshore fisheries management, including that of coral-reef fisheries in the southern part of the country, has evolved into a system based firmly on ancient tradition, but has changed substantially in functional response to modern circumstances (Ruddle, 1990). Elsewhere, new forms of local authority, such as elected village fisheries management councils to replace hereditary authority, may emerge as a means of adapting to contemporary conditions, especially if governments are supportive.

Although traditional marine environmental knowledge and related management practices in the Pacific Islands have been in gradual decline ever since the region was first colonized by Western powers (Johannes, 1978), efforts are accelerating today to record local knowledge and understand local management systems, to integrate them with Western knowledge, and to help reinvigorate or remodel local authority in an effort to better respond to reef fisheries management needs.

We are in the very early stages of these efforts in some areas (Hviding and Baines, 1992). In other areas, the process has not yet begun. The need to study traditional knowledge and management systems and to integrate them better into future government marine resource management programs grows more pressing every year. The loss of such information is accelerating as many of the cultures that possess it change under the pressures of Westernization.

Integrating traditional and modern approaches to marine resource management offers no panacea for the accelerating decline of coral-reef resources around the world. But in many regions it does offer a substantial improvement over past management efforts imposed on local reef fishing cultures by biologists ignorant of what local people have to contribute and overly optimistic about what Western science can accomplish unaided by local advice.

17

Resource Use: Conflicts and Management Solutions

Gregor Hodgson

Coral-reef management solutions require more than just a knowledge of coral-reef biology. The direct and indirect effects of human activities on coral reefs can also cause several types of socioeconomic conflicts that must be considered in the management strategy. We will take a detailed look at a case study from the Philippines that shows how the techniques of ecology and economics can be combined to provide answers to questions of resource use. Then we will examine the methods used for coastal zone and marine park planning and management, and discuss the potential for successful implementation in developed and developing nations.

Coral-reef resources have long been utilized for food, building materials, medicines, jewelry, and curios. Building materials have been extracted from coral reefs for centuries. Visitors to Cebu City, the second largest city in the Philippines, are usually shown Santo Niño, one of the country's oldest Catholic churches (Fig. 17–1). The meter-thick walls of the 400-year-old church were built almost entirely of blocks cut from large coral "heads" that were mined from nearby reefs (Fig. 17–2). Historical examples of coral-block construction are common in many other locations around the world. In recent decades, labor-intensive coral-block construction has become less common, while coral continues to be dredged, crushed, and burned to make lime, used directly as fill, or mixed with cement to make concrete for construction of roads, seawalls, and other structures (Schlapak and Herbich, 1978).

A visit to Beijing's Forbidden City reveals that the imperial emperors who ruled China for several thousand years were particularly fond of coral-reef resources such as giant clams, black, red, and blue coral, as well as mother-of-pearl from oysters. Room after room is filled with displays of intricate sculptures carved from coral (Fig. 17–3) and decorated with precious gems. For the emperors holding court in cold Beijing, colorful artwork crafted from tropical treasures must

Figure 17–1. Santo Niño Church in Cebu City, Philippines.

have seemed delightfully exotic. Coral-reef organisms continue to be harvested in many parts of the world to be sold as curios (Fig. 17–4).

Since many of the resources of coral reefs are generated by living organisms that can reproduce themselves, why should we worry about using these renewable resources? One major problem is that humans tend to exploit some reefs more than others, particularly those that are nearby. This uneven pressure on coral-reef resources can easily lead to overexploitation, for example, overfishing. Unfortunately, the complexity of the coral-reef ecosystem has been a barrier to our scientific understanding of how such systems respond to exploitation (Sorokin, 1993). There are large gaps in our scientific knowledge of subjects such as estimation of sustainable rates of fish or coral harvest.

In cases where suitable data are available, they have revealed that despite high productivity, many reef systems are easily overexploited and would require some type of management if a sustainable yield is desired (section 18.1; McManus,

Figure 17–2. Close-up of building blocks in walls of the Santo Niño Church cut from coral heads.

Figure 17–3. Coral carvings were popular with the emperors of ancient China and are now displayed in Beijing's Forbidden City.

Figure 17–4. Warehouse of coral-reef seashells in Cebu City.

1980; Wells, 1981; Grigg, 1984; Marten and Polovina, 1982; Kenchington and Hudson, 1987; Salvat, 1987; McManus et al., 1992).

In addition to direct uses of coral reefs, many human activities, often far from the ocean, are known to damage coral-reef ecology (Salvat, 1987; Kenchington, 1990). Important examples of causes of damage are chemical pollution from industrial discharges, oil spills, toxic runoff from golf courses or farms; sediment pollution from poorly managed land clearing; nitrification from discharge of poorly treated human or livestock sewage.

Unfortunately for coral reefs, the ecological effects of both direct and indirect influences are usually additive and sometimes synergistic; therefore, reefs located near human population centers are often subject to multiple direct and indirect influences. In some cases, human activities have led to the local extirpation of coral reefs, for example, in Tolo Harbour, Hong Kong (Scott, 1984), and Jakarta Bay, Indonesia (Tomascik et al., 1993).

Faced with this increasing degradation of coral reefs and depletion of resources, it might be tempting to recommend a prohibition of all activities on coral reefs. But erecting regulatory "fences" that prohibit entry to all but researchers and enforcers is obviously not a practical or desirable goal. What is a useful goal for coral-reef management?

17.1. Sustainable Use

The World (Brundtland) Commission on Environment and Development (De La Court, 1990) defined sustainable development as "providing for the needs of the present without jeopardizing the needs of the future." Sustainable development is a worthy goal of coral-reef management because it accepts the reality that humans have need for coral-reef resources, and yet it provides a guideline for how far the utilization in the short term should be allowed to proceed so as not to destroy the ability to utilize in the long term. Although this goal is not easily attainable, particularly for reefs located near human population centers, there are several examples where the goal has been reached (Grigg, 1977; Ross, 1984; Johannes and MacFarlane, 1991).

Since human activities are among the most important factors in the life and death of many coral reefs, and yet the data are insufficient to allow scientists to accurately predict long-term effects, it is prudent to follow the precautionary principle and to adopt a conservative approach to exploitation of coral-reef resources (Kenchington, 1990). Therefore, if we are in doubt about the effects of a proposed activity, we should err on the side of caution when considering how to regulate that activity. In Australian law, the National Environmental Protection Council Act of 1994 states that lack of full scientific certainty that serious or irreversible environmental damage will occur cannot be used as a reason for postponing measures to prevent environmental degradation. The onus is on the entrepreneur to prove that resource degradation will not occur to the loss of the general public before a venture is undertaken.

In addition to a management goal, we need planning and management tools. Since the implicit rationale for allowing human activities on coral reefs is generally economic, to produce goods and services, it follows that any attempts to regulate these activities should be based not only on sound scientific knowledge, but also on a firm understanding of the socioeconomics of resource use. For a given location, an analysis of socioeconomic systems allows us to translate measurable ecological effects, such as loss of abundance and species diversity, into economic losses and social disruption. One way to study socioeconomic systems is to examine resource-use conflicts.

17.2. Resource-Use Conflicts

Often, the use of a resource for one commercial purpose may have a direct or indirect effect on the use of the same or another resource by a different industry or noncommercial user. Economists call this situation a resource-use conflict. Some resource-use conflicts involving coral reefs are given in Table 17–1.

One example of a resource-use conflict occurs in Sri Lanka where a major cause of coastal erosion is coral mining. The coral is used to manufacture lime for cement and other products. But by removing the offshore coral barrier, storm waves are allowed to batter full force against Sri Lanka's west and southwest coasts, particularly during the southwest monsoon. Wave erosion causes losses of between 17.5 and 28.5 hectares of land each year, and damages houses, hotels, roads, and other coastal structures, causing economic losses of millions of rupees annually (Lowry and Wickremeratne, 1987). Aside from seriously damaging the ecology of coral reefs, coral mining is a commercial resource use that deleteriously affects other industries such as tourism and transportation. Tourism faces losses resulting from direct wave damage to land, hotels, and restaurants, and indirectly from reduced environmental quality for scuba tours. Transportation losses result from direct damage to roads and railroads.

Biologists are likely to be most concerned about the loss of natural resources. But what about government policymakers and implementers? Would they be more interested in the resources or the economic losses? By comparing the economic gains from coral mining with the losses it causes, natural resource managers, regardless of their background, will have a much stronger argument for regulating coral mining than if they rely on the "damaged ecology" argument alone.

Another example of a resource associated with coral reefs is crude oil. It is believed there are large oil reserves lying under coral reefs such as the Great Barrier Reef in Australia (Kenchington, 1990) and Pratas Reef in the South China Sea (Valencia, 1985). In this case, the resource is not a part of the living coral reef, but simply lies beneath it. Exploitation of the crude oil will conflict with the other uses listed in Table 17–1 by causing operational or accidental damage.

Table 17.1. Some Resource-Use Conflicts Involving Coral Reefs

Coral-Reef Resource	Primary Use	Conflicting Uses
Coral	Mining	Conservation Research/Education Tourism/Recreation Fisheries Coastal Protection
Fish	Commercial fishing	Recreational and Artisanal Fishing Tourism/Conservation
Crude oil	Oil extraction	Conservation Research Tourism/Recreation Education Fisheries

Investigating potential resource-use conflicts is a difficult task that may require expertise in diverse fields such as economics, ecology, law, and sociology. A detailed assessment of resource-use conflicts requires a set of analytical techniques developed within a new field called ecological economics that emphasizes an interdisciplinary approach (Costanza, 1991). Ecological economics builds on and refines many of the concepts developed within the fields of natural resource economics and environmental economics.

A major goal of ecological economics is to monetize as many "downstream" effects as possible in the economic analysis of a planned project's viability. In the past, many downstream effects such as pollution have been considered external to economic analyses. The cost of such externalities in terms of damage to the environment and human health has been excluded traditionally from economic balance sheets. Several theoretical techniques have been developed to internalize these costs and they are finding increasing use, particularly by international banks (Dixon and Hufschmidt, 1986; Dixon and Sherman, 1990).

17.3. Case Study: Logging Versus Fisheries and Tourism in the Philippines

Although cost-benefit analysis has been in use for many years, its application to ecological valuation is relatively new. A simple ecological-economic analysis is presented in the following case study. This will demonstrate how cost-benefit analysis can be used to translate ecological processes into economic values to guide management decisions.

Fieldwork for this case study of resource-use conflicts was carried out in Palawan Island, Philippines, in 1985 and 1986 (Hodgson and Dixon, 1988, 1992). In this case, two industries—tourism and fisheries—were competing with a third industry—logging. The competition was indirect, through crucial ecosystem links; erosion from coastal logging in the terrestrial ecosystem produced negative effects on the marine ecosystem via sedimentation, an economic externality to logging that posed a threat to the viability of the tourism and fishery industries. In order to introduce the study, some basic economic data and information on the logging industry are presented.

17.3.1. Logging and Siltation

The detrimental side effects of logging on watersheds have been well documented throughout the world (BIOTROP, 1978; Hamilton and King, 1983) and include damage to young trees via unplanned felling, soil degradation, and soil loss through increased soil exposure to wind and rain. Although tree cutting exposes underlying soil to the direct effects of wind and rain, by removing protective layers of leaf canopy, the major cause of erosion resulting from either selective or clear-cut logging operations in the tropics has been shown to be road building (O'Loughlin, 1985; Hodgson and Dixon, 1988, 1992).

Although the links between watershed erosion, sediment-laden rivers, and siltation of the coastal marine environment appear obvious, it is not a trivial problem to determine the precise origin of sediment delivered many kilometers to the sea or to quantify the rate at which it is delivered. Multiple sediment sources and sinks, formation and erosion of bedload deposits, bank erosion, and seasonal entry of marine sediment to estuaries are some examples of the complex processes affecting this seemingly simple transport of sediment from mountain to sea. Agronomists, foresters, and hydrologists have studied the effects of erosion on terrestrial ecology, but few have looked beyond the freshwater systems to the sea (Williams and Hamilton, 1982; O'Loughlin and Pearce, 1984). Numerous marine scientists have studied the effects of sedimentation on tropical coastal marine organisms. However, the precise source and cause of the sediment erosion has often been assumed but rarely documented (Salvat, 1987). The Palawan study was unusual because it was a "mountaintop to seabed" investigation that documented the source of sediment, the cause and mechanism of sediment erosion, and quantified sediment transport and the ecological effects on downstream coral reefs.

Sedimentation in the sea is a natural process resulting from erosion of land and transport of soil to the sea, or from resuspension of sediment previously deposited along coastal margins or on the seabed. Many human activities, particularly land-clearing associated with farming, logging, and road construction, accelerate erosion and subsequent marine sedimentation (Gomez et al., 1982; White, 1987; Salvat, 1987). This anthropogenic sedimentation is a widespread form of pollution that is increasing in severity (Wells and Hanna, 1992; Hodgson, 1993). The effects of coastal marine pollution are especially serious in developing countries such as the Philippines where the rapidly expanding human population depends heavily on marine species to meet protein requirements.

The site of the Philippine sedimentation study was Bacuit Bay, located near the village of El Nido at the northern tip of Palawan Island (Fig. 17–5). This deep-water bay encompasses an area of 120 km^2, contains five islands, and has nine islands located on its seaward shelf. Flourishing coral reefs surround each island and form a continuous band along the bay coastline interrupted only by river passes. On land, the Bacuit Bay drainage basin covers 78.3 km^2 and extends inland to the central Palawan dividing range. Prior to the initiation of logging in 1985, 53% of the basin was composed of primary forest. A breakdown of the 1986 land-use Bacuit Bay drainage area is listed in Table 17–2.

Logging within the drainage basin was carried out in 1985 and temporarily ceased between January and December 1986, the period of the study. Many different fishing methods were practiced in the bay to catch a diverse array of species including resident coral-reef fish and fish indirectly dependent on reefs. The tourism industry was dependent on local and foreign scuba divers who pay

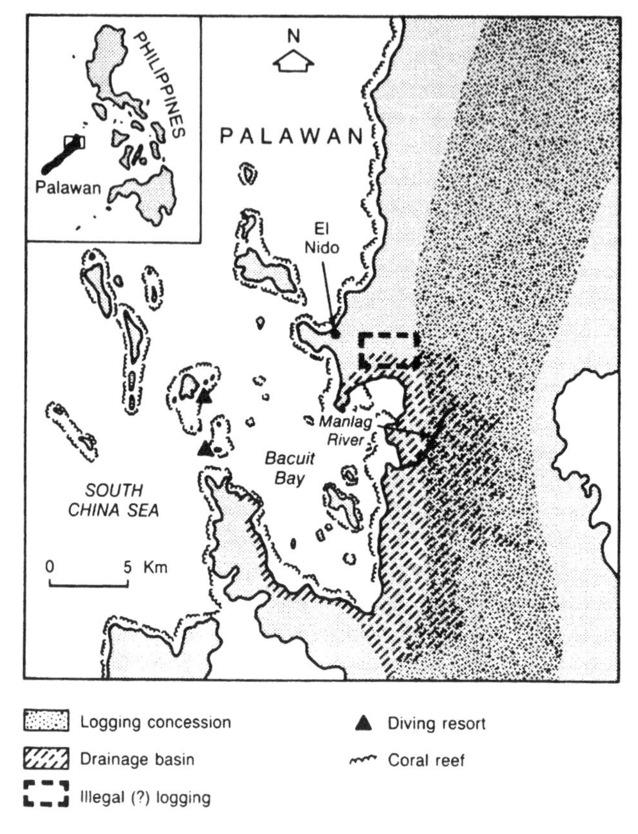

Figure 17–5. Map of Philippines (inset) and the Study Site, Bacuit Bay, near El Nido on Palawan Island. By January 1986, about 9% of the original forest area drained by the Manlag River had been cut.

a high premium to fly to this remote location to dive in clear water and to see coral reefs in near pristine condition. It was assumed that degradation of the natural beauty of the underwater or terrestrial environment would make tourism less profitable.

17.3.2. An Integrated Assessment

The El Nido study was "integrated" in that it involved investigations of links among ecosystems—terrestrial, riparian, and marine—and included both socio-economic and ecological perspectives. A study of the terrestrial ecosystem was made to determine if logging significantly increased erosion above natural levels, and if so, whether this eroded soil was transported to the marine ecosystem or was stored in river systems. A simultaneous study of Bacuit Bay measured

Table 17.2. Land Use in the Bascuit Bay Drainage Basin (1986)

Land Use	Area (km^2)	% of Total
Forest		
Primary Dipterocarp forest	37.0	47.3
Scrub, secondary forest	27.1	34.6
Logged forest	4.8	6.1
Mangrove forest	3.9	5.0
Subtotal	72.8	93.0
Agriculture		
Swidden/cashew	3.6	4.6
Rice paddy	1.1	1.4
Coconut plantation	0.8	1.0
Subtotal	5.5	7.0
Total	78.3	100.0

sediment transport and deposition, and the effects of sediment deposition on marine life. A separate evaluation was made of the socioeconomics of two El Nido industries dependent on coral reefs: dive tourism and fisheries.

Erosion plots were built in three areas: road, cut forest, and uncut forest. Runoff and eroded sediment was trapped and measured directly. The erosion plot studies showed that sheet erosion per hectare of logged area (cut forest and logging roads combined) during 1986 was about 240 times greater than from primary (uncut) forest plots. Although roads only covered approximately 3% of the forest area, they contributed 84% of the erosion (Figs. 17–6 and 17–7).

River discharge and suspended sediment levels were monitored in two adjacent rivers: one draining the logged forest (Manlag River) and the second draining uncut primary forest (Balangoyen River). Sediment transport in the Manlag River was 16 times greater than sediment transport in the Balangoyen River. Along with other physical evidence, this confirmed that sediment transport to Bacuit Bay was significantly increased as a result of logging.

Suspended sediment concentrations were monitored in Bacuit Bay, and the size and depth of sediment plumes were measured. Sediment plumes from the Manlag River spread out over the inner portion of Bacuit Bay, increasing sediment deposition by as much as 60 times the monthly deposition rate measured at control reefs located near the bay entrance, which were unaffected by terrestrial sedimentation.

17.3.3. Coral-Reef Damage

Laboratory and field experiments on the effects of sedimentation on 50 species of Indo-Pacific reef corals from 14 families revealed that there is a hierarchy of tolerance to sedimentation that is related to each species' growth form, corallite

Figure 17–6. Erosion from a logging road.

size, and ability to extend polyps above the colony surface (Hodgson and Dixon, 1988, 1992). Sediment deposition of 20 mg/cm^2/day (0.2 kg/m^2/day) was sufficient to injure and kill many species; those species with small corallites, poor polyp extensibility, and ramose growth form were the most susceptible to damage.

Between January and December 1986, the amount of living coral cover and

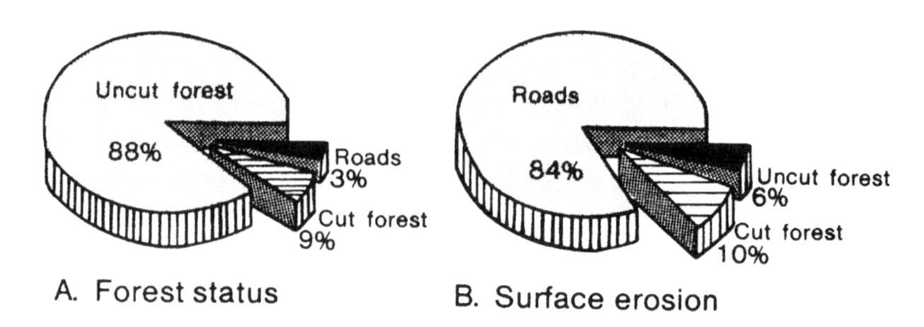

Figure 17–7. Although comprising a small fraction of the original forest area, logging roads contributed most of the surface erosion.

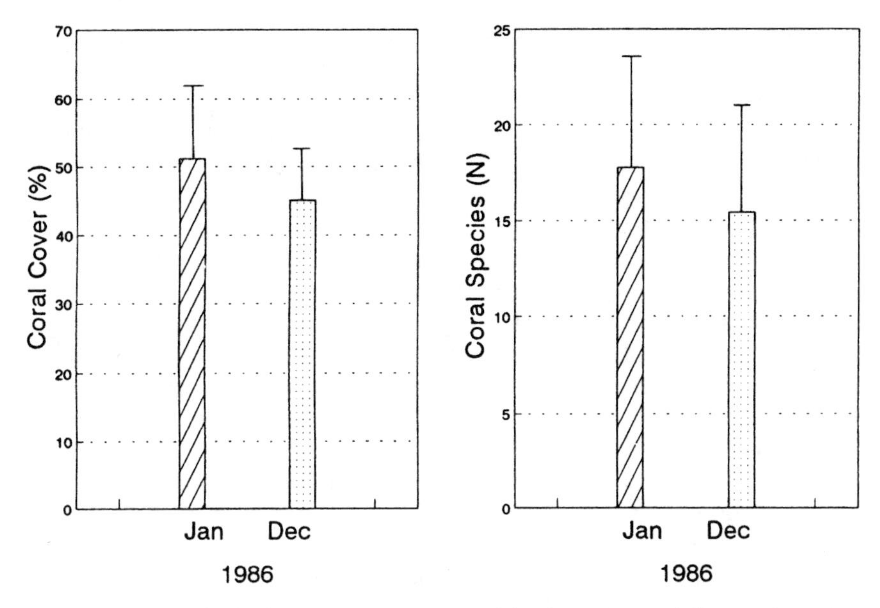

Figure 17–8. Reduction in coral cover and number of species on eight reefs in Bacuit Bay exposed to sedimentation due to logging in 1986. Error bar gives SD for eight stations; five 10 m transects per station.

number of species declined significantly at all seven stations affected by increased sedimentation ($p < 0.0005$; Fig. 17–8). The cover of recently killed coral increased significantly at the station nearest the river mouth ($p < 0.01$). There was a significant positive relationship between loss of live coral cover at each station and mean sediment deposition ($p < 0.02$; $r^2 = 0.62$; $y = 6.28x + 1.3$).

17.3.4. Effects of Sediment on Fish

Adult fish, in contrast with corals, can escape from localized high turbidity by swimming away. In laboratory tests of numerous (mainly temperate) species, very high suspended sediment concentrations (about 1,000–50,000 mg/l) were required to kill adult fish via gill clogging and oxygen starvation (Peddicord et al., 1975; Alabaster and Lloyd, 1984). Although fish larvae may be less tolerant, this extreme range of suspended sediment concentration is rarely found in nature. For example, the highest single turbidity reading from the Manlag River (draining the logged area) was 3,000 mg/l, and from Bacuit Bay, 1,000 mg/l. Therefore, changes in abundance and diversity of coral-reef and pelagic fishes were not expected to be caused by fish mortality as a direct result of sedimentation.

Numerous studies have documented direct and indirect dependence of coral-reef fish on the coral-reef community (Chapter 11; Carpenter et al., 1982; Sano et al., 1987). Large decreases in fish diversity and abundance have been measured when the coral-reef community is damaged. In one case, the complete destruction of a coral reef at Iriomote Island, Japan, resulted in the loss of 90% of fish abundance within 2 years (Sano et al., 1987). Although some coral-reef fish feed on coral directly (Reese, 1981; Cox, 1983; Tricas, 1986), most depend on prey organisms for their food supply, for example, crustaceans, other fish, or algae living in association with coral reefs (Sale, 1991). In general, major shifts in fish diversity and abundance are expected to result from damage or removal of components of coral-reef community structure that reef fish depend on for food, shelter, reproduction, and recruitment of juvenile fish (Randall, 1974; Sale, 1991).

The abundance, diversity, and biomass of fish at the Bacuit Bay reefs were estimated. The dependency of fishes on the living coral reef of Bacuit Bay was examined via regression analysis of changes in coral cover and diversity versus fish biomass on coral-reef transects surveyed at the beginning and end of the one year of logging (1986).

The different levels of sediment deposition measured during 1986 at eight stations produced linearly proportionate reductions in coral cover and diversity at these sites. Linear relationships between coral cover and coral diversity, respectively, and fish biomass were used to predict changes in coral and fish ecological parameters resulting from predicted increases in sediment deposition.

Analysis of changes measured in Bacuit Bay coral and fish populations during 1986 produced the following relationships:

1. There was an annual decrease in coral cover of 1% for every additional 400 tons/km^2 of annual sediment deposition in Bacuit Bay.
2. There was an annual decrease of one coral species (local extinction) in Bacuit Bay per increase of 100 tons/km^2 of annual sediment deposition.
3. Fish biomass decreased by 2.43% for each 1% annual decrease in coral cover.
4. For each annual decrease of one coral species associated with coral-cover loss, fish biomass decreases by 0.8%.

Using these relationships it was possible to estimate the long-term results of the loss of either coral cover or diversity on fish biomass and thus fish catch in Bacuit Bay.

Based on the predicted ecological effects, the economic consequences of logging on fisheries and tourism were estimated over a 10-year time frame based on two possible development options: (1) a complete stop to logging in the affected area, or (2) continued logging of the watershed with resulting reduction in projected revenues of the fisheries and tourism industries.

17.3.5. Assumptions in Estimation of Effects

To predict the potential effects of the alternative development options on Bacuit Bay drainage basin for 10 years, several simplifying assumptions were made.

The first assumptions, based on available evidence, were that logging of the remaining 88.5% of the drainage basin would be finished within 5 years beginning in January 1986 and that the resulting erosion and sedimentation would be proportional to the area logged.

To account for forest recovery and reduced erosion following logging, the sediment output (and deposition)/logged-area ratio was assumed to remain constant until one year after the drainage-area forest had been completely logged (1993). Sediment output was then reduced by 30% per year through the tenth year after logging began. This model is consistent with empirical results from Malaysia and California (O'Loughlin, 1985; Erman and Mahoney, 1983).

In order to calculate the predicted sediment deposition for each succeeding year of the 10-year time horizon, the Manlag River sediment output in 1986 was used and was increased in proportion to the yearly increase of the newly logged area. Using the Manlag River sediment discharge, rather than the higher sediment deposition figure measured in 1986, resulted in a conservative estimate of actual sediment output caused by logging.

To reduce the complexity of the model, sediment deposition and ecological effects were averaged over the entire bay. In reality, the continuous gradients measured for sediment deposition and sedimentation damage varied temporally and spatially throughout the bay. This simplifying assumption was made in order to emphasize the significance of major trends.

The third assumption was that the fish catch for the entire bay and shelf would decline at a rate set by the regression equations of coral ecological parameters on fish parameters beginning one year after logging was restarted. The one-year delay was used to account for potential delay inherent in the predominantly indirect effects of coral-reef degradation on fish population dynamics (Sano et al., 1987). A precondition for the validity of this assumption is that fish catch is approximately proportional to fish abundance given a constant level of fishing effort. Reliable long-term data on fishing effort (or stock size) in and around Bacuit Bay were not available. Rather than attempting to guess the future of these complex relationships (Pauly and Murphy, 1982), a constant effort and catch was used for the analysis of Bacuit Bay fisheries. Based on these simplifying assumptions, the Bacuit Bay model predicted decreased coral cover and fish catch as shown in Figure 17–9.

17.3.6. Economics of Management Alternatives

The Bacuit Bay ecosystem is an example of a complex natural system that provides a variety of resources. Multiple uses exist for the products of the system; some uses can coexist easily and others conflict. The fisheries industry and tourism, for example, both depend on a healthy marine environment. The undisturbed bay ecosystem, including algal, coral-reef, plankton, and fish populations, produces the products desired by both the tourism and the fishing industries.

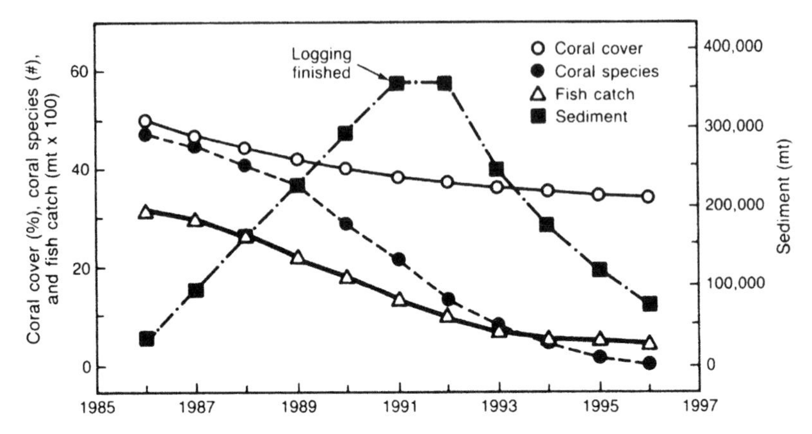

Figure 17–9. The results of the Bacuit Bay model predicted declines in coral cover, coral species, and fish catch due to increased sedimentation from logging.

There was good cooperation between the tourism and fishery industries in terms of law enforcement and the operation of marine preserves and fish-feeding stations for scuba divers. Assuming that this level of cooperation would continue, the fisheries industry was not expected to have a negative effect on the tourism industry.

The logging industry, on the other hand, is concerned only with timber extraction and has little or no economic interest in the marine environment. A reduction in marine resources will have no direct effect on timber production; logging activities, however, can have a major negative influence on the marine resource–dependent sectors.

In order to compare the benefits generated by each activity under the two development options, the present value of gross revenues over a 10-year period was calculated. Present-value analysis is used to take into account the time factor in revenue production and the existence of private and social discount rates (Hufschmidt et al., 1983; Dixon and Hufschmidt, 1986). That is, a dollar promised tomorrow is worth more than a dollar promised in 2 years. The dollar received tomorrow can be used for investment or to earn interest over time. Present-value analysis is useful for comparing, for example, the value of a tree sold today for lumber versus a tree that we expect can be sold several years hence. In this analysis, two discount rates were used, 10% and 15%, the range of rates typically used in analysis of government and private projects (Dixon and Hufschmidt, 1986). A summary of the results of the analysis is presented in Table 17–3.

The results are striking; the gross revenue total under the Stop Logging Option (1) is more than double the gross revenue under the Continued Logging Option (2). Since Option 1 will prevent further logging in the Bacuit Bay drainage basin, the gross revenues from logging are zero. Fisheries and tourism, however, generate

Table 17.3. Economic Analysis of Two Development Options

	Option 1	Option 2	Option 1–2
Gross Revenue			
Tourism	$47,415	$ 8,178	$39,237
Fisheries (with tuna)	28,070	12,844	15,226
	(46,070)	(21,471)	(24,599)
Logging	0	11,375	–11,375
Total	$75,485	$32,397	$43,088
Present Value (10%)			
Tourism	$25,336	$ 6,279	$19,057
Fisheries (with tuna)	17,248	9,166	8,082
	(28,308)	(15,125)	(13,183)
Logging	0	8,624	–8,624
Total	$42,584	$24,069	$18,515
Present Value (15%)			
Tourism	$19,410	$ 5,591	$13,819
Fisheries (with tuna)	14,458	7,895	6,563
	(23,122)	(13,083)	(10,039)
Logging	0	7,626	–7,626
Total	$33,868	$21,112	$12,756

Notes: Gross revenues, present values of gross revenues (× $1,000) using 10% and 15% discount rates for each industry sector under Option 1 (logging ban) and Option 2 (continued logging). The last column lists the difference obtained when Option 2 values are subtracted from Option 1 values. The values for the fisheries sector when tuna catch is included are shown in parentheses, but they are not included in the totals.

large and continuing benefits. The size of the tourism benefits is expected to grow over time as demand and market increase. The fisheries benefit will remain constant. In contrast, Option 2, which allows continued logging, generates smaller and decreasing benefits. After 5 years the logs will be gone as well as a significant part of the tourism and fishery sectors. The modest logging revenues generated under Option 2 are more than offset by decreased tourism and fisheries income.

The present value of gross revenues under Option 1, calculated using a 10% discount rate, is almost double that under Option 2. Even at the higher, 15% discount rate, the Option 1 total present value is still 1.6 times the Option 2 present value.

17.3.7. Intergenerational Equity and Sustainable Development

The conflict between purely economic priorities (e.g., maximizing present value) and socioecological priorities (sustaining resource yield) is illustrated by examining the present value of renewable resources to be harvested 30, 50, or 100 years in the future. The present value of such future harvests rapidly approaches zero with increasing time, even when moderate discount rates are used. Since future harvests of natural resources clearly do have value, the application of present-value analysis to slow-growing natural resources has been controversial. Some economists have argued that near-zero discount rates would be appropriate when calculating present value of natural resources (Finney and Western, 1986). Others have suggested that when projects span more than one human generation, discounting is inappropriate (Norgaard and Howarth, 1991). An intermediate approach has been to use variable discount rates to take into account changes in time preference when project duration is long relative to human life span (Crocker and Shogren, 1993).

One way to clarify the limitations of present-value analysis and approximately reflect longer-run concerns would be to examine the process of equitable distribution among individuals and between generations of people (Norgaard and Howarth, 1991). In the case of Bacuit Bay (El Nido), residents are highly dependent on fish catch. Many individuals who stand to benefit from future fish catch or who could be hurt by diminishing fish catch may not be alive to enjoy any possible long-term economic benefits from the logging operations that were predicted to damage the fishery. In fact, some researchers have argued that the Philippine logging industry is exploitative, leaving little sustained economic growth behind (Tadem et al., 1984). Certainly, if most of the profits of logging went to nonresidents of El Nido, it is unlikely that logging would do much to create future income-generating opportunities for El Nido residents. Thus, from the point of view of distribution of revenues between generations, a logging ban was the preferred development option.

One way to address concern about intergenerational equity is to focus on preservation of future options rather than on generation of present and future benefits and their distribution. In this approach, the optimal development alternative is the one that maintains maximum sustainable resource utilization in perpetuity. In the Bacuit Bay case, the development option resulting in sustainable use of the fisheries and tourism resources will be favored as it seems unlikely that logging could be developed sustainably (see Warford, 1986, and Barbier, 1987, for detailed discussions of sustainable economic development).

Additional factors that remain externalities in this analysis are employment, job training, infrastructure development, risk of flooding, and biodiversity preservation. In all cases except infrastructure development, the Stop Logging Option would be preferred.

The results of the El Nido study indicated that continued logging of the drainage basin would result in environmental damage and high economic costs compared to the effects of the alternative, prohibiting further logging in the specified area. Instituting a ban on future logging in the Bacuit Bay drainage basin was the alternative of choice.

17.3.8. El Nido Today

Along with other information, the El Nido study was used by the Philippine government to help plan the future of northern Palawan. Large-scale commercial logging was stopped in 1987; therefore, it is not possible to compare the ecological and economic predictions of the model with the actual outcome. The El Nido area was designated as a marine park. Although no follow-up study has been made, local residents claim that sediment pollution continued to damage coral reefs in the inner section of Bacuit Bay for several years following the end of logging operations. Tourism has expanded, bringing new challenges to the implementation of conservation plans for the area.

The results of the ecological-economics analysis of development alternatives for El Nido suggest several points of interest for those responsible for coral-reef management. The results indicate that sedimentation pollution could seriously degrade coastal marine fisheries, and reduce tourism potential and recreation options in the tropics by damaging coral reefs. Those concerned with maintaining a sustainable marine fisheries catch might consider increasing the role of sedimentation pollution monitoring and control, both for rivers and the coastal zone, particularly in areas of suspected value to the fisheries trophic system. More basic research needs to be carried out to investigate the direct and indirect effects of sedimentation on commercially important tropical marine fish species. Higher priority should be given to examining possible effects on valuable reef fish such as grouper, as well as species such as tuna that have traditionally been considered entirely pelagic (and therefore not affected by coastal pollution), but that appear to be dependent on the nearshore marine environment for at least part of their life cycle (Leis et al., 1991).

From the perspective of coastal zone management and development planning, the El Nido study showed that it is possible to obtain clear, cost-effective answers to development questions from integrated ecological research and economic evaluation. The primary features of the integrated approach are that data on both ecological and socioeconomic aspects are collected and analyzed, and the study area is considered to be composed of a series of linked ecosystems extending from mountaintop to seafloor. The dynamic links between these ecosystems are regarded as equal in importance to the processes occurring solely within each ecosystem. Natural systems are linked by freshwater flows, tidal currents, and the movements of organisms, particularly planktonic larvae. It is more realistic to conceptualize the coastal zone as a series of linked ecosystems rather than independent entities as is commonly done.

17.4. Marine Parks

Throughout the world, integrated coastal zone planning and management have been recognized as necessary steps on the way to sustainable development (Sorensen et al., 1984; Kelleher and Kenchington, 1991). The establishment of marine-protected areas is one aspect of coastal zone planning and management that was rapidly adopted in many tropical countries, particularly with respect to tourism and fisheries development in coral-reef areas.

Unfortunately, in an effort to rapidly protect these precious resources, some developing nations established marine reserves and parks before management plans had been completed. In particular, some parks were established without consultation with the local inhabitants or consideration of their historical rights to resources, such as fish, within park boundaries. This resulted in the declaration of a number of poorly planned "paper parks"—parks that exist on paper but provide no conservation value (Hodgson, 1992). For example, over 100 marine reserves have been created in Indonesia, Malaysia, the Philippines, and Thailand (White, 1988b), but few are operated in a way that will ensure conservation over the long term.

The design of marine parks has evolved from a philosophy of total protection of small areas to one allowing multiple uses of large areas with an integrated management system that takes account of all resource management goals, including total protection (Kelleher, 1993). A number of useful guidelines have been proposed for setting up marine-protected areas that, if followed, would reduce the chances of more paper parks being established (Salm and Clark, 1984; Kenchington and Hudson, 1987; White, 1988; Kenchington, 1990; Kelleher and Kenchington, 1991). The guidelines recommend:

- Integrated consideration of terrestrial and aquatic ecosystems
- Early and open public consultation
- Provision of alternative resources to those potentially deprived by the implementation of restrictions
- Formulation of clear goals for the park
- Use of management tools such as zoning to provide for multiple uses that are consistent with the park's goals
- Establishment of a management plan that specifies how the goals will be achieved
- Provision of legislation and regulations that are enforceable, implementable, efficient, comprehensive, and in concert with established local, national, and international policies

It is not an easy task to establish and maintain a marine park, and different problems will be faced in different countries. In developed countries such as Australia, powerful commercial interests are often the major potential threat to conservation, for example, drilling for oil on the Great Barrier Reef. In less-developed countries, the major threat to conservation may be an impoverished local population simply trying to survive (section 18.4.3).

An important aspect of park design is the division of the area into zones for different uses. The application of zoning will depend on the special physical and biological characteristics of each potential park area. One zoning system that has proved workable is placement of a core zone of high conservation value inside a buffer zone that will protect the core from outside disturbance.

There are coral-reef marine parks in several countries that are successfully achieving their multiple-use goals of conservation, education, and recreation. The most ambitious project was begun in 1975 with the establishment of the 344,000 km^2 Great Barrier Reef Marine Park. The Great Barrier Reef Marine Park is a multiple-use protected area, not a national park (Kelleher, 1993).

The Great Barrier Reef Marine Park Authority (GBRMPA) was established to manage the area and marine park rangers were authorized to enforce regulations such as no oil exploration, mining, spearfishing with scuba, or littering (Kenchington, 1990).

Regardless of the economic status of a given country, the success of a marine park is directly attributable to early and transparent public consultation. GBRMPA solicits public input when designing management plans. Whenever a zoning plan is in preparation, reef users are invited to comment. Once potential conflicts are resolved, the draft zoning plan is advertised and the public is again invited to comment. To ensure that the plans remain consistent with current conditions, they are revised about every 5 years. Some of the types of zones used in the park are listed in Table 17–4.

In addition to these activity-control zones, periodic restrictions may be placed on specified areas to protect breeding grounds, to allow heavily fished areas to recover, to allow small areas to remain totally protected, or to allow scientific research to be carried out without disturbance. Proponents of planned developments near the park are required to carry out a detailed Environmental Impact

Table 17.4. Zoning of Cairns and Cormorant Pass GBRMP

Zone Type	Area (km^2)	Allowed Uses
General Use "A"	25,800	Most activities including fishing, tourism,* shipping, defense, collecting,* research,* traditional hunting and fishing
General Use "B"	7,700	Same as "A" except no trawling or shipping
Marine National Park "A"	672	Research,* traditional hunting and fishing," tourism," recreation (no collecting)
Marine National Park "B"	251	Research,* traditional hunting and fishing,* nonextractive tourism,* recreation
Marine National Park Buffer	305	Same as above, but allow troll fishing
Scientific Research	191	Scientific research,* traditional hunting and fishing*
Preservation	68	Scientific research* only in exceptional circumstances

Source: After Woodley, 1985. *Require permits.

Assessment (EIA) to predict the likely effects should the development proceed. Guidelines for EIAs focusing on coral reefs are given by Carpenter and Maragos (1989) and Maragos (1993).

Along with the Great Barrier Reef Marine Park in Australia, the Virgin Islands National Park on St. John, and Pennekamp State Park in Florida stand as examples of how preservation of coral reefs through coastal zone management and planning can complement other forms of development and create vigorous tourism industries. Although small areas of park can be set aside for the purpose of pure preservation, carefully planned ecotourism and recreation development offer the best potential incentives for the conservation of coral reefs. When marine parks are established in areas that previously supported subsistence or commercial fisheries, tourism also provides alternative income-generating opportunities for local inhabitants who are suddenly cut off from their traditional means of support (Hodgson and Dixon, 1992).

Uncontrolled tourism has its own negative effects. Large numbers of ecotourists can degrade parks; a combination of education and regulation must be used to prevent divers, reef walkers, and boaters from inadvertently damaging the fragile reefs (Rogers et al., 1988). The negative effects of poorly planned tourism development extend well beyond physical damage, and in extreme cases, may result in localized self-destruction of the industry (Miller and Auyong, 1991).

One argument in favor of establishing marine-protected areas is that they should enhance fisheries by serving as the marine equivalent of terrestrial "seed banks" (section 18.3). Theoretically, restricting fishermen's access to certain portions of a coral reef allows increased reproductive success and larval recruitment, leading to enhanced fish catch on nearby reefs (Roberts and Polunin, 1991). Such a system is a form of traditional fisheries management (Johannes and MacFarlane, 1991) that is currently used to regulate reef fisheries in Hawaii. It appears that one of the goals of setting up marine reserves, to enhance fish populations within and nearby the reserves, is achievable.

17.5. Implementation of Sustainable Development

The best management plans in the world will achieve little if they cannot be properly implemented. The ingredients for successful implementation of coastal zone management plans are political will, adequate long-term funding, and public support. If any one of these ingredients is lacking, the programs will probably fail.

Numerous skeletons of failed programs can be seen around the world. Topping the list of failures are foreign-aid-funded paper parks in developing countries that lacked local public support and long-term funding. In such cases, as soon as the foreign aid runs out, the "park" simply falls apart. The second common type of failures are those projects that, although adequately funded, lack political support, so that a formal government structure is never established and the money is spent on nonproductive activities.

Some support for marine resource planning and management can be gleaned from international agreements. There are numerous regional and international agreements that aim to provide protection for coral-reef resources (Groombridge, 1992). For example, the United Nations Convention on Law of the Sea provides a legal basis for stewardship of marine resources. The convention includes the text, "States have the obligation to protect and preserve the marine environment," but it provides a mechanism (the exclusive economic zone) to protect marine resources within 200 nautical miles of the coast only from foreign exploitation. It does not provide a mechanism for protection against exploitation from within a country.

For political will, motivation of the local residents, and effective enforcement of management policies of the nearshore resources of coral reefs, the management should be based in the local villages (section 18.5), but this requires the support of the government. Without backing by the central government, village-based management of marine resources is susceptible to being overwhelmed by foreign and/or large money interests.

There is no international police force to enforce environmental laws or to ensure that coastal zone planning and management are properly carried out. Sadly, it is primarily through negative feedback from management crises that governments have learned that sustainable development is not an idealistic goal but rather a necessity. As unsustainable development continues, some consequences may be dramatic and obvious, whereas others may be subtle, appearing only gradually.

Pattaya, located south of Bangkok, was once the most popular beach resort in Thailand. The main attractions were clean white-sand beaches, blue water, and scuba diving on nearby coral reefs. Exponential and haphazard development throughout the 1980s created widespread, severe water pollution as well as social problems. Environmental degradation was slow enough that many did not recognize the problems until it was too late. The result has been a serious slump in the tourism industry. The Thai government has recognized the problem, but appropriate remedies are difficult to implement and the recovery of damaged coral reefs will take many years (Chancharaswat, 1991).

In Hong Kong, sustainable development has never been a high priority. In 1991, large-scale dredging of inshore seabeds began to supply 400,000 m^3 of sand fill for reclamation associated with port and airport expansion. Although few tourists see them, coral reefs fringe the eastern shores of the colony. A side effect of the dredging has been sedimentation damage to coral reefs at the Ninepin Islands, a popular dive spot for local scuba enthusiasts (Hodgson, 1993). In this case, the damage occurred over a few months, and resulted in a significant reduction in coral cover and number of species at the reefs. Because a high proportion of damaged corals were decades-old colonies of tabletop *Acropora,* barring further damage, it will take 50 years for the reef to return to its predamage condition.

Figure 17–10. Stuffed endangered sea turtles, protected by CITES, but for sale in a curio shop in Vietnam.

When unsustainable resource use continues and some species are overexploited, their populations may decline to such low numbers that authorities consider their extinction to be imminent, in which case they may be placed on the endangered species list compiled by the International Union for the Conservation of Nature (IUCN) or another organization. In such cases, the rules are relatively well established. Trade in threatened and endangered species of plants and animals has been banned by the 1975 Convention on International Trade in Endangered Species of Wild Fauna and Flora (CITES), an international agreement signed by over 100 countries. For example, it is now illegal to transport items such as jewelry made from the shells of the endangered green turtle into or out of CITES signatory countries. Both publicity about CITES and enforcement of its regulations have played crucial roles in protecting numerous marine species.

Unfortunately, some countries with large numbers of native rare and endangered marine species find it difficult to enforce their own conservation regulations, and curios made from endangered marine species are still openly displayed in shops in many countries (Fig. 17–10; section 18.4.1). This problem is particularly acute in socialist countries, which have recently opened their markets, such as China, Myanmar, and Vietnam, thus dramatically increasing the incentives to exploit marine resources (section 18.4.1).

The lack of political will, administrative structures, and funding available to regulate the use of marine resources has led frustrated scientists and managers in some countries to recommend the banning of all resource use. For example, both China and the Philippines periodically place a total ban on the harvesting of hard corals for the curio trade, with severe penalties for those caught breaking the law. Such laws may be justified when exploitation of certain species, such as giant clams, has clearly soared out of control or when little is known about their population biology.

But more often, and certainly in the case of hard corals, they are an admission of defeat by government, and a step backward in that they deny the scientific reality of sustainable use of living resources (Ross, 1984). In practice, such laws are expensive to enforce and give corrupt law enforcers another avenue of extracting money from poor villagers. The laws are also incongruous given those countries' unregulated trade in rare shells also collected from coral reefs. Rather than admitting defeat, a sustainable multiple-use management plan should be designed based on scientific investigation of population biology. Ideally, by allowing controlled exploitation, implementation of the plan can be funded by taxes on the export of the resources.

17.6. Conclusion

Coral reefs are renewable resources that have a long history of exploitation by humans and that continue to serve as a direct source of a wide variety of products. The results of unsustainable use of coral reefs can be seen throughout the world: reefs with few fish, edible molluscs, or sea cucumbers and corals damaged by dynamite fishing. Inadvertent damage by household and industrial chemical pollution, eutrophication, and restricted water circulation associated with coastal construction activities, and from sedimentation from poorly planned clearing of land, are also serious threats to the health of coral reefs.

In general, coral reefs located near large human population centers have suffered the most serious degradation. However, few truly pristine coral reefs remain anywhere. Coral reefs that were previously protected from human predation by deep water and distance are steadily becoming exposed to human exploitation (section 18.4.1). In the cumulative equation that describes the life and death of coral reefs, anthropogenic damage is a critical factor that must be added to damage by disease, storms, and other natural forces.

Anthropogenic exploitation and inadvertent damage to coral reefs can be managed. The primary goal of management is sustainable use. Integrated coastal zone planning and management provide an avenue for sustainable use of marine resources including coral reefs. As part of this process, establishment of multiple-use marine-protected areas is essential for research, education, recreation, and preservation of biodiversity. Given careful planning and management, some products can continue to be harvested from coral reefs at a sustainable rate, although perhaps not at rates sufficient to supply all demands, especially as human populations increase. Inadvertent "downstream" damage to coral reefs, as a result of development of other resources, can be avoided.

Coral-reef management will only achieve its goals if it is backed by public support, political will, and adequate funding. This level of support will only be achieved by advertising the spiritual, ecological, and economic value of coral reefs. Widespread recognition that coral reefs supply many goods and services offers the best protection against potential damage from competing resource-use schemes, for example, oil extraction. The recently developed field of ecological economics has been shown to offer useful methods for drawing attention to the economic value of marine resources in the political arena. Greater use of the techniques of ecological economics by those with an interest in coral reefs will help generate the public support and political will needed to implement appropriate management strategies for coral reefs.

18

Implications for Resource Management

Charles Birkeland

Coral reefs are immediately perceived as diverse, abundant, and productive in resources. It is ironic that these impressions attract endeavors that strike at the most vulnerable aspects of reef communities. Although coral reefs are the most productive of marine communities, they are also the most vulnerable to overexploitation. The vulnerability is because of characteristics of coral reefs from the level of the ecosystem to the level of individual species.

The standing stocks of coral-reef fishes encourage the development of fishery-based economies. Fisheries have been estimated to be as high as 160–200 metric tons km^{-2} in the Atlantic (Randall, 1963; Munro, 1983) and as high as 93–239 metric tons km^{-2} in the Pacific (Goldman and Talbot, 1976; Williams and Hatcher, 1983). The mean harvests on reefs of some small islands in the central Philippines over a 5-year period have been found to be 11.4 and 16.5 metric tons km^{-2} yr^{-1} (Alcala and Luchavez, 1982). The shoreline fishery of American Samoa has yielded an average of 18 metric tons km^{-2} yr^{-1} with some catches as high as 26.6 metric tons km^{-2} yr^{-1} (Wass, 1982). Two hundred to three hundred species of fishes are collected for food from certain coral reefs in the western Pacific (McManus et al., 1992; Wells and Hanna, 1992; Myers, 1993). In the Caribbean region, about 180 species of reef-associated fishes are marketed (Munro, 1983). The standing stocks of coral-reef fishes are about 30–40 times greater than standing stocks on demersal fishing grounds in Southeast Asia, the Mediterranean, or other temperate regions (Russ, 1984a).

Coral reefs are the basis of livelihoods for perhaps tens of millions of people in the tropics (Salvat, 1992a), coral reefs provide most of the protein in the diets of some societies of islanders, and coral reefs have sustained subsistence fisheries in some of the islands of Oceania for hundreds or possibly thousands of years. About 60,000 small-scale fishery operations harvest coral reefs in the Caribbean and 14,000 people fish the reefs of the Gulf of Lingayen in the Philippines (Wells and Hanna, 1992). Calculations and extrapolations (although recognized to be speculative and from incomplete inventories) have indicated that coral-reef-related fisheries may have a potential yield approaching 9% of the total marine fisheries of the world (Smith, 1978). Yet coral reefs have provided few, if any, sustained commercial export fisheries.

18.1. Export Fisheries Not Sustained

Although the diversity, standing stock, and yield of coral-reef fishery resources are spectacular, the initial impressions of a cornucopia have not been substantiated. Most coral-reef fisheries previously touted with exemplary high yields have not been sustained when exploited commercially or when the local human population grew substantially. The harvest from the reefs of American Samoa has been recorded to be as high as 26.6 metric tons km^{-2} yr^{-1} (Wass, 1982), and this figure is commonly used as one of the greater examples of the potential yield of coral-reef fisheries (e.g., Russ, 1984a; Munro and Williams, 1988). But there was a 70% drop in CPUE (catch per unit effort) at American Samoa from 1979 to 1994 (Craig, 1994), the number of fishes per hectare (17 most common species) decreased by 75% from 1980 to 1994, and the relative abundances among species changed drastically, with a decrease in preferred species (Craig et al., 1995).

Likewise, the potential harvest by Jamaican fishermen has been estimated "on a conservative basis" to be 16,000 tons per year using traps (Munro, 1983), but the CPUE at Southeast Pedro Bank near Jamaica declined by 82% over a 15-year period (Koslow et al., 1988).

In the Philippines, up to 16.5 metric tons km^{-2} yr^{-1} have been harvested (Alcala and Luchavez, 1982). However, in a study in which Philippine reef-fish populations were monitored over a 4-year period (McManus et al., 1992), the numbers of adult fish declined by 80% and the number of species known to reach adulthood declined by 33%. As adult fish become harder to find, a greater proportion of the catch is made up of juveniles. The more prized species disappear faster than the less desired species and the fishermen turn to nonselective fishing techniques such as dynamite and cyanide. These nonselective techniques are often damaging to the coral reef as a habitat for fishes. The living coral cover on the reef flat and in the lagoon were reduced by 60% by these fishing methods (McManus et al., 1992). Fishermen using these nonselective methods are in direct competition with fishermen using more desirable fishing techniques, and the destruction of habitat reduces the success of future harvest. This accelerates the conversion of other fishermen to less selective and more destructive fishing methods (McManus et al., 1992).

Even reefs that are "lightly fished" have failed to sustain their promised yield. The fishing effort on Bermuda was considered "low" (Russ, 1984a), but the total catch dropped from a maximum of 678,000 kg in 1986 to about 380,000 in 1990.

The differences noted by the public were not so much in the 50% drop in total yield, but rather in the changes in species composition. In the 1950s, about 70% of the catch was grouper and less than 1% was parrotfish and other "miscellaneous fishes" that were less desired as food fish. The proportion of grouper declined over the years until it made up 13.7% of the catch in 1990–1991 while the miscellaneous fishes made up over 30%.

Likewise, the fishery on Guam had "relatively good statistics" but was "so lightly exploited that it is difficult to draw any reliable conclusions about the limits of coral-reef fish productivity" (Munro and Williams, 1988, p. 546). Nevertheless, there was an increase in commercial fishing on the reef flats of Guam in 1984. At this time, the CPUE started to decrease so dramatically that many fishermen quit fishing. Since 1984, both the number of active fishermen and the CPUE have continued to decrease simultaneously. In an attempt to sustain subsistence and recreational fishing on these "lightly exploited" reefs, the government is preparing legislation to prohibit commercial fishing on the reefs of Guam.

Coral-reef populations do not always recover within decades after being exploited. An offshore pinnacle was discovered near Guam in 1967 and harvested down within 6 months (Ikehara et al., 1970). The fish have still not returned after nearly 30 years.

During 14 years prior to World War II, well over 30 million sea cucumbers were exported from Micronesia, especially from Chuuk (formerly called Truk). Although 59 years have passed, recent surveys at a number of sites around Chuuk indicate that the sea-cucumber populations have not yet recovered (Chapter 8).

In a few days, Taiwanese fishing boats harvested 15,000 giant clams from Helen's Reef south of Palau, but it may take up to 20 years for the population to recover (Wexler, 1994), in part because of the high natural rate of predation on the juveniles.

Even some relatively unfished areas downstream from heavily fished areas demonstrate major variations in stock size and substantial reduction in catch rates of targeted fish species. This might be attributed to chronic reductions in recruitment because of the overfishing on upstream reefs (Koslow et al., 1988). Munro (1983) determined that about half of 37 species harvested in traps become catchable long before they reach sexual maturity. He concluded that in heavily fished areas, such as the north coast of Jamaica, very few fishes survive long enough to spawn and so recruitment depends on larvae drifting in from other regions. It may be that the populations that do not recover for decades have also been overexploited at the areas depended upon as the source of larval recruits.

18.2. High Productivity, Low Yield

Although coral reefs have among the greatest gross primary productivities of any ecosystem in the sea (Crisp, 1975; Lewis, 1977), the fisheries yield and net system productivity are relatively low (Fig. 18–1). The ratio of fisheries yield per unit gross primary productivity for the Peruvian upwelling is about 20–60 times greater than for coral reefs (Chapter 12; Fig. 18–1; Ryther, 1969; Nixon, 1982). For coral-reef systems, in contrast to systems with abundant nutrient input such as upwelling and estuaries, the net system productivity averages very close to zero (Kinsey, 1983). Far less than 1% of the gross primary productivity is converted to production that is meaningful for human consumption (Chapter 7).

Part of this low net system yield might result from a large number of trophic levels in coral reefs (section 12.3.2 in Chapter 12). Ryther (1963) assigned one or two steps between phytoplankton and humans to upwelling systems and acknowledged that the upwelling regions of the world have the shortest food webs in terms of the major bulk of the biomass. Grigg et al. (1984) assigned six trophic levels to coral reefs, probably the most complex of marine ecosystems. Much of the assimilated energy is lost in respiration at each step in the food web so that it seems possible that there could be less excess production in a system with substantial portions of the biomass in six trophic levels than in a system with most of the biomass concentrated in the lower two.

It also seems logical that greater yield could be obtained from a system if lower trophic levels were harvested. Much of the great increases in marine fisheries yield of the world in the 1950s and 1960s came from the shift in emphasis in fisheries from the predatory fishes to the more plentiful fishes lower

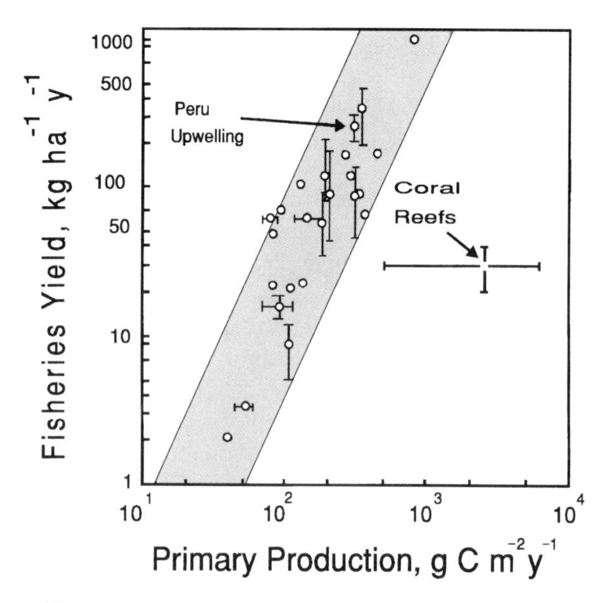

Figure 18–1. Differences between coral reefs and other marine ecosystems in the relationships of fisheries yield to gross primary production. The non-coral-reef relationship is based on 49 studies at 25 sites (redrawn from Nixon, 1982).

in the food web. The catch of anchoveta *(Engraulis ringens)* in the Peru upwelling increased from about 200,000 metric tons in 1955 to 12 million tons in 1970, a 60-fold increase (Larkin, 1978). In view of these successes, a first principle of resource management of coral reefs has been "To maximize efficiency, harvest at levels of sustained yield as low on the food chain as possible. Cropping predators would increase potential yield" (Grigg et al., 1984).

But in using for coral reefs the principles of management strategies allegedly successful in upwelling regions, it is important to keep in mind that there are fundamental differences among these ecosystems. Oceanographic processes are the major driving force in large-scale current ecosystems, while species interactions (e.g., predation) are the driving force in coral-reef ecosystems (Birkeland, 1987, 1988b; Sherman, 1994). Predation or overfishing does not influence the process of upwelling. On coral reefs, in contrast, the movements of fishes may cause enough movement of nutrients in coral-reef ecosystems to influence the growth of corals (Meyer et al., 1983), and overfishing can have large-scale ecosystem-level effects (Chapter 9; Hughes, 1994; Roberts, 1995; Andres and Witman, 1995; McClanahan et al., 1996).

Large-scale changes in fisheries in coastal-current ecosystems are determined primarily by physical environmental changes and only secondarily by intensive fishing (Sherman, 1994). There is a general positive association between gross primary productivity of the ecosystem and fisheries yield among most of the world's fisheries, both marine and freshwater (Fig. 18–1; Nixon, 1982). Reproductive success and year-class strength depend on river discharge and upwelling (Aleem, 1972; Sutcliffe, 1972; Wolff et al., 1987). Whenever El Niño causes the reduction of nutrient input by upwelling, the effects of a failure of upwelling and nutrient input are manifested at all levels in the food web (Chapter 12; Glynn, 1990a).

In coral-reef ecosystems, the resident population of fishes affects the recruitment of fishes and corals and thereby the structure of the coral-reef community at the site (Chapter 9; Shulman et al., 1983; Hixon, 1991). The decrease in percent cover of living coral along the north coast of Jamaica from about 70% in the 1970s to less than 5% in the early 1990s is substantially influenced by overfishing of herbivorous fishes and triggerfishes (Hughes, 1994). Individual species of predators, such as the crown-of-thorns seastar *Acanthaster planci* in the Pacific, the sea urchin *Diadema antillarum* in the Atlantic, and *Eucidaris thouarsii* in the eastern tropical Pacific, have ecosystem-level effects on the geology and ecology of coral-reef communities (Chapter 9).

18.2.1. Recycling

Physiological and life-history characteristics of animals exploited for food from coral reefs, which are most strikingly different from the characteristics of animals from regions of nutrient input (Chapter 12), and which make them especially vulnerable to overexploitation (section 18.2.3), are most directly explained as adaptations to survival in a system in which food and space are in limited supply and are largely recycled. Of the estimated 2–3% of gross primary productivity that becomes net primary productivity from a coral-reef community, perhaps 75% (Chapter 7) of the net system yield (E) is in forms unusable by humans (e.g., dissolved organic materials and detritus). About 15% accumulates in reef structures. This leaves about 10% of the 2–3% of the total gross productivity of a coral reef, far less than 1%, that is usable secondary production (Hatcher, 1996). Of this, much is algae and invertebrates.

The degree of recycling of nutrients within coral-reef communities is extreme. The diet of "herbivorous" zooplankton of coral reefs is almost entirely detrital material. Chlorophyll has been found to be present in only 2–6% of the food mass, and most of this was from detrital benthic algal fragments rather than phytoplankton (Gerber and Marshall, 1974). A majority of the zooplankton itself is resident, dwelling within or near the substratum, but periodically ascending into the water column (Porter and Porter, 1977; Hammer and Carleton, 1979; Walter and Pasamonte, 1982; Grimm and Clayshulte, 1982). In a sense, even much of the zooplankton is "recycled" rather than drifting through the system.

Fish feces have been observed to be fed upon by corals (McCloskey and Chesher, 1971), and Robertson (1982) deduced that some fecal material from fishes may be eaten and recycled through five fishes before it reaches the seafloor to be consumed by corals or other invertebrates. Much of the detrital material eventually reaches the sediment, from which it is again consumed by deposit feeders, in contrast to pelagic regions where it is exported to below the photic zone.

At the physiological level, nutrients are recycled between zooxanthellae (algal cells), *Prochloron* (a genus of prokaryote with chlorophyll), or cyanobacteria and the tissues of their animal hosts (including foraminiferans, sponges, stony corals, soft corals, fire corals, sea fans, anemones, giant clams, jellyfish, sea slugs, tunicates, and other invertebrates). The symbioses between single-celled algae within the tissues of animals of numerous phyla provide a major portion of the primary production on coral reefs.

18.2.2. Constraints of Complexity

In regions of abundant nutrient input from upwelling, river discharge, or terrestrial runoff, the diatoms, naked flagellates, and other small, relatively simple and free-floating, single-celled phytoplankters can respond rapidly to nutrient input and increase in population density (Geider et al., 1986). The feeding rates of herbivorous copepods also increase as the phytoplankton becomes more densely available (Mayzaud and Poulet, 1978), although it is possible for saturation to be reached if a lot of food is introduced within a very short period. The secondary production by the small herbivorous zooplankters is generally associated with the distribution of primary production. Gross growth efficiencies of herbivorous marine plankton in upwelling regions are at least twice the 10% generally quoted for terrestrial communities (Pomeroy, 1979; Barnes and Hughes, 1988).

In contrast to the rapid and substantial response of pelagic communities to nutrient input, the density of single-celled algae in the tissues of corals is relatively constant within any particular set of environmental conditions associated with depth (Chapter 5). The complex physiological accommodations between the host and the symbiotic algae can be upset if there is a rapid proliferation of the intracellular algae. There is some indication that high levels of dissolved nitrogen (exceeding about 10 μm) destabilize the symbiosis of corals and intracellular algae by enhancing the growth rates of the zooxanthellae (Chapter 5). Although this matter has not been finally resolved, it appears that if a rapid increase of intracellular algal density upsets the physiological balance in the symbiosis, the corals may respond by expelling the "excess" zooxanthellae. Because a relatively large proportion of the primary production in coral-reef ecosystems is from intracellular symbioses rather than from phytoplankton (Johannes et al., 1970; Scott and Jitts, 1977; Lewis, 1982), it is likely that regulatory processes within the complex physiological adaptations tend to dampen the response of coral-reef communities to pulses of nutrient input. These complex interactions may also lead to a dampening of a positive reponse of the coral-reef system to increased harvesting.

18.2.3. Life-History Traits

The life-history adaptations of the species themselves may have a more immediate effect on the ability of the system to accommodate an increase in fishing pressure. Populations of animals in regions of rich nutrient input are characterized by periodically abundant recruitment, rapid growth, early sexual maturity, dense aggregations, and rapid population turnover (Birkeland, 1988a, b). These life-history characteristics facilitate population responses to increased fishing pressure. In regions of less concentrated nutrient input where coral reefs are usually found, the dominant animals and especially those targeted by fishermen are characterized by irregular recruitment success, high natural mortality of juveniles, slow growth, low natural adult mortality, postponed first reproduction, increased fecundity with age (Fig. 18–2), sedentary postsettlement life-history stages, long adult life, and multiple reproduction (Bohnsack, 1992). A number of coral-reef fishes are sequential hermaphrodites. These life-history characteristics make successful reproduction in these populations especially vulnerable to the targeting of large individuals (Bannerot et al., 1987).

Predation is intense on juvenile fishes on coral reefs. The proportion of juveniles surviving their first year on coral reefs is sometimes less than 0.008 (Shulman and Ogden, 1987) or 0.007 (Craig, 1995). Coral reefs have a disproportionate diversity and prevalence of predators, and the ratio of juveniles to adults in populations on coral reefs is much less than in neighboring habitats (Birkeland

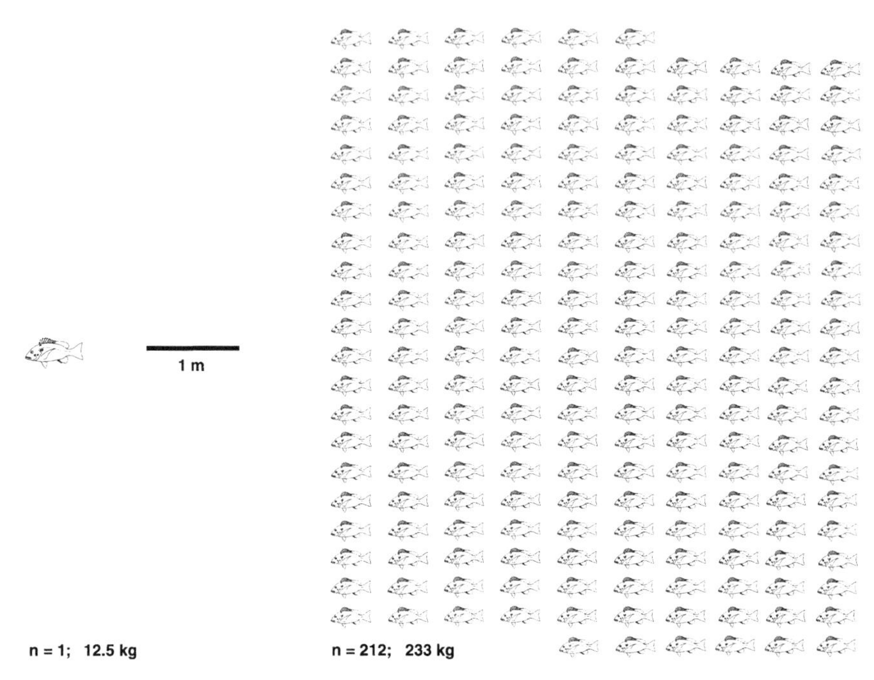

Figure 18–2. Fecundity increases exponentially with body size in fish. One 61 cm female red snapper (lutjanid) produces the same number of eggs as 212 females at 42 cm each, or 12.5 kg of large snapper produces the equivalent number of eggs as 233 kg of snappers; each smaller snapper is 69% the length of the larger one (Plan Development Team, 1990). The reproductive potential of a population is disproportionately affected when fishermen target large individuals. This is especially true for protandrous species.

and Amesbury, 1988). The tremendous standing stock of animals, the low yield from coral-reef systems, and intensity of predation suggest that the natural system already approaches maximum sustainable yield in the biomass supported. The standing stock of animals adapted to multiple reproductive efforts (e.g., giant clams, sea cucumbers, sharks, groupers) has usually accumulated over a number of years. The catch per unit effort (CPUE) of large fishes should start to decline very soon after a substantial increase in fishing pressure. If the system is already at maximum sustainable yield, further extraction from the system, for example, by commercial export fisheries, must lead to overharvesting.

18.3. Reserves for Breeding Stock

Empirical evidence demonstrates that coral-reef reserves increase the abundance of fishes both inside and outside the area of protection (Fig. 18–3; Alcala and Russ, 1990; Bohnsack, 1994). By reducing mortality from fishing, reserves or sanctuaries provide an area in which a spawning-stock population and even some large individuals are allowed to survive. This substantially increases the spawning potential (Fig. 18–2) and genetic variability (Fig. 18–4) of the targeted species. Furthermore, undisturbed spawning grounds are also provided. These provide potential long-term maintenance or even enhancement of fisheries yield to areas outside the reserves. A model shows that for snappers *(Lutjanus campechanus)*, protection of 20% of the area increases egg production by 1,200% (Bohnsack, 1994). Reproductive output from reserves helps replenish fished areas outside the reserves by natural larval dispersal and/or by movements of adult fishes out of the reserves into the less crowded areas.

Fishery reserves in the Philippines (Alcala and Russ, 1990) and in the Caribbean (Polunin and Roberts, 1993) have been found to allow increases in abundances of both targeted and nontargeted species. Alcala and Russ (1990) documented that nearly twice as many fishes could be caught on 75% of the reef as compared to the entire reef, presumably because of movements of fishes from the high-population densities on the reserve to areas with more resources and space outside the protected area (Fig. 18–3). The reason that nontargeted species also benefit may be due to nonselective fishing methods in the nonprotected areas.

Reserves protect the genetic diversity of targeted fish populations, whereas catch quotas based on optimal yield calculations do not. The fecundity of fishes increases exponentially with body size (Fig. 18–2); thus, in natural populations in low-nutrient conditions, where predation on larvae and juveniles is intense and successful recruitment is sparse and uncertain, the selective pressure strongly favors large body size, long life, and multiple reproduction. When the species that formerly was top predator and had a refuge in size is now the prey of humans, however, the probability of living long enough to reproduce at a large size becomes small; and the selective pressure strongly favors rapid growth and early reproduction (Fig. 18–4).

It is not just the species with larger individuals that are especially vulnerable on coral reefs and that can be favored with breeding-stock protection. For example, the yellowstripe goatfish *Mulloides flavolineatus* is a schooling predator of small, soft-bodied invertebrates in the sand. It reaches sexual maturity within one year after hatching and has a long period of repeated spawnings each year. The local fishermen on Guam probably do not realize that fishing pressure has been drastically reducing the reproductive potential of the goatfish each year over about a 7-year period. By 1991, the reproductive potential of *M. flavolineatus* was only 5% of its unfished potential (Davis, 1992). By harvesting the larger individuals of even the smaller, more rapidly maturing fishes, a greater portion of the breeding stock is taken than is perceived by the fishermen. Fishery reserves protect against both recruitment overfishing and growth overfishing.

It is also not just the territorial or benthic species that are vulnerable to overfishing and benefit from refuges. The motile caesionids, carangids, and scombrids decreased by 64% following the breakdown in protection of a reserve in

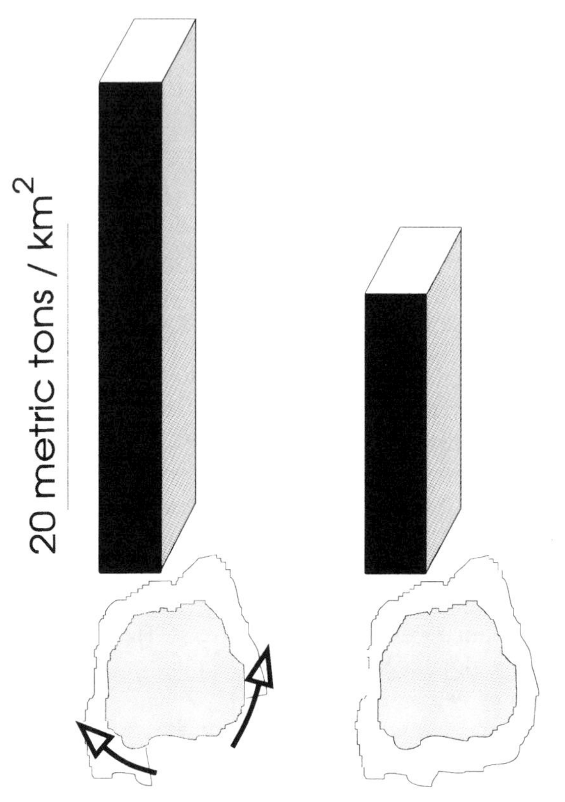

Figure 18–3. Fisheries yield on Sumilon Reef, Philippines, declined 54% when a reserve was opened to fishing and the breeding stock was taken (Alcala and Russ, 1990). The arrows indicate movement of fishes and larvae from the reserve out to neighboring reefs.

the Philippines (Alcala and Russ, 1990). It is possible that a decrease in food supply made the refuge lose its attraction to the more wide-ranging fishes when protection broke down. The holistic approach of refuges provides insights and poses questions not available with a species-by-species modeling approach. Fishery reserves also have the potential of generating revenue from tourism, providing relatively undisturbed habitats and fish populations for scientific study and educational use.

The complex interactions at all levels on coral reefs (Chapter 11), the great diversity of species (Chapter 14), the life-history characteristics of animals adapted to systems with low concentrations of nutrient input (Chapter 12), the patchiness of reef populations, and the vagaries of recruitment from the plankton

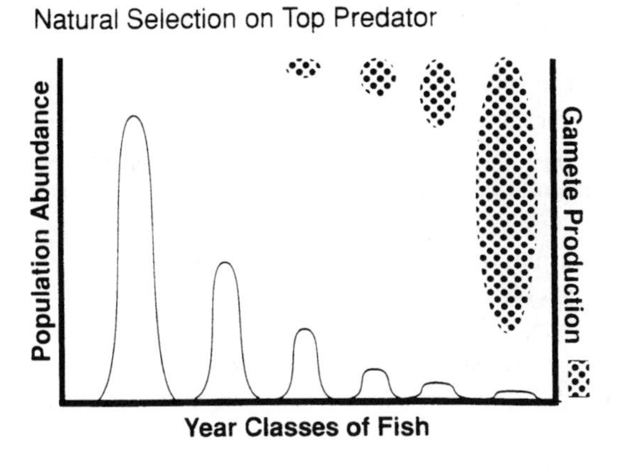

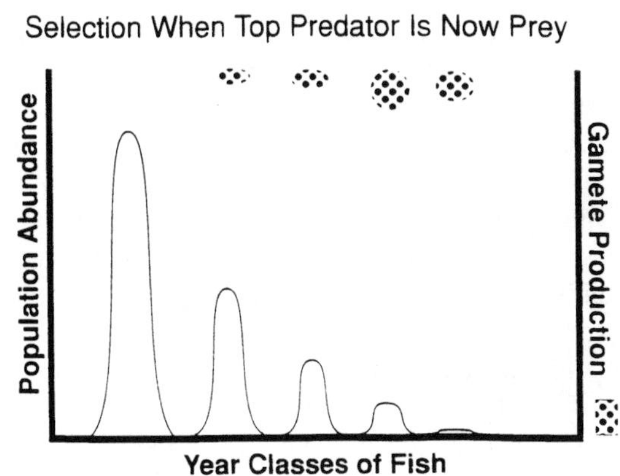

Figure 18–4. There is a strong selective advantage for a top predator to grow large because fecundity increases exponentially with body size. When the top predator becomes prey to humans, selection favors early reproduction. Reserves protect the intraspecific genetic diversity of natural populations (Plan Development Team, 1990).

make the reductionist approach in calculating optimal sustainable yields less reliable in preserving breeding stocks than is the holistic approach of establishing refuges. Fishery reserves provide a backup in case of management failure, they are more readily understood by the public than the limitations imposed by optimal-yield calculations, they simplify enforcement, and they affect the users of the fishery more equitably than do more complex regulations (Plan Development Team, 1990). Considering the rates of growth of human populations in coastal regions of the tropics (section 18.5), reserves may be the only practical means of maintaining viable spawning-stock populations with any potential of sustaining fisheries.

18.4. Economics of Harvesting Diverse Resources

Smaller yields per unit gross primary productivity of coral reefs compared with upwelling areas (Fig. 18–1) are partially a result of the economic inefficiency of harvesting and processing a diverse assemblage of species using a variety of gear in topographically complex coral reefs. A few people on a purse seiner in pelagic upwelling areas or on a trawler on soft-bottom areas can harvest tons of fish, while the same number of fishermen swimming around coral heads or leaving traps overnight can collect only kilograms of fish. But these factors that explain the low rates of extraction per fisherman would also tend to protect the fishes on coral reefs from economic overexploitation. Why are coral-reef fisheries especially vulnerable to overfishing despite these natural constraints to economic overfishing?

It has been suggested that economic overfishing occurs before biological overfishing. The individuals become rare enough that it is no longer profitable to fish for them and so the fishermen stop before a population nears extinction. For example, the bounty system has failed to control crown-of-thorns starfish *Acanthaster planci* because it is no longer profitable to search for the starfish when the population density is low. Since each female *A. planci* can produce millions of eggs per spawning, outbreaks can start from very small remnant populations. It has been suggested that the bounty system actually promotes the persistence of *A. planci* by thinning the population so that *A. planci* would not overexploit its own food supply (Yamaguchi, 1986).

But the bounty for the crown-of-thorns starfish is at a low value (usually $0.10–0.20 per starfish). The opposite is the case for coral-reef resources. Biological overfishing is taking place on oceanic Pacific reefs long before protective constraints of economic overfishing come into effect because (1) the tremendous dollar value in Asia of some coral-reef species encourages fishing effort even after the targeted species have become too rare to sustain a viable reproductive population; (2) economic subsidies from developed countries to local fishermen (sometimes to buy the rights from island governments to offshore pelagic fishing in the Exclusive Economic Zone) provide money to sustain fishing on coral reefs without necessity of economic return; and (3) the diverse assemblage of the fishes themselves provides a variety of alternative species to catch and sell. These alternative, initially nontargeted species sustain the fishery and keep the fishermen active in the area. The more prized species can therefore continue to be captured as they are encountered, even though they may become too rare to sustain a fishery by themselves. The targeted species can become biologically overfished before the fishermen are economically constrained.

In fact, as the species becomes more rare, it becomes more highly valued and the price increases. In the past, dugongs and sea turtles would be hunted as needed for food or for a special occasion. In more recent times, rare items are taken at any opportunity, whether needed or not and even if illegal, because if they are not taken as the situation occurs, they may be taken by someone else and the opportunity may not happen again. Unlike the bounty on *A. planci,* where it becomes increasingly uneconomical to harvest a species as it becomes rare, the effort becomes more focused on valuable species as they become rare.

18.4.1. High Values

China's population is growing at 14 million per year and its economic growth over the past decade breaks all records, exceeding even the greatest of Japanese economic growth records (*New Scientist,* 7 January 1995). More than 80 million mainland Chinese now earn between $10,000 and $40,000 per year (Naisbitt, 1996). The increase in disposable income in Asia has increased the demand for protein, medicines, and aphrodisiacs. There are 3.8 million marine fishermen working out of China, 200,000 out of Japan, and many out of Korea and Taiwan. Ting Hong Corporation alone has about 500–600 ships operating in the tropical western Pacific. As demand and dollar value increase, and as marine resources are being depleted near the continent, fishing boats from many Asian countries are ranging across Oceania.

An estimated 20,000–25,000 tonnes of live reef fish (valued at about $1 billion) are being imported to Asia (Johannes and Riepen, 1995). To collect these live fish economically, the entrepreneurs facilitate the use of sodium cyanide by the fishermen. The cyanide is very toxic to the corals and other reef organisms and so the reefs are not expected to recover for several years. The cost of transporting live fish aboard large ships makes a one-time sweep more profitable than development of a sustainable yield. Therefore, a single pass by the Asian fishing boats provides the islanders with quick cash, but eliminates large spawning aggregations of fishes that have previously sustained coastal villagers for centuries (Johannes and Riepen, 1995).

In 1991, $95 million worth of sharkfin (conservatively calculated as 17,000–28,000 tons dry weight) was imported to Hong Kong alone, with additional millions of dollars in sharkfins being imported to other Asian countries (Bonfil, 1994). Sharks have relatively low fecundity and are adapted to long life and multiple reproduction. In Guam, a small bowl of sharkfin soup costs $80, a medium-sized living crab costs $60, and small live groupers are bought wholesale at $60 each.

In Hong Kong and Singapore, a half-kilogram live grouper can sell for over $100 and some species can sell for $180 per kilogram. The lips of the wrasse *Cheilinus undulatus* can sell for $225 per plate (Dayton, 1995). One 40 kg *Cheilinus undulatus* was valued at $1,800.

As the disposable income increases in Asian countries, the expenditures on mythological cures and aphrodisiacs create new revenue for exploiting animals that were previously ignored. An estimated 60–70 tonnes of dried seahorses (approximately 20 million individuals) were imported into Asian countries in

1994 (Prein, 1995). The low fecundity, lengthy parental care, low natural adult mortality, and sparse populations make seahorses vulnerable to such intensive harvesting. Seahorses are now as high as $850 per kilogram, but this price is bound to rise as they become harder to supply for their alleged aphrodisiac and medicinal potential.

High monetary values on protein from coral-reef resources are putting pressures on many coral-reef management policies, traditions, and laws. Sea turtles are protected by the Convention on International Trade in Endangered Species of Wild Fauna and Flora (CITES), but this regulation is often ignored. In Naha (Okinawa) small (30–45 cm carapace) sea turtles are publicly displayed for sale at the equivalent of $1,000–$1,400. It takes 20–25 years for sea turtles to reach sexual maturity while the money earned from the sale of the turtle carapaces can gain interest at a much greater rate, about 8% per annum. When interest rates on money are far greater than return on natural capital, it is in the interests of many to "liquidate" the natural populations, publicly ignoring management policies, traditions, and national and international law.

The distances of islands in Oceania from Asia make a one-time harvest of standing stocks of slow-growing, long-lived species more economical than a sustained fishery. For example, divers from a Taiwanese fishing boat illegally cut the adductor muscles from 15,000 giant clams on remote Helen's Reef (between Palau and Papua New Guinea) before they were caught (Wexler, 1994). It has been estimated that it will take at least two decades for the giant clam population to recover. Adductor muscles from giant clams sold for up to $100 per kilogram; the rest of the meat was left to rot on the reef. The several years necessary for giant clams to grow to sexual maturity make a "one-pass" economic strategy preferable for boats from distant ports.

The high dollar value makes it worthwhile to take the risk of arrest of the crew and forfeiture of the boat. It has been estimated that 406,000 *Tridacna derasa* were taken by poachers from foreign fishing boats between 1969 and 1977 in the Swain Reef section of the Great Barrier Reef alone (Pearson, 1977). Prior to 1977, 22 Taiwanese fishing boats were apprehended for poaching (Pearson, 1977), but the return on successful trips is more than adequate to absorb the cost of losing ships and transporting the crew back to Taiwan.

Trochus niloticus was harvested in a sustainable manner throughout eastern Indonesia for most of this century (Reid, 1992). However, commercial export was begun in the 1960s and it was then that the catch rate increased dramatically. As the stocks became overharvested in Indonesia and the price of trochus shell increased, the fishermen would take long voyages to Australian waters, risking arrest, imprisonment, and the destruction of their boat (Reid, 1992).

18.4.2. Subsidies

The large standing stocks of fishes and initial catches from stocks that have built up over a number of years have sometimes led to the conclusion that coral-reef fisheries have a great potential that is underutilized. Fishery agencies have sometimes been given the development of commercial harvest of marine resources as their mandate to boost the local economy. International development agencies have subsidized such efforts.

Japan AID has given Micronesian nations subsidies to build up the potential for harvesting coral-reef resources on some years in which Japan has been granted access to the pelagic fisheries in the Economic Exclusion Zone of these nations. In one year, Japan AID provided 70 fishing boats with outboard motors, fuel tanks, and fishing gear to the small island of Kosrae (Federated States of Micronesia). The Marine Resources Division of the Kosrae State Government advised against accepting the boats and equipment on the basis that the fish populations on the small fringing reef could not sustain this much fishing pressure. But the acquisition of free boats and equipment was a conspicuous and potential vote-getting accomplishment for the governor's office. The stocks of fishes were seriously depleted within a few months after the additional boats were put to use, and the reefs around Kosrae continued to remain depleted of large fishes, at least for a decade through 1991.

Coral-reef fisheries are also naturally subsidized by the diverse assemblage of the fishes themselves. The more highly prized species can be biologically overfished while the alternative species continue support of the fishery. The more prized species continue to be taken occasionally as they are encountered, no matter how rare they become, while the commercial fisheries are sustained by alternative, less highly priced species of fishes. A moratorium on fishing that would allow the recovery of the more prized fish populations is prevented by the fisheries operations being continuously active. Records from the marine resources agencies of Guam and American Samoa indicate that there is a shift in species prevalent in the catches before there is a drastic drop in CPUE.

18.4.3. "Ratchet Effect" from Population Growth

In developed countries, a rich harvest in good years encourages additional investment in equipment in order to increase the volume of catch for increased profit. The commitments and investments during rich years compel the fishermen to continue to extract fishes at the same level during poor years in order to maintain payments on their investments and to avoid layoffs of employees (Ludwig et al., 1993). The biological resource populations do not have a chance to rebound when intense exploitation is continued while the reproductive stock is low. This is called the *ratchet effect* (Ludwig et al., 1993).

An analogous ratchet effect can be seen with human population growth and urbanization in less developed countries. This is also a positive-feedback process. As the populations of fishermen increase through growing numbers of people and/or decreasing availability of jobs, the populations of fishes become severely overharvested. Fishermen might be aware that they will catch larger fish with less effort if they restrain from fishing for a year or two. But they have families to feed and they cannot wait a year. In desperate, last-measure attempts to obtain enough smaller fish to feed their families, the fishermen turn to methods such as dynamite fishing and muro-ami that are destructive to the habitat and to the recruits of the fish population: the smaller the fishes, the greater the numbers that need to be caught to feed the families, and the less the chance for the fish population to recover.

18.5. Village-Based Control

In regions with human populations small enough to be sustained directly by reef resources, traditional fishery rights often involve limited entry (Johannes 1978, 1981). When resource use is controlled by a village, clan, or chief, then it is usually in the best interest of those in control to manage the resources sustainably (Chapter 16). The authority to control the resource gives hope for the future, and this provides the motivation to manage the resource wisely. Protective management by a local village has been demonstrated to substantially increase both yield and CPUE (Alcala and Russ, 1990).

In contrast to local village management, in situations where there is open access and the resources can be sold, the fisherman is motivated to catch all he can. Even though he is probably aware that it is theoretically in his long-term best interests to leave some stock for reproduction, the catch he leaves will likely be taken by someone else, so he should take it now. This philosophy is called the tragedy of the commons (Hardin, 1968; Hardin and Badin, 1977).

As resources near the continental waters of Asia became depleted, numerous large ships extended their range of operations farther east among the oceanic islands (Johannes and Riepen, 1995). The costs of large-scale operations, large ships, and long-distance transportation prompts entrepreneurs to develop a higher profit margin by organizing a one-time "clean sweep" rather than a sustainable fishery. It is in the best interest of the large-scale businesses to collect entire spawning aggregations of targeted fishes such as groupers. This is analogous to clear-cutting forests for a higher immediate profit.

In contrast, for local residents involved in subsistence and small-scale commercial fisheries, it is beneficial to develop sustainable fisheries. However, even if the village has legal rights to close or regulate the harvest of reef resources by outsiders, it may not have the power to effectively enforce these rights against the persuasive powers of large amounts of cash from Asia, especially if the negotiations for access to local resources are made with higher government officials. It is essential that local governments give the villagers training, support, and explicit legal authority to manage their resources and regulate access to their resources. This will increase the proportion of the profits retained in the country, reduce the use of destructive fishing techniques, and increase the incentive for sustainability as a goal of management (Johannes and Riepen, 1995).

Another contemporary process that is undermining traditional management practices is immigration of foreign cultures. Alien workers occasionally run into conflict because of incompatibility with islanders in terms of harvesting procedures. Palauans traditionally spread out their search for food over a wide area in order to avoid concentrating the pressure on reef populations and harvesting any one area too severely. They also occasionally pick up organisms such as giant clams and bring them back to the reef flat near their homes where they are kept as a nearby supply in case of emergency, for example, when stormy seas prevent them from ranging far. Whether it is difference in culture, a lack of understanding of Palauan property rights, or a lack of access to transportation by boat to reefs farther offshore, conflicts occasionally arise because immigrant labors tend to thoroughly glean the nearby reefs.

Western ideals sometimes find local traditions of limited access to reefs unjust and discriminatory, but the biological realities of the coral-reef systems indicate that reef communities are not the cornucopias that they appeared to be and that regulation of harvests is probably necessary for sustainability. The population of Yap is now about 7,000; in the past it was about 40,000. It is thought that the traditional regulations, taboos, and restrictions of access and times of harvest had the effect of guarding against the resources being severely overharvested and thereby allowed a larger population of humans to be supported.

Rapid urbanization may rend the social fabric. Some of the greatest rates of human population growth in the world are in Oceania. The Marshalls are growing at 4.2% per annum, Federated States of Micronesia at 3.6%, Solomon Islands at 3.5%, and French Polynesia at 3.0%. In most nations of Oceania, the rates of urbanization are substantially greater than rates of population growth (Marshalls 5.7%, Solomon Islands 5.3%, Tuvalu 4.8%, and Vanuatu 4.5% [Bryant, 1994]). It is probably difficult for an elite few to retain traditional authority over marine resources near areas of urbanization where there are many immigrants and insufficient employment opportunities, and cash is needed to acquire attractive new consumer goods.

As human populations grow beyond the carrying capacity of coral reefs, as the dollar value of reef resources overrides management, tradition, and law, and as urbanization and immigration of people change the culture from subsistence to participation in the world economy, compelling and lucrative alternative sources of income must be found for the majority of fishermen living near coral reefs. The life-history characteristics of coral-reef organisms and the relatively closed nature of the ecosystem make it unrealistic to expect reef resources to continue to meet increased demands. Radically new approaches to resource management have been developed by Palau and Bermuda for sustainable income from coral reefs. Their nonexportive approaches are compatible with the attributes of coral-reef ecosystems and the demands of the growing pressures of human activities.

18.6. Exemplary Cases: Bermuda and Palau

There was a widespread concern in Bermuda that the coral reefs were deteriorating because of commercial fishing, even though the fisheries economy and number of people employed in extractive fisheries was small compared with the number employed by tourism. Tourism and recreation grossed more than $9 million in Bermuda in 1988. The hotel owners, charter-boat fishermen, dive- and tour-boat operators, and other businesses felt the incomes of many were threatened by the activities of a few fishermen. In 1990, considering evidence that the catches were declining substantially and the reef communities were deteriorating, the Government of Bermuda offered payment of up to $75,000 per fishermen, in addition to compensating them for hardware such as pots and winches, in order to close the coral reefs of Bermuda to pot fishing (Butler et al., 1993). This was not popular with commercial fishermen and was considered by some to be "draconian." But with the establishment of a number of offshore fish-aggregation devices, some fishermen found alternative livelihoods with recreational charter fishing, offshore longlining for pelagic fishes, and deep-water trapping for crabs and prawns. Unlike fishing on coral reefs, which is naturally subsidized by alternative species when the preferred species become scarce (section 18.4.2), the alternative fisheries each targeted certain types of fish and therefore were relatively self-limiting, requiring healthy stocks in order to have a viable fishery.

Japan AID contributed 11 diesel fishing boats and other fishing gear to Palau to promote the exportation of reef fishes. After a year, thousands of dollars were being exchanged in the marketing of reef fishes, but after boat operation expenses, as well as marketing and processing the fish, the fishermen were netting only $0.50 per fish. In the export business, fishermen would have to sell about 5 kg of reef fish to purchase 0.2 kg of canned tuna at the supermarket. It was clear that exporting fishes was far less profitable but in competition with recreational fishing, tourism, and selling fishes to local restaurants and hotels. It was also in competition with subsistence fishing.

Diving activities by tourists brought in over $12 million to Palau in 1992 (not counting additional expenditures such as hotels, meals, and transportation). From diving tours on only a few hundred meters of reef at Blue Corner and nearby Nemelis Dropoff alone, Palau earns about $7,000 per day on the average, and up to $14,000 per day in peak season (C. Cook, The Nature Conservancy, Palau Office, pers. comm.). Extractive fishing from a small area of reef would not be this profitable or sustainable. The section of reef called Blue Corner and its fish population seem exemplary for coral-reef-exploited resources in that they have "improved" in money-making quality over recent years. The fishes may have become accustomed to large numbers of human observers. Even dog-tooth tuna, humphead wrasse, and gray reef sharks will now come close to photographers at Blue Corner.

In view of the more sustainable economic use of resources in subsistence and tourism than in export of biomass, the Government of Palau has passed legislation called the Marine Protection Act of 1994 to gradually phase out the export of fishes. "The purpose of this act is to promote sustainability and develop marine resources of the republic while also preserving the livelihood of the commercial fishermen of the republic." At this time, the commercial export from Palau of fishes is illegal from March to July, which are the spawning seasons for most species. The local fishermen are encouraged to switch completely to other forms of endeavor, for example, chartering for sport fishing or selling fish locally to hotels and restaurants. The money stays in Palau when the fish are sold locally, the fisherman can make a lot more money per fish by cutting the costs of shipping and middlemen, and the quality of fish in the local hotels and restaurants is greater when there is no competition with the export market.

It has been said that "Once a fish is dead, it doesn't matter whether it is used for subsistence or commercial export. Biologically, there is no difference between subsistence and export fisheries." But it takes at least an order of magnitude of more fish to support a few commercial fishermen than it does to support a population by subsistence because most of the proceeds from the sale of fishes go to supporting the marketing and transportation infrastructure. Coral-reef resources can sustain a much greater number of people or careers in the world economy if used in nonextractive enterprises. The nonexportive, service-oriented approaches of Bermuda and Palau are compatible with the attributes of coral-reef ecosystems and species, and with the demands of the growing human populations, growing economies, and urbanization.

18.7. Nonextractive Commercial Uses of Coral-Reef Resources

As populations of humans grow beyond the carrying capacity of the local coral reefs, and when the society wishes to participate in the world economy, attractive and lucrative alternative sources of income must be found for the majority of fishermen living near coral reefs. Alternatives that have potential for some locations are tourism and aquaculture. Just as coral reefs can support a large biomass, diversity, and energy flow with relatively little exported materials from the natural system, the numbers of jobs, financial support for infrastructure, and financial flow are much more substantial to local economies if the coral reef is exploited by nonextractive methods (section 8.6 in Chapter 8).

There is far more potential in the tropical coastal communities for tourism than for marine fisheries. The money exchanged in world tourism in 1992 was $1.9 trillion, over 27 times the $70 billion of the world marine fisheries revenue

(Weber, 1993). To participate in the world economy, the islands of Oceania can be highly competitive for the high spenders in tourism, while geography (transportation costs), ecology (low yield, effects of overfishing on ecosystem processes), and biology (life-history characteristics of the species) work against commercial export fisheries. Furthermore, the infrastructure for commercial fisheries (processing, refrigeration, transportation) is more expensive and risky than the commitment to ecotourism. The cost of airfares is paid by the tourists themselves, in contrast to the subtraction of air transport costs from the profit on exported fish sales. For example, charter-boat sport fishing of offshore pelagics brings in considerably more money per fish on Guam than direct sale of the fish itself. Some commercial fishermen on Guam feel their destiny is to be fishermen, but they cannot make a living by extractive fishing because of the costs of operation and marketing. In order to be fishermen, they have to subsidize their commercial fishing endeavors by chartering for sport fishing or tourism.

Of course, tourism can also be damaging to coral reefs if not properly managed. Tourist facilities should not allow pollution from sewage, detergents, excessive freshwater, sediments, or garbage. The constuction of piers and other structures should be done in a manner that affects the current and sediment patterns as little as possible (Carpenter and Maragos, 1989). Mooring buoys and drift dives are preferable dive-boat logistics to anchoring because of the damage anchors do to corals. Dive resorts and hotels generally distribute guidelines for proper diving practices, such as buoyancy control rather than standing on corals, and photography rather than collecting specimens for souvenirs. Resorts should keep the number of divers at a site within appropriate limits.

Even though tourism can damage a coral reef without proper management, tourism is not as intrinsically adverse to the characteristics of the coral-reef ecosystem and to the natural histories of the coral-reef species as is commercial extraction of biomass. Tourism can be sustainable if properly managed.

Another source of money that does not require chronic extraction of biological resources directly from coral-reef populations is aquaculture. Aquaculture is a possible alternative to extraction of species for food, for the aquarium trade, and for pharmaceuticals. Although coral-reef species are characteristically vulnerable to overexploitation in the reef community, some species are adaptable to culturing in aquaria. Larvae and juveniles of some species can live in cement or plastic tanks in dense concentrations, they do not require feeding after the first month, and they grow relatively fast. Giant clams are an example.

In the natural environment of the reef, recruits to populations of giant clams run a gamut of predators and competitors. Adult tridacnids can spawn hundreds of millions of eggs, but the probability of any individual surviving to reproduction in the complex coral-reef community is on the order of 10^{-9}. Few small tridacnids survive when transplanted to the reef community, apparently because of predation. As evidenced by the local extinctions of the larger species of tridacnids, the population on the coral reef cannot support the added pressure of commercial exploitation. The larger species of giant clams (*Tridacna gigas* and *T. derasa*) are on the list of threatened species of the International Union for the Conservation of Nature (IUCN).

Although *Tridacna derasa, T. gigas,* and *Hippopus hippopus* cannot be commercially harvested sustainably from coral reefs, a few individuals may be used as a broodstock for a commercial enterprise in the realm of aquaculture. The populations in aquaculture are protected from natural predators and so they produce a tremendously greater yield than populations in nature, and with no known harm to the environment. The Micronesian Mariculture Demonstration Center (MMDC) in Palau has produced giant clams as a successful venture for 15 years (1989–1994) under the direction of Gerald Heslinga. On a 1-hectare subtidal reef farm, MMDC maintained a standing stock of over 100 tons of cultured clam biomass, while donating about one ton of clams per month to local conservation programs, and exporting sufficient quantities to cover the operating costs and salaries of the hatchery staff. The hatchery raised over 20,000 tridacnids to sexual maturity and these were used as breeding stock for susequent production of several hundred thousand F2 and F3 giant clam offspring each year. MMDC was consistently able to ship 50 kg per week (200 kg per month) for a year to Japan to fulfill a contract for adductor muscles and mantle meat (without viscera or gills), and this was only a fraction of the biomass exported that year. Cultured clam products were successfully marketed in the sashimi trade in Japan, Saipan, and Palau, in the aquarium trade in the United States and Europe, and in the local handicraft trade.

The natural populations of giant clams are not able to meet similar quotas over extended periods of time because of the high rates of mortality of juveniles. The Marine Protection Act of 1994 of Palau prohibits the commercial export of any giant clam (tridacnid) caught in the wild, but allows for the export of "cultured" individuals (raised in aquaculture). The law applies the same rules to other marine animals, including many species of holothuroids, crustaceans, and fishes in Palau.

Pearl culture from broodstocks of *Pinctada margaritifera* and *P. maxima* is bringing over $60 million annually to French Polynesia and other Pacific islands. The exporting of pearls does not cause stress to coral-reef ecosystems because $CaCO_3$ is in rich supply compared to the mass of pearls that could ever be exported.

Major factors in the failure of natural populations to recover from overharvesting are often irregular recruitment and intense predation on recruits. These problems can be overcome for some species with aquaculture, but the establishment of an aquaculture facility must be carefully planned and managed if it is to be sustainable and if it is not to damage the surrounding habitats and livelihoods. The developers of aquaculture must not focus on the species being raised or their specific enterprise, but rather they must keep the entire system in perspective, including the economics of the region, the social acceptability of the enterprise, the availability of a nearby airport, the accessibility of the market, and the general availability of appropriate space and water. Establishing the aquaculture facility must not involve the removal of mangroves or other abuses to the coastal ecosystem. It must not cause red tides or anaerobic conditions in surrounding waters by eutrophication, or introduce alien species, or draw too heavily on the local supply of freshwater. There is danger in shipping live animals for broodstock from facility to facility because the animals may carry diseases, parasites, or predators (Chapter 6). Doty (1983) advances "coral-reef diversified farming," which alters the natural habitat only minimally and requires manual effort rather than capital investment.

The marine aquarium trade for coral-reef organisms involves hundreds of millions of dollars in the United States alone. When collecting small fishes from the reef is not carefully regulated, it can put a stress on particular populations of prized species. But there is a major drive to develop aquaculture of corals, soft corals, and coral-reef fishes for the aquarium trade. The Pacific Aquaculture Development Program of the Pacific Aquaculture Association has put priority on giant clam, pearl oyster, trochus, marine shrimp, sponges, algae, rabbitfish, mullet, grouper, and aquarium species.

Because the biodiversity on coral reefs is especially strong at the phyletic level (Chapter 14), the genetic diversity on coral reefs may have the greatest potential for new pharmaceuticals. It is unlikely, however, that extraction of pharmaceuticals will pose a threat to coral reefs. Biodiversity partially protects the reef community from economically feasible harvest for pharmaceuticals. This is because the most interesting chemicals are usually species specific, and the species may be rare or patchily distributed, and the natural production of the chemical may vary in space and time. Once the chemical structures are clinically tested and found to be useful, pharmaceutical companies find it more economical to synthesize the compounds. If synthesis is especially difficult or economically impractical, the aquaculture of source organisms is usually more reliable than attempting to harvest them from the reef. For example, bryostatin from a bryozoan species of *Bugula* has undergone preclinical development, but cannot be synthesized yet. Therefore, the potential for aquaculture of *Bugula* is being investigated.

18.8. Integrative Coastal-Area Management

Proper land management is a "win-win" endeavor. To control excessive sedimentation and nutrient input is of benefit to habitats and commercial enterprises both at the source and at the destination, both on the land and on the coral reef. Soil and nutrients are valuable to the farms and other terrestrial enterprises, and sediment and nutrients are damaging to fisheries, tourism, and other coral-reef enterprises (Chapter 17).

The Great Barrier Reef (GBR) is 40–80 km offshore of the Queensland coast, but the chairperson of the Great Barrier Reef Marine Park Authority (GBRMPA) nevertheless considers protection of the GBR from increasing nutrient levels to be the greatest challenge facing the GBRMPA in the next two decades (Kelleher, 1990). Approximately a quarter of the research budget of the GBRMPA is committed to working with farmers on nonwasteful methods of irrigation and application of fertilizer.

Sedimentation from incompetent land management is the form of pollution most harmful to coral reefs (Johannes, 1975; Hatcher et al., 1989; Rogers, 1990). The complexity and importance of the influences of neighboring terrestrial and marine habitats on each other (Chapter 13) make development of a science of integrated coastal-area management most urgent. Salvat (1987), Maragos (1992, 1993), and Carpenter and Maragos (1989) provide guidelines for engineering procedures that mitigate the harmful effects of coastal construction projects.

A major block to development of an operative integrated coastal-area management program is that administrative law often demands precisely defined boundaries, usually near the shoreline. But natural law and ecological processes operate through distances that vary in space and time. For example, the Aswan Dam has a tremendous influence on the coastal ecosystems of the southeastern Mediterranean, so it must be included in a coastal-area management plan, although it lies about 900 km inland from the seashore. From the legal perspective, natural processes must adhere to stationary boundaries that are well defined on maps. A challenge to the legal profession is to develop an operational legal system based on definitions of critical processes rather than stationary boundaries. A guidebook to contemporary multidisciplinary issues of planning and management of marine environments and resources is provided by Kenchington (1990).

18.9. Conclusion

The guidebook information suggests that we need to develop a new paradigm for the exploitation of coral reefs, a new perspective that might also be a framework for our management of the Earth as a whole as humans become overcrowded and resources become scarce. In many situations, the coral-reef ecosystem is characterized by high gross primary productivity but low net system yield (Figs. 7–6 and 18–1). It is often considered remarkable that coral reefs in regions of low nutrient input, such as on atolls in the Pacific gyre, are able to support such high biomass and highly diverse communities of organisms.

Populations of coral-reef organisms are most often characterized by irregular recruitment (Chapter 8) and coral-reef ecosystems are relatively closed in terms of nutrient recycling (Chapter 7). These traits make coral-reef ecosystems fundamentally different than systems driven by nutrient input from external sources such as regions of upwelling, estuaries, floodplains, or river deltas (Chapter 12). Coral reefs do occur in areas of nutrient input, such as in lagoons around high islands and near continents, and there may be some export from these reef systems. But the coral-reef organisms targeted for extraction, such as giant clams and groupers, are nevertheless adapted to the complex communities under oligotrophic conditions (section 12.2 in Chapter 12) and are vulnerable to overharvest.

Physical oceanographic processes drive systems based on major currents and upwelling while species interactions have a major influence on coral-reef ecosystems (Birkeland, 1987, 1988b; Sherman, 1994; Roberts, 1995). The new paradigm for coral-reef management is that it may be more profitable and sustainable to make nonextractive use of coral-reef resources. Rather than have maximum sustainable yield as the major objective for all fisheries, we should focus on harvesting biomass mainly from habitats driven by nutrient input from outside sources.

Marine communities can have functions other than biomass production that are useful for humans (section 1.1 in Chapter 1). Coral reefs function as protection against waves for other coastal habitats, sources of food for pelagic fishes, sources of genetic diversity for aquaculture and pharmaceuticals, and sources of income from tourism and recreation. The general reaction to this proposal for nonextractive exploitation of reefs has been one of pessimism, because it is assumed that human nature does not allow people to broaden their short-term perspective and use resources rationally for the long-term profit. It has been suggested that it is necessary to allow fisheries to be overfished because people will not take advice on management during times of plenty, while the reefs are still perceived as a cornucopia. This approach must be used cautiously in view of the fact that some populations may never recover.

Humans can manage farms by tilling the soil and planting, forests by reseeding and selective logging, wildlife by stocking and culling, and small lakes by stocking of brood stock and control of water quality. We speak of managing coral reefs, but the more we learn about the coral-reef ecosystem, the planktonic nature of population replenishment, the recycling of nutrients at all levels in the system, the complexity of interactions between species and between habitats, the more we realize that we cannot effectively control populations of reef organisms in the foreseeable future. We must manage human activities that affect the reefs and adjust our methods of exploitation rather than manage the reefs themselves.

The management of human activities through regulations imposed by central administration, whether based on quotas or on fishing effort, has not been working (section 17.5 in Chapter 17). Regulations require enforcement, and this is impractical along the extensive coastal regions of the tropics. A stronger approach is to have governments compel self-control by local villages, clans, or families through retracting open access and legalizing limited access (Chapter 16). Placing the management of resources back into the domain of local residents strengthens the motivation by those harvesting the resources to keep the future of the resources secure and makes enforcement practical.

The primary recommendations of this book are that (1) in order to develop an economy based on ecological sustainability, the commercial use of coral-reef resources should be generally nonextractive (although moderate subsistence or recreational use of coral-reef resources can be sustainable if carefully managed), exploiting instead the ecological services of coral reefs through tourism and genetic information of the reef for production of foods and pharmaceuticals through aquaculture; and that (2) management plans for coastal areas must be integrative, flexible, and based on environmental processes and habitat characteristics. Irregular recruitment and complex interactions among the diverse natural components of the coral-reef ecosystem require that management strategies must include economic, social, and political aspects of the human components as well as the ecological factors.

Pacific islanders who depended on the resources of coral reefs were able to recognize the limits of their domain and their resources because they lived on small landmasses surrounded by finite reefs (Johannes, 1977). They often developed territorial jurisdictions and traditional harvest methods that promoted sustainability. They "discovered the cornerstone of sound fisheries management . . . centuries before any form of marine fisheries management was seriously considered in the west" (Johannes, 1977). These management customs are being undermined by objectives accompanying westernization, that is, free equalitarian access and optimal biomass export for monetary gain. This is happening in the places and times where and when it is most inappropriate.

A greater number of people and diversity of careers can be supported sustainably by coral reefs if economies are based on the natural attributes of the species and ecosystem, that is, recycling and low amounts of export. The rational way to exploit the coral-reef ecosystem is with subsistence fishing for food by resident human populations of moderate and sustainable densities, and with the monetary economy based on nonextractive enterprises such as tourism and aquaculture. In Bermuda and Palau, the local leaders have been promoting the support of more jobs, larger revenue, and a good standard of living by developing an economy based on nonextractive (nonexport) uses of coral-reef resources, with the fisheries yield for subsistence, recreation, and selling to local hotels and restaurants.

At this time, as the human population is approaching, if not exceeding, the carrying capacity of the world, we might consider the coral-reef system as a model for the world, with an economy based on quality of life and sustainability, with a strong role of mutualistic symbioses, rather than an economy based on perpetual growth, export, and colonization of new regions of abundant resources.

References Cited

Abbott, R.T., and S.P. Dance. 1982. Compendium of seashells. R.P. Dutton, New York.

Abe, N. 1937. Postlarval development of the coral *Fungia actiniformis* var. *palawensis* Doderlein. Palao Trop. Biol. Stn. Stud. 1: 73–93.

Abele, L.G. 1972. Comparative habitat diversity and faunal relationships between the Pacific and Caribbean Panamanian decapod Crustacea: a preliminary report, with some remarks on the crustacean fauna of Panamá. Bull. Biol. Soc. Wash. 2: 125–138.

Abele, L.G. 1974. Species diversity of decapod crustaceans in marine habitats. Ecology 55: 156–161.

Abrams, P.A. 1984. Foraging time optimization and interactions in food webs. Am. Nat. 124: 80–96.

Abrams, P.A. 1987. Indirect interactions between species that share a predator: varieties of indirect effects. *In* W.C. Kerfoot and A. Sih (eds.), Predation: direct and indirect impacts on aquatic communities, pages 38–54. United Press of New England, Hanover, New Hampshire, 386 p.

Abrams, P.A. 1992. Predators that benefit prey and prey that harm predators: unusual effects of interacting foraging adaptations. Am. Nat. 140: 573–600.

Abrams, R.W., M.D. Abrams, and M.W. Schein. 1983. Diurnal observations on the behavioral ecology of *Gymnothorax moringa* (Cuvier) and *Muraena miliaris* (Kaup) on a Caribbean coral reef. Coral Reefs 1: 185–192.

Adams, C.A., M.J. Oesterling, S.C. Snedaker, and W. Seaman. 1973. Quantitative dietary analyses for selected dominant fishes of the ten thousand islands, Florida. *In* S.C. Snedaker and A.E. Lugo (eds.), The role of mangrove ecosystems in the maintenance of environmental quality and a high productivity of desirable fisheries, pages H-1–H-56. Final Report submitted to the Bureau of Sport Fisheries and Wildlife.

Adams, C.G., J. Butterlin, and B. Samanta. 1986. Larger foraminifera and events at the Eocene–Oligocene boundary in the Indo-West Pacific region. *In* C. Pomerol and I. Premoli Silva (eds.), Terminal Eocene events, pages 237–252. Elsevier, Amsterdam.

Adams, C.G., A.W. Genry, and P.J. Whybrow. 1983. Dating the terminal Tethyan event. Utrecht Micropaleontological Bull. 30: 273–298.

Adams, C.G., D.E. Lee, and B.R. Rosen. 1990. Conflicting isotopic and biotic evidence for tropical sea-surface temperatures during the Tertiary. Palaeogeography, Palaeoclimatology, Palaeoecology 77: 289–313.

Adams, S.M. (ed.). 1990. Biological indicators of stress in fish. Amer. Fish. Symp. 8. Bethesda, Maryland.

Adey, W.H. 1978. Coral reef morphogenesis: a multidimensional model. Science 202: 831–837.

Adey, W.H., and R. Burke. 1976. Holocene bioherms (algal ridges and bank-barrier reefs) of the eastern Caribbean. Geol. Soc. Amer. Bull. 87: 95–109.

Adey, W.H., and R.B. Burke. 1977. Holocene bioherms of the Lesser Antilles—geologic control of development. *In* S.H. Frost, M.P. Weiss, and J.B. Saunders (eds.), Reefs and related carbonates—ecology and sedimentology, pages 67–82. Amer. Assoc. Petr. Geol., Studies in Geology 4.

Adey, W.H., I.G. Macintyre, R. Stuckenrath, and R.F. Dill. 1977. Relict barrier reef system off St. Croix: its implications with respect to late Cenezoic coral reef development in the western Atlantic. Proc. 3rd Internat. Coral Reef Symp., Miami 2. Geology: 15–21.

Adey, W.H., and R.S. Steneck. 1985. Highly productive eastern Caribbean reefs: synergistic effects of biological, chemical, physical, and geological factors. *In* M.L. Reaka (ed.), The ecology of coral reefs, pages 163–187. NOAA Symp. Ser. Undersea Res. 3(1), 208 p.

Aeby, G.S. 1991. Behavioral and ecological relationships of a parasite and its hosts within a coral reef system. Pac. Sci. 45: 263–269.

Agassiz, L. 1852. Coral reefs. *In* Annual Report of the Superintendent of the Coast Survey, showing the progress of that work during the year ending November, 1851, pages 153–154. 32d Congr., 1st Sess. Senate. Ex. Doc. No.3, R. Armstrong, Printer.

Aharon, P., S.L. Goldstein, C.W. Wheeler, and G. Jacobson. 1993. Sea-level events in the South Pacific linked with the Messinian salinity crisis. Geology 21: 771–775.

Ahmadjian, V., and S. Paracer. 1986. Symbiosis: an introduction to biological associations. University Press of New England, Hanover, New Hampshire.

Ahr, W.M., and R.J. Stanton, Jr. 1973. The sedimentologic and paleoecologic significance of *Lithotrya,* a rock-boring barnacle. J. Sed. Petrol. 43: 20–23.

Alabaster, J.S., and R. Lloyd. 1984. Water quality criteria for freshwater fisheries. Butterworths, London.

Alcala, A.C., and T. Luchavez. 1982. Fish yields of the coral reef surrounding Apo Island, Negros Occidental, Central Visayas, Philippines. Proc. 4th Internat. Coral Reef Symp., Manila 1: 69–73.

Alcala, A.C., and G.R. Russ. 1990. A direct test of the effects of protective management on abundance and yield of tropical marine resources. J. Conserv. Internat. Explor. Mer 46: 40–47.

Alder, J., and R. Braley. 1989. Serious mortality in populations of giant clams on reefs surrounding Lizard Island, Great Barrier Reef. Austr. J. Mar. Freshw. Res. 40: 205–213.

Aleem, A.A. 1972. Effect of river outflow management on marine life. Mar. Biol. 15: 200–208.

Alino, P.M., P.W. Sammarco, and J.C. Coll. 1992. Competitive strategies in soft corals (Coelenterata, Octocorallia). IV. Environmentally induced reversals in competitive superiority. Mar. Ecol. Progr. Ser. 81: 129–145.

Allen, G.R. 1979, 1981. Butterfly and angelfishes of the world. Mergus Publishers, Hans A. Baensch, Melle, Germany.

Allen, G.R. 1991. Damselfishes of the world. Mergus Publishers, Hans A. Baensch, Melle, Germany.

Allen, G.R., and D.R. Robertson. 1994. Fishes of the tropical eastern Pacific. University of Hawaii Press, Honolulu.

Allen, G.R., and R. Steene. 1994. Indo-Pacific coral reef field guide. Tropical Reef Research, Singapore.

Allmon, W.D., G. Rosenberg, R.W. Portell, and K.S. Schindler. 1993. Diversity of Atlantic coastal plain mollusks since the Pliocene. Science 260: 1626–1629.

Alongi, D.M. 1988. Detritus in coral reef ecosystems: fluxes and fates. Proc. 6th Internat. Coral Reef Symp., Townsville 1: 29–36.

Al-Sofyani, A., and P. Spencer Davies. 1993. Seasonal variation in production and respiration of Red Sea Corals. Proc. 7th Internat. Coral Reef Symp., Guam 1: 351–357.

Alvarez, L.W., W. Alvarez, F. Asaro, and H.V. Michel. 1980. Extraterrestrial cause for the Cretaceous–Tertiary extinction. Science 208: 1095–1108.

Alvarez, W., J. Smit, W. Lowrie, F. Asaro, S.V. Margolis, P. Claeys, M. Kastner, and A.R. Hildebrand. 1992. Proximal impact deposits at the Cretaceous–Tertiary boundary in the Gulf of Mexico: a restudy of DSDP Leg 77 Sites 536 and 540. Geology 20: 697–700.

Anderson, R.M. 1986. Genetic variability in resistance to parasitic invasion: population implications for invertebrate host species. Symp. Zool. Soc. London 56: 239–274.

Andres, N.G., and J.D. Witman. 1995. Trends in community structure on a Jamaican reef. Mar. Ecol. Progr. Ser. 118: 305–310.

Andrews, J.C. 1979. Pathology of seaweeds: current status and future prospects. Experientia 35: 429–450.

Andrews, J.C., and H. Muller. 1983. Space–time variability of nutrients in a lagoonal patch reef. Limnol. Oceanogr. 28: 215–227.

Andrews, J.C., and G.L. Pickard. 1990. The physical oceanography of coral-reef systems. *In* Z. Dubinsky (ed.), Ecosystems of the world, vol. 25, Coral reefs, pages 11–48. Elsevier Science Publishing, New York, 550 p.

Antonius, A. 1985. Coral diseases in the Indo-Pacific: a first record. P.S.Z.N.I. Mar. Ecol. 6: 197–218.

Atkinson, M.J. 1988. Are coral reefs nutrient limited? Proc. 6th Internat. Coral Reef Symp., Townsville 1: 157–166.

Atkinson, M.J. 1992. Productivity of Eniwetak Atoll reef flats predicted from mass transfer relationships. Cont. Shelf Res. 12: 799–807.

Atkinson, M.J., and R.W. Grigg. 1984. Model of a coral reef ecosystem II. Gross and net primary production at French Frigate Shoals, Hawaii. Coral Reefs 3: 13–22.

Atkinson, M.J., and S.V. Smith. 1983. C:N:P ratios of benthic marine plants. Limnol. Oceanogr. 28: 568–574.

Atkinson, S., and M.J. Atkinson 1992. Detection of estradiol-17β during a mass coral spawn. Coral Reefs 11: 33–35.

Atoda, K. 1947a. The larva and postlarval development of some reef-building corals. I. *Pocillopora damicornis* (Dana). Sci. Rep. Tohoku Univ. 18: 24–47.

Atoda, K. 1947b. The larva and postlarval development of some reef-building corals. II. *Stylophora pistillata* (Esper). Sci. Rep. Tohoku Univ. 18: 48–64.

Atoda, K. 1951a. The larva and postlarval development of some reef-building corals. III. *Acropora brueggemanni* (Brook). J. Morph. 89: 1–13.

Atoda, K. 1951b. The larva and postlarval development of some reef-building corals. IV. *Galaxea aspera* (Quelch). J. Morph. 89: 17–35.

Atoda, K. 1951c. The larva and postlarval development of some reef-building corals. V. *Seriatopora hystrix* (Dana). Sci. Rep. Tohoku Univ. 19: 33–39.

Axelrod, D.I. 1992. What is an equable climate? Palaeogeography, Palaeoclimatology, Palaeoecology 91: 1–12.

Ayling, A.M., and A.L. Ayling. 1987. Ningaloo Marine Park: preliminary fish density assessment and habitat survey with information on coral damage due to *Drupella cornus* grazing. Department of Conservation and Land Management, Western Australia.

Ayling, A.M., and A.L. Ayling. 1992. Preliminary information on the effects of *Drupella* spp. grazing on the Great Barrier Reef. *In* S. Turner (ed.), *Drupella cornus:* a synopsis, pages 37–42. CALM Occasional Paper 3/92, Department of Conservation and Land Management, Western Australia.

Babcock, R.C. 1984. Reproduction and distribution of two species of *Goniastrea* (Scleractinia) from the Great Barrier Reef province. Coral Reefs 2: 187–204.

Babcock, R.C. 1985. Growth and mortality in juvenile corals (*Goniastrea, Platygyra* and *Acropora*): the first year. Proc. 5th Internat. Coral Reef Congr., Tahiti 4: 355–360.

Babcock, R.C. 1988a. Fine-scale spatial and temporal patterns in coral settlement. Proc. 6th Internat. Coral Reef Symp., Townsville 2: 635–639.

Babcock, R.C. 1988b. Age-structure, survivorship, and fecundity in populations of massive corals. Proc. 6th Internat. Coral Reef Symp., Townsville 2: 625–633.

Babcock, R.C. 1990. Reproduction and development of the blue coral *Heliopora coerulea* (Alcyonaria: Coenothecalia). Mar. Biol. 104: 475–481.

Babcock, R.C., G.D. Bull, P.L. Harrison, A.J. Heyward, J.K. Oliver, C.C. Wallace, and B.L. Willis. 1986. Synchronous spawning of 105 scleractinian coral species on the Great Barrier Reef. Mar. Biol. 90: 379–394.

Bak, H.P., and S. Nojima. 1980. Immigration of a tropical sea urchin, *Astropyga radiata*

(Leske) in a temperate eel grass, *Zostera marina* L., patch: its feeding habit and grazing effect on the patch. Publ. Amakusa Mar. Biol. Lab. 5: 153–169.

Bak, R.P.M. 1985. Recruitment patterns and mass mortalities in the sea urchin *Diadema antillarum.* Proc. 5th Internat. Coral Reef Congr. Tahiti 5:267–272.

Bak, R.P.M. 1990. Patterns of echinoid bioerosion in two Pacific coral reef lagoons. Mar. Ecol. Progr. Ser. 66: 267–272.

Bak, R.P.M., and M.S. Engel. 1979. Distribution, abundance, and survival of juvenile hermatypic corals (Scleractinia) and the importance of life history stages in the parent coral community. Mar. Biol. 54: 341–352.

Bak, R.P.M., and G. van Eys. 1975. Predation of the sea urchin *Diadema antillarum* Philippi on living coral. Oecologia 20: 111–115.

Balazs, G.H., and S.G. Pooley (eds.). 1991. Research plan for marine turtle fibropapilloma. Results of a December 1990 workshop, Nat. Mar. Fish. Serv. Tech. Memorandum, NOAA-TM-NMFS-SWFSC-156. U.S. Dept. Commerce, NOAA.

Bannerot, S.P., W.W. Fox, Jr., and J.E. Powers. 1987. Reproductive strategies and the management of snappers and groupers in the Gulf of Mexico and Caribbean. *In* J.J. Polovina and S. Ralston (eds.), Tropical snappers and groupers: biology and fisheries management, pages 561–603. Westview, Boulder, Colorado.

Barbier, E.B. 1987. The concept of sustainable economic development. Env. Conserv. 14 (2): 101–110.

Bardach, J.E. 1959. The summer standing crop of fish on a shallow Bermuda reef. Limnol. Oceanogr. 4: 77–85.

Bardach, J.E. 1961. Transport of calcareous fragments by reef fishes. Science 133: 98–99.

Barlow, G.W. 1974. Extraspecific imposition of social grazing among surgeonfishes (Pisces: Acanthuridae). J. Zool. Lond. 174: 333–340.

Barnard, J.L. 1991. Amphipodological agreement with Platnick. J. Nat. Hist. 25: 1675–1676.

Barnard, J.L., and G.S. Karaman. 1991. The families and genera of marine gammaridean Amphipoda (except marine gammaroids). Records Austr. Mus., Suppl. 13, 866 p.

Barnes, D.J. (ed.). 1983. Perspectives on coral reefs. Brian Clouston Publ., Manuka, Australia, 277 p.

Barnes, D.J., and B.E. Chalker, 1990. Calcification and photosynthesis in reef-building corals and algae. *In* Z. Dubinsky (ed.), Ecosystems of the world, vol. 25: Coral reefs, pages 109–131. Elsevier Science Publishing, New York, 550 p.

Barnes, D.J., and M.J. Devereux. 1984. Productivity and calcification of a coral reef: a survey using pH and oxygen electrode techniques. J. Exp. Mar. Biol. Ecol. 79: 213–231.

Barnes, H. 1956. *Balanus balanoides* (L.) in the Firth of Clyde: the development and annual variation of the larval population, and the causative factors. J. Anim. Ecol. 25: 72–84.

Barnes, R.D. 1980. Invertebrate zoology. Saunders College Publ.

Barnes, R.S.K., and R.N. Hughes. 1988. An introduction to marine ecology. Blackwell Scientific Publications 351 p.

Barnola, J.M., D. Raynaud, D. Korotkevich, and C. Lorius. 1987. Vostok ice core provides 160,000 year record of atmospheric CO_2. Nature 329: 408–414.

Basile, L.L., R.J. Cuffey, and D.F. Kosich. 1984. Sclerosponges, pharetronids, and sphinctozoans (relict cryptic hard-bodied Porifera) in the modern reefs of Enewetak atoll. J. Paleontology 58: 636–650.

Baskin, Y. 1994. Ecologists dare to ask: how much does diversity matter? Science 264: 202–203.

Bathurst, R.G.C. 1976. Carbonate sediments and their diagenesis. Elsevier, Amsterdam.

Bauer, J.C. 1980. Observation on geographic variations in population density of the echinoid *Diadema antillarum* within the western Atlantic. Bull. Mar. Sci. 30: 509–514.

Bayer, F.M. 1961. The shallow-water Octocorallia of the West Indian region. Martinus Nijhoff, The Hague.

Bayer, F.M. (ed.). 1974. Prostaglandins from *Plexaura homomalla:* ecology, utilization and conservation of a major medical marine resource, a symposium. Stud. Trop. Oceanogr. No. 12. University of Miami Press, Coral Gables, Florida, 165 p.

Bayer, F.M. 1981. Key to the genera of Octocorallia exclusive of Pennatulacea (Coelenterata: Anthozoa), with diagnoses of new taxa. Proc. Biol. Soc. Wash. 94: 902–947.

Bayer, F.M., G.L. Voss, and C.R. Robins. 1970. Bioenvironmental and radiological safety feasibility studies Atlantic–Pacific interoceanic canal. Rosenstiel School of Marine and Atmospheric Sciences, University of Miami.

Beckvar, N. 1981. Cultivation, spawning, and growth of the giant clams *Tridacna gigas, T. derasa,* and *T. squamosa* in Palau, Caroline Islands. Aquaculture 24: 21–30.

Bell, J.D., M. Harmelin-Vivien, and R. Galzin. 1985. Large scale spatial variation in the abundance of butterflyfishes (Chaetodontidae) on Polynesian reefs. Proc. 5th Internat. Coral Reef Congr., Tahiti 5: 421–426.

Bellwood, D.R., and J.H. Choat. 1990. A functional analysis of grazing in parrotfishes (family Scaridae): the ecological implications. Environ. Biol. Fish. 28: 189–214.

Benayahu, Y. 1993. Onset of zooxanthellae acquisition in course of ontogenesis of broadcasting and brooding soft corals. Proc. 7th Internat. Coral Reef Symp., Guam 1: 500.

Bender, E.A., T.J. Case, and M.E. Gilpin. 1984. Perturbation experiments in community ecology: theory and practice. Ecology 65: 1–13.

Bennett, K.D. 1990. Milankovitch cycles and their effects on species in ecological and evolutionary time. Paleobiology 16: 11–21.

Benzie, J.A.H. 1992. Review of genetics, dispersal and recruitment of crown-of-thorns starfish *(Acanthaster planci).* Austr. J. Mar. Freshw. Res. 43: 597–610.

Benzie, J.A.H., and J.A. Stoddart. 1992. Genetic structure of crown-of-thorns starfish *(Acanthaster planci)* in Australia. Mar. Biol. 112 (4): 631–639.

Berger, W.H. 1982. Increase of carbon dioxide in the atmosphere during deglaciation: The coral reef hypothesis. Naturwissenschaften 69: 87–88.

Berggren, W.A. 1982. A Cenozoic time scale—some implications for regional geology and paleobiogeography. Lethaia 5: 195–215.

Berggren, W.A. 1984. Role of ocean gateways in climate change. *In* Climate and earth history. National Academy of Sciences, Washington, D.C.

Bergquist, P.R. 1965. The sponges of Micronesia, Part I. The Palau archipelago. Pac. Sci. 19: 123–204.

Bermas, N.A., P.M. Aliño, M.P. Atrigenio, and A. Uychiaoco. 1993. Observations on the reproduction of scleractinian and soft corals in the Philippines. Proc. 7th Internat. Coral Reef Symp., Guam 1: 443–447.

Bernard, P.A. 1984. Coquillages du Gabon (Shells of Gabon). Private publication, Libreville, Gabon.

Berner, R.A., A.C. Lasaga, and R.M. Garrels. 1983. The carbonate-silicate geochemical cycle and its effects on atmospheric carbon dioxide over the past 100 million years. Amer. J. Sci. 283: 641–683.

Berry, F.H., and W.J. Baldwin. 1966. Triggerfishes (Balistidae) of the Eastern Pacific. Proc. Cal. Acad. Sci., Fourth Series 34: 429–474.

Bert, T.M. 1986. Speciation in western Atlantic stone crabs (genus *Menippe*): the role of ecological processes and climatic events in the formation and distribution of species. Mar. Biol. 93: 157–170.

Bertsch, H., and S. Johnson. 1981. Hawaiian nudibranchs. Oriental Publishing Co., Honolulu.

Bilger, R.W., and M.J. Atkinson. 1995. Effects of nutrient loading on mass transfer rates to a coral reef community. Limnol. Oceanogr. 40: 279–289.

BIOTROP. 1978. Proceedings of the Symposium on the Long-Term Effects of Logging in Southeast Asia, R.S. Suparto (ed.). SEAMEO Regional Center for Tropical Biology, Bogor, Indonesia.

Birkeland, C. 1977. The importance of rate of biomass accumulation in early successional stages of benthic communities to the survival of coral recruits. Proc. 3rd Internat. Coral Reef Symp., Miami 1. Biology: 15–21.

Birkeland, C. 1981. *Acanthaster* in the cultures of high islands. Atoll Res. Bull. 255: 55–58.

Birkeland, C. 1982. Terrestrial runoff as a cause of outbreaks of *Acanthaster planci* (Echinodermata: Asteroidea). Mar. Biol. 69: 175–185.

Birkeland, C. 1985. Ecological interactions between mangroves, seagrass beds and coral reefs. *In* Ecological interactions between tropical coastal systems, pages 1–26. UNEP Regional Seas Reports and Studies No. 73.

Birkeland, C. 1987. Nutrient availability as a major determinant of differences among coastal hard-substratum communities in different regions of the tropics. *In* C. Birkeland (ed.), Comparison between Atlantic and Pacific tropical marine coastal ecosystems: community structure, ecological processes, and productivity, pages 45–97. UNESCO Reports in Marine Science 46. 262 p.

Birkeland, C. 1988a. Second-order ecological effects of nutrient input into coral communities. Galaxea 7: 91–100.

Birkeland, C. 1988b. Geographic comparisons of coral-reef community processes. Proc. 6th Internat. Coral Reef Symp., Townsville 1: 211–220.

Birkeland, C. 1989a. The influence of echinoderms on coral-reef communities. *In* M. Jangoux and J.M. Lawrence (eds.), Echinoderm studies, pages 1–79. A.A. Balkema, Rotterdam.

Birkeland, C. 1989b. The Faustian traits of the crown-of-thorns starfish. Amer. Sci. 77: 154–164.

Birkeland, C. 1996. Why some species are especially influential on coral-reef communities and others are not. Galaxea 13: 77–84.

Birkeland, C., and S.S. Amesbury. 1988. Fish-transect surveys to determine the influences of neighboring habitats on fish community structure in the tropical Pacific. *In* A.L. Dahl (ed.), Regional co-operation on environmental protection of the marine and coastal areas of the Pacific, pages 195–202. UNEP Regional Seas Reports and Studies No. 97.

Birkeland, C., and J.S. Lucas. 1990. *Acanthaster planci:* major management problem of coral reefs. CRC Press, Boca Raton, Florida, 257 p.

Birkeland, C., S.G. Nelson, S. Wilkins, and P. Gates. 1985. Effects of grazing by herbivorous fishes on coral reef community metabolism. Proc. 5th Internat. Coral Reef Congr., Tahiti 4: 47–51.

Birkeland, C., and R.H. Randall. 1979. Report on the *Acanthaster planci (Alamea)* studies on Tutuila, American Samoa. Report submitted to Office of Marine Resources, Government of American Samoa, 69 p.

Birkeland, C., D. Rowley, and R.H. Randall. 1982. Coral recruitment patterns at Guam. Proc. 4th Internat. Coral Reef Symp., Manila 2: 339–344.

Blank, R.J., and R.K. Trench. 1985. Speciation and symbiotic dinoflagellates. Science 229: 656–658.

Blum, S.D. 1989. Biogeography of the Chaetodontidae: an analysis of allopatry among closely related species. Env. Biol. Fish. 25: 9–31.

Boekschoten, G.J., and M.B. Best. 1988. Fossil and recent shallow water corals from the Atlantic Islands off Western Africa. Zoologische Mededelingen 62: 99–112.

Boesch, D.F., N.E. Armstrong, C.F. D'Elia, N.G. Maynard, H.W. Paerl, and S.L. Williams. 1993. Deterioration of the Florida Bay ecosystem: an evaluation of the scientific evidence. Report to the Interagency Coodinating Group, Department of the Interior, Washington, D.C., 27 p.

Böhlke, J.E., and C.C.G. Chaplin. 1993. Fishes of the Bahamas and adjacent tropical waters. University of Texas Press.

Bohnsack, J.A. 1992. Reef resource habitat protection: the forgotton factor. *In* R.H. Stroud (ed.), Stemming the tide of coastal fish habitat loss, pages 117–129. Mar. Recreational Fisheries 14.

Bohnsack, J.A. 1994. How marine fishery reserves can improve reef fisheries. Proc. 43rd Gulf and Caribb. Fisheries Inst.: 217–241.

Bonfil, R. 1994. Overview of world elasmobranch fisheries. FAO Fisheries Tech. Paper No. 341. 119 p.

Borowitzka, M.A. 1977. Algal calcification. Oceanogr. Mar. Biol. Ann. Rev. 15: 189–223.

Borowitzka, M.A., and A.W.D. Larkum. 1986. Reef algae. Oceanus 29: 49–54.

Boschma, H. 1925. On the feeding reactions and digestion in the coral polyp *Astrangia danae,* with notes on its symbiosis with zooxanthellae. Biol. Bull. 49: 407–439.

Boschma, H. 1929. On the post-larval development of the coral *Maendra areolata.* Pap. Tortugas Lab. 26: 129–147.

Bosscher, H., and W. Schlager. 1992. Computer simulation of reef growth. Sedimentology 39: 503–512.

Bothwell, A.M. 1982. Fragmentation, a means of asexual reproduction and dispersal in the coral genus Acropora (Scleractinia: Astrocoeniida: Acroporidae)—a preliminary report. Proc. 4th Internat. Coral Reef Symp., Manila 2: 137–144.

Boto, K.G., and J.S. Bunt. 1981. Tidal export of particulate organic matter from a northern Australian mangrove system. Est. Coast. Shelf Sci. 13: 247–255.

Boto, K.G., and J.T. Wellington. 1984. Soil characteristics and nutrient status in a northern Australian mangrove forest. Estuaries 7: 61–69.

Bouchet, P. 1979. How many molluscan species in New Caledonia? Hawaiian Shell News 27 (6): 10.

Bouchet, P., and G. Poppe. 1988. Deep water volutes from the New Caledonian region, with a discussion on biogeography. Venus 47: 15–32.

Bouchon-Navaro, Y., and C. Bouchon. 1989. Correlations between chaetodontid fishes and coral communities of the Gulf of Aqaba (Red Sea). Env. Biol. Fish. 25: 47–60.

Bouchon-Navaro, Y., C. Bouchon, and M.L. Harmelin-Vivien. 1985. Impact of coral degradation on a chaetodontid fish assemblage (Moorea, French Polynesia). Proc. 5th Int. Coral Reef Congr., Tahiti 5: 427–432.

Bowen, B.W., A.B. Meylan, and J.C. Avise. 1989. An odyssey of the green sea turtle: Ascension Island revisited. Proc. Nat. Acad. Sci. 86: 573.

Bradbury, R.H. 1991. Understanding *Acanthaster.* Coenoses 6 (3): 121–126.

Branch, M., and G. Branch. 1981. The living shores of Southern Africa. C. Struik Publishers, Cape Town, 272 p.

Brawley, S.H. 1992. Mesoherbivore. *In* D.M. John, S.J. Hawkins, and J.H. Price (eds.), Plant-animal interactions in the marine benthos, pages 235–263. Syst. Assoc. Spec. 46.

Brawley, S.H., and W.H. Adey. 1977. Territorial behavior of three spot damselfish *(Eupomacentrus planifrons)* increases free algal biomass and productivity. Env. Biol. Fish. 2: 45–51.

Brawley, S.H., and W.H. Adey. 1981. The effect of micrograzers on algal community structure in a coral-free microcosm. Mar. Biol. 61: 167–177.

Briand, F. 1983. Environmental control of food web structure. Ecology 64: 253–263.

Briggs, J.C. 1974. Marine zoogeography. McGraw-Hill, New York.

Briggs, J.C. 1987. Amphitropical distribution and evolution in the Indo-West Pacific Ocean. Syst. Zool. 36: 237–247.

Briggs, J.C. 1992. The marine East Indies: centre of origin? Global Ecology and Biogeography Letters 2: 149–156.

Brock, R.E. 1979. An experimental study on the effects of grazing by parrotfishes and role of refuges in benthic community structure. Mar. Biol. 51: 381–388.

Brodie, J. 1995. The problems of nutrients and eutrophication in the Australian marine environment. *In* L.P. Zann and D.C. Sutton (eds.), The State of the Marine Environment Report for Australia, Technical Annex 2, pages 1–29. Great Barrier Reef Marine Park Authority, Townsville, 93 p.

Broecker, W.S., and C. Peng. 1987. The role of $CaCO_3$ compensation in the glacial to interglacial atmosphere. Global Biogeochemical Cycles 1: 15–29.

Brooks, J.L., and S.I. Dodson. 1965. Predation, body size, and composition of plankton. Science 150: 28–35.

Brostoff, W.N. 1988. Seaweed community structure and productivity: the role of mesograzers. Proc. 6th Internat. Coral Reef Symp., Townsville 2: 1–6.

Brown, B.E., and R.P. Dunne. 1988. The impact of coral mining on coral reefs in the Maldives. Env. Conserv. 15: 159–165.

Brown, B.E., and L.S. Howard. 1985. Assessing the effects of "stress" on reef corals. Advances in Mar. Biol. 22: 1–63.

Brown, B.E., M.D.A. LeTissier, R.P. Dunne, and T.P. Scoffin. 1993. Natural and anthropogenic disturbances on intertidal reefs of SE Phuket, Thailand 1979–1992. *In* R.N. Ginsburg (compiler), Proceedings of the Colloquium and Forum on Global Aspects of Coral Reefs: Health, Hazards, and History. Rosenstiel School of Marine and Atmospheric Sciences, University of Miami, 420 p.

Brown, B.E., M.D.A. LeTissier, T.P. Scoffin, and A.W. Tudhope. 1990. Evaluation of the environmental impact of dredging on intertidal coral reefs and physiological parameters. Mar. Ecol. Progr. Ser. 65: 273–281.

Brown, B.E., and J.C. Ogden. 1992. Coral bleaching. Sci. Amer. 268: 64–70.

Brown, J.H. 1988. Species diversity. *In* A.A. Myers and P.S. Giller (eds.), Analytical biogeography, pages 57–89. Chapman and Hall, London.

Brown, J.H., D.W. Davidson, J.C. Munger, and R.S. Inouye. 1986. Experimental community ecology: the desert granivore system. *In* J. Diamond and T.J. Case (eds.), Community ecology, pages 41–61. Harper and Row, New York, 665 p.

Bruce, A.J. 1976. Coral reef Caridea and "commensalism." Micronesica 12: 83–98.

Brusca, R.C. 1980. Common intertidal invertebrates of the Gulf of California. University of Arizona, Tucson.

Bryan, J.R. 1991. A Paleocene coral-algal-sponge reef from southwestern Alabama and the ecology of Early Tertiary reefs. Lethaia 24: 423–438.

Bryant, J.J. 1996. The impact of human settlements on marine/coastal biodiversity in the tropical Pacific. *In* Marine and coastal biodiversity in the tropical island Pacific Region 2: population, development and conservation priorities. East-West Center and Pacific Science Association, Honolulu.

Bryant, J.P., J. Tuomi, and P. Niemala. 1988. Environmental constraint of constitutive and long-term inducible defenses in woody plants. *In* K.C. Spencer (ed.), Chemical mediation of coevolution. Academic Press, San Diego, 609 p.

Budd, A.F. 1989. Biogeography of Neogene Caribbean reef corals and its implications for the ancestry of eastern Pacific reef corals. Mem. Assoc. Australasian Palaeontologists 8: 219–230.

Budd, A.F., and H.M. Guzmán. 1994. *Siderastrea glynni,* a new species of scleractinian coral (Cnidaria: Anthozoa) from the Eastern Pacific. Proc. Biol. Soc. Wash. 107: 591–599.

Budd, A.F., T.A. Stemann, and K.G. Johnson. 1994. Stratigraphic distributions of genera and species of Neogene to Recent Caribbean reef corals. J. Paleontology 68: 951–977.

Budd, A.F., T.A. Stemann, and R.H. Stewart. 1992. Eocene Caribbean reef corals: a unique fauna from the Gatuncillo Formation of Panama. J. Paleontology 66: 570–594.

Buddemeier, R.W., and D.G. Fautin. 1993. Coral bleaching as an adaptive mechanism. BioScience 43: 320–326.

Buddemeier, R.W., and D. Hopley. 1988. Turn-ons and turn-offs: causes and mechanisms of the initiation and termination of coral reef growth. Proc. 6th Internat. Coral Reef Symp., Townsville 1: 253–261.

Buddemeier, R.W., J.E. Maragos, and D.W. Knutson. 1974. Radiographic studies of reef coral exoskeletons: rates and patterns of coral growth. J. Exp. Mar. Biol. and Ecol. 14: 179–199.

Buddemeier, R.W., and J.A. Oberdorfer. 1986. Internal hydrology and geochemistry of coral reefs and atoll islands: key to diagenetic variations. *In* J.H. Shroeder and B.H. Purser (eds.), Reef diagenesis, pages 91–111, Springer-Verlag, Berlin.

Buddemeier, R.W., and S.V. Smith. 1988. Coral reef growth in an era of rapidly rising sealevels: predictions and suggestions for long-term research. Coral Reefs 7: 51–56.

Burgess, C.M. 1985. Cowries of the world. Gordon Verhoef, Seacomber Publications, Cape Town, South Africa.

Burukovskii, R.N. 1982. Key to shrimps and lobsters. Oxonian Press, New Delhi.

Butler, J.N., J. Burnett-Herkes, J.A. Barnes, and J. Ward. 1993. The Bermuda fisheries: a tragedy of the commons averted? Environment 35: 7–41.

Bythell, J., and C.R.C. Sheppard. 1993. Mass mortality of Caribbean shallow corals. Mar. Pollut. Bull. 26: 296–297.

Cairns, S.D. 1977. Guide to the commoner shallow-water gorgonians (seawhips, seafeathers and seafans) of Florida, the Gulf of Mexico, and the Caribbean region, SeaGrant Field Guide Series No. 6. University of Miami Seagrant Program.

Cairns, S.D. 1982. Stony corals (Cnidaria: Hydrozoa, Scleractinia) of Carrie Bow Cay and vicinity, Belize. *In* K. Rützler and I.G. Macintyre (eds.), The Atlantic barrier reef ecosystem at Carrie Bow Cay, Belize I: structure and communities, pages 271–302. Smithson. Contr. Mar. Sci. 12, 539 p. Smithsonian Institution Press, Washington, D.C.

Cairns, S.D. 1983. A generic revision of the Stylasteridae (Coelenterata: Hydrozoa). Part 1. Description of the genera. Bull. Mar. Sci. 33: 427–508.

Cairns, S.D. 1991a. A generic revision of the Stylasteridae (Coelenterata: Hydrozoa). Part 3. Keys to the genera. Bull. Mar. Sci. 49: 538–545.

Cairns, S.D. 1991b. A revision of the ahermatypic Scleractinia of the Galápagos and Cocos Islands. Smithson. Contr. Zool. 504: 1–32.

Cairns, S.D., and S.A. Parker. 1992. Review of the recent Scleractinia (stony corals) of South Australia, Victoria and Tasmania. Records South Austr. Mus. Monogr. Ser. 3: 1–82.

Cairns, S.D., J.K.L. Dickson, and A.W. Maki. 1978. Estimating the hazard of chemical substances to aquatic life. ASTM, Philadelphia.

Calgren, O. 1949. A survey of the Ptychodactiaria, Corallimorpharia and Actiniaria. Kgl. Svenska Vetenspkapasakad. Handl., Ser. 41: 1–121.

California Coastal Zone Conservation Commission. 1975. California Coastal Plan. State of California documents and publications, Sacramento.

Cameron, A.M. 1977. *Acanthaster* and coral reefs: population outbreaks of a rare and specialized carnivore in a complex high-diversity system. Proc. 3rd Internat. Coral Reef Symp., Miami 1. Biology: 193–199.

Cameron, A.M., and R. Endean. 1982. Renewed population outbreaks of a rare and specialized carnivore (the starfish *Acanthaster planci*) in a complex high-diversity system (the Great Barrier Reef). Proc. 4th Internat. Coral Reef Symp., Manila 2: 593–596.

Cameron, A.M., R. Endean, and L.M. DeVantier. 1991. Predation on massive corals: are devastating population outbreaks of *Acanthaster planci* novel events? Mar. Ecol. Prog. Ser. 75: 251–258.

Camp, D.K., S.P. Cobb, and J.F. van Breedveld. 1973. Overgrazing of seagrasses by a regular urchin, *Lytechinus variegatus*. BioScience 23: 37–38.

Campbell, A.C., J.K. Dart, S.M. Head, and R.F.G. Ormond. 1973. The feeding activity of *Echinostrephus molaris* (deBlainville) in the central Red Sea. Mar. Behav. Physiol. 2: 155–169.

Cannon, L.R.G. 1986. Turbellaria of the world. A guide to families and genera. Queensland Museum, Brisbane.

Capone, D.G. 1983. Benthic nitrogen fixation. *In* E.J. Carpenter and D.G. Capone (eds.), Nitrogen in the marine environment, pages 105–137. Academic Press, New York.

Capone, D.G., P. Penhale, R. Oremland, and B.F. Taylor. 1979. Relationship between productivity and $N_2(C_2H_2)$ fixation in a *Thalassia testudinum* community. Limnol. Oceanogr. 24: 117–124.

Carefoot, T.H. 1987. *Aplysia:* its biology and ecology. Oceanogr. Mar. Biol. Ann. Rev. 25: 167–284.

Carlton, J.T. 1985. Transoceanic and interoceanic dispersal of coastal marine organisms: the biology of ballast water. Oceanogr. Mar. Biol. Ann. Rev. 23: 313–371.

Carlton, J.T. 1987. Patterns of transoceanic marine biological invasions in the Pacific Ocean. Bull. Mar. Sci. 41: 452–465.

Carlton, J.T. 1992. Dispersal of living organisms into aquatic ecosystems: the mechanisms of dispersal as mediated by aquaculture and fisheries activities. *In* A. Rosenfield and

R. Mann (eds.), Dispersal of living organisms into aquatic ecosystems, pages 13–45. University of Maryland, College Park.

Carlton, J.T., and J.B. Geller. 1993. Ecological roulette: the global transport of nonindigenous marine organisms. Science 261: 78–82.

Carlton, R.G., and L.L. Richardson. 1995. Oxygen and sulfide dynamics in a horizontally migrating cyanobacterial mat: black band disease of corals. FEMS Micobiology Ecology 18: 155–162.

Carlton, J.T., J.K. Thompson, L.E. Schemel, and F.H. Nichols. 1990. Remarkable invasion of San Francisco Bay (California, USA) by the Asian clam *Potamocorbula amurensis.* I. Introduction and dispersal. Mar. Ecol. Progr. Ser. 66: 81–84.

Carlton, J.T., G.J. Vermeij, D.R. Lindberg, D.A. Carlton, and E.C. Dudley. 1991. The first historical extinction of a marine invertebrate in an ocean basin: the demise of the eelgrass limpet *Lottia alveus.* Biol. Bull. 180: 72–80.

Carpenter, K.E., R.I. Miclat, V.D. Albaladejo, and V.T. Corpuz. 1982. The influence of substrate structure on the local abundance and diversity of Philippine reef fishes. Proc. 4th Internat. Coral Reef Symp., Manila 2: 497–502.

Carpenter, R.A., and J.E. Maragos. 1989. How to assess environmental impacts on tropical islands and coastal areas. East-West Center, Honolulu, 345 p.

Carpenter, R.C. 1981. Grazing by *Diadema antillarum* (Philippi) and its effects on the benthic algal community. J. Mar. Res. 39: 749–765.

Carpenter, R.C. 1984. Predator and population density control of homing behavior in the Caribbean echinoid *Diadema antillarum* Philippi. Mar. Biol. 82: 101–108.

Carpenter, R.C. 1985. Sea urchin mass-mortality: effects on reef algal abundance, species composition, and metabolism and other coral reef herbivores. Proc. 5th Internat. Coral Reef Congr., Tahiti 4: 53–60.

Carpenter, R.C. 1986a. Partitioning herbivory and its effects on coral reef algal communities. Ecol. Monogr. 56: 345–363.

Carpenter, R.C. 1986b. Relationships between primary production and irradiance in coral reef algal communities. Limnol. Oceanogr. 30: 784–793.

Carpenter, R.C. 1990a. Mass mortality of *Diadema antillarum* I. Long-term effects on sea urchin population-dynamics and coral reef algal communities. Mar. Biol. 104: 67–77.

Carpenter, R.C. 1990b. Mass mortality of *Diadema antillarum* II. Effects on population densities and grazing intensity of parrotfishes and surgeonfishes. Mar. Biol. 104: 79–86.

Carriker, M.R., E.H. Smith, and R.T. Wilce (eds.). 1969. Penetration of calcium carbonate substrates by lower plants and invertebrates. Am. Zool. 9: 629–1020.

Carson, H.L. 1983. Chromosomal sequences and interisland colonizations in Hawaiian *Drosophila.* Genetics 103: 465–482.

Carter, J.G. 1990. Skeletal biomineralization: patterns, processes and evolutionary trends. Van Nostrand Reinhold, New York.

Cernohorsky, W.O. 1972. Marine shells of the Pacific. Pacific Publications, Sydney.

Cernohorsky, W.O. 1978. Tropical Pacific marine shells. Pacific Publications, Sydney.

Chace, F.A., Jr. 1983–1988. The caridean shrimps (Crustacea: Decapoda) of the ALBATROSS Philippine Expedition 1907–1910, Parts 1–5. Smithson. Contr. Zool. 381, 397, 411, 432, 466.

Chace, F.A. Jr., and A.J. Bruce. 1993. The caridean shrimps (Crustacea: Decapoda) of the ALBATROSS Philippine Expedition 1907–1910, Part 6: Superfamily Palaemonoidea. Smithson. Contr. Zool. 543.

Chancharaswat, K. 1991. The role of environmental protection and coastal resources management in national development planning in Thailand. *In* T.E. Chua and L.F. Scura (eds.), Managing ASEAN's coastal resources for sustainable development: roles of policymakers, scientists, donors, media and communities, pages 49–57. ICLARM Conference Proceedings 30. ICLARM, Manila, 125 p.

Chapman, G.A. 1995. Sea urchin sperm cell test. *In* G.M. Rand (ed.), Fundamentals of aquatic toxicology: effects, environmental fate, and risk assessment, pages 189–205. Taylor and Francis, Washington D.C.

Charpy, L., and C. Charpy-Robaud. 1988. Phosphorus cycle and budget in an atoll lagoon. Proc. 6th Internat. Coral Reef Symp., Townsville 2: 547–550.

Chave, K.E., S.V. Smith, and K.J. Roy. 1972. Carbonate production by coral reefs. Mar. Geol. 12: 123–140.

Chesher, R.H. 1969. Destruction of Pacific corals by the seastar *Acanthaster planci.* Science 165: 280–283.

Chevalier, J.P. 1966. Contribution à l'étude des Madréporaires des côtes occidentales del' Afrique tropicale. Bull. del'I.F.A.N., Sér. A28: 912–975, 1356–1405.

Chevalier, J.P. 1982. Reef Scleractinia of French Polynesia. Proc. 4th Internat. Coral Reef Symp., Manila 2: 177–182.

Choat, J.H. 1991. The biology of herbivorous fishes on coral reefs. *In* P.F. Sale (ed.), The ecology of fishes on coral reefs, pages 120–155. Academic Press, San Diego, 754 p.

Choat, J.H., and D.R. Bellwood. 1991. Reef fishes: their history and evolution. *In* P.F. Sale (ed.), The ecology of fishes on coral reefs, pages 39–66. Academic Press, San Diego, 754 p.

Chornesky, E.A., and E.C. Peters. 1987. Sexual reproduction and colony growth in the scleractinian coral *Porites astreoides.* Biol. Bull. 172: 161–177.

Chua, T.E., and D. Pauly. 1989. Coastal area management in Southeast Asia: policies, management strategies and case studies. ICLARM, Manila, 254 p.

Chua, T.E., and L.F. Scura (eds.). 1991. Managing ASEAN's coastal resources for sustainable development: roles of policy makers, scientists, donors, media and communities. ICLARM Conference Proceedings 30. ICLARM, Manila, 125 p.

Cimino, G., S. DeRosa, S. DeStefano, and G. Sodano. 1982. The chemical defense of four Mediterranean nudibranchs. Compar. Biochem. Physiol. 73B: 471–474.

Clark, A.M., and J. Courtman-Stock. 1976. The echinoderms of southern Africa. British Museum (Natural History), London.

Clark, A.M., and M.E. Downey. 1992. Starfishes of the Atlantic. Chapman and Hall, London.

Clark, A.M., and F.W.E. Rowe. 1971. Monograph of shallow-water Indo-West Pacific echinoderms. British Museum (Natural History) London, 238 p.

Clarke, A. 1992. Is the real latitudinal diversity cline in the sea? Trends Ecol. Evol. 7: 286–287.

Cloud, P.E., Jr. 1959. Geology of Saipan, Mariana Islands, Part 4. Submarine topography and shoal-water ecology. Geol. Surv. Prof. Pap. 280–K: 361–445.

Cloud, P.E., Jr., and M.G. Glaessner. 1982. The ediacarian period and system: Metazoa inherit the Earth. Science 217: 783–792.

Coates, A.G., J.B.C. Jackson, and L.S. Collins. 1992. Closure of the Isthmus of Panama: the nearshore marine record of Costa Rica and Western Panama. Geol. Soc. Amer. Bull. 104: 814–828.

Coen, L.D. 1988a. Herbivory by crabs and the control of algal epibionts on Caribbean host corals. Oecologia 75: 198–203.

Coen, L.D. 1988b. Herbivory by Caribbean majid crabs: feeding ecology and plant susceptibility. J. Exp. Mar. Biol. Ecol. 122: 257–276.

Cohen, D.M. 1970. How many recent fishes are there? Proc. Calif. Acad. Sci. 38: 341–346.

Coley, P.D., J.P. Bryant, and F.S. Chapin III. 1985. Resource availability and plant antiherbivore defense. Science 230: 895–899.

Colgan, M.W. 1982. Succession and recovery of a coral reef after predation by *Acanthaster planci* (L.). Proc. 4th Internat. Coral Reef Symp., Manila 2: 333–338.

Colgan, M.W. 1984. The Cretaceous coral *Heliopora* (Octocorallia, Coenothecalia)—a common Indo-Pacific reef builder. *In* N. Eldredge and S.M. Stanley (eds.), Living fossils, pages 226–271 Springer-Verlag, New York.

Colgan, M.W. 1987. Coral reef recovery on Guam (Micronesia) after catastrophic predation by *Acanthaster planci:* a study of community development. Ecology 68: 1592–1605.

Colin, P.L. 1988. Marine invertebrates and plants of the living reef. Tropical Fish Hobbyist, Neptune City, New Jersey.

Colin, P.L., and C. Arneson. 1995. Tropical Pacific invertebrates. Coral Reef Press, Koror, Palau.

Coll, J.C., B.F. Bowden, P.M. Alino, A. Heaton, G.M. Konig, R. DeNys, and R. Willis. 1989. Chemically mediated interactions between marine organisms. Chemica Scripta 29: 383–388.

Coll, J.C., B.F. Bowden, and M.N. Clayton. 1990. Chemistry and coral reproduction. Chemistry in Britain, August 1990: 761–763.

Coll, J.C., B.F. Bowden, G.V. Meehan, G.M. Konig, A.R. Carroll, D.M. Tapiolas, P.M. Alino, A. Heaton, R. DeNys, P.A. Leone, M. Maida, T.L. Aceret, R.H. Willis, R.C. Babcock, B.L. Willis, Z. Florian, M.N. Clayton, and R.L. Miller. 1994. Chemical aspects of mass spawning in corals. I. Sperm-attractant molecules in the eggs of the scleractinian coral *Montipora digitata.* Mar. Biol. 118: 177–182.

Condie, K.C. 1989. Origin of the Earth's crust. Palaeogeography, Palaeoclimatology, Palaeoecology 75: 57–81.

Connell, D.W., and G.J. Miller. 1984. Chemistry and ecotoxicology of pollution. Wiley, New York, 444 p.

Connell, J.H. 1978. Diversity in tropical rain forests and coral reefs. Science 199: 1302–1310

Connell, J.H. 1983. On the prevalence and relative importance of interspecific competition: evidence from field experiments. Am. Nat. 122: 661–678.

Conway Morris, S. 1993. The fossil record and the early evolution of the Metazoa. Nature 361: 219–225.

Cook, C.B., and C.F. D'Elia. 1987. Are natural populations of zooxanthellae ever nutrient-limited? Symbiosis 4: 199–212.

Cook, C.B., A. Logan, J. Ward, B. Luckhurst, and C.J. Berg, Jr. 1990. Elevated temperatures and bleaching on a high latitude coral reef: the 1988 Bermuda event. Coral Reefs 9: 45–49.

Copland, J.W., and J.S. Lucas (eds.). 1988. Giant clams in Asia and the Pacific. Australian Centre for International Agricultural Research, Canberra.

Copper, P. 1988. Ecological succession in Phanerozoic reef ecosystems: is it real? Palaios 3: 136–151.

Copper, P. 1994. Ancient reef ecosystem expansion and collapse. Coral Reefs 13: 3–11.

Cornell, J.H. 1978. Diversity in tropical rainforests and coral reefs. Science 199: 1302–1310.

Corrêa, D.D. 1961. Nemerteans from Florida and Virgin Islands. Bull. Mar. Sci. 11: 1–44.

Cortés, J. 1985. Preliminary observations of *Alpheus simus* Guerin-Meneville, 1856 (Crustacea: Alphaeidae): a little-known Caribbean bioeroder. Proc. 5th Int. Coral Reef Cong., Tahiti 5: 351–53.

Cortés, J., and M.J. Risk. 1985. Are reefs under siltation stress: Cahuita, Costa Rica. Bull. Mar. Sci. 36: 339–356.

Costanza, R. 1991. Ecological economics. Columbia University Press, New York.

Couch, J.A., and J.W. Fournie (eds.). 1993. Pathobiology of marine and estuarine organisms. CRC Press, Boca Raton, Florida.

Coudray, J., and L. Montaggioni. 1982. Coraux et récifs coralliens de la province indo-pacifique: répartition géographique et altitudinale en relation avec la tectonique globale. Bull. de la Société Géologique de France 24: 981–993.

Cox, E.F. 1983. Aspects of corallivory by *Chaetodon unimaculatus* in Kaneohe Bay, Oahu. Master's thesis, Zoology Department, University of Hawaii, Honolulu.

Cox, E.F. 1986. The effects of a selective corallivore on growth rates and competition for space between two species of Hawaiian corals. J. Exp. Mar. Biol. Ecol. 101: 161–174.

Craig, P. 1994. Coral reefs in American Samoa. Department of Marine Wildlife Resources 12/94.

Craig, P. 1995. Life history and harvest of the surgeonfish *Acanthurus lineatus* in American Samoa, Biol. Rept. Ser. No. 77. Department of Marine Wildlife Resources, American Samoa, 30 p.

Craig, P., A. Green, and S. Saucerman. 1995. Coral reef troubles in American Samoa. SPC Fisheries Newsletter No. 72: 33–34.

Craik, W., R. Kenchington, and G. Kelleher. 1990. Coral-reef management. *In* Z. Dubinsky (ed.), Ecosystems of the world, vol. 25: Coral reefs, pages 453–467. Elsevier Science Publishing, New York, 550 p.

Crevello, P.D., J.L. Wilson, J.F. Sarg, and J.F. Read (eds.). 1989. Controls on carbonate platform and basin development. Spec. Publ. No. 44. Society of Economic Paleontologists and Mineralogists, Tulsa, Oklahoma.

Crisp, D.J. 1975. Secondary productivity in the sea. *In* D.E. Reichle, J.E. Franklin, and D.W. Goodall (eds.), Proceedings of the Symposium on Productivity of World Ecosystems, pages 71–89. National Academy of Sciences, Washington, D.C.

Crocker, T.D., and J.F. Shogren. 1993. Dynamic inconsistency in valuing environmental goods. Ecological Economics 7: 239–254.

Crossland, C.J. 1988. Latitudinal comparisons of coral reef structure and function. Proc. 6th Internat. Coral Reef Symp., Townsville 1: 221–226.

Crossland, C.J., and D.J. Barnes. 1983. Dissolved nutrients and organic particulates in water flowing over coral reefs at Lizard Island. Austr. J. Mar. Freshw. Res. 34: 835–844.

Crossland, C.J., B.G. Hatcher, and S.V. Smith. 1991. The role of coral reefs in global carbon production. Coral Reefs 10: 55–64.

Culotta, W.A., and G.V. Pickwell. 1993. The venomous sea snakes: a comprehensive bibliography. Krieger Publishing Company, Malabar, Florida.

Cumming, R.L. 1988. Pyramidellid parasites of giant clam mariculture systems. *In* J.W. Copland and J.S. Lucas (eds.), Giant clams in Asia and the Pacific, pages 231–236. Australian Centre for International Agricultural Research, Canberra.

Dahl, A.L. 1973. Surface area in ecological analysis: quantification of benthic coral-reef algae. Mar. Biol. 23: 239–249.

Dai, A., and S. Yang. 1991. Crabs of the China Seas. China Ocean Press, Beijing; Springer-Verlag, Heidelberg, 682 p.

Dai, C.F. 1990. Interspecific competition in Taiwanese corals with special reference to interactions between alcyonaceans and scleractinians. Mar. Ecol. Progr. Ser. 60: 291–297.

Dai, C.F., K. Soong, and T.Y. Fan. 1993. Sexual reproduction in corals in northern and southern Taiwan. Proc. 7th Internat. Coral Reef Symp., Guam 1: 448–455.

Daly, R.A. 1915. The glacial-control theory of coral reefs. Proc. Amer. Acad. Arts and Sci. 51: 155–251.

Dana, T.F. 1975. Development of contemporary Eastern Pacific coral reefs. Mar. Biol. 33: 355–374.

Dana, T.F., W.A. Newman, and E.W. Fager. 1972. Acanthaster aggregations: interpreted as primary responses to natural phenomena. Pac. Sci. 26: 355–372.

Dart, J.K.G. 1972. Echinoids, algal lawn and coral recolonization. Nature 239: 50–51.

Darwin, C. 1842. The structure and distribution of coral reefs. Smith, Elder & Company, London, 214 p.

Davies, P.J. 1988. Evolution of the Great Barrier Reef—reductionist dream or expansionist vision? Proc. 6th Internat. Coral Reef Symp., Townsville 1: 9–18.

Davies, P.J., and P.A. Hutchings. 1983. Initial colonization, erosion and accretion on coral substrate. Experimental results, Lizard Island, Great Barrier Reef. Coral Reefs 2: 27–35.

Davies, P.J., and L. Montaggioni. 1985. Reef growth and sea-level change: the environmental signature. Proc. 5th Internat. Coral Reef Congr., Tahiti 3: 477–511.

Davies, P.J., P.A. Symonds, D.A. Feary, and C.J. Pigram. 1987. Horizontal plate motion: a key allocyclic factor in the evolution of the Great Barrier Reef. Science 238: 1697–1700.

Davis, G.W. 1992. Biology of *Mulloides flavolineatus.* Study 3 in studies of Guam's recreationally-important fish, Job Progress Report for Project No. FW-2R-28. 10 p.

Davis, W.M. 1923. The marginal belts of the coral seas. Proc. Natl. Acad. Sci. 9: 292–296.

Dawson, E.Y. 1954. Marine plants in the vicinity of the Institut Océanographique de Nha Trang, Viêt Nam. Pac. Sci. 8: 373–481.

Dawson, E.Y. 1956. Some marine algae of the Southern Marshall Islands. Pac. Sci. 10: 25–66.

Dawson, E.Y. 1957. An annotated list of marine algae from Eniwetok Atoll, Marshall Islands. Pac. Sci. 11: 92–132.

Day, J.H. 1967. A monograph of the Polychaeta of Southern Africa. British Museum (Natural History), London.

Day, R.W. 1977. Two contrasting effects of predation on species richness in coral reef habitats. Mar. Biol. 44: 1–5.

Dayton, L. 1995. The killing reefs. The New Scientist 148: 14–15.

Dayton, P.K. 1985. The structure and regulation of some South American kelp communities. Ecol. Monogr. 55: 447–468.

De La Court, T. 1990. Beyond Brundtland: green development in the 1990s. Zed Books, London.

deLaubenfels, M.W. 1954. The sponges of the West-Central Pacific. Oregon State College, Corvalis, Oregon.

D'Elia, C.F., R.W. Buddemeier, and S.V. Smith. 1991. Workshop on coral bleaching, coral reef ecosystems and global change: report of proceedings. Maryland Sea Grant UM-SG-TS-91-03. University of Maryland, College Park, Maryland, 49 p.

D'Elia, C.F., J.G. Sanders, and W.R. Boynton. 1986. Nutrient enrichment studies in a coastal plain estuary: phytoplankton growth in large-scale continuous cultures. Can. J. Fish. Aquat. Sci. 43: 397–406.

D'Elia, C.F., and W.J. Wiebe. 1990. Biochemical nutrient cycles in coral reef ecosystems. *In* Z. Dubinsky (ed.), Ecosystems of the world, vol. 25: Coral reefs, pages 49–74. Elsevier Science Publishing, New York, 550 p.

Delmas, R.J. 1992. Environmental information for ice cores. Rev. Geophysics 30: 1–21.

Derr, M. 1992. Raiders of the reef. Audubon 94 (2): 48–56.

de Ruyter van Steveninck, E.D., and R.P.M. Bak. 1986. Changes in abundance of coral-reef bottom components related to mass mortality of the sea urchin *Diadema antillarum.* Mar. Ecol. Progr. Ser. 34: 87–94.

Dethier, M.N., and D.O. Duggins. 1984. An "indirect commensalism" between marine herbivores and the importance of competitive hierarchies. Am. Nat. 124: 205–219.

Devaney, D.M., and L.G. Eldredge (eds.). 1977. Reef and shore fauna of Hawaii. Section 1: Protozoa through Ctenophora. Bishop Museum Special Publication 64 (1). Bishop Museum Press, Honolulu.

Devaney, D.M., and L.G. Eldredge (eds.). 1987. Reef and shore fauna of Hawaii. Section 2: Platyhleminthes through Phoronida, and Section 3: Sipuncula through Annelida [bound together]. Bishop Museum Special Publication 64 (2 and 3). Bishop Museum Press, Honolulu.

DeVantier, L.M., R.E. Reichelt, and R.H. Bradbury. 1986. Does *Spirobranchus giganteus* protect host *Porites* from predation by *Acanthaster planci:* predation pressure as a mechanism of coevolution? Mar. Ecol. Progr. Ser. 32: 307–310.

DeVoe, M.R. 1992. Introductions and transfers of marine species: achieving a balance between economic development and resource Protection. Proceedings of the Conference and Workshop, October 30–November 2, Hilton Head, South Carolina. South Carolina Sea Grant Consortium, Charleston.

Dhondt, A.V. 1992. Palaeogeographic distribution of Cretaceous Tethyan non-rudist bivalves. *In* Schriftenreihe der Erdwissen schaft lichen Kommissionen der Österreichischen Akademieder Wissenschaften (New aspects on Tethyan Cretaceous fossil assemblages), pages 75–94. Österreichischen Akademie der Wissenschaften, Austria 9.

Diamond, J.M. 1977. Continental and insular speciation in Pacific land birds. Syst. Zool. 26: 263–268.

Diamond, J.M., and A.G. Marshall. 1976. Origin of the New Hebridean avifauna. Emu 76: 187–200.

Dight, I.J., L. Bode, and M.K. James. 1990a. Modelling the larval dispersal of *Acanthaster planci* I. Large scale hydrodynamics, Cairns Section, Great Barrier Reef Marine Park. Coral Reefs 9: 115–123.

Dight, I.J., M.K. James, and L. Bode. 1990b. Modelling the larval dispersal of *Acanthaster planci* II. Patterns of reef connectivity. Coral Reefs 9: 125–134.

Dill, R.F., E.A. Shinn, A.T. Jones, K. Kelly, and R.P. Steinen. 1986. Giant subtidal stromatolites forming in normal salinity waters. Nature 324: 55–58.

DiSalvo, L.H. 1969. Isolation of bacteria from the corallum of *Porites lobata* (Vaughn) and its possible significance. Am. Zool. 9: 735–40.

Dixon, J.A., and M.M. Hufschmidt. 1986. Economic valuation techniques for the environment. Johns Hopkins University Press, Baltimore.

Dixon, J.A., and P.B. Sherman. 1990. Economics of protected areas: a new look at benefits and costs. Earthscan, London.

Dodge, R.E., and J.C. Lang. 1983. Environmental correlates of hermatypic coral *(Montastrea annularis)* growth on the East Flower Gardens Bank, northwest Gulf of Mexico. Limnol. Oceanogr. 28(2): 228–240.

Dodge, R.E., and A.M. Szmant-Froelich. 1985. Effects of drilling fluids on reef corals: a review. *In* I.W. Duedall, D.R. Kester, P.K. Park, B.H. Ketchum (eds.), Wastes in the ocean, vol. 4, pages 341–364. Wiley (Interscience), New York.

Dodge, R.E., and J.R. Vaisnys. 1977. Coral populations and growth patterns: responses to sedimentation and turbidity associated with dredging. J. Mar. Res. 35: 715–730.

Doherty, P.J. 1991. Spatial and temporal patterns in recruitment. *In* P.J. Sale (ed.), The ecology of fishes on coral reefs, pages 261–293. Academic Press, San Diego, 754 p.

Doherty, P.J., and J. Davidson. 1988. Monitoring the distribution and abundance of juvenile *Acanthaster planci* in the central Great Barrier Reef. Proc. 6th Int. Coral Reef Symp., Townsville 2: 131–136.

Doherty, P.J., and D.M. Williams. 1988. The replenishment of coral reef fish populations. Oceanogr. Mar. Biol. Ann. Rev. 26: 487–551.

Dollar, S.J., and R.W. Grigg. 1981. Impact of a kaolin clay spill on a coral reef in Hawaii. Mar. Biol. 65: 269–276.

Done, T.J. 1985. Effects of two *Acanthaster* outbreaks on coral community structure: the meaning of devastation. Proc. 5th Internat. Coral Reef Congr., Tahiti 5: 315–320.

Done, T.J. 1987. Simulation of the effects of *Acanthaster planci* on the population structure of massive corals in the genus Porites: evidence of population resilience? Coral Reefs 6: 75–90.

Done, T.J. 1988. Simulation of recovery of pre-disturbance size structure in populations of Porites spp. damaged by the crown-of-thorns starfish *Acanthaster planci.* Mar. Biol. 100: 51–61.

Done, T.J. 1992. Phase shifts in coral reef communities and their ecological signficance. Hydrobiologia 247: 121–132.

Done, T.J., A.E. Dayton, and R. Steger. 1991. Regional and local variability in recovery of shallow coral communities: Moorea, French Polynesia and central Great Barrier Reef. Coral Reefs 9: 183–192.

Done, T.J., J.C. Ogden, W.W. Wiebe, and BIOCORE Working Group. 1996. Biodiversity and ecosystem function of coral reefs. *In* H. Cushman and H. Mooney (eds.), Global biodiversity assessment, pages 381–387. United Nations Scientific Committee on Problems in the Environment (SCOPE), Washington, D.C.

Donn, T.F., and M.R. Boardman. 1988. Bioerosion of rocky carbonate coastlines on Andros Island, Bahamas. J. Coastal Res. 4: 381–394.

Doolette, J.B., and W.B. Magrath (eds.). 1990. Watershed development in Asia. Strategies and technologies. World Bank Tech. Paper 127. 227 p.

Doty, M.S. 1983. Coral reef diversified farming. *In* C.K. Tseng (ed.), Proceedings of the Joint China–U.S. Phycol. Symp, pages 437–478. Science Press, Beijing.

Downing, N., and C.R. El-Zahr. 1987. Gut evacuation and filling rates in the rock-boring sea urchin, *Echinometra mathaei.* Bull. Mar. Sci. 41: 579–584.

Dubin, R.E. 1982. Behavioral interactions between Caribbean reef fish and eels (Muraenidae and Opichthidae). Copeia 1982: 229–232.

Dubois, R., and E.L. Towle. 1985. Coral harvesting and mining management practices.

In J.R. Clarke (ed.), *Coasts-coastal resource management: development case studies,* pages 203–283. Coastal Publication No. 3. Research Planning Institute, Columbia, South Carolina.

Ducklow, H.W. 1990. The biomass, production and fate of bacteria in coral reefs. *In* Z. Dubinsky (ed.), Ecosystems of the world, vol. 25: Coral reefs, pages 265–290. Elsevier Science Publishing, New York, 550 p.

Duerden, J.E. 1902. West Indian madreporarian polyps. Mem. Natl. Acad. Sci. 8: 401–597.

Dugdale, R.C., and J.J. Goering. 1967. Uptake of new and regenerated forms of nitrogen in primary productivity. Limnol. Oceanogr. 23: 196–206.

Dungan, M.L. 1986. Three-way interactions: barnacles, limpets, and algae in a Sonoran desert rocky intertidal zone. Am. Nat. 127: 292–316.

Dungan, M.L. 1987. Indirect mutualism: complementary effects of grazing and predation in a rocky intertidal community. *In* W.C. Kerfoot and A. Sih (eds.), Predation: direct and indirect impacts on aquatic communities, pages 188–200. University Press of New England, Hanover, New Hampshire, 386 p.

Durako, M.J., and K.M. Kuss. 1994. Effects of *Labyrinthula* infection on the photosynthetic capacity of *Thalassia testudinum.* Bull. Mar. Sci. 54(3): 727–732.

Durham, J.W., and J.L. Barnard. 1952. Stony corals of the Eastern Pacific collected by the Velero III and Velero IV. Allan Hancock Pac. Exped. 16: 1–110.

Dustan, P. 1975. Growth and form in the reef-building coral *Montastrea annularis.* Mar. Biol. 33: 101–107.

Dustan, P. 1985. Studies on the bio-optics of coral reefs. *In* M.L. Reaka (ed.), The ecology of coral reefs, pages 189–198. NOAA Symp. Ser. Undersea Res. 3(1). 208 p.

Dykens, J.A. and J.M. Shick. 1984. Photobiology of the symbiotic sea anemone, *Anthopleura elegantissima:* defenses against photo dynamic effects, and seasonal photo acclimatization. Biol. Bull. 167: 683–697.

Eakin, C.M. 1987. Damselfishes and their algal lawns: a case of plural mutualism. Symbiosis 4: 275–288.

Eakin, C.M. 1993. Post-El Niño Panamanian reefs: less accretion, more erosion and damselfish protection. Proc. 7th Internat. Coral Reef Symp., Guam 1: 387–396.

Earle, S.A. 1972. A review of the marine plants of Panama. Bull. Biol. Soc. Wash. 2: 69–87.

Ebbs, N.K. 1966. The coral-inhabiting polychaetes of the northern Florida reef tract. Bull. Mar. Sci. 16: 485–555.

Ebeling, A.W., and M.A. Hixon. 1991. Tropical and temperate reef fishes: comparison of community structures. *In* P.F. Sale (ed.), The ecology of fishes on coral reefs, pages 509–563. Academic Press, San Diego. 754 p.

Ebert, T.A. 1971. A preliminary quantitative survey of the echinoid fauna of Kealakekua and Honaunau Bays, Hawaii. Pac. Sci. 25: 112–131.

Ebert, T.A. 1980. Relative growth of sea urchins jaws: an example of plastic resource allocation. Bull. Mar. Sci. 30: 467–474.

Ebert, T.A. 1983. Recruitment in echinoderms. Echinoderm Studies 1: 169–203.

Edinger, E.N. 1991. Mass extinction of Caribbean corals at the Oligocene–Miocene boundary: paleoecology, paleoceanography, paleobiogeography. M.Sc. thesis, McMaster University, Hamilton, Ontario.

Edinger, E.N., and M.J. Risk. 1994. Oligocene–Miocene extinction and geographic restriction of Caribbean corals: roles of turbidity, temperature, and nutrients. Palaios 9: 576–598.

Edmondson, C.H. 1929. Growth of Hawaiian corals. Bull. Bernice P. Bishop Mus. 58: 1–38.

Edmondson, C.H. 1946. Behavior of coral planulae under altered saline and thermal conditions. Occ. Pap. Bernice P. Bishop Mus. 18: 283–304.

Edmunds, P.J. 1991. Extent and effect of black band disease on a Caribbean reef. Coral Reefs 10: 161–165.

Ehrhardt, H., and H. Moosleitner. 1995. Meerwasser atlas, bands 1–3. Mergus, Melle, Germany.

Ekman, S. 1953. Zoogeography of the sea. Sidgwick and Jackson, London.

Eldredge, L.G. 1988. Case studies of the impacts of introduced animal species on renewable resources in the U.S.-affiliated Pacific islands. *In* B.D. Smith (ed.), Topic reviews in insular development and management in the Pacific U.S.-affiliated islands, pages 118–146. Tech. Rept No. 88. University of Guam Marine Laboratory, Guam.

Eldredge, L.G., and S.E. Miller. 1995. How many species are there in Hawaii? Occ. Pap. Bernice P. Bishop Mus. 41: 1–18.

Emerson, W.K. 1991. First records for *Cymatium mundum* (Gould) in the eastern Pacific Ocean, with comments on the zoogeography of the tropical trans-Pacific tonnacean and non-tonnacean prosobranch gastropods with Indo-Pacific faunal affinities in west American waters. Nautilus 105: 62–80.

Emerson, W.K., and H.W. Chaney. 1995. A zoogeographic review of the Cypraeidae (Mollusca: Gastropoda) occurring in the eastern Pacific Ocean. The Veliger 38: 8–21.

Emlen, J.M. 1973. Ecology: an evolutionary approach. Addison-Wesley, Reading, Massachusetts.

Emson, R.H., P.V. Mladenov, and I.C. Wilkie. 1985. Studies of the biology of the West Indian copepod *Ophiopsyllus reductus* (Siphonostomatoida: Cancerillidae) parasitic upon the brittlestar *Ophiocomella ophiactoides*. J. Nat. Hist. 19: 151–171.

Endean, R. 1969. Report on investigations made into aspects of the current *Acanthaster planci* (crown-of-thorns) infestations of reefs of the Great Barrier Reef. Fisheries Branch, Queensland Department of Primary Industries, Brisbane, 35 p.

Endean, R. 1971. The recovery of coral reefs devastated by catastrophic events with particular reference to current *Acanthaster planci* plagues in the tropical Indo-West Pacific region. J. Mar. Biol. Assoc. India 13: 1–13.

Endean, R. 1973. Population explosions of *Acanthaster planci* on the Great Barrier Reef and associated destruction of hermatypic corals in the Indo-West Pacific region. *In* O.A. Jones and R.E. Endean (eds.), Biology and geology of coral reefs. II. Biology 1, pages 389–438. Academic Press, New York.

Endean, R. 1976. Destruction and recovery of coral reef communities. *In* O.A. Jones and

R. Endean (eds.), Biology and geology of coral reefs. III. Biology 2, pages 215–254. Academic Press, New York.

Endean, R. 1977. *Acanthaster planci* infestations of reefs of the Great Barrier Reef. Proc. 3rd Internat. Coral Reef Symp., Miami 1. Biology: 185–191.

Endean, R. 1982. Crown-of-thorns starfish on the Great Barrier Reef. Endeavour 6: 10–14.

Endean, R., and A.M. Cameron. 1990a. Trends and new perspectives in coral-reef ecology. *In* Z. Dubinsky (ed.), Ecosystems of the world, vol. 25: Coral reefs, pages 469–492. Elsevier Science Publishing, New York, 550 p.

Endean, R., and A.M. Cameron. 1990b. *Acanthaster planci* population outbreaks. *In* Z. Dubinsky (ed.), Ecosystems of the world, vol. 25: Coral reefs, pages 419–437. Elsevier Science Publishing, New York, 550 p.

Endean, R., and W. Stablum. 1973. A study of some aspects of the crown-of-thorns starfish *(Acanthaster planci)* infestations of Australia's Great Barrier Reef. Atoll Res. Bull. 167: 1–60.

Erez, J. 1990. On the importance of food source in coral-reef ecosystems. *In* Z. Dubinsky (ed.), Ecosystems of the world, vol. 25: Coral reefs, pages 411–418. Elsevier Science Publishing, New York, 550 p.

Erman, D.C., and D. Mahoney. 1983. Recovery after logging in streams with and without buffer strips in Northern California. Calif. Water Resources Center Contrib. 186. University of California, Davis.

Fabricius, K.E., and F.H. Fabricius. 1992. Re-assessment of ossicle frequency patterns in sediment cores: rate of sedimentation related to *Acanthaster planci.* Coral Reefs 11: 109–114.

Fadlallah, Y.H. 1983. Sexual reproduction, development and larval biology in scleractinian corals. A review. Coral Reefs 2: 129–150.

Fadlallah, Y.H. 1985. Reproduction in the coral *Pocillopora verrucosa* on the reefs adjacent to the industrial city of Yanbu, Red Sea, Saudi Arabia. Proc. 5th Internat. Coral Reef Congr., Tahiti 4: 314–318.

Faeth, S.H. 1988. Plant-mediated interactions between seasonal herbivores: enough for evolution or coevolution? *In* K.C. Spencer (ed.), Chemical mediation of coevolution. Academic Press, San Diego, 609 p.

Fagerstrom, J.A. 1987. The evolution of reef communities. Wiley, New York, 600 p.

Fairbridge, R., and C. Teichert. 1948. The low isles of the Great Barrier Reefs, a new analysis. Geogr. J. 3: 67–88.

Fairweather, P.G. 1990. Is predation capable of interacting with other community processes on rocky reefs? Austr. J. Ecol. 15: 453–464.

Falkowski, P.G., Z. Dubinsky, L. Muscatine, and L.R. McCloskey. 1993. Population control in symbiotic corals. BioScience 43: 606–611.

Fauchald, K. 1977. The polychaete worms. Nat. Hist. Mus. Los Angeles County, Science Series 28. 188 p.

Fauchald, K., and P.A. Jumars. 1979. The diet of worms: a study of polychaete feeding guilds. Ann. Rev. Oceanogr. Mar. Biol. 17: 193–284.

Faulkner, D.J. 1984a. Marine natural products: metabolites of marine algae and herbivorous marine molluscs. Natural Products Report 1: 251–280.

Faulkner, D.J. 1984b. Marine natural products: metabolites of marine invertebrates. Natural Products Report 1: 551–598.

Faulkner, D.J. 1986. Marine natural products. Natural Products Report 3: 2–33.

Faulkner, D.J. 1987. Marine natural products. Natural Products Report 4: 539–576.

Faulkner, D.J. 1988. Feeding deterrents in molluscs. *In* D.G. Fautin (ed.), Biomedical importance of marine organisms, pages 29–36. Mem. Cal. Acad. Sci. 13.

Faulkner, D.J. 1992. Chemical defenses in marine molluscs. *In* V.J. Paul (ed.), Ecological roles of marine secondary metabolites. Cornell University Press, Ithaca, New York.

Faulkner, D.J., T.F. Molinski, R.J. Andersen, E.J. Dumdei, and E. Dilip DeSilva. 1990. Geographical variation in defensive chemicals from Pacific coastdorid nudibranchs and some related marine molluscs. Compar. Biochem. Physiol. 97C: 233–240.

Fautin, D.G., and G.R. Allen. 1992. Field guide to anemone fishes and their host anemones. Western Australia Museum, Perth.

Fautin, D.G., and R.W. Buddemeier. 1992. Geochemistry and scleractinian evolution. Am. Zool. 32: 135A.

Fautin, D.G., and J.M. Lowenstein. 1993. Phylogenetic relationships among scleractinians, actinians, and corallimorpharians (Coelenterata: Anthozoa). Proc. 7th Internat. Coral Reef Symp., Guam 2: 665–670.

Feder, H.M. 1966. Cleaning symbiosis in the marine environment. *In* S.M. Henry (ed.), Symbiosis 1. Academic Press, New York.

Fischer, A.G. 1965. Fossils, early life, and atmospheric history. Proc. Natl. Acad. Sci. 53: 1205–1215.

Fischer, A.G., and M.A. Arthur. 1977. Secular variations in the pelagic realm. *In* H.E. Cook and P. Enos (eds.), Deep-water carbonate environments, pages 19–50. Spec. Publ. 25. Society of Economic Paleontologists and Mineralogists, Tulsa, Oklahoma.

Fisher, J.L. 1969. Starfish infestation: hypothesis (letter). Science 165: 645.

Fisher, M.R., and S.C. Hand. 1984. Chemoautotrophic symbionts in the bivalve *Lucina floridana* from seagrass beds. Biol. Bull. 167: 445–459.

Fisher, W.S. (ed.) 1988. Disease processes in marine bivalve molluscs. Spec. Publ. 18. American Fisheries Society, Bethesda, Maryland.

Fitzgerald, W.J., Jr., and S.G. Nelson. 1979. Development of aquaculture in an island community (Guam, Mariana Islands). Proc. World Mariculture Soc. 10: 39–50.

Fleminger, A. 1986. The Pleistocene equatorial barrier between the Indian and Pacific Oceans and a likely cause for Wallace's line. *In* A.C. Pierrot-Bults, B.J.S. vander Spoel, B.J. Zahuranec, and R.K. Johnson (eds.), Pelagic biogeography, pages 84–97. UNESCO Tech. Papers Mar. Sci. 49.

Focke, J.W. 1978. Limestone cliff morphology on Curaçao (Netherlands Antilles) with special attention to the origin of notches and vermetid/coralline algal surf benches ("cornices," "trottoirs"). Z. Geomorphol. 22: 329–349.

Forest, J., and D. Guinot. 1961. Crustacés décapods brachyoures de Tahiti et des Tuamotu. Expédition Française sur les récifs coralliens de la Nouvelle-Calédonie. Volume Préliminaire. Éditions de la Fondation Singer-Polignac, Paris.

Foster, B.A. 1980. Shallow water barnacles from Hong Kong. *In* B.S. Morton and C.K. Tseng (eds.), Proc. First International Marine Biological Workshop: The marine flora and fauna of Hong Kong and Southern China, pages 207–232. Hong Kong University Press, Hong Kong.

Foster, S.A. 1985. Group for aging by a coral reef fish: a mechanism for gaining access to defended resources. Anim. Behav. 33: 782–792.

Foster, S.A. 1987. Territoriality of the dusky damselfish: influence on algal biomass and on the relative impacts of grazing by fishes and *Diadema.* Oikos 50: 153–160.

Frakes, L.A., and J.E. Francis. 1988. A guide to Phanerozoic cold polar climates from high-latitude ice-rafting in the Cretaceous. Nature 333: 547–548.

Frankel, E. 1977. Previous *Acanthaster* aggregations in the Great Barrier Reef. Proc. 3rd Internat. Coral Reef Symp., Miami 1. Biology: 201–208.

Frankel, E. 1978. Evidence from the Great Barrier Reef of ancient *Acanthaster* aggregations. Atoll Res. Bull. 220: 75–86.

Frost, S.H. 1977. Cenozoic reef systems of Caribbean-prospects for paleoecologic synthesis. *In* S.H. Frost, M.P. Weiss, and J.B. Saunders (eds.), Reefs and related carbonates—ecology and sedimentology, pages 93–110. Stud. Geol. No. 4. American Association of Petroleum Geologists, Tulsa, Oklahoma.

Frydl, P., and G.W. Stearn. 1978. Rate of bioerosion by parrotfish in Barbados reef environments. J. Sediment. Petrol. 48: 1149–1157.

Fujioka, Y., and K. Yamazato. 1983. Host selection of some Okinawan coral associated gastropods belonging to the genera *Drupella, Coralliophila* and *Quoyula.* Galaxea 2: 59–73.

Fuller, R.W., J.H. Cardellina II, Y. Kato, L.S. Brinen, J. Clardy, K.M. Snader, and M.R. Boyd. 1992. A pentahalogenated monoterpene from the red alga *Portieria hornemanni* produces a novel cytotoxicity profile against a diverse panel of human tumor cell lines. J. Med. Chem. 35: 3007–3011.

Futuyma, D. 1986. Evolutionary biology. Sinauer, New York.

Garth, J.S. 1974. On the occurrence in the eastern tropical Pacific of Indo-west Pacific decapod crustaceans commensal with reef-building corals. Proc. 2nd Internat. Coral Reef Symp., Brisbane 1: 397–404.

Gawel, M.J. 1977. The common shallow-water soft corals (Alcyonacea) of Guam. M.Sc. thesis, University of Guam, 201 p.

Geider, R.J., T. Platt, and J.A. Raven. 1986. Size dependence of growth and photosynthesis in diatoms: a synthesis. Mar. Ecol. Progr. Ser. 30: 93–104.

Geister, J. 1977. The influence of wave exposure on the ecological zonation of Caribbean reefs. Proc. 3rd Internat. Coral Reef Symp., Miami 1. Biology: 23–29.

Gerber, R.P., and N. Marshall. 1974a. Ingestion of detritus by the lagoon pelagic community at Eniwetok Atoll. Limnol. Oceanogr. 19: 815–824.

Gerber, R.P., and N. Marshall. 1974b. Reef pseudoplankton in lagoon trophic systems. Proc. 2nd Internat. Coral Reef Symp., Brisbane 1: 105–107.

Gibson, R. 1979–1983. Nemerteans of the Great Barrier Reef. 1: Anopla Palaeonemertea; 2: Anopla Heteronemertea (Baseodiscidae); 3: Anopla Heteronemertea (Lineidae); 4: Anopla Heteronemertea (Valenciniidae); 5: Enopla Hoplonemertea (Monostilifera); 6: Enopla (Polystylifera: Reptantia). Zool. J. Linn. Soc. 65: 307–337; 66: 137–160; 71: 171–235; 72: 165–174; 75: 269–296; 78: 73–104.

Gibson, R. 1995. Nemertean genera and species of the world: an annotated checklist of original names and description citations, synonyms, current taxonomic status, habitats and recorded zoogeographic distribution. J. Nat. Hist. 29: 271–562.

Gilchrist, S. 1985. Hermit crab corallivore activity. Proc. 5th Internat. Coral Reef Congr., Tahiti 5: 211–214.

Gilliland, R.L. 1989. Solar evolution. Palaeogeography, Palaeoclimatology, Palaeoecology 75: 35–55.

Ginsburg, R.N. 1983. Geological and biological roles of cavities in coral reefs. *In* D.J. Barnes (ed.), Perspectives on coral reefs, pages 148–153. Brian Clouston Publ., Manuka, Australia, 277 p.

Ginsburg, R.N. (compiler). 1993. Proceedings of the Colloquium and Forum on Global Aspects of Coral Reefs: Health, Hazards, and History. Rosenstiel School of Marine and Atmospheric Sciences, University of Miami, 420 p.

Ginsburg, R.N., and E.A. Shinn. 1993. Preferential distribution of reefs in the Florida reef tract: the past is a key to the present. *In* R.N. Ginsburg (compiler), Proceedings of the Colloquium and Forum on Global Aspects of Coral Reefs: Health, Hazards, and History, pages 21–26. Rosenstiel School of Marine and Atmospheric Sciences, University of Miami, 420 p.

Gladfelter, W.G. 1982. Whiteband disease in *Acropora palmata:* implications for the structure and growth of shallow reefs. Bull. Mar. Sci. 32: 639–643.

Glazebrook, J.S., and R.S.F. Campbell. 1990a. A survey of the diseases of marine turtles in northern Australia. I. Farmed turtles. Diseases Aquat. Org. 9: 83–95.

Glazebrook, J.S., and R.S.F. Campbell. 1990b. A survey of the diseases of marine turtles in northern Australia. II. Oceanarium-reared and wild turtles. Diseases Aquat. Org. 9: 97–104.

Glynn, P.W. 1970. On the ecology of the Caribbean chitons *Acanthopleura granulata* Gmelin and *Chiton tuberculatus* Linné: density, mortality, feeding, reproduction, and growth. Smithson. Contr. Zool. 66: 1–21.

Glynn, P.W. 1972. Observations on the ecology of the Caribbean and Pacific coasts of Panama. Bull. Biol. Soc. Wash. 2: 13–20.

Glynn, P.W. 1973. Ecology of a Caribbean coral reef. The *Porites* reef-flat biotope. II. Plankton community with evidence for depletion. Mar. Biol. 20: 297–318.

Glynn, P.W. 1976. Some physical and biological determinants of coral community structure in the eastern Pacific. Ecol. Monogr. 46: 431–456.

Glynn, P.W. 1980. Defense by symbiotic crustacea of host corals elicited by chemical cues from predator. Oecologia 47: 287–290.

Glynn, P.W. 1982. *Acanthaster* population regulation by shrimp and a worm. Proc. 4th Internat. Coral Reef Symp., Manila 2: 607–612.

Glynn, P.W. 1983a. Crustacean symbionts and the defense of corals: coevolution on the reef? *In* M.H. Nitecki (ed.), Coevolution, pages 111–178. University of Chicago Press, Chicago.

Glynn, P.W. 1983b. Increased survivorship in corals harboring crustacean symbionts. Mar. Biol. Lett. 4: 105–111.

Glynn, P.W. 1984. Widespread coral mortality and the 1982–1983 El Niño warming event. Env. Conserv. 11: 133–146.

Glynn, P.W. 1985. El Niño-associated disturbance to coral reefs and post disturbance mortality by *Acanthaster planci.* Mar. Ecol. Progr. Ser. 26: 295–300.

Glynn, P.W. 1987. Some ecological consequences of coral-crustacean guard mutualisms in the Indian and Pacific Oceans. Symbiosis 4: 301–324.

Glynn, P.W. 1988a. El Niño warming, coral mortality and reef framework destruction by echinoid bioerosion in the eastern Pacific. Galaxea 7: 129–160.

Glynn, P.W. 1988b. Predation on coral reefs: some key processes, concepts and research directions. Proc. 6th Internat. Coral Reef Symp., Townsville 1: 51–62.

Glynn, P.W. (ed.) 1990a. Global ecological consequences of the 1982–1983 El Niño—Southern oscillation. Elsevier Oceanography Series, Amsterdam, 563 p.

Glynn, P.W. 1990b. Feeding ecology of selected coral-reef macro consumers: patterns and effects on coral community structure. *In* Z. Dubinsky (ed.), Ecosystems of the world, vol. 25: Coral reefs, pages 365–400. Elsevier Science Publishing, New York, 550 p.

Glynn, P.W. 1991. Coral reef bleaching in the 1980s and possible connections with global warming. Trends Ecol. Evol. 6: 175–179.

Glynn, P.W. 1993. Coral reef bleaching: ecological perspectives. Coral Reefs 12: 1–17.

Glynn, P.W., and M.W. Colgan. 1988. Defense of corals and enhancement of coral diversity by territorial damselfishes. Proc. 6th Internat. Coral Reef Symp., Townsville 2: 157–163.

Glynn, P.W., and M.W. Colgan. 1992. Sporadic disturbances in fluctuating coral reef environments: El Niño and coral reef development in the Eastern Pacific. Amer. Zool. 32: 707–718.

Glynn, P.W., and L. D'Croz. 1990. Experimental evidence for high temperature stress as the cause of El Niño—coincident coral mortality. Coral Reefs 8: 181–191.

Glynn, P.W., and W.H. de Weerdt. 1991. Elimination of two reef-building hydro corals following the 1982–83 El Niño warming event. Science 253: 69–71.

Glynn, P.W., and J.S. Feingold. 1992. Hydrocoral species not extinct. Science 257: 1845.

Glynn, P.W., N.J. Gassman, C.M. Eakin, J. Cortes, D.B. Smith, and H. Guzman. 1991. Reef coral reproduction in the eastern Pacific: Costa Rica, Panama, and Galapagos Islands (Ecuador) I. Pocilloporidae. Mar. Biol. 109: 355–368.

Glynn, P.W., and D.A. Krupp. 1986. Feeding biology of a Hawaiian seastar corallivore *Culcita novaeguineae* Muller and Troschel. J. Exp. Mar. Biol. Ecol. 96: 75–96.

Glynn, P.W., and R.H. Stewart. 1973. Distribution of coral reefs in the Pearl Islands (Gulf of Panama) in relation to thermal conditions. Limnol. Oceanogr. 18: 367–378.

Glynn, P.W., R.H. Stewart, and J.E. McCosker. 1972. Pacific coral reefs of Panamá: structure, distribution and predators. Geol. Rundschau 61: 483–519.

Glynn, P.W., A.M. Szmant, E.F. Corcoran, and S.V. Cofer-Shabica. 1989. Condition of coral reef cnidarians from the northern Florida reef tract: pesticides, heavy metals, and histopathological examination. Mar. Pollut. Bull. 20: 568–576.

Glynn, P.W., J.E.N. Veron, and G.M. Wellington. 1996. Clipperton Atoll (eastern Pacific): oceanography, geomorphology, reef-building coral ecology and biogeography. Coral Reefs 15: 71–99.

Glynn, P.W., and G.M. Wellington. 1983. Corals and coral reefs of the Galapagos Islands (with an annotated list of the scleractinian corals of the Galapagos by J.W. Wells). University of California Press, Berkeley, 330 p.

Glynn, P.W., G.M. Wellington, and C. Birkeland. 1979. Coral growth in the Galapagos: limitation by sea urchins. Science 203: 47–49.

Gofas, S., J. Pinto Afonso, and M. Brandao. 1981. Conchase Moluscos de Angola/ Coquillages et Mollusques d'Angola. Elf Aquitaine Angola, Universida de Agostinho Neto.

Goggin, C.L., and R.J.G. Lester. 1987. Occurrence of *Perkinsus* species (Protozoa, Apicomplexa) in bivalves from the Great Barrier Reef. Diseases Aquat. Org. 3: 113–117.

Goldberg, D.E. 1990. Components of resource competition in plant communities. *In* J.B. Graceand D. Tilman (eds.), Perspectives on plant competition. Academic Press, San Diego.

Goldberg, D.E., and A.M. Barton. 1992. Patterns and consequences of interspecific competition in natural communities: a review of field experiments with plants. Am. Nat. 139: 771–801.

Goldman, B., and F.H. Talbot. 1976. Aspects of the ecology of coral reef fishes. *In* O.A. Jones and R. Endean, (eds.), Biology and geology of coral reefs. III. Biology 2, pages 125–254. Academic Press, New York.

Golley, F., H.T. Odum, and R.F. Wilson. 1962. The structure and metabolism of a Puerto Rican redman grove forest in May. Ecology 43: 9–19.

Golubic, S., R.D. Perkins, and K.J. Lukas. 1975. Boring microorganisms and microborings in carbonate substrates. *In* R.W. Frey (ed.), The study of trace fossils, pages 229–259. Springer-Verlag, New York.

Gomez, E.D., A.C. Alcala, and A.C. San Diego. 1982. Status of Philippine coral reefs–1981. Proc. 4th Internat. Coral Reef Symp., Manila 1: 275–282.

Goreau, T.F. 1964. On the predation of coral by the spiny starfish *Acanthaster planci* (L.) in the southern Red Sea. Bull. Mar. Sci. 23: 399–464.

Gorlick, D.L., P.D. Atkins, and G.S. Losey, Jr. 1978. Cleaning stations as water holes, garbage dumps, and sites for the evolution of reciprocal altruism? Am. Nat. 112: 341–353.

Gorlick, D.L., P.D. Atkins, and G.S. Losey. 1987. Effect of cleaning by *Labroides dimidia-*

tus (Labridae) on ectoparasite population infecting *Pomacentrus vaiuli* (Pomacentridae) at Enewetak Atoll. Copeia 1987: 41–45.

Gosliner, T. 1987a. Biogeography of the opisthobranch gastropod fauna of Southern Africa. Amer. Malacol. Bull. 5: 243–258.

Gosliner, T. 1987b. Nudibranchs of southern Africa. Sea Challengers, Monterey, California.

Gosliner, T.M., and R. Darheim. 1996. Indo-Pacific opisthobranch gastropod biogeography: how do we know what we don't know. Amer. Malacological Bull. 12: in press.

Gosliner, T., D. Behrens, and G. Williams. 1996. Indo-Pacific reef invertebrates. Sea Challengers, Monterey, California.

Grant, J., and G. Gust. 1987. Prediction of coastal sediment stability from photo pigment content of mats of purple sulfur bacteria. Nature 330: 244–246.

Grassle, J.F. 1973. Variety in coral reef communities. *In* O.A. Jones and R. Endean (eds.), Biology and geology of coral reefs. II. Biology 1, pages 247–270. Academic Press, New York.

Grassle, J.F., and N.J. Maciolek. 1992. Deep-sea species richness: regional and local diversity estimates from quantitative bottom samples. Am. Nat. 139: 313–341.

Grauss, R.R., and I.G. Macintyre. 1982. Variations in growth forms of the reef coral *Montastrea annularis:* a quantitative evaluation of growth response to light distribution using computer simulation. *In* K. Rützler and I.G. Macintyre (eds.), The Atlantic barrier reef ecosystem at Carrie Bow Cay, Belize I. Structure and communities, pages 441–464. Smithson. Contr. Mar. Sci. 12. 539 p. Smithsonian Institution Press, Wash., D.C.

Grauss, R., I.G. Macintyre, and B.E. Herchenroder. 1984. Computer simulation of the reef zonation at Discovery Bay, Jamaica: hurricane disruption and long-term physical oceanographic controls. Coral Reefs 3: 59–68.

Greenway, M. 1976. The grazing of *Thalassia testudinum* in Kingston Harbor, Jamaica. Aquat. Bot. 2: 117–126.

Grigg, R.W. 1977. Fishery management of precious corals in Hawaii. Proc. 3rd Internat. Coral Reef Symp., Miami 2, Geology: 609–616.

Grigg, R.W. 1982. Darwin Point: a threshold for atoll formation. Coral Reefs 1: 29–35.

Grigg, R.W. 1984. Resource management of precious corals: a review and application to shallow water reef building corals. Mar. Ecol. 5 (1): 57–74.

Grigg, R.W., and S.J. Dollar. 1990. Natural and anthropogenic disturbance on coral reefs. *In* Z. Dubinsky (ed.), Ecosystems of the world, vol. 25: Coral reefs, pages 439–452. Elsevier Science Publishing, New York, 550 p.

Grigg, R.W., and R. Hey. 1992. Paleoceanography of the tropical eastern Pacific Ocean. Science 255: 172–178.

Grigg, R.W., J.J. Polovina, and M.J. Atkinson 1984. Model of a coral reef ecosystem. III. Resource limitation, community regulation, fisheries yield and resource management. Coral Reefs 3: 23–27.

Grigg, R.W., J.M. Wells, and C. Wallace. 1981. *Acropora* in Hawaii, Part 1: history of the scientific record, systematics and ecology. Pac. Sci. 35: 1–13.

Grimm, G.R., and R.N. Clayshulte. 1982. Demersal plankton from Western Shoals, Apra Harbor, Guam. Proc. 4th Internat. Coral Reef Symp., Manila 1: 454 (Abstr.).

Groombridge, B. 1992. Global biodiversity: status of Earth's living resources. Chapman and Hall, London.

Grotzinger, J.P. 1989. Facies and evolution of Precambrian carbonate depositionals systems: emergence of the modern platform archetype. *In* P.D. Crevello, J.L. Wison, J.F. Sarg, and J.F. Read (eds.), Controls on carbonate platform and basin development, pages 79–106. SEPM Special Publication No. 44. Society of Economic Paleontologists and Mineralogists, Tulsa, Oklahoma.

Guille, A., P. Laboute, and J.L. Menou. 1986. Guide des étioles de mer, oursins et autres échinodermes du lagonde Nouvelle Calédonie. Éditions del' ORSTOM, Paris.

Gulland, J.A. 1976. Production and catches of fish in the sea. *In* D.H. Cushing and J.J. Walsh (eds.), The ecology of the seas, pages 283–314. W.B. Saunders Co., Philadelphia.

Guzman, H.M., and I. Holst. 1993. Effects of chronic oil-sediment pollution on the reproduction of the Caribbean coral *Siderastrea siderea*. Mar. Pollut. Bull. 26: 276–282.

Habe, T. 1964. Shells of the Western Pacific in color, II. Koikusha, Osaka, Japan.

Hackney, J.M., R.C. Carpenter, and W.H. Adey. 1989. Characteristic adaptations to grazing among algal turfs on a Caribbean coral reef. Phycologia 28: 109–119.

Hada, Y. 1932. A note of earlier stage of colony formation with the coral *Pocillopora caespitosa*. Sci. Rep. Tohuku Univ. 6: 425–431.

Hadfield, M.G., and J.T. Pennington. 1990. Nature of the metamorphic signal and its internal transduction in larvae of the nudibranch *Phestilla sibogae*. Bull. Mar. Sci. 46: 455–465.

Haigler, S.A. 1969. Boring mechanism of *Polydora websteri* inhabiting *Crassostrea virginica*. Am. Zool. 9: 821–828.

Hairston, N.G., F.E. Smith, and L.B. Slobodkin. 1960. Community structure, population control and competition. Am. Nat. 94: 421–425.

Hallam, A. 1985. Jurassic molluscan migration and evolution in relation to sealevel change. *In* G.M. Friedman (ed.), Sedimentary and evolutionary cycles, pages 4–5. Springer-Verlag, Berlin.

Hallock, P. 1987. Fluctuations in the trophic resource continuum: a factor in global biodiversity cycles. Palaeoceanography 2: 457–471.

Hallock, P. 1988a. The role of nutrient availability in bioerosion: consequences to carbonate buildups. Palaeogeography, Palaeoclimatology, Palaeoecology 62: 275–291.

Hallock, P. 1988b. Interoceanic differences in Foraminifera with symbiotic algae: a result of nutrient supplies? Proc. 6th Internat. Coral Reef Symp., Townsville 3: 251–255.

Hallock, P., A.C. Hine, G.A. Vargo, J.A. Elrod, and W.C. Jaap. 1988. Platforms of the Nicaraguan Rise: examples of the sensitivity of carbonate sedimentation to excess trophic resources. Geology 16: 1104–1107.

Hallock, P., and M.W. Peebles. 1993. Foraminifera with chlorophyte endosymbionts: habitats of six species in the Florida Keys. Mar. Micropaleontol. 20: 277–292.

Hallock, P., I. Premoli Silva, and A. Boersma. 1991. Similarities between planktonic and larger foraminiferal evolutionary trends through Paleogene paleoceanographic changes. Palaeogeography, Palaeoclimatology, Palaeoecology 83: 49–64.

Hallock, P., F.E. Müller-Karger, and J.C. Halas. 1993. Coral reef decline. Natl. Geogr. Res. Explor. 9: 358–378.

Hallock, P., and W. Schlager. 1986. Nutrient excess and the demise of coral reefs and carbonate platforms. Palaios 1: 389–398.

Hamilton, L., and P.N. King. 1983. Tropical forested watersheds. Westview, Boulder, Colorado.

Hammond, L.S. 1983. Nutrition of deposit-feeding holothuroids and echinoids (Echinodermata) from a shallow reef lagoon, Discovery bay, Jamaiaca. Mar. Ecol. Prog. Ser. 10: 297–305.

Hamner, W.M., and J.H. Carleton. 1979. Copepod swarms: attributes and role in coral reef ecosystems. Limnol. Oceanogr. 24: 1–14.

Hamner, W.M., and E. Wolanski. 1988. Hydrodynamic forcing functions and biological processes on coral reefs: a status review. Proc. 6th Internat. Coral Reef Symp., Townsville 1: 103–113.

Hand, C. 1966. On the evolution of the Actiniaria. *In* W.J. Rees (ed.), The Cnidaria and their evolution, pages 135–140 (Symp. Zool. Soc. London 16) Academic Press, London.

Hansen, J.A., D.M. Alongi, D.J.W. Moriarty, and P.C. Pollard. 1987. The dynamics of benthic microbial communities at Davies Reef, central Great Barrier Reef. Coral Reefs 6: 63–70.

Hansen, T. 1980. Influence of larval dispersal and geographic distribution on species longevity in neogastropods. Paleobiology 6: 193–207.

Haq, B.U., J. Hardenbol, and P.R. Vail. 1987. Chronology of fluctuating sealevels since the Triassic (250 million years ago to present). Science 235: 1156–1167.

Haq, B.U., and F.W.B. Van Eysinga. 1987. Geologic time table. Elsevier Science Publishing, Amsterdam.

Hardin, G. 1968. The tragedy of the commons. Science 162: 1243–1248.

Hardin, G., and J. Baden. 1977. Managing the commons. W.H. Freeman, New York, 294 p.

Harland, A.D., and N.R. Nganro. 1990. Copper uptake by the sea anemone *Anemoni aviridis* and the role of zooxanthellae in metal regulation. Mar. Biol. 104: 297–301.

Harmelin-Vivien, M.L. 1989. Reef fish community structure: an Indo-Pacific comparison. *In* M.L. Harmelin-Vivien and F. Bourliere (eds.), Vertebrates in complex tropical systems, pages 21–60. Springer-Verlag, New York.

Harmelin-Vivien, M.L., and Y. Bouchon-Navaro. 1982. Trophic relationships among chaetodontid fishes in the Gulf of Aqaba (Red Sea). Proc. 4th Internat. Coral Reef Symp., Manila 2: 537–544.

Harmelin-Vivien, M.L., and Y. Bouchon-Navaro. 1983. Feeding diets and significance of coral feeding among chaetodontid fishes in Moorea (French Polynesia). Coral Reefs 2: 119–127.

Harmelin-Vivien, M.L., and P. Laborute. 1986. Catastrophic impact of hurricanes on atoll reef slopes in the Tuamotu (French Polynesia). Coral Reefs 5: 55–62.

Harmelin-Vivien, M.L., M. Peyrot-Clausade, and J.C. Romano. 1992. Transformation of algal turf by echinoids and scarid fishes on French Polynesian coral reefs. Coral Reefs 11: 45–50.

Harrigan, J.F. 1972. The planula larva of *Pocillopora damicornis,* lunar periodicity of swarming and substratum selection behavior. Ph.D. dissertation, University of Hawaii.

Harriot, V.J. 1983a. Reproductive ecology of four scleractinian species at Lizard Island, Great Barrier Reef. Coral Reefs 2: 9–18.

Harriot, V.J. 1983b. Reproductive seasonality, settlement, and post-settlement mortality of *Pocillopora damicornis* (Linnaeus), at Lizard Island, Great Barrier Reef. Coral Reefs 2: 151–157.

Harrison, P.L. 1985. Sexual characteristics of scleractinian corals: systematic and evolutionary implications. Proc. 5th Internat. Coral Reef Congr., Tahiti 4: 337–342.

Harrison, P.L., R.C. Babcock, G.D. Bull, J.K. Oliver, C.C. Wallace, and B.L. Willis. 1984. Mass spawning in tropical reef corals. Science 223: 1186–1189.

Harrison, P.L., and C.C. Wallace. 1990. Coral reproduction. *In* Z. Dubinsky (ed.), Ecosystems of the world, vol. 25: Coral reefs, pages 133–208. Elsevier Science Publishing, New York, 550 p.

Harrison, R.G. 1990. Hybrid zones: windows on evolutionary process. *In* D. Futuyma and J. Antonovics (eds.), Oxford surveys in evolutionary biology, vol. 7, pages 69–118. Oxford University Press, New York.

Hartman, W.D. 1982. Porifera. *In* S.P. Parker (ed.), Synopsis and classification of living organisms, pages 640–666. McGraw-Hill, New York.

Hartman, W.D., and T.F. Goreau. 1975. A Pacific tabulate sponge, living representative of a new order of sclerosponges. Postilla. Peabody Museum (Yale Univ.) 167: 1–14, figs. 1–15.

Harvell, C.D. 1984. Predator-induced defense in a marine bryozoan. Science 224: 1357–1359.

Harvell, C.D. 1986. The ecology and evolution of inducible defenses in a marine bryozoan: cues, costs, and consequences. Am. Nat. 128: 810–823.

Harvell, C.D. 1990. The ecology and evolution of inducible defenses. Quart. Rev. Biol. 65: 323–340.

Hatcher, A.I., and C.A. Frith. 1985. The control of nitrate and ammonium concentrations in a coral reef lagoon. Coral Reefs 4: 101–110.

Hatcher, B.G. 1982. The interaction between grazing organisms and the epilithic algal community of a coral reef: a quantitative assessment. Proc. 4th Internat. Coral Reef Symp., Manila 2: 515–524.

Hatcher, B.G. 1984. A maritime accident provides evidence for alternate stable states in benthic communities on coral reefs. Coral Reefs 3: 199–204.

Hatcher, B.G. 1988. The primary productivity of coral reefs: a beggar's banquet. Trends Ecol. Evol. 3: 106–111.

Hatcher, B.G. 1990. Coral reef primary productivity: a hierarchy of pattern and process. Trends Ecol. Evol. 5: 149–155.

Hatcher, B.G., J. Imberger, and S.V. Smith. 1987. Scaling analysis of coral reef systems: an approach to problems of scale. Coral Reefs 5: 171–181.

Hatcher, B.G., R.E. Johannes, and A.I. Robertson. 1989. Review of research relevant to the conservation of shallow tropical marine systems. Oceanogr. Mar. Biol. Ann. Rev. 27: 337–414.

Hatcher, B.G., and A.W.D. Larkum. 1983. An experimental analysis of factors controlling the standing crop of the pilithic algal community on a coral reef. J. Exp. Mar. Biol. Ecol. 69: 61–84.

Havel, J.E. 1987. Predator-induced defenses: a review. Pages 263–278 *In* W.C. Kerfoot and A. Sih (eds.), Predation: direct and indirect impacts on aquatic communities, University Press of New England, Hanover, New Hampshire, 386 p.

Hay, M.E. 1981a. Spatial patterns of grazing intensity on a Caribbean barrier reef: herbivory and algal distribution. Aquat. Bot. 11: 97–109.

Hay, M.E. 1981b. Herbivory, algal distribution, and the maintenance of between-habitat diversity on a tropical fringing reef. Am. Nat. 118: 520–540.

Hay, M.E. 1984a. Predictable spatial escapes from herbivory: how do these affect the evolution of herbivore resistance in tropical marine communities? Oecologia 64: 396–407.

Hay, M.E. 1984b. Patterns of fish and urchin grazing on Caribbean coral reefs: are previous results typical? Ecology 65: 446–454.

Hay, M.E. 1986. Associational plant defenses and the maintenance of species diversity: turning competitors into accomplices. Am. Nat. 128: 617–641.

Hay, M.E. 1991. Fish–seaweed interactions on coral reefs: effects of herbivorous fishes and adaptations of their prey. *In* P.F. Sale (ed.), The ecology of fishes on coral reefs, pages 96–119. Academic Press, San Diego, 754 p.

Hay, M.E. 1992. The role of seaweed chemical defenses in the evolution of feeding specialization and in the mediation of complex interactions. *In* V.J. Paul (ed.), Ecological roles of marine natural products. Comstock Publishing Associates, Ithaca, New York.

Hay, M.E., T. Colburn, and D. Downing. 1983. Spatial and temporal patterns in herbivory on a Caribbean fringing reef: the effects on plant distribution. Oecologia 58: 299–308.

Hay, M.E., J.E. Duffy, and W. Fenical. 1990a. Host-plant specialization decreases predation on a marine amphipod: an herbivore in plant's clothing. Ecology 71: 733–743.

Hay, M.E., J.E. Duffy, V.J. Paul, P.E. Renaud, and W. Fenical. 1990b. Specialist herbivores reduce their susceptibility to predation by feeding on the chemically-defended seaweed *Avrainvillea longicaulis*. Limnol. Oceanogr. 35: 1734–1743.

Hay, M.E., and W. Fenical. 1988. Marine plant–herbivore interactions: the ecology of chemical defense. Ann. Rev. Ecol. Syst. 19: 111–145.

Hay, M.E., J.R. Pawlik, J.E. Duffy, and W. Fenical. 1989. Seaweed-herbivore-predator interactions: host-plant specialization reduces predation on small herbivores. Oecologia 81: 418–427.

Hay, M.E., and P.R. Taylor. 1985. Competition between herbivorous fishes and urchins on Caribbean reefs. Oecologia 65: 591–598.

Hayami, I., and T. Kase. 1993. Submarine cave Bivalvia from the Ryukyu Islands: systematics and evolutionary significance. Univ. Mus., Univ. Tokyo, Bull. 35: 1–133.

Hayashibara, T., K. Shimoike, T. Kimura, S. Hosaka, A. Heyward, P. Harrison, K. Kudo, and M. Omori. 1993. Patterns of coral spawning at Akajima Island, Okinawa, Japan. Mar. Ecol. Progr. Ser. 101: 253–262.

Hayes, J.A. 1990a. Distribution, movement and impact of the corallivorous gastropod *Coralliophila abbreviata* (Lamarck) on a Panamanian patch reef. J. Exp. Mar. Biol. Ecol. 142: 25–42.

Hayes, J.A. 1990b. Prey preference in a caribbean corallivore, *Coralliophila abbreviata* (Lamarck) (Gastropoda, Coralliophilidae). Bull. Mar. Sci. 47: 557–560.

Heatwole, H. 1987. Sea snakes. The New South Wales University Press.

Heckel, P.H. 1974. Carbonate buildups in the geologic record: a review. *In* L.F. Laporte (ed.), Reefs in time and space, pages 90–154. SEPM Spec. Pub. 18. Society of Economic Paleontologists and Mineralogists, Tulsa, Oklahoma.

Hedgpeth, J.W. 1993. Foreign invaders. Science 261: 34–35.

Heinmann, M., C.D. Keeling, and C.J. Tucker. 1989. A three-dimensional model of atmospheric CO_2 transport based on observed winds 3. Seasonal cycle and synoptic time scale variations. Geophysical Monographs 55: 277–303.

Hellebust, J.A. 1974. Extracellular products. *In* W.D.P. Stewart (ed.), Algal physiology and biochemistry, pages 838–863. Blackwell Scientific, Oxford.

Henderson, R.A. 1992. Assessment of crown-of-thorns skeletal elements in surface sediment of the Great Barrier Reef. Coral Reefs 11: 103–108.

Henderson, R.A., and P.D. Walbran. 1992. Interpretation of the fossil record of *Acanthaster planci* from the Great Barrier Reef: a reply to criticism. Coral Reefs 11: 95–101.

Hendler, G., J.E. Miller, D.L. Pawson, and P.M. Kier. 1995. Seastars, sea urchins, and allies. Echinoderms of Florida and the Caribbean. Smithsonian Press, Washington D.C.

Herring, P.J. 1972. Observation on the distribution and feeding habits of some littoral echinoids from Zanzibar. J. Nat. Hist. 6: 169–175.

Heyward, A.J. 1985. Sexual reproduction in five species of the coral *Montipora.* Tech. Rep. Hawaii Inst. Mar. Biol. 37: 170–178.

Heyward, A.J. 1989. Reproductive status of some Guam corals. Micronesica 21: 272–274.

Heyward, A.J., and R. Babcock. 1986. Self- and cross fertilization in scleractinian corals. Mar. Biol. 90: 191–195.

Heyward, A.J., and J.D. Collins. 1985. Growth and sexual reproduction in the scleractinian coral *Montipora digitata* (Dana). Austr. J. Mar. Freshw. Res. 36: 441–446.

Heyward, A.J., K. Yamazato, T. Yemin, and M. Minei. 1987. Sexual reproduction of corals in Okinawa. Galaxea 6: 331–343.

Higgins, R.P., and H. Thiele (eds.) 1988. Introduction to the study of meiofauna. Smithsonian, Washington, D.C.

Highsmith, R.C. 1980. Geographic patterns of coral bioerosion: a productivity hypothesis. J. Exp. Mar. Biol. Ecol. 46: 177–196.

Highsmith, R.C. 1981. Lime-boring algae in hermatypic coral skeletons. J. Exp. Mar. Biol. Ecol. 55: 267–281.

Highsmith, R.C. 1982. Reproduction by fragmentation in corals. Mar. Ecol. Progr. Ser. 7: 207–226.

Hinde, R. 1988. Symbiotic nutrition and nutrient limitation. Proc. 6th Internat. Coral Reef Symp., Townsville 1: 199–204.

Hinds, P.A., and D.L. Ballantine. 1987. Effects of the Caribbean threespot damselfish, *Stegastes planifrons* (Cuvier), on algal lawn composition. Aquat. Bot. 27: 299–308.

Hine, A.C., P. Hallock, M.W. Harris, H.T. Mullins, D.F. Belknap, and W.C. Jaap. 1988. *Halimeda* bioherms along an open seaway: Miskito Channel, Nicaraguan Rise, SW Caribbean Sea. Coral Reefs 6: 173–178.

Hine, A.C., and A.C. Neumann. 1977. Shallow carbonate bank margin growth and structure, Little Bahama Bank. Amer. Assoc. Petr. Geol. Bull. 61: 376–406.

Hixon, M.A. 1983. Fish grazing and community structure of reef corals and algae: a synthesis of recent studies. *In* M.L. Reaka (ed.), The ecology of deep and shallow coral reefs, pages 79–87. NOAA Symp. Ser. Undersea Res. 1 (1). 149 p.

Hixon, M.A. 1986. Fish predation and local prey diversity. *In* C.A. Simenstad and G.M. Cailliet (eds.), Contemporary studies on fish feeding, pages 235–257. Junk Publ., Dordrecht.

Hixon, M.A. 1991. Predation as a process structuring coral reef fish communities. *In* P.F. Sale (ed.), The ecology of fishes on coral reefs, pages 437–508. Academic Press, San Diego, 754 p.

Hixon, M.A., and J.P. Beets. 1993. Predation, prey refuges, and the structure of coral-reef fish assemblages. Ecol. Monogr. 63: 77–101.

Hixon, M.A., and W.N. Brostoff. 1983. Damselfish as keystone species in reverse: intermediate disturbance and diversity of reef algae. Science 220: 511–513.

Hixon, M.A., and W.N. Brostoff. 1985. Substrate characteristics, fish grazing, and epibenthic reef assemblages off Hawaii. Bull. Mar. Sci. 37: 200–213.

Hixon, M.A., and W.N. Brostoff. 1996. Succession and herbivory: effects of differential fish grazing on Hawaiian coral-reef algae. Ecol. Monogr. 66: 67–90.

Hixon, M.A., and B.A. Menge. 1991. Species diversity: prey refuges modify the interactive effects of predation and competition. Theor. Pop. Biol. 39: 178–200.

Hobson, E.S. 1991. Trophic relationships of fishes specialized to feed on zooplankters above coral reefs. *In* P.J. Sale (ed.), The ecology of fishes on coral reefs, pages 69–95. Academic Press, San Diego, 754 p.

Hodgson, G. 1989. The effects of sedimentation on Indo-Pacific reef corals. Ph.D. dissertation. University of Hawaii, Honolulu.

Hodgson, G. 1990. Tetracycline reduces sedimentation damage to corals. Mar. Biol. 104: 493–496.

Hodgson, G. 1992. An alternative to "paperparks." *In* S.K. Yap and S.W. Lee (eds.), Proceedings of the International Conference on Conservation of Tropical Biodiversity,

Kuala Lumpur, 12–16 June 1990, pages 158–165. Malayan Nature Society, Kuala Lumpur, Malaysia, 656 p.

Hodgson, G. 1993a. Sedimentation damage to reef corals. *In* Ginsburg, R.N. (compiler), Proceedings of the Colloquium and Forum on Global Aspects of Coral Reefs: Health, Hazards, and History, pages 520–525. Rosenstiel School of Marine and Atmospheric Sciences, University of Miami, 420 p.

Hodgson, G. 1993b. The environmental impact of marine dredging in Hong Kong. Paper presented at the 2nd International Conference on the Marine Biology of the South China Sea, Guangzhou, Peoples Republic of China, 3–7 April 1993.

Hodgson, G., and J.A. Dixon. 1988. Logging versus fisheries and tourism: economic dimensions. Occasional Paper. East–West Center Environment and Policy Institute, Honolulu.

Hodgson, G., and J.A. Dixon. 1992. Sedimentation damage to marine resources: environmental and economic analysis. *In* J.B. Marsh (ed.), Resources and environment in Asia's Marine Sector, pages 421–446. Taylor & Francis, Washington, D.C., 457 p.

Hoegh-Guldberg, O. 1994. Population dynamics of symbiotic zooxanthellae in the coral *Pocillopora damicornis* exposed to elevated ammonium [$(NH_4)_2SO_4$] concentrations. Pac. Sci. 48: 263–272.

Hoeksema, B.W. 1989. Taxonomy, phylogeny and biogeography of mushroom corals (Scleractinia: Fungiidae). Zoologische Verhandelingen 254: 1–295.

Hoeksema, B.W., and M.B. Best. 1991. New observations on scleractinian corals from Indonesia 2. Sipunculan-associated species belonging to the genera *Heterocyathus* and *Heteropsammia.* Zoologische Mededelingen 65: 221–245.

Holt, R.D. 1977. Predation, apparent competition, and the structure of prey communities. Theor. Pop. Biol. 11: 197–229.

Holt, R.D. 1984. Spatial heterogeneity, indirect interactions, and the coexistence of prey species. Am. Nat. 124: 377–406.

Holt, R.D., and B.P. Kotler. 1987. Short-term apparent competition. Am. Nat. 130: 412–430.

Hooper, J.N.A., and C. Lévi. 1994. Biogeography of Indo-west Pacific sponges: Microcionidae, Raspailiidae, Axinellidae. *In* R.W.M. Soestvan, T.M.G. Kempenvan, and J.C. Brakeman (eds.), Sponges in time and space: biology, chemistry, paleontology, pages 191–212. Proc. 4th Internat. Porifera Congress, Amsterdam, Netherlands. A.A. Balkema, Rotterdam.

Hooper, J.N.A., and F. Wiedenmayer. 1994. Porifera. *In* W.W.D. Houston (ed.), Zoological catalogue of Australia. CSIRO Australia, Melbourne 12.

Hopkinson, C.S., B.F. Sherr, and H.W. Ducklow. 1987. Microbial regeneration in the water column of Davies Reef, Australia. Mar. Ecol. Progr. Ser. 41: 147–153.

Hopley, D. 1982. The geomorphology of the Great Barrier Reef. Wiley, New York, 453 p.

Horn, M.H. 1989. Biology of marine herbivorous fishes. Oceanogr. Mar. Biol. Ann. Rev. 27: 167–272.

Houbrick, R.S. 1978. The family Cerithiidae in the Indo-Pacific. Part 1: The genera *Rhinoclavis, Pseudovertagus* and *Clavocerithium.* Monogr. Mar. Mollusca 1: 1–130.

Hourigan, T.F. 1989. Environmental determinants of butterflyfish social systems. Env. Biol. Fish. 25: 61–78.

Houston, W.W.D. 1983. (ed.) Zoological catalogue of Australia. CSIRO Australia, Melbourne.

Howarth, F.G. 1991. Environmental impacts of classical biological control. Ann. Rev. Entomol. 36: 485–509.

Huang, C., and A. Sih. 1990. Experimental studies on behaviorally mediated, indirect interactions through a shared predator. Ecology 71: 1515–1522.

Hubbard, D.K. 1986. Sedimentation as a control of reef development. Coral Reefs 5: 117–125.

Hubbard, D.K. 1987. A general review of sedimentation as it relates to environmental stress in the Virgin Islands Biosphere Reserve and the eastern Caribbean in general. Biosphere Reserve Research Report 20. U.S. National Park Service, 42 p.

Hubbard, D.K. 1989. Modern carbonate environments of St. Croix and the Caribbean: a general overview. *In* D.K. Hubbard (ed.), Terrestrial and marine geology of St. Croix, U.S. Virgin Islands, pages 85–94. Spec. Pub. No. 8. West Indies Laboratory, St. Croix, USVI.

Hubbard, D.K. 1992a. A modern example of reef backstepping from eastern St. Croix, USVI. Abstracts and Program, AAPG Annual Meeting, Calgary, Canada, 57.

Hubbard, D.K. 1992b. Hurricane-induced sediment transport in open-shelf tropical systems—an example from St. Croix, U.S. Virgin Islands. J. Sedimentary Petrology 62: 946–960.

Hubbard, D.K., R.B. Burke, and I.P. Gill. 1986. Styles of reef accretion along a steep, shelf-edge reef, St. Croix, U.S. Virgin Islands. J. Sedimentary Petrology 56: 848–861.

Hubbard, D.K., A.I. Miller, and D. Scaturo. 1990. Production and cycling of calcium carbonate in a shelf-edge reef system (St. Croix, U.S. Virgin Islands): applications to the nature of reef systems in the fossil record. J. Sedimentary Petrology 60: 335–360.

Hubbard, D.K., and D. Scaturo. 1985. Growth rates of seven species of scleractinean corals from Cane Bay and Salt River, St. Croix, U.S. Virgin Islands. Bull. Mar. Sci. 36: 325–338.

Hubbard, D.K., J.D. Stump, and B. Carter. 1987. Sedimentation and reef development in Hawksnest, Fish and Reef Bays, St. John, U.S. Virgin Islands. Biosphere Reserve Research Report 21. Virgin Islands Resource Management Cooperative/National Park Service, 99 p.

Hubbard, D.K., L.G. Ward, D.M. FitzGerald, and A.C. Hine. 1974. Bank margin morphology and sedimentation, Lucaya, Grand Bahama Island. Tech. Report No. 7—CRD. Department of Geology, University of South Carolina, Columbia, 36 p.

Hubbard, J.A.E.B., and Y. Pocock. 1974. Sediment rejection by recent scleractinean corals: a key to paleoenvironmental reconstruction. Geol. Rundschau 61: 598–626.

Hudson, J.H. 1981. Growth rates in *Montastrea annularis:* a record of environmental change in Key Largo coral reef marine sanctuary, Florida. Bull. Mar. Sci. 31: 444–459.

Hudson, J.H., and R. Diaz. 1988. Damage survey and restoration of M/V WELLWOOD grounding site, Molasses Reef, Key Largo National Marine Sanctuary, Florida. Proc. 6th Internat. Coral Reef Symp., Townsville 2: 231–236.

Hufschmidt, M.M., D.E. James, A.D. Meister, B.T. Bower, and J.A. Dixon. 1983. Environment, natural systems, and development. Johns Hopkins University Press, Baltimore, 338 p.

Huggett, R.J., R.A. Kimerle, P.M. Mehrle, Jr., and H.L. Bergman. 1992. Biomarkers: biochemical, physiological, and histological markers of anthropogenic stress. Lewis Publishers, Boca Raton, Florida.

Hughes, T.P. 1984. Population dynamics based on individual size rather than age: a general model with a reef coral example. Am. Nat. 123: 778–795.

Hughes, T.P. 1985. Life histories and population dynamics of early successional corals. Proc. 5th Internat. Coral Reef Congr., Tahiti 4: 101–106.

Hughes, T.P. 1989. Community structure and diversity of coral reefs: the role of history. Ecology 70: 275–279.

Hughes, T.P. 1994. Catastrophes, phase shifts, and large-scale degradation of a Caribbean coral reef. Science 265: 1547–1551.

Hughes, T.P., and J.B.C. Jackson. 1980. Do corals lie about their age? Some demographic consequences of partial mortality, fission and fusion. Science 209: 713–715.

Hughes, T.P., D.C. Reed, and M.J. Boyle. 1987. Herbivory on coral reefs: community structure following mass mortalities of sea urchins. J. Exp. Mar. Biol. Ecol. 113: 39–59.

Humann, P. 1989. Reef fish identification. New World Publications, Jacksonville, Florida.

Humann, P. 1992. Reef creature identification. New World Publications, Jacksonville, Florida.

Humann, P. 1993. Reef coral identification. New World Publications, Jacksonville, Florida.

Humfrey, M. 1975. Seashells of the West Indies. A guide to the marine molluscs of the Caribbean. Taplinger, New York.

Humphrey, J.D. 1988. Disease risks associated with translocation of shellfish, with special reference to the giant clam *Tridacna gigas*. *In* J.W. Copland and J.S. Lucas (eds.), Giant clams in Asia and the Pacific, pages 241–244. Australian Centre for International Agricultural Research, Canberra.

Hunte, W., and D. Younglao. 1988. Recruitment and population recovery in the black sea urchin *Diadema antillarum* in Barbados. Mar. Ecol. Prog. Ser. 45: 109–119.

Hunter, C.L. 1988. Genotypic diversity and population structure of the Hawaiian reef coral, *Porites compressa*. Ph.D. dissertation, Univ. of Hawaii, Honolulu.

Hunter, C.L. 1988. Environmental cues controlling spawning in two Hawaiian corals: *Montipora verrucosa* and *M. dilatata*. Proc. 6th Internat. Coral Reef Symp., Townsville 2: 727–732.

Huston, M.A. 1985. Patterns of species diversity on coral reefs. Ann. Rev. Ecol. Syst. 16: 149–177.

Hutchings, P.A. 1986. Biological destruction of coral reefs. Coral Reefs 4: 239–252.

Hviding, E., and G.B.K. Baines. 1992. Fisheries management in the Pacific: tradition and the challenges of development in Marovo, Solomon Islands. Discussion Paper 32. U.N. Res. Instit. Social Development, Geneva.

Ikehara, I.I., H.T. Kami, and R.K. Sakamoto. 1970. Exploratory fishing survey of the inshore fisheries resources of Guam. Proc. 2nd CSK Symp., Tokyo: 425–437.

Iliffe, T.M., H. Wilkens, J. Parzefall and D. Williams. 1984. Marine lava cave fauna: composition, biogeography, and origins. Science 225: 309–311.

Ingersoll, C.G. 1995. Sediment tests. *In* G.M. Rand (ed.), Fundamentals of aquatic toxicology: effects, environmental fate, and risk assessment, pages 231–255. Taylor and Francis, Washington, D.C.

Isdale, P.J. 1981. Geographical variation in the growth rate of the hermatypic coral *Porites* in the Great Barrier Reef Province. Unpublished Ph.D. dissertation, James Cook University, North Queensland, Australia.

Isdale, P.J. 1984a. Geographical patterns in coral growth rates on the Great Barrier Reef. *In* Baker, J.T., R.M. Carter, P.W. Sammarco, and K.P. Stark (eds.), Proceedings of the Great Barrier Reef Conference, pages 327–330. Townsville, Australia.

Isdale, P.J. 1984b. Fluorescent bands in massive corals record centuries of coastal rainfall. Nature 310: 578–579.

Itzkowitz, M. 1979. The feeding strategies of a facultative cleaner fish, *Thalassoma bifasciatum* (Pisces: Labridae). J. Zool. London. 187: 403–413.

Ivlev, V.S. 1961. Experimental ecology of the feeding of fishes. Yale Univ. Press, New Haven, Connecticut.

Jaap, W.C. 1984. The ecology of the south Florida reefs: a community profile. FWS/OBS-82/08. U.S. Fish and Wildlife Service, Washington, D.C.

Jablonski, D. 1986. Causes and consequences of mass extinctions: a comparative approach. *In* D.K. Elliott (ed.), Dynamics of extinction, pages 183–229. Wiley, New York.

Jablonski, D. 1993. The tropics as a source of evolutionary novelty through geological time. Nature 364: 142–144.

Jackson, J.B.C. 1991. Adaptation and diversity of reef corals. BioScience 41: 475–482.

Jackson, J.B.C. 1992. Pleistocene perspectives on coral reef community structure. Am. Zool. 32: 719–731.

Jackson, J.B.C. 1994. Constancy and change of life in the sea. Philosophical Trans. Roy. Soc. London B. Biol. Sci. 344: 55–60.

Jackson, J.B.C., J.D. Cubit, B.D. Keller, V. Batista, K. Burns, H.M. Caffey, R.L. Caldwell, S.D. Garrity, C.D. Gelter, C. Gonzalez, H.M. Guzman, K.W. Kaufmann, A.H. Knap, S.C. Levings, M.J. Marshall, R. Sleger, R.C. Thampson, and E. Weit. 1989. Ecological effects of a major oil spill on Panamanian coastal marine communities. Ecology 243: 37–44.

Jackson, J.B.C., T.F. Goreau, and W.D. Hartman. 1971. Recent brachiopod-coralline sponge communities and their paleoecological significance. Science 173: 623–625.

Jackson, J.B.C., P. Jung, A.G. Coates, and L.S. Collins. 1993. Diversity and extinction of tropical American mollusks and emergence of the Isthmus of Panama. Science 260: 1624–1626.

Jackson, J.B.C., J.E. Winston, and A.G. Coates. 1985. Niche breadth, geographic range, and extinction of Caribbean reef-associated cheilostome Bryozoa and Scleractinia. Proc. 5th Internat. Coral Reef Congr., Tahiti 4: 151–158.

Jacobsen, E.R., C. Buergelt, B. Williams, and R.K. Harris. 1991. Herpes virus in cutaneous fibropapillomas of the green turtle *Chelonias mydas*. Diseases Aquat. Org. 12: 1–6.

James, M.J. ed. 1991. Galápagos marine invertebrates. Plenum, New York.

James, N.P. 1983. Reefs in carbonate depositional environments. *In* P.A. Scholle, D.G. Bebout, and C.H. Moore (eds.), AAPG Memoir 33, pages 345–462. American Association of Petroleum Geologists, Tulsa, Oklahoma.

Jastrow, R., and M.H. Thompson. 1972. Astronomy: fundamentals and frontiers. Wiley, New York.

Johannes, R.E. 1975. Pollution and degradation of coral reef communities. *In* E.J. Ferguson Wood and R.E. Johannes (eds.), Tropical marine pollution, pages 13–51. Elsevier Oceanogr. Ser. 12. Elsevier, Amsterdam, 192 p.

Johannes, R.E. 1978a. Traditional marine conservation methods in Oceania and their demise. Ann. Rev. Ecol. Syst. 9: 349–364.

Johannes, R.E. 1978b. Reproductive strategies of coastal marine fishes in the tropics. Env. Biol. Fish. 3: 65–84.

Johannes, R.E. 1980a. Using knowledge of the reproductive behaviour of reef and lagoon fishes to improve yields. *In* J. Bardach, J. Magnuson, R. May, and J. Reinhart (eds.), Fish behavior and fisheries management (capture and culture), pages 247–270. ICLARM, Manila.

Johannes, R.E. 1980b. The ecological significance of the submarine discharge of groundwater. Mar. Ecol. Prog. Ser. 3: 365–373.

Johannes, R.E. 1981. Words of the lagoon: fishing and marine lore in the Palau District of Micronesia. University of California Press.

Johannes, R.E. 1993. Integrating traditional ecological knowledge and management with environmental impact assessment. *In* J.T. Inglis (ed.), Traditional ecological knowledge: concepts and cases, pages 33–40. International Progress on Traditional Ecological Knowledge and International Development Research Centre.

Johannes, R.E. 1994. Government-supported, village-based management of marine resources in Vanuatu Report #94/2. South Pacific Forum Fisheries Agency.

Johannes, R.E., and J.W. MacFarlane. 1990. Assessing traditional fishing rights systems in the context of fisheries management: a Torres Strait example. *In* K. Ruddle and R.E. Johannes (eds.), Traditional management of coastal systems in the Asia and the Pacific: a compendium, pages 241–261. UNESCO, Jakarta.

Johannes, R.E., and J.W. MacFarlane. 1991. Traditional fishing in the Torres Strait Islands. CSIRO, 210 p.

Johannes, R.E., and M. Riepen. 1995. Environmental, economic, and social implications of the live reef fish trade in Asia and the western Pacific. Report to The Nature Conservancy and the South Pacific Forum Fisheries Agency. 80 p.

Johannes, R.E., S.L. Coles, and N.T. Kuenzel. 1970. The role of zooplankton in the nutrition of some scleractinian corals. Limnol. Oceanogr. 15: 579–586.

Johnson, D. 1988. Development of *Mytilus edulis* embryos: a bioassay for polluted waters. Mar. Ecol. Prog. Ser. 46: 135–138.

Johnson, J.H. 1971. An introduction to the study of organic limestones. Quarterly of the Colorado School of Mines 66: 1–185.

Johnson, K.G., A.F. Budd, and T.A. Stemann. 1995. Extinction selectivity and ecology of Neogene Caribbean reef corals. Paleobiology 21: 52–73.

Johnson, M.W. 1974. On the dispersal of lobster larvae into the East Pacific barrier (Decapoda, Palinuridea). Fish. Bull. 72: 639–647.

Johnston, I.S. 1980. The ultrastructure of skeletogenesis in hermatypic corals. Internat. Rev. Cytology 67: 171–214.

Jokiel, P.L. 1985. Lunar periodicity of planula release in the reef coral *Pocillopora damicornis* in relation to various environmental factors. Proc. 5th Internat. Coral Reef Congr., Tahiti 4: 307–312.

Jokiel, P.L. 1986. Growth of the reef coral *Porites compressa* on the Coconut Island reef, Kaneohe Bay. *In* P.L. Jokiel, R.H. Richmond, and R.A. Rogers (eds.), Coral reef population biology, pages 101–110. Tech. Rept. No. 37. Hawaii Institute of Marine Biology.

Jokiel, P.L., and S.L. Coles. 1990. Response of Hawaiian and other Indo-Pacific reef corals to elevated temperature. Coral Reefs 8: 155–162.

Jokiel, P.L., C.L. Hunter, S. Taguchi, and L. Watarai. 1993. Ecological impact of a freshwater "reef kill" in Kaneohe Bay, Oahu, Hawaii. Coral Reefs 12: 177–184.

Jokiel, P.L., R.Y. Ito, and P.M. Liu. 1985. Night irradiance and synchronization of lunar release of planula larvae in the reef coral *Pocillopora damicornis.* Mar. Biol. 88: 167–174.

Jokiel, P.L., and J.E. Maragos. 1978. Reef corals of Canton Atoll: II. Local distribution. Atoll Res. Bull. 221: 71–98.

Jokiel, P., and F.J. Martinelli. 1992. The vortex model of coral reef biogeography. J. Biogeogr. 19: 449–458.

Jokiel, P.L., and J.I. Morrissey. 1986. Influence of size on primary production in the reef coral *Pocillopora damicornis* and the macroalga *Acanthophora spicifera.* Mar. Biol. 91: 15–26.

Jokiel, P.L., and R.H. York. 1982. Solar ultraviolet photobiology of the reef coral *Pocillopora damicornis* and symbiotic zooxanthellae. Bull. Mar. Sci. 32: 301–315.

Jokiel, P.L., and R.H. York. 1984. Importance of ultraviolet in microalgae growth photo inhibition. Limnol. Oceanogr. 29: 192–199.

Jones, G.P., D.J. Ferrell, and P.F. Sale. 1991. Fish predation and its impact on the invertebrates of coral reefs and adjacent sediments. *In* P.F. Sale (ed.), The ecology of fishes on coral reefs, pages 156–179. Academic Press, San Diego, 754 p.

Jones, R.S. 1968. Ecological relationships in Hawaiian and Johnston Island Acanthuridae (surgeonfishes). Micronesica 4: 309–361.

Josselyn, M.N., G.M. Cailliet, T.M. Niesen, R. Cowen, A.C. Hurley, J. Connor, and S. Hawes. 1983. Composition, export and faunal utilization of drift vegetation in the Salt River submarine canyon. Est. Coastal Shelf Sci. 17: 447–465.

Kabat, A.R. 1996. Biogeography of the genera of Naticidae (Gastropoda) in the Indo-Pacific. Amer. Malacol. Bull. 13: in press.

Kamura, S., and S. Choonhabandit. 1986. Distribution of benthic marine algae on the coasts of Khang Khao and Thai Ta Mun, Sichang Islands, the Gulf of Thailand. Galaxea 5: 97–114.

Kaneshiro, K.Y. 1987. The dynamics of sexual selection and its pleiotropic effects. Behav. Genet. 17: 559–569.

Karlson, R.H., and D.R. Levitan. 1989. Recruitment-limitation in open populations of *Diadema antillarum:* an evaluation. Oecologia 82: 40–44.

Karplus, I. 1978. A feeding association between the grouper *Epinephelus fasciatus* and the moray eel *Gymnothorax griseus.* Copeia 1978: 164.

Karuso, P. 1987. Chemical ecology of the nudibranchs. *In* P.J. Scheuer (ed.) Bioorganic marine chemistry, vol. 1, pages 31–60. Springer-Verlag, Berlin.

Kase, T., and I. Hayami. 1992. Unique submarine cave mollusc fauna: composition, origin and adaptation. J. Molluscan Stud. 58: 446–449.

Kauffman, E.G., and C.C. Johnson. 1988. The morphological and ecological evolution of middle and upper Cretaceous reef building rudistids. Palaios 3: 194–126.

Kaufman, L. 1977. The threespot damselfish: effects on benthic biota of Caribbean coral reefs. Proc. 3rd Internat. Coral Reef Symp., Miami 1: 559–564.

Kawaguti, S. 1940. An abundance of reef coral planulae in plankton. Zool. Mag. (Tokyo) 52: 31.

Kawaguti, S. 1941. On the physiology of reef corals. V. Tropisms of coral planulae considered as a factor of distribution on the reef. Palao Trop. Biol. Stn. Stud. 2: 319–328.

Kawanabe, H., J.E. Cohen, and K. Iwasaki (eds.) 1993. Mutualism and community organization: behavioural, theoretical, and food-web approaches. Oxford University Press, New York, 426 p.

Kay, E.A. 1967. The composition and relationships of marine molluscan fauna of the Hawaiian Islands. Venus 25: 94–104.

Kay, E.A. 1979a. Hawaiian marine shells. Reef and shore fauna of Hawaii. Section 4: Mollusca. Bishop Museum Press, Honolulu.

Kay, E.A. 1979b. Little worlds of the Pacific. An essay on Pacific Basin biogeography. Univ. Hawaii Harold L. Lyon Arboretum Lecture Number 9: 1–40.

Kay, E.A. 1984. Patterns of speciation in the Indo-West Pacific. *In* F.J. Radovsky, P.H. Raven, and S.H. Sohmer (eds.), Biogeography of the tropical Pacific, pages 15–31. Honolulu, Spec. Publ. No. 72. Bishop Museum Press, Honolulu.

Kay, E.A. 1990. Cypraeidae of the Indo-Pacific: Cenozoic fossil history and biogeography. Bull. Mar. Sci. 47: 23–34.

Kay, E.A., and S.R. Palumbi. 1987. Endemism and evolution in Hawaii and marine invertebrates. Trends Ecol. Evol. 2: 183–186.

Kaye, C. 1959. Shoreline features and Quaternary shoreline changes, Puerto Rico. USGS Prof. Paper 317-B. U.S. Govt. Printing Office, Washington, D.C., 140 p.

Keen, A.M. 1971. Seashells of tropical West America. Stanford University Press, Stanford.

Keesing, J.K., R.H. Bradbury, L.M. DeVantier, M.J. Riddle, and G. De'ath. 1992. Geological evidence for recurring outbreaks of the crown-of thorns starfish: a reassessment from an ecological perspective. Coral Reefs 11: 79–85.

Keesing, J.K., and J.S. Lucas. 1992. Field measurement of feeding and movement rates of the crown-of-thorns starfish *Acanthaster planci* (L.). J. Exp. Mar. Biol. Ecol. 156(1): 89–104.

Keigwin, L.D., Jr. 1982b. Isotopic paleoceanography of the Caribbean and East Pacific: role of Panama uplift in Late Neogene time. Science 217: 350–353.

Keigwin, L.D., Jr. 1982a. Neogene planktonic foraminifers from Deep Sea Drilling Project Sites 502 and 503. Initial Reports of the Deep Sea Drilling Project 68: 269–288.

Kelleher, G.K. 1990. The reef and conservation strategies. Reeflections No. 25: 4–5.

Kelleher, G.K. 1993. Can the Great Barrier Reef model of protected a areas save reefs worldwide? P28–P34 *In* R.N. Ginsburg (compiler), Proceedings of the Colloquium and Forum on Global Aspects of Coral Reefs: Health, Hazards, and History, pages P28–P34. University of Miami, Rosenstiel School of Marine and Atmospheric Sciences, 420 p.

Kelleher, G.K., and R.A. Kenchington. 1991. Guidelines for establishing marine protected areas. A marine conservation and development report. IUCN, Gland, Switzerland.

Kelly-Borges, M., and C. Valentine. 1995. The sponges of the tropical island region of Oceania: a taxonomic review. *In* J.E. Maragos, M.N.A. Peterson, L.G. Eldredge, J.E. Bardach, and H.F. Takeuchi (eds.), Marine and coastal biodiversity in the tropical island Pacific region, vol. 1. Species systematics and in formation management priorities, pages 83–120. East-West Center, University of Hawaii, Honolulu.

Kenchington, R.A. 1990. Managing marine environments. Taylor and Francis, New York, 248 p.

Kenchington, R.A., and B.E.T. Hudson (eds.) 1987. UNESCO coral reef management handbook. Unesco, Jakarta, 321 p.

Kenchington, R., and G. Kelleher. 1992. Crown-of thorns starfish management conundrums. Coral Reefs 11: 53–56.

Kendrick, B., M.J. Risk, J. Michaelides, and K. Bergman. 1982. Amphibious microborers: bioeroding fungi isolated from live corals. Bull. Mar. Sci. 32: 862–867.

Kennett, J.P. 1982. Marine geology. Prentice-Hall, Englewood Cliffs, New Jersey.

Kenny, D., and C. Loehle. 1991. Are foodwebs randomly connected? Ecology 72: 1794–1799.

Kensley, B., and M. Schotte. 1989. Guide to the marine isopod crustaceans of the Caribbean. Smithsonian, Washington, D.C.

Kenyon, J.C. 1993. Chromosome number in ten species of the coral genus *Acropora.* Proc. 7th Internat. Coral Reef Symp., Guam 1: 471–475.

Kenyon, J.C. 1995. Latitudinal differences between Palau and Yap in coral reproductive synchrony. Pac. Sci. 49: 156–164.

Kerfoot, W.C., and A. Sih. 1987. Predation: direct and indirect impacts on aquatic communities. University Press of New England, Hanover, New Hampshire, 386 p.

Kerstitch, A. 1989. Sea of Cortez marine invertebrates. Sea Challengers, Monterey, California.

Kiene, W.E. 1988. A model of bioerosion on the Great Barrier Reef. Proc. 6th Internat. Coral Reef Symp., Townsville 3: 449–454.

Kilar, J.A., and J.N. Norris. 1988. Composition, export, and import of drift vegetation on a tropical, plant-dominated, fringing reef platform (Caribbean Panama). Coral Reefs 7: 93–103.

Kimmerer, W.J., E. Gartside, and J.J. Orsi. 1994. Predation by an introduced clam as the likely cause of substantial declines in zooplankton of San Francisco Bay. Mar. Ecol. Progr. Ser. 113: 81–93.

Kinne, O. (ed) 1980. Diseases of marine animals. I. General aspects, Protozoa to Gastropoda. Wiley, New York.

Kinsey, D.W. 1978. Productivity and calcification estimates using slack-water periods and field enclosures. *In* D.R. Stoddart and R.E. Johannes (eds.), Coral reefs: research methods, pages 439–468. UNESCO, Paris.

Kinsey, D.W. 1983. Standards of performance in coral reef primary production and carbonate turnover. *In* D.J. Barnes (ed.), Perspectives on coral reefs, pages 209–220. Brian Clouston Publ., Manuka, Australia, 277 p.

Kinsey, D.W. 1985a. Metabolism, calcification and carbon production. I. Systems level studies. Proc. 5th Internat. Coral Reef Congr., Tahiti 4: 505–526.

Kinsey, D.W. 1985b. The functional role of back reef and lagoonal systems in the central great Barrier Reef. Proc. 5th Internat. Coral Reef Congr., Tahiti 6: 223–228.

Kinsey, D.W. 1988. Coral reef system response to some natural and anthropogenic stresses. Galaxea 7: 113–128.

Kinsey, D.W., and P.J. Davies. 1979. Effects of elevated nitrogen and phosphorous on coral reef growth. Limnol. Oceanogr. 24: 935–940.

Kinsey, D.W., and D. Hopley. 1991. The significance of coral reefs as global carbon sinks—response to greenhouse. Palaeogeography, Palaeoclimatology, Palaeoecology 89: 363–377.

Kinzie, R.A. 1993. Spawning in the reef corals *Pocillopora verrucosa* and *P. eydouxi* at Sesoko Island, Okinawa. Galaxea 11: 93–105.

Kira, T. 1962. Shells of the Western Pacific in color. Koikusha, Osaka, Japan.

Klumpp, D.W., and A.D. McKinnon. 1992. Community structure, biomass and productivity of epilithic algal communities on the Great Barrier Reef: dynamics at different spatial scales. Mar. Ecol. Progr. Ser. 86: 77–89.

Klumpp, D.W., A.D. McKinnon, and P. Daniel. 1987. Damselfish territories: zones of high productivity on coral reefs. Mar. Ecol. Progr. Ser. 40: 41–51.

Klumpp, D.W., A.D. McKinnon, and C.N. Mundy. 1988. Motile cryptofauna of a coral reef: abundance, distribution and trophic potential. Mar. Ecol. Progr. Ser. 45: 95–108.

Klumpp, D.W., and N.V.C. Polunin. 1989. Partitioning among grazers of food resources within damselfish territories on a coral reef. J. Exp. Mar. Biol. Ecol. 125: 145–169.

Klumpp, D.W., and A. Pulfrich. 1989. Trophic significance of herbivorous macro invertebrates on the central Great Barrier Reef. Coral Reefs 8: 135–144.

Knowlton, N. 1992. Thresholds and multiple stable states in coral reef community dynamics. Am. Zool. 32: 674–682.

Knowlton, N. 1993. Sibling species in the sea. Ann. Rev. Ecol. Syst. 24: 189–216.

Knowlton, N., and J.B.C. Jackson. 1994. New taxonomy and niche partitioning on coral reefs: jack of all trades or master of some? Trends Ecol. Evol. 9: 7–9.

Knowlton, N., J.C. Lang, and B.D. Keller. 1990. Case study of natural population collapse: post-hurricane predation on Jamaican staghorn corals. Smithson. Contr. Mar. Sci. 31: 1–25.

Knowlton, N., J.C. Lang, M.C. Rooney, and P. Clifford. 1981. Evidence for delayed mortality in hurricane-damaged Jamaican staghorn corals. Nature 294: 251–252.

Knowlton, N., L.A. Weigt, L.A. Solórzano, D.K. Mills, and E. Bermingham. 1993. Divergence in proteins, mitochondrial DNA, and the reproductive compatibility across the Isthmus of Panama. Science 260: 1629–1632.

Knowlton, N., E. Weil, L.A. Weigt, and H.M. Guzman. 1992. Sibling species in *Montastrea annularis,* coral bleaching, and the coral climate record. Science 255: 330–333.

Kohn, A.J. 1959. The ecology of *Conus* in Hawaii. Ecol. Monogr. 29: 47–90.

Kohn, A.J. 1971. Diversity, utilization of resources, and adaptive radiation in shallow-water marine invertebrates of tropical oceanic islands. Limnol. Oceanogr. 16: 332–348.

Kohn, A.J. 1980. *Conus kahiko,* a new Pleistocene gastropod from Oahu, Hawaii. J. Paleontology 54: 534–541.

Kohn, A.J. 1990. Tempo and mode of evolution in Conidae. Malacologia 32: 55–67.

Kohn, A.J., and F.E. Perron. 1994. Life history and biogeography patterns in *Conus.* Clarendon Press, Oxford.

Kohn, A.J., and J.K. White. 1977. Polychaete annelids of an intertidal reef limestone platform at Tanguisson, Guam. Micronesia 13: 199–215.

Kojis, B.L., and N.J. Quinn. 1981. Aspects of sexual reproduction and larval development in the shallow water hermatypic coral *Goniastrea australensis.* Bull. Mar. Sci. 31: 558–573.

Kojis, B.L., and N.J. Quinn. 1982a. Reproductive strategies in four species of *Porites* (Scleractinia). Proc. 4th Internat. Coral Reef Symp., Manila 2: 145–151.

Kojis, B.L., and N.J. Quinn. 1982b. Reproductive ecology of two Faviid corals (Coelenterata: Scleractinia). Mar. Ecol. Prog. Ser. 8: 251–255.

Kojis, B.L., and N.J. Quinn. 1984. Seasonal and depth variation in fecundity of *Acropora palifera* at two reefs in Papua New Guinea. Coral Reefs 3: 165–172.

Kojis, B.L., and N.J. Quinn. 1985. Puberty in *Goniastrea favulus.* Age or size limited? Proc. 5th Internat. Coral Reef Congr., Tahiti 4: 289–293.

Kojis, B.L., and N.J. Quinn. 1993. Biological limits to Caribbean reef recovery, a comparison with western South Pacific reefs. *In* Ginsburg, R.N. (compiler), Proceedings of the Colloquium and Forum on Global Aspects of Coral Reefs: Health, Hazards, and History, pages P35–P41. Rosenstiel School of Marine and Atmospheric Sciences, University of Miami, 420 p.

Koop, K., and A.W.D. Larkum. 1987. Deposition of organic material in a coral reef lagoon, One Tree Island, Great Barrier Reef. Est. Coastal Shelf Sci. 25: 1–9.

Koslow, J.A., F. Hanley, and R. Wicklund. 1988. Effects of fishing on reef fish communities at Pedro Bank and Port Royal Cays, Jamaica. Mar. Ecol. Progr. Ser. 43: 201–212.

Kott, P. 1985. The Australian Ascidiacea Part 1, Phlebobranchia and Stolidobranchia. Mem. Queensl. Mus. 23: 1–440.

Kott, P. 1990a. The Australian Ascidiacea Part 2, Aplousobranchia (1). Mem. Queensl. Mus. 29: 1–266.

Kott, P. 1990b. The Australian Ascidiacea, Phlebobranchia and Stolidobranchia, Supplement. Mem. Queensl. Mus. 29: 267–298.

Kozloff, E.N. 1987. Marine invertebrates of the Pacific Northwest. University of Washington Press, Seattle, 511 p.

Kramp, P.L. 1961. Synopsis of the medusae of the world. J. Mar. Biol. Assoc. U.K. 40: 1–469.

Krupp, D.A. 1983. Sexual reproduction and early development of the solitary coral *Fungia scutaria* (Anthozoa: Scleractinia). Coral Reefs 2: 159–164.

Krupp, D.A., P.L. Jokiel, and T.S. Chartrand. 1993. Asexual reproduction by the solitary scleractinian coral *Fungia scutaria* on dead parent coralla in Kaneohe Bay, Oahu, Hawaiian Islands. Proc. 7th Internat. Coral Reef Symp., Guam 1: 527–534.

Kühlmann, D.H.H. 1985. Living coral reefs of the world. Arco Publishing, New York.

Kühlmann, D.H.H. 1988. The sensitivity of coral reefs to environmental pollution. Ambio 17: 13–21.

Labeyrie, J., C. Lalou, and G. Delibrias. 1969. Étude des transgressions marines sur l'atoll de Mururoa par la datation des differents niveaux de corail. Cahiers Pacifique 13: 59–68.

Laborel, J. 1974. West African reef corals, an hypothesis on their origin. Proc. 2nd Internat. Coral Reef Symp., Brisbane 1: 425–443.

Ladd, H.S. 1960. Origin of the Pacific island molluscan fauna. Amer. J. Sci. 258A: 137–150.

Ladd, H.S. 1961. Reef building. Science 134: 703–715.

Ladd, H.S., and S.O. Schlanger. 1960. Drilling operations on Enewetak Atoll. USGS Prof. Paper 260-Y. U.S. Govt. Printing Office, Washington, D.C.

Lamprell, K., and T. Whitehead. 1992. Bivalves of Australia, vol. 1. Crawford House Press, Bathurst, NSW.

Land, L.S. 1979. The fate of reef-derived sediment on the north Jamaican island slope. Mar. Geol. 29: 55–71.

Lapointe, B.E., M.M. Littler, and D.S. Littler. 1987. A comparison of nutrient-limited productivity in macroalgae from a Caribbean barrier reef and from a mangrove ecosystem. Aquat. Bot. 28: 243–255.

Larkin, P.A. 1978. Fisheries management—an essay for ecologists. Ann. Rev. Ecol. Syst. 9: 57–73.

Larkum, A.W.D. 1983. Primary productivity of plant communities on coral reefs. *In* D.J. Barnes (ed.), Perspectives on coral reefs, pages 221–230. Brian Clouston Publ., Manuka, Australia, 277 p.

Larkum, A.W.D., and A.D.L. Steven. 1994. ENCORE: The effect of nutrient enrichment on coral reefs: 1. Experimental design and research programme. Mar. Pollut. Bull. 29: 112–120.

Lassuy, D.R. 1980. Effects of "farming" behavior by *Eupomacentrus lividus* and *Hemiglyphidodon plagiometopon* on algal community structure. Bull. Mar. Sci. 30: 304–312.

Lawlor, L.R. 1979. Direct and indirect effects of n-species competition. Oecologia 43: 355–364.

Lawrence, J.M. 1975. On the relationships between marine plants and sea urchins. Oceanogr. Mar. Biol. Ann. Rev. 13: 213–286.

Ledua, E., and T.J.H. Adams. 1988. Quarantine aspects of the reintroduction of *Tridacna gigas* to Fiji. *In* J.W. Copland and J.S. Lucas (eds.), Giant clams in Asia and the Pacific, pages 237–240. Australian Centre for International Agricultural Research, Canberra.

Lee, J.J., and O.R. Anderson. 1991. Symbiosis in Foraminifera. *In* J.J. Lee and O.R. Anderson (eds.), Biology of Foraminifera, pages 157–220. Academic Press, London.

Lee, J.J., S.H. Hunter, and E.C. Bovee (eds.) 1985. An illustrated guide to the Protozoa. Society of Protozoologists, Lawrence, Kansas.

Lees, A. 1975. Possible influence of salinity and temperature on modern shelf carbonate sedimentation. Mar. Geol. 19: 159–198.

Legendre, L., S. Demers, B. Delesalle, and C. Harnois. 1988. Biomass and photosynthetic activity of phototrophic picoplankton in coral reef waters (Moorea Island, French Polynesia). Mar. Ecol. Progr. Ser. 47: 153–160.

Leis, J.M. 1983. Coral reef fish larvae (Labridae) in the East Pacific Barrier. Copeia 1983: 826–828.

Leis, J.M., T. Trnski, M. Harmelin-Vivien, J.P. Renon, V. Dufour, M.K. El Moudni, and R. Galzin. 1991. High concentration of tuna larvae (Pisces: Scombridae) in near-reef waters of French Polynesia (Society and Tuamotu Islands). Bull. Mar. Sci. 48: 150–158.

Lessios, H.A. 1988. Mass mortality of *Diadema antillarum* in the Caribbean: What have we learned? Ann. Rev. Ecol. Syst. 19: 371–393.

Lessios, H.A. 1995. *Diadema antillarum* 10 years after mass mortality: still rare, despite help from a competitor. Proc. Roy. Soc. Lond. B. 259: 331–337.

Lessios, H.A., D.R. Robertson, and J.D. Cubit. 1984. Spread of *Diadema* mass mortality through the Caribbean. Science 226: 335–337.

Levine, S.H. 1976. Competitive interactions in ecosystems. Am. Nat. 110: 903–910.

Levitan, D.R. 1988. Algal-urchin biomass responses following mass mortality of *Diadema antillarum* Philippi at Saint John, U.S. Virgin Islands. J. Exp. Mar. Biol. Ecol. 119: 167–178.

Levitan, D.R. 1991. Skeletal changes in the test and jaws of the sea urchin *Diadema antillarum* in response to food limitation. Mar. Biol. 111: 431–435.

Levitan, D.R. 1992. Community structure in times past: influence of human fishing pressure on algal-urchin interactions. Ecology 73: 1597–1605.

Levitan, D.R., M.A. Sewell, and F. Chia. 1992. How distribution and abundance influence fertilization success in the sea urchin *Strongylocentrotus franciscanus*. Ecology 73: 248–254.

Lewin, R.A., L. Cheng, and R.A. Alberte. 1983. *Prochloron*-ascidian symbioses: photosynthetic potential and productivity. Micronesica 19: 165–170.

Lewis, D.H., and D.C. Smith. 1971. The autotrophic nutrition of marine coelenterates with special reference to hermatypic corals. I. movement of photosynthetic products between the symbionts. Proc. Roy. Soc. Lond. B. 178: 111–129.

Lewis, J.B. 1966. Growth and breeding in the tropical echinoid *Diadema antillarum*. Bull. Mar. Sci. 16: 151–158.

Lewis, J.B. 1977. Processes of organic production of reef corals. Biol. Rev. 52: 305–347.

Lewis, J.B. 1981. Coral reef ecosystems. *In* A.R. Lewin (ed.), Analysis of marine ecosystems, pages 127–158. Academic Press, London.

Lewis, J.B. 1982. Estimates of secondary production of reef corals. Proc. 4th Internat. Coral Reef Symp., Manila 2: 369–374.

Lewis, S.M. 1985. Herbivory on coral reefs: algal susceptibility to herbivorous fishes. Oecologia 65: 370–375.

Lewis, S.M. 1986. The role of herbivorous fishes in the organization of a Caribbean reef community. Ecol. Monogr. 56: 183–200.

Lewis, S.M., J.N. Norris, and R.B. Searles. 1987. The regulation of morphological plasticity in tropical reef algae by herbivory. Ecology 68: 636–641.

Lewis, S.M., and P.C. Wainwright. 1985. Herbivore abundance and grazing intensity on a Caribbean coral reef. J. Exp. Mar. Biol. Ecol. 87: 215–228.

Liddell, W.D., and S.L. Ohlhorst. 1986. Changes in benthic community composition following the mass mortality of *Diadema* at Jamaica. J. Exp. Mar. Biol. Ecol. 95: 271–278.

Liddell, W.D., and S.L. Ohlhorst. 1988. Comparison of Western Atlantic coral reef communities. Proc. 6th Internat. Coral Reef Symp., Townsville 3: 281–286.

Lieber, M. 1994. More than a living: fishing and the social order in a Polynesian Atoll. Westview Press, Boulder, Colorado.

Lieske, E., and R.M. Myers. 1994. Collins' hand guide to the coral reef fishes of the world. Harper-Collins, London.

Limbaugh, C. 1961. Cleaning symbiosis. Sci. Amer. 205: 42–49.

Littler, D.S., M.M. Littler, K.E. Bucher, and J.N. Norris. 1989. Marine plants of the

Caribbean: a field guide from Florida to Brazil. Smithsonian Press, Washington, D.C., 263 p.

Littler, M.M., and D.S. Littler. 1980. The evolution of thallus form and survival strategies in benthic marine macroalgae: field and laboratory tests of a functional form model. Am. Nat. 116: 25–44.

Littler, M.M., and D.S. Littler. 1984. Models of tropical reef biogenesis: the contribution of algae. Progr. Phycol. Res. 3: 323–364.

Littler, M.M., and D.S. Littler. 1988. Structure and role of algae in tropical reef communities. *In* C.A. Lemmbi and J.R. Waaland (eds.), Algae and human affairs, pages 29–56. Cambridge Univ. Press,

Littler, M.M., and D.S. Littler. 1994. A pathogen of reef-building algae discovered in the South Pacific. Coral Reefs 13: 202.

Littler, M.M., and D.S. Littler. 1995. Impact of CLOD pathogen on Pacific coral reefs. Science 267: 1356–1360.

Littler, M.M., D.S. Littler, and P.R. Taylor. 1983a. Evolutionary strategies in a tropical barrier reef system: functional-form groups of marine macroalgae. J. Phycol. 19: 229–237.

Littler, M.M., D.S. Littler, and P.R. Taylor. 1987. Animal-plant defense associations: effects on the distribution and abundance of tropical reef macrophytes. J. Exp. Mar. Biol. Ecol. 105: 107–121.

Littler, M.M., P.R. Taylor, and D.S. Littler. 1983b. Algal resistance to herbivory on a Caribbean barrier reef. Coral Reefs 2: 111–118.

Littler, M.M., P.R. Taylor, and D.S. Littler. 1986. Plant defense associations in the marine environment. Coral Reefs 5: 63–71.

Littler, M.M., P.R. Taylor, and D.S. Littler. 1989. Complex interactions in the control of coral zonation on a Caribbean reef flat. Oecologia 80: 331–340.

Lively, C.M. 1986. Predator-induced shell dimorphism in the acorn barnacle *Chthamalus anisopoma.* Evolution 40: 232–242.

LMER Coordinating Committee. 1993. Understanding changes in coastal environments: the LMER Program. EOS 73: 481–485.

Lobel, P.S. 1980. Herbivory by damselfishes and their role in coral reef community ecology. Bull. Mar. Sci. 30: 273–289.

Loeblich, A.R.J., and H. Tappan. 1988. Foraminiferal genera and their classification, 2 volumes. Van Nostrand Reinhold, New York.

Loehle, C. 1990. Indirect effects: a critique and alternate methods. Ecology 71: 2382–2386.

Logan, B.W., J.F. Read, G.M. Hagan, P.F. Hoffman, R.G. Brown, P.J. Woods, and C.D. Gebelein. 1974. Evolution and diagenesis of Quaternary carbonate sequences, Shark Bay, Western Australia. Amer. Assoc. Petrol. Geol. Mem. 22: 1–358.

Longman, M.W. 1981. A process approach to recognizing facies of reef complexes. *In* D.F. Toomey (ed.), European fossil reef models, pages 9–40. SEPM Spec. Publ. No. 30. Society of Economic Paleontologists and Mineralogists, Tulsa, Oklahoma.

Losey, G.S., Jr. 1971. Communication between fishes in cleaning symbiosis. *In* T.C. Cheng (ed.), Aspects of the biology of symbiosis. University Park Press, Baltimore.

Losey, G.S., Jr. 1972. The ecological importance of cleaning symbiosis. Copeia 1972: 820–833.

Losey, G.S., Jr. 1974. Cleaning symbiosis in Puerto Rico with comparison to the tropical Pacific. Copeia 1974: 960–970.

Losey, G.S., Jr. 1979. Fish cleaning symbiosis: proximate causes of host behavior. Anim. Behav. 27: 669–685.

Lovelock, J.E. 1988. The ages of Gaia. Norton, New York.

Low, R.M. 1971. Interspecific territoriality in a pomacentrid reef fish, *Pomacentrus flavicauda* Whitley. Ecology 52: 648–654.

Lowenstam, H.A. 1974. Impact of life on chemical and physical processes. *In* E.D. Goldberg (ed.), The sea, vol. 5, Marine chemistry, pages 715–796. Wiley, New York.

Lowenstam, H.A., and S. Weiner. 1989. On biomineralization. Oxford University Press, New York.

Lowry, K., and H.J.M. Wickremeratne. 1987. Coastal area management in Sri Lanka. Ocean Year Book 7: 263–293.

Loya, Y. 1976. The Red Sea coral *Stylophora pistillata* is an r-strategist. Nature 259: 478–480.

Loya, Y., and B. Rinkevich. 1979. Abortion effect in corals induced by oil pollution. Mar. Ecol. Progr. Ser. 1: 77–80.

Loya, Y., and B. Rinkevich. 1980. Effects of oil pollution on coral reef communities. Mar. Ecol. Progr. Ser. 3: 167–180.

Lubchenco, J. 1978. Plant species diversity in a marine intertidal community: importance of herbivore food preference and algal competitive abilities. Am. Nat. 112: 23–39.

Lucas, J.S. 1973. Reproductive and larval biology of *Acanthaster planci* (L.) in Great Barrier Reef waters. Micronesica 9: 197–203.

Lucas, J.S. 1975. Environmental influences on the early development of *Acanthaster planci* (L.). *In* Crown-of-thorns starfish Seminar Proceedings, pages 109–121. Brisbane, Australian Government Publ. Service, Canberra.

Lucas, J.S. 1982. Quantitative studies of feeding and nutrition during larval development of the coral reef asteroid *Acanthaster planci* (L.). J. Exp. Mar. Biol. Ecol. 65: 173–193.

Lucas, J.S. 1984. Growth, maturation and effects of diet in *Acanthaster planci* (L.) (Asteroidea) and hybrids reared in the laboratory. J. Exp. Mar. Biol. Ecol. 79: 129–147.

Ludwig, D., R. Hilborn, and C. Walters. 1993. Uncertainty, resource exploitation, and conservation: lessons from history. Science 260: 17–18.

Lugo, A.E.G., M.M. Evink, A. Brinson, A. Broce, and S.C. Snedaker. 1973. Diurnal rates of photosynthesis, respiration and transpiration in mangrove forests of south Florida. *In* S.C. Snedaker and A.E. Lugo (eds.), The role of mangrove ecosystems in the maintenance of environmental quality and a high productivity of desirable fisheries, pages D-1–D-45. Final report submitted to the Bureau of Sport Fisheries and Wildlife.

Luyendyk, B.P., D. Forsyth, and J.D. Phillips. 1972. Experimental approach to paleocirculation of oceanic surface waters. Geol. Soc. Amer. Bull. 83: 2649–2664.

MacGruder, W.H., and J.W. Hunt. 1979. Seaweeds of Hawaii. Oriental Publishing Co., Honolulu.

Macintyre, I.G. 1975. A diver operated hydraulic drill for coring submerged substrates. Atoll Res. Bull. 185: 21–25.

Macintyre, I.G. 1984. Preburial and shallow-subsurface alteration of modern scleractinian corals. *In* W.A. Oliver., Jr., W.J. Sando, S.D. Cairns, A.G. Coates, I.G. Macintyre, F.M. Bayer, and J.E. Sorauf (eds.), Recent advances in the paleobiology And geology of the Cnidaria, pages 229–244. Palaeontographica Americana 54.

Macintyre, I.G., and P.W. Glynn. 1976. Evolution of modern Caribbean fringing reef, Galtea Point, Panama. Amer. Assoc. Petr. Geol. Bull. 60: 1054–1072.

MacKenzie, F.T., and J.W. Morse. 1992. Sedimentary carbonates through Phanerozoic time. Geochem. Cosmochem. Acta 56: 3281–3295.

Maclean, J. 1984. Red tide—a growing problem in the Indo-Pacific region. ICLARM Newsletter 7 (4): 20.

Madsen, F.J. 1955. A note on the sea-star genus *Acanthaster.* Vidensk, Meddrdan sknaturh. Foren. 117: 179–192.

Maier-Reimer, E., U. Mikalojewicz, and T. Crowley. 1990. Ocean GCM sensitivity experiment with an open Central American isthmus. Paleoceanography 5: 349–366.

Maluf, L.Y. 1988. Biogeography of the central eastern Pacific shelf echinoderms. *In* R.D. Burke, P.V. Mladenov, P. Lambert, and R.L. Parsley (eds.), Echinoderm biology, pages 389–398. A.A. Balkema, Rotterdam.

Manning, R.B., and L.B. Holthuis. 1981. West African brachyuran crabs (Crustacea: Decapoda). Smithson. Contr. Zool. 306: 1–379.

Maragos, J. 1972. A study of the ecology of Hawaiian coral reefs. Ph.D. dissertation, University Hawaii, Honolulu, 280 p.

Maragos, J.E. 1992. Restoring coral reefs with emphasis on Pacific reefs. *In* G.W. Thayer (ed.), Restoring the nations's marine environment, pages 141–221. Maryland Sea Grant Book, College Park.

Maragos, J.E. 1993. Impact of coastal construction on coral reefs in the U.S.-affiliated Pacific Islands. Coastal Management 21: 235–269.

Maragos, J.E., and M.E. Elliott. 1985. Coastal resource inventories in Hawaii, Samoa, and Micronesia. Proc. 5th Internat. Coral Reef Congr., Tahiti 5: 577–582.

Maragos, J.E., and K.Z. Meier. 1993. Reefs and corals of the main Palau Islands. *In* Rapid ecological assessment of Palau II. The main Palau Islands.

Margulis, L. 1993. Symbiosis and cell evolution. Freeman, New York.

Margulis, L., and R. Fester. 1991. Symbiosis as a source of evolutionary innovation: speciation and morphogenesis. Proc. Bellagio Conference.

Margulis, L., and D. Sagan. 1986. Microcosmos. Summit Books,

Marsh, J.A., Jr. 1977. Terrestrial inputs of nitrogen and phosphorous on fringing reefs of Guam. Proc. 3rd Internat. Coral Reef Symp., Miami 1. Biology: 331–336.

Marsh, J.A., Jr., and S.V. Smith. 1978. Productivity measurements of coral reefs in flowing water. *In* D.R. Stoddart and R.E. Johannes (eds.), Coral reefs: research methods, pages 361–378. UNESCO, Paris.

Marshall, N. 1985. Ecological sustainable yield (fisheries potential) of coral reef areas, as related to physiographic features of coral reef environments. Proc. 5th Internat. Coral Reef Congr. Tahiti 5: 525–530.

Marshall, S.M., and T.A. Stephenson. 1933. The breeding of reef animals. Part I. The corals. Sci. Rep. Great Barrier Reef Exped. 3: 219–245.

Marszalek, D.S. 1982. Impact of dredging on a subtropical reef community, southeast Florida, U.S.A. Proc. Fourth Internat. Coral Reef Symp., Manila 1: 147–153.

Marten, G.G., and J.J. Polovina. 1982. A comparative study of fish yields from various tropical ecosystems. *In* D. Pauly and G.I. Murphy (eds.), Theory and management of tropical fisheries, pages 255–286. ICLARM, Manila, Philippines.

Martin, J.H. 1990. Glacial–interglacial CO_2 exchange: their own hypothesis. Paleoceanography 5: 1–14.

Mathais, J.A., and N.P.E. Langham. 1978. Coral reefs. *In* T.E. Chua and N.P.E. Langham (eds.), Coastal resources of west Sabah, pages 122–151. University of Sains Malaysia, Penang.

Mather, P., and I. Bennett (eds.) 1993. A coral reef handbook. Australian Coral Reef Society, Brisbane, 264 p.

Mattson, J.S., and J.A. DeFoor, II. 1985. Natural resource damages: restitution as a mechanism to slow destruction of Florida's natural resources. J. Land Use and Env. Law 1 (3): 295–319.

May, J.A., I.G. Macintyre, and R.D. Perkins. 1982. Distribution of microborers within planted substrates along a barrier reef transect, Carrie Bow Cay, Belize. *In* K. Rützler and I.G. Macintyre (eds.), The Atlantic barrier reef ecosystem at Carrie Bow Cay, Belize I: structure and communities, pages 93–107. Smithson. Contrib. Mar. Sci. 12. 539 p.

May, R.M. 1988. Personnal communication, BBC radio interview.

May, R.M. 1994. Biological diversity: differences between land and sea. Philosophical Trans. Roy. Soc. London B 343: 105–111.

Mayor, J.W. 1915. On the development of the coral *Agaricia fragilis* (Dana). Proc. Amer. Acad. Arts Sci. 51: 485–511.

Mayr, E. 1971. Populations, species and evolution. Harvard University Press, Cambridge, Massachusetts.

Mayzaud, P., and S.A. Poulet. 1978. The importance of the time factor in the response of zooplankton to varying concentrations of naturally occurring particulate matter. Limnol. Oceanogr. 23: 1144–1154.

McCall, J., B. Rosen, and J. Darrell. 1994. Carbonate deposition in accretionary prism settings: Early Miocene coral limestones and corals of the Makran Mountain range in Southern Iran. Facies 31: 141–178.

McCarthy, J.F., and L.R. Shugart. 1980. Biomarkers of environmental contamination. Lewis Publ., Boca Raton, Florida.

McClanahan, T.R. 1989. Kenyan coral reef-associated gastropod fauna: a comparison between protected and unprotected reefs. Mar. Ecol. Progr. Ser. 53: 11–20.

McClanahan, T.R. 1990. Kenyan coral reef-associated gastropod assemblages: distribution and diversity patterns. Coral Reefs 9: 63–74.

McClanahan, T.R. 1992. Resource utilization, competition, and predation: a model and example from coral reef grazers. Ecological Modelling 61: 195–215.

McClanahan, T.R., A.T. Kamukuru, N.A. Muthiga, M. Gilagabher Yebio, and D. Obura. 1996. Effect of sea urchin reductions on algae, coral, and fish populations. Conserv. Biol. 10 (1): 136–154.

McClanahan, T.R., and J.D. Kurtis. 1991. Population regulation of the rock-boring sea urchin *Echinometra mathaei* (de Blainville). J. Exp. Mar. Biol. Ecol. 147: 121–146.

McClanahan, T.R., and N.A. Muthiga. 1988. Changes in Kenya coral reef community structure and function due to exploitation. Hydrobiologia 166: 269–276.

McClanahan, T.R., and N.A. Muthiga. 1989. Patterns of predation on a sea urchin, *Echinometra mathaei* (de Blainville), on Kenyan coral reefs. J. Exp. Mar. Biol. Ecol. 126: 77–94.

McClanahan, T.R., and S.H. Shafir. 1990. Causes and consequences of sea urchin abundance and diversity in Kenyan coral reef lagoons. Oecologia 83: 362–370.

McCloskey, L.R., and R.H. Chesher. 1971. Effects of man-made pollution on the dynamics of coral reefs. *In* J.W. Miller, J.G. VanDerwalker, and R.A. Waller (eds.), Scientists in the sea. Tektite II, pages VI-229–VI-237. U.S. Department of the Interior, Washington, D.C.

McConnaughey, T.A. 1989. Biomineralization mechanisms. *In* R.E. Crick (ed.), The origin, evolution, and modern aspects of biomineralization in animals and plants. Plenum Press, New York.

McDonald, J.H., and R.K. Koehn. 1988. The mussels *Mytilus galloprovincialis* and *M. trossulus* on the Pacific coast of North America. Mar. Biol. 99: 111–118.

McFarland, W.N. 1979. Observations on recruitment in haemulid fishes. Proc. Gulf and Carib. Fisheries Inst. 32: 132–138.

McFarland, W.N. 1991. The visual world of coral reef fishes. *In* P.J. Sale (ed.), The ecology of fishes on coral reefs, pages 16–38. Academic Press, San Diego, 754 p.

McFarland, W.N., E.B. Brothers, J.C. Ogden, M.J. Shulman, E.L. Bermingham, and N.M. Kotchian-Prentiss. 1985. Recruitment patterns in young French grunts, *Haemulon flavolineatum* (Family Haemulidae), at St. Croix, Virgin Islands. Fish. Bull. 83: 413–426.

McIvor, C.C., J.A. Ley, and R.D. Bjork. 1994. Changes in freshwater in flow from the Everglades to Florida Bay including effects on biota and biotic processes: a review. *In* S.M. Davis and J.C. Ogden (eds.), Everglades: The ecosystem and its restoration, pages 117–146. St. Lucie Press, Deray Beach Florida, 826 p.

McLean, R.F. 1967. Measurement of beach rock erosion by some tropical marine gastropods. Bull. Mar. Sci. 17: 551–561.

McManus, J.W. 1980. Philippine coral exports: the coral drain. ICLARM Newsletter, January: 18–20.

McManus, J.W. 1985. Marine speciation, tectonics and sea-level changes in Southeast Asia. Proc. 5th Internat. Coral Reef Congr., Tahiti 4: 133–138.

McManus, J.W., C.L. Nañola, R.B. Reyes, Jr., and K.N. Kesner. 1992. Resource ecology of the Bolinao coral reef system. ICLARM Stud. Rev. 22. 117 p.

McRoy, C.P. 1983. Nutrient cycles in Caribbean seagrass ecosystems. *In* J.C. Ogden and E.H. Gladfelter (eds.), Coral reefs, seagrass beds and mangroves: their interaction in the coastal zones of the Caribbean, pages 69–73. Reports in Marine Science 23. UNESCO, 133 p.

McRoy, C.P., J.J. Goering, and B. Eheney. 1973. Nitrogen fixation associated with seagrasses. Limnol. Oceanogr. 18: 998–1002.

McRoy, C.P., and D.S. Lloyd. 1982. Comparative function and stability of macrophyte based ecosystems. *In* A.R. Longhursted (ed.), Analysis of marine ecosystems, pages 473–490. Academic Press, London, 741 p.

Meehan, B.W., J.T. Carlton, and R. Wenne. 1989. Genetic affinities of the bivalve *Macoma balthica* from the Pacific coast of North America: evidence for recent introduction and historical distribution. Mar. Biol. 102: 235–241.

Meyer, J.L., and E.T. Schultz. 1985a. Migrating haemulid fishes as a source of nutrients and organic matter on coral reefs. Limnol. Oceanogr. 30: 146–156.

Meyer, J.L., and E.T. Schultz. 1985b. Tissue condition and growth rate of corals associated with schooling fish. Limnol. Oceanogr. 30: 157–166.

Meyer, J.L., E.T. Schultz, and G.S. Helfman. 1983. Fish schools: an asset to corals. Science 220: 1047–1049.

Millard, N.A.H. 1975. Monograph of the Hydroida of Southern Africa. Annals of the South African Museum 68: 1–513.

Miller, J.A. 1991. Washington watch. Does coral bleaching mean global warming? BioScience 41: 77.

Miller, M.L., and J. Auyong. 1991. Coastal zone tourism. Marine Policy, March: 75–99.

Miller, T.E., and W.C. Kerfoot. 1987. Redefining indirect effects. *In* W.C. Kerfoot and A. Sih (eds.), Predation: direct and indirect impacts on aquatic communities, pages 33–37. University Press of New England, Hanover, New Hampshire, 386 p.

Millero, F.J. 1979. The thermodynamics of the carbonate system in seawater. Geochem. Cosmochem. Acta 43: 1651–1661.

Milliman, J.D. 1974. Marine carbonates. Springer-Verlag, Berlin.

Milliman, J.D. 1992. River sediment discharge to the sea: new analysis of old data. Coastal systems studies and sustainable development. UNESCO Tech. Pap. Mar. Sci. 64: 56–66.

Milliman, J.D., and R.H. Meade. 1983. World-wide delivery of river sediment to the oceans. J. Geol. 91: 1–21.

Milliman, J.D., Y.S. Qin, M.E. Ren, and Y. Saito. 1987. Man's influence on the erosion and transport of sediment by Asian rivers: the Yellow River (Huanghe) example. J. Geol. 95: 751–762.

Millis, N.F. 1981. Marine microbiology. *In* M.N. Clayton and R.J. King (eds.), Marine botany: an Australasian perspective, pages 36–60. Longman Cheshire, Melbourne.

Mills, E.L., J.L. Forney, and K.J. Wagner. 1987. Fish predation and its cascading effect on the Oneida Lake food chain. *In* W.C. Kerfoot and A. Sih, (eds.), Predation: direct and indirect impacts on aquatic communities, pages 118–131. University Press of New England, Hanover, New Hampshire, 386 p.

Mitchell, R., and I. Chet. 1975. Bacterial attack of corals in polluted seawater. Microbial Ecology 2: 227–233.

Miyake, S. 1983. Japanese crustacean decapods and stomatopods in color, 2 vols. (in Japanese). Hoikusha Publishing Co., Japan.

Mojsisovics, E.M.V. 1879. Die Dolomit-Riffe von Sud Tirol und Venetien. A. Holden Co., Wien, 551 p.

Monniot, C., and F. Monniot. 1996. Ascidians of the Western Pacific Ocean Islands. Micronesica 29: in press.

Monniot, C., F. Monniot, and P. Laboute. 1985. Ascidies du port de Papeete (Polynésie française): relation save clemilieu nature let apports intercontinent aux parla navigation. Bull. du Muséum National d'Histoire Naturelle 4 esérie, section A (Zoologie, Biologieet Écologie Animales) 7: 481–495.

Monniot, C., F. Monniot, and P. Laboute. 1991. Coral reef ascidians. Editions del' ORS-TOM, Paris.

Montgomery, W.L. 1980. The impact of non-selective grazing by the giant blue damselfish, *Microspathodon dorsalis,* on algal communities in the Gulf of California, Mexico. Bull. Mar. Sci. 30: 290–303.

Montgomery, W.L., and S.D. Gerking. 1980. Marine macroalgae as foods for fishes: an evaluation of potential food quality. Env. Biol. Fish. 5: 143–153.

Moore, C.H., and W.W. Shedd. 1977. Effective rates of sponge bioerosion as a function of carbonate production. Proc. 3rd Internat. Coral Reef Symp., Miami 2. Geology: 499–506.

Moore, R.C. (ed.) 1953–. Treatise on invertebrate paleontology. University of Kansas, Lawrence, Kansas.

Moosleitner, V.H. 1980. Cleaner fish and cleaner shrimps in the Mediterranean. Zool. Anz. Jena 205: 219–240.

Moran, P.J. 1986. The *Acanthaster* phenomenon. Oceanogr. Mar. Biol. Ann. Rev. 24: 379–480.

Moran, P.J., R.E. Reichelt, and R.H. Bradbury. 1986. An assessment of the geological evidence for previous *Acanthaster* outbreaks. Coral Reefs 4: 235–238.

Morelock, J., K. Boulon, and G. Galler. 1979. Sediment stress and coral reefs. *In* Lopez (ed.), Proceedings of the Symposium on Energy and Industry in the Marine Environment in Guayanilla Bay, pages 46–58. University of Puerto Rico.

Moretzsohn, F., and E.A. Kay. 1995. Hawaiian marine mollusks—an update to Kay, 1979. Pages distributed at 1995 American Malacological Union Meetings, Hilo, University of Hawai'i at Manoa, Honolulu.

Moriarty, D.J.W., P.C. Pollard, D.M. Alongi, C.R. Wilkinson, and J.S. Gray. 1985.

Bacterial productivity and trophic relationships with consumers on a coral reef (MEC-ORI). Proc. 5th Internat. Coral Reef Congr. Tahiti 3: 457–462.

Morris, I., A.E. Smith, and H.E. Glover. 1981. Products of photosynthesis in phytoplankton off the Orinoco River and the Caribbean Sea. Limnol. Oceanogr. 26: 1034–1044.

Morris, R.H., D.P. Abbott, and E.C. Haderlie (eds.). 1980. Intertidal invertebrates of California. Stanford University Press, Stanford.

Morrison, D. 1984. Mass mortality of *Diadema antillarum* on a Jamaican coral reef: effect on the algal community. *In* Advances in Reef Science, pages 85–86. Rosenstiel School of Marine and Atmospheric Sciences, University of Miami.

Morrison, D. 1988. Comparing fish and urchin grazing in shallow and deeper coral reef algal communities. Ecology 69: 1367–1382.

Morse, D.E., and A.N.C. Morse. 1991. Enzymatic characterization of the morphogen recognized by *Agaricia humilis* larvae. Biol. Bull. 181: 104–122.

Morse, D.E., and A.N.C. Morse. 1993. Sulfated polysaccharide induces settlement and metamorphosis of *Agaricia humilis* larvae on specific crustose algae. Proc. 7th Internat. Coral Reef Symp., Guam 1: 502 (Abstr.).

Morse, D.E., A.N.C. Morse, P.T. Raimondi, and N. Hooker. 1994. Morphogen-based chemical fly paper for *Agaricia humilis* coral larvae. Biol. Bull. 186: 172–181.

Morton, B. 1987. Recent marine introductions into Hong Kong. Bull. Mar. Sci. 41: 503–513.

Moyer, J.T., W.K. Emerson, and M. Ross. 1982. Massive destruction of scleractinian corals by the muricid gastropod, *Drupella,* in Japan and the Philippines. The Nautilus 96: 69–82.

Muir, J. 1911. My first summer in the Sierra. Houghton Mifflin, Boston.

Muller-Parker, G., C.B. Cook, and C.F. D'Elia. 1994. Elemental composition of the coral *Pocillopora damicornis* exposed to elevated seawater ammonium. Pac. Sci. 48: 234–246.

Munro, J.L. (ed.). 1983. Caribbean coral reef fishery resources. ICLARM Studies and Reviews 7, 276 p.

Munro, J.L. 1984. Coral reef fisheries and world fish production. ICLARM Newsletter 7(4): 3–4.

Munro, J.L., and D.M. Williams. 1985. Assessment and management of coral reef fisheries: biological, environmental and socio-economic aspects. Proc. 5th Internat. Coral Reef Congr., Tahiti 4: 543–581.

Murray, J., E. Murray, M.S. Johnson, and B. Clarke. 1988. The extinction of *Partula* on Moorea. Pac. Sci. 42: 150–153.

Muscatine, L. 1990. The role of symbiotic algae in carbon and energy flux in reef corals. *In* Z. Dubinsky (ed.), Ecosystems of the world, vol. 25: Coral reefs, pages 75–87. Elsevier Science Publishing, New York, 550 p.

Muscatine, L., and J.W. Porter. 1977. Reef corals: mutualistic symbiosis adapted to nutrient-poor environments. BioScience 27: 454–460.

Muscatine, L., and V. Weis. 1991. Productivity of zooxanthellae and biogeochemical cycles. *In* P.G. Falkowski and A.D. Woodhead (eds.), Primary productivity and biogeochemical cycles in the sea, pages 257–271. Plenum Press, New York.

Muthiga, N.A., and T.R. McClanahan. 1987. Population changes of a sea urchin *(Echinometra mathaei)* on an exploited fringing reef. Afr. J. Ecol. 25: 1–8.

Myers, A.A. 1991. How did Hawaii accumulate its biota? A test from the Amphipoda. Global Ecology and Biogeography Letters 1: 24–29.

Myers, J.H. 1988. The induced defense hypothesis: does it apply to the population dynamics of insects? *In* K.C. Spencer (ed.), Chemical mediation of coevolution. Academic Press, San Diego, 609 p.

Myers, R.F. 1993. Micronesian reef fishes. Coral Graphics, Guam.

Naisbitt, J. 1996. Megatrends Asia. Simon and Schuster, New York, 320 p.

Nedwell, D.B. 1975. Inorganic nitrogen metabolism in a eutrophicated mangrove estuary. Water Res. 9: 221–231.

Nelson, J.S. 1984. Fishes of the world. Wiley-Interscience, New York.

Nelson, S.G., and R.N. Tsutsui. 1982. Browsing by herbivorous reef fishes on the agarophyte *Gracilaria edulis* (Rhodophyta) at Guam, Mariana Islands. Proc. 4th Internat. Coral Reef Symp., Manila 2: 503–506.

Neudecker, S. 1979. Effects of grazing and browsing fishes on the zonation of corals in Guam. Ecology 60: 666–672.

Neumann, A.C. 1966. Observations on coastal erosion in Bermuda and measurements of the boring rates of the sponge *Cliona lampa.* Limnol. Oceanogr. 11: 92–108.

Neumann, A.C., and I. Macintyre. 1985. Reef response to sealevel rise: keep-up, catch-up, or give-up. Proc. 5th Internat. Coral Reef Congr., Tahiti 3: 105–110.

Neumann, C.J., G.W. Cry, E.L. Caso, and B.R. Jarvinen. 1981. Tropical cyclones of the North Atlantic Ocean, 1871–1980 (updated annually). National Climatic Center Publ., 174 p.

Newman, W.A. 1986. Origin of the Hawaiian marine fauna: dispersal and vicariance as indicated by barnacles and other organisms. *In* Crustacean biogeography, pages 21–49. Crustacean Issues 4. A.A. Balkema, Boston.

Newman, W.A., and B.A. Foster. 1987. Southern hemisphere endemism among the barnacles: explained in part by extinction of amphitropical taxa? Bull. Mar. Sci. 41: 361–377.

New Scientist, 7 January 1995. "A new cultural revolution?" vol. 145, issue 1959.

Nishida, M., and J.S. Lucas. 1988. Genetic differences between geographic populations of the crown-of-thorns starfish throughout the Pacific region. Mar. Biol. 98: 359–368.

Nixon, S.W. 1982. Nutrient dynamics, primary production and fisheries yields of lagoons. Oceanologica Acta SP: 357–371.

Nixon, S.W., B.N. Furnas, V. Lee, N. Marshall, O. Jim-Eong, W. Chee-Hoong, G. Wooi-Khoon, and A. Sasekumar. 1984. The role of mangroves in the carbon and nutrient dynamics of Malaysia estuaries. *In* Proc. As. Symp. Mang. Env. Res. and Manag., pages 534–544.

Norgaard, R.B., and R.B. Howarth. 1991. Sustainability and discounting the future. *In* R.

Constanza (ed.), Ecological economics, pages 88–101. Columbia University Press, New York.

Nys, R. de, J.C. Coll, and I.R. Price. 1991. Chemically mediated interactions between the red alga *Plocamium hamatum* (Rhodophyta) and the octocoral *Sinularia cruciata* (Alcyonacea). Mar. Biol. 108: 315–320.

Odum, E.P., E.J. Keunzler, and M.X. Bunt. 1958. Uptake of P^{32} and primary productivity in marine benthic algae. Limnol. Oceanogr. 3: 340–345.

Odum, H.H., and E.P. Odum. 1955. Trophic structure and productivity of a windward coral reef community on Eniwetok Atoll. Ecol. Monogr. 25: 291–320.

Officer, C.B., and C.L. Drake. 1985. Terminal Cretaceous environmental events. Science 227: 1161–1167.

Ogden, J.C. 1976. Some aspects of herbivore-plant relationships on Caribbean reefs and seagrass beds. Aquat. Bot. 2: 103–116.

Ogden, J.C. 1977. Carbonate-sediment production by parrotfish and sea urchins on Caribbean reefs. *In* S.H. Frost, M.P. Weiss, and J.B. Saunders (eds.), Reefs and related carbonates—ecology and sedimentology, pages 281–288. Studies in Geology 4. American Association of Petroleum Geologists, Tulsa, Oklahoma.

Ogden, J.C. 1987. Cooperative coastal ecology at Caribbean marine laboratories. Oceanus 30: 9–15.

Ogden, J.C. 1988. The influence of adjacent systems on the structure and function of coral reefs. Proc. 6th Internat. Coral Reef Symp., Townsville 1: 123–129.

Ogden, J.C. 1992. The impact of Hurricane Andrew on the ecosystems of South Florida. Conserv. Biol. 6: 488–490.

Ogden, J.C., R.A. Brown, and N. Salesky. 1973. Grazing by the echinoid *Diadema antillarum* Philippi: formation of halos around West Indian patch reefs. Science 182: 715–717.

Ogden, J.C., and N.S. Buckman. 1973. Movements, foraging groups, and diurnal migrations of the striped parrotfish *Scarus croicensis* Block (Scaridae). Ecology 54: 589–596.

Ogden, J.C., and R.C. Carpenter. 1987. Long-spined black sea urchin. Species profiles: life histories and environmental requirements of coastal fish and invertebrates (South Florida). Biol. Report 82. 18 p.

Ogden, J.C., and P.R. Ehrlich. 1977. The behavior of heterotypic resting schools of juvenile grunts (Pomadasyidae). Mar. Biol. 42: 273–280.

Ogden, J.C., and E.H. Gladfelter (eds.) 1983. Coral reefs, seagrass beds, and mangroves: their interaction in the coastal zones of the Caribbean. Reports in Mar. Sci. 23. UNESCO, 133 p.

Ogden, J.C., and E.H. Gladfelter (eds.) 1986. Caribbean coastal marine productivity. Reports in Mar. Sci. 41. UNESCO, 59 p.

Ogden, J.C., and P.S. Lobel. 1978. The role of herbivorous fishes and urchins in coral reef communities. Env. Biol. Fish. 3: 49–63.

Ogden, J.C., J.W. Porter, N.P. Smith, A.M. Szmant, W.C. Jaap, and D. Forcucci. 1994.

A long-term interdisciplinary study of the Florida Keys seascape. Bull. Mar. Sci. 54: 1059–1071.

Ogden, J.C., and T.P. Quinn. 1984. Migration in coral reef fishes: ecological significance and orientation mechanisms. *In* J.D. McCleave, G.P. Arnold, J.J. Dodson, and W.H. Neilleds (eds.), Mechanisms of migration in fishes, pages 293–308. Plenum, New York.

Ogden, J.C., and R.L. Wicklund (eds.) 1988. Mass bleaching of coral reefs in the Caribbean: a research strategy. Res. Rept. 88-2. NOAA National Undersea Research Program, 51 p.

Ogden, J.C., and J.C. Zieman. 1977. Ecological aspects of coral reef-seagrass bed contacts in the Caribbean. Proc. 3rd Internat. Coral Reef Symp., Miami. 1. Biology: 377–382.

Oliver, J.K., R.C. Babcock, P.L. Harrison, and B.L. Willis. 1988. Geographical extent of mass coral spawning: clues to ultimate causal factors. Proc. 6th Internat. Coral Reef Symp., Townsville 2: 803–810.

Oliver, J.K., and B.L. Willis. 1987. Coral-spawn slicks in the Great Barrier Reef: preliminary observations. Mar. Biol. 94: 521–529.

Oliver, P.G. 1992. Bivalved seashells of the Red Sea. Verlag Christa Hemmen, Wiesbaden, Germany.

Oliver, P.G., and R. von Cosel. 1992. Taxonomy of tropical West African bivalves IV. Arcidae. Bulletin du Muséum National d' Histoire Naturelle 4 esérie, section A (Zoologie, Biologieet Écologie Animales) 14: 293–381.

O'Loughlin, C. 1985. The influence of forest roads on erosion and stream sedimentation—comparisons between temperate and tropical forests. Working paper. Environment and Policy Institute, East-West Center, Honolulu.

O'Loughlin, C.L., and A.J. Pearce. 1984. Proceedings: Symposium on the effects of forest land use on erosion and slope stability, 7–11 May 1984. East-West Center, Honolulu.

Olsen, D.A., and J.A. LaPlace. 1979. A study of a Virgin Islands grouper fishery based on a breeding aggregation. *In* Proceedings of the 31st Annual Gulf and Caribbean Fisheries Institute, pages 130–144.

Olson, R.R. 1987. In situ culturing as a test of the larval starvation hypothesis for the crown-of-thorns starfish, *Acanthaster planci.* Limnol. Oceanogr. 32: 895–904.

Opdyke, B.N., and J.C.G. Walker. 1992. Return of the coral reef hypothesis: basin to shelf partitioning of $CaCO_3$ and its effect on atmospheric CO_2. Geology 20: 733–736.

Ormond, R.F.G., and A.C. Campbell. 1971. Observations on *Acanthaster planci* and other coral reef echinoderms in the Sudanese Red Sea. Symp. Zool. Soc. Lond. 28: 433–454.

Ormond, R.F.G., N.H. Hanscomb, and D.H. Beach. 1976. Foods election and learning in the crown-of-thorns starfish, *Acanthaster planci* (L.). Mar. Behav. Physiol. 4: 93–105.

Ormond, R.F.G., S.H. Head, R.J. Moore, P.R. Rainbow, and A.P. Saunders. 1973. Formation and breakdown of aggregations of the crown-of-thorns starfish, *Acanthaster planci* (L.). Nature 246: 167–169.

Ott, B., and J.B. Lewis. 1972. The importance of the gastropod *Coralliophila abbreviata* (Lamarck) and the polychaete *Hermodice carunculata* (Pallas) as coral reef predators. Can. J. Zool. 50: 1651–1656.

Page, H.M., and D.M. Hubbard. 1987. Temporal and spatial patterns of growth in mussels

Mytilus edulis on an offshore platform: relationships to water temperature and food availability. J. Exp. Mar. Biol. Ecol. 111: 159–179.

Paine, R.T. 1966. Foodweb complexity and species diversity. Am. Nat. 100:65–75.

Paine, R.T. 1980. Foodwebs: linkage, interaction strength and community infrastructure. J. Anim. Ecol. 49: 667–685.

Paine, R.T. 1988. Foodwebs: roadmaps of interactions or grist for theoretical development? Ecology 69: 1648–1654.

Paine, R.T. 1992. Food-web analysis through field measurement of per capita interaction strength. Nature 355: 73–75.

Palumbi, S.R. 1992. Marine speciation on a small planet. Trends Ecol. Evol. 7: 114–118.

Palumbi, S.R. 1994. Genetic divergence, reproductive isolation, and marine speciation. Ann. Rev. Ecol. Syst. 25: 547–572.

Palumbi, S.R., and E.C. Metz. 1991. Strong reproductive isolation between closely related tropical sea urchins (genus *Echinometra*). Molecular Biology and Evolution 8: 227–239.

Panayatou, T. 1982. Management concepts for small scale fisheries: economic and social aspects. FAO Fisheries Tech. Pap. 228.

Pandolfi, J.M. 1992. Successive isolation rather than evolutionary centres for the origination of Indo-Pacific reef corals. J. Biogeogr. 19: 593–609.

Pannier, F. 1979. Mangroves impacted by human-induced disturbances: a case study of the Orinoco delta mangrove ecosystem. Env. Manag. 3: 205–216.

Parker, S.P. (ed.). 1982. Synopsis and classification of living organisms. McGraw-Hill, New York.

Pastorak, R.A., and G.R. Bilyard. 1985. Effects of sewage pollution on coral reef communities. Mar. Ecol. Progr. Ser. 21: 175–189.

Patten, B.C. 1990. Environ theory and indirect effects: a reply to Loehle. Ecology 71: 2386–2393.

Patton, W.K. 1974. Community structure among the animals inhabiting the coral *Pocillopora damicornis* at Heron Island, Australia. *In* S.H. Frost, M.P. Weissand, and J.B. Saunders (eds.), Symbiosis in the sea, pages 219–243. Univ. of S. Carolina Press, Columbia, South Carolina.

Paul, V.J. (ed.). 1992. Ecological roles of marine natural products. Cornell University Press, Ithaca, New York.

Paul, V.J., N. Lindquist, and W. Fenical. 1990. Chemical defenses of the tropical ascidian *Atapozoa* sp. and its nudibranch predators *Nembrotha* spp. Mar. Ecol. Progr. Ser. 59: 109–118.

Paul, V.J., and S.C. Pennings. 1991. Diet-derived chemical defenses in the seahare *Stylocheilus longicauda*. J. Exp. Mar. Biol. Ecol. 151: 227–243.

Paul, V.J., and K.L. Van Alstyne. 1988. Use of ingested algal diterpenoids by *Elysia halimedae* Macnae (Opisthobranchia: Ascoglossa) as antipredator defenses. J. Exp. Mar. Biol. Ecol. 119: 15–29.

Paulay, G. 1989. Marine invertebrates of the Pitcairn Islands: species composition and biogeography of corals, molluscs, and echinoderms. Atoll Res. Bull. 326: 1–28.

Paulay, G. 1991. Late Cenozoic sealevel fluctuations and the diversity and species composition of insular shallowwater marine faunas. *In* E.C. Dudley (ed.), The unity of evolutionary biology, pages 184–193. Proc. Fourth Internat. Congr. Syst. Evol. Biol. Dioscorides, Portland, Oregon.

Paulay, G. 1994. Biodiversity on oceanic islands: its origin and extinction. Am. Zool. 34: 134–144.

Paulay, G. 1996a. Dynamic clams: historical changes in the bivalve fauna of Pacific islands as a result of sealevel fluctuations. Amer. Malacol. Bull. 13: in press.

Paulay, G. 1996b. Coral species and evolution. Review of: Veron, J.E.N. 1995. Corals in space and time. The biogeography and evolution of the Scleractinia. University of New South Wales Press, Sydney. J. Biogeogr. 23: in press.

Paulay, G., and L.R. McEdward. 1990. A simulation model of island reef morphology: the effects of sea level fluctuations, growth, subsidence and erosion. Coral Reefs 9: 51–62.

Pauly, D., and G.I. Murphy. 1982. Theory and management of tropical fisheries. ICLARM, Manila, Philippines, 360 p.

Pawlik, J.R., and M.G. Hadfield. 1990. A symposium on chemical factors that influence the settlement and metamorphosis of marine invertebrate larvae: introduction and perspective. Bull. Mar. Sci. 46: 450–454.

Pawlik, J.R., M.R. Kernan, T.F. Molinski, M.K. Harper, and D.J. Faulkner. 1988. Defensive chemicals of the Spanish dancer nudibranch *Hexabranchus sanguineus* and its egg ribbons: macrolides derived from a sponge diet. J. Exp. Mar. Biol. Ecol. 119: 99–109.

Pearson, R.G. 1975. Coral reefs, unpredictable climatic factors and *Acanthaster. In* Crown-of-thorns starfish seminar proceedings, pages 131–134. Australian Govt. Publishing Service, Canberra.

Pearson, R.G. 1977. Impact of illegal giant clam fishing being investigated. Austr. Fisheries 36: 22.

Pearson, R.G. 1981. Recovery and recolonization of coral reefs. Mar. Ecol. Progr. Ser. 4: 105–122.

Pearson, R.G., and R. Endean. 1969. A preliminary study of the coral predator *Acanthaster planci* (L.) (Asteroidea) on the Great Barrier Reef. Fish. Notes Dept. Harbours and Marine, Qld. 3: 27–55.

Peck, S.B. 1993. Galapagos species diversity: is it on the land or in the sea? Noticias de Galapagos 52: 18–21.

Peddicord, R.K., V.A. McFarland, D. Pbelfiori, and T.E. Byrd. 1975. Effects of suspended solids on San Francisco Bay organisms. Physical impact study. Dredge disposal study San Francisco Bay and estuary. U.S. Army Corps of Engineers.

Peet, R.K. 1974. Measurement of species diversity. Ann. Rev. Ecol. Syst. 5: 285–307.

Pennings, S.C. 1990. Multiple factors promoting narrow host range in the seahare, *Aplysia californica.* Oecologia 82: 192–200.

Pennings, S.C., and V. J. Paul. 1993. Sequestration of dietary secondary metabolites by three species of seahares: location, specificity and dynamics. Mar. Biol. 117: 535–546.

Pentecost, A. 1991. Calcification processes in algae and cyanobacteria. *In* R. Riding (ed.), Calcareous algae and stromatolites, pages 3–20. Springer-Verlag, New York.

Peters, E.C. 1984. A survey of cellular reactions to environmental stress and disease in Caribbean scleractinian corals. Helgoländer Meeresuntersuchungen 37: 113–137.

Peters, E.C. 1988. Symbiosis to pathology: are the roles of microorganisms as pathogens of coral reef organisms predictable from existing knowledge? Proc. 6th Internat. Coral Reef Symp., Townsville 1: 205–209.

Peters, E.C. 1993. Diseases of other invertebrate phyla: Porifera, Cnidaria, Ctenophora, Annelida, Echinodermata. *In* J.A. Couch and J.W. Fournie (eds.), Pathobiology of marine and estuarine organisms, pages 388–444. CRC Press, Boca Raton, Florida.

Petersen, R.C., Jr., B.L. Madsen, M.A. Wilzbach, C.H.D. Magadza, A. Paarlberg, A. Kullberg, and K.W. Cummins. 1987. Stream management: emerging global similarities. Ambio 16: 166–179.

Petuch, E.J. 1981a. A volutid species radiation from northern Honduras, with notes on the Honduran Caloosahatchian secondary relict pocket. Proc. Biol. Soc. Wash. 94: 1110–1130.

Petuch, E.J. 1981b. A relict Neogene caenogastropod fauna from northern South America. Malacologia 20: 307–347.

Petuch, E.J. 1982. Geographical heterochrony: comtemporaneous [sic] coexistence of Neogene and Recent molluscan fauna in the Americas. Palaeogeography, Palaeoclimatology, Palaeoecology 37: 277–312.

Pfister, C.A., and M.E. Hay. 1988. Associational plant refuges: convergent patterns in marine and terrestrial communities result from differing mechanisms. Oecologia 77: 118–129.

Phillips, R.C., and E.G. Meñez. 1988. Seagrasses. Smithson. Contr. Mar. Sci. 34.

Pianka, E.R. 1966. Latitudinal gradients in species diversity: a review of concepts. Am. Nat. 100: 33–46.

Pianka, E.R. 1988. Evolutionary ecology. Harper and Row, New York.

Peters, E.C., J.C. Halas, H.B. McCarty. 1986. Calicoblastic neoplasms in *Acropora palmata,* with a review of reports on anomalies of growth and form in corals. Journal of the National Cancer Institute 76: 895–912.

Pimentel, D., C. Harvey, P. Resosudarmo, K. Sinclair, D. Kurz, M. McNair, S. Crist, L. Shpritz, L. Fitton, R. Saffouri, and R. Blair. 1995. Environmental and economic costs of soil erosion and conservation benefits. Science 267: 1117–1123.

Pinkerton, E. (ed.). 1989. Co-operative management of local fisheries. University of British Columbia Press, Vancouver.

Pitcher, T.J. 1986. Predators and food are the keys to understanding fish shoals: a review of recent experiments. Naturaliste Can. Ann. Rev. Ecol. Syst. 113: 225–233.

Pitcher, T.J., and A.E. Magurran. 1983. Shoal size, patch profitability and information exchange in foraging goldfish. Anim. Behav. 31: 546–555.

Pitcher, T.J., A.E. Magurran, and I.J. Winfield. 1982. Fish in larger shoals find food faster. Behav. Ecol. Sociobiol. 10: 149–151.

Plan Development Team. 1990. The potential of marine fishery reserves for reef fish management in the U.S. southern Atlantic. Tech. Memorandum NMFS-SEFC-261. NOAA, 40 p.

Platt, T., P. Jauhari, and S. Sathyendranath. 1991. The importance and measurement of new production. *In* P.G. Falkowski and A.D. Woodhead (eds.), Primary productivity and biogeochemical cycles in the sea, pages 273–284. Plenum Press, New York.

Platt, T., M. Lewis, and R. Geider. 1984. Thermodynamics of the pelagic ecosystem: elementary closure conditions for biological production in the open ocean. *In* M.J.R. Fasham (ed.), Flows of energy and materials in marine ecosystems, pages 49–84. Plenum Press, New York.

Plaziat, J.C., and C. Perrin. 1992. Multi kilometer sized reefs built by foraminiferans (*Solenomeris*) from the early Eocene of the Pyreneandomain (S France, N Spain): paleoecologic relations with coral reefs. Palaeogeography, Palaeoclimatology, Paleoecology 96: 195–232.

Polis, G.A. 1991. Complex trophic interactions in deserts: an empirical critique of food-web theory. Am. Nat. 138: 123–155.

Polovina, J.J. 1984. Model of a coral reef ecosystem 1. The ECOPATH model and its application to French Frigate Shoals. Coral Reefs 3: 1–11.

Polunin, N.V.C. 1988. Efficient uptake of algal production by a single resident herbivorous fish on a reef. J. Exp. Mar. Biol. Ecol. 123: 61–76.

Polunin, N.V.C., and I. Koike. 1987. Temporal focusing of nitrogen release by a periodically feeding reef fish. J. Exp. Mar. Biol. Ecol. 111: 285–296.

Polunin, N.V.C., and C.M. Roberts. 1993. Greater biomass and value of target coral-reef fishes in two small Caribbean marine reserves. Mar. Ecol. Progr. Ser. 100: 167–176.

Pomeroy, L.R. 1979. Secondary production mechanisms of continental shelf communities. *In* R.J. Livingston (ed.), Ecological processes in coastal and marine systems, pages 233–264. Plenum Press, New York.

Pomponi, S.A. 1979. Ultrastructure and cytochemistry of the etching area of boring sponges. *In* C. Levi and N. Boury-Esnault (eds.), Biologiedes Spongiaires, pages 317–323. Colloques Internationaux du Centre National de la Recherche Scientifique 291.

Poppe, G.T., and Y. Goto. 1992. Volutes. L'Infromatore Piceno, Ancona.

Porter, J.W., J.F. Battey, and G.J. Smith. 1982. Perturbation and change in coral reef communities. Proc. Natl. Acad. Sci. 79: 1678–1681.

Porter, J.W., L. Muscatine, Z. Dubinsky, and P.G. Falkowski. 1984. Primary production and photo adaptation on light and shade-adapted colonies of the symbiotic coral *Stylophora pistillata.* Proc. Roy. Soc. London B. 222: 161–180.

Porter, J.W., and K.G. Porter. 1977. Quantitative sampling of demersal plankton migrating from different coral reef substrates. Limnol. Oceanogr. 22: 553–556.

Potts, D.C. 1977. Suppression of coral populations by filamentous algae within damselfish territories. J. Exp. Mar. Biol. Ecol. 38: 207–216.

Potts, D.C. 1981. Crown-of-thorns starfish—man-induced pest or natural phenomenon?

In R.L. Kitching and R.E. Jones (eds.), The ecology of pests. Some Australian case histories, pages 55–86. CSIRO, Melbourne.

Potts, D.C., and R.L. Garthwaite. 1991. Evolution of reef building corals during periods of rapid global change. *In* E.C. Dudley (ed.), The unity of evolutionary biology, pages 170–178. Proc. Fourth Internat. Congr. Syst. Evol. Biol. Dioscorides. Portland, Oregon.

Potts, D.C., R.L. Garthwaite, and A.F. Budd. 1993. Speciation in the coral genus *Porites.* Proc. 7th Internat. Coral Reef Symp., Guam 2: 679 (Abstr.).

Power, M.E. 1984. Depth distributions of armored catfish: predator-induced resource avoidance? Ecology 65: 523–528.

Power, M.E. 1987. Predator avoidance by grazing fishes in temperate and tropical streams: importance of stream depth and prey size. *In* W.C. Kerfoot and A. Sih (eds.), Predation: direct and indirect impacts on aquatic communities, pages 333–351. University Press of New England, Hanover, New Hampshire, 386 p.

Power, M.E. 1990. Effects of fish in river food webs. Science 250: 811–814.

Power, M.E., and W.J. Matthews. 1983. Algae-grazing minnows *(Campostoma anomalum),* piscivorous bass (*Micropterus* spp.) and the distribution of attached algae in a small prairie-margin stream. Oecologia 60: 328–332.

Power, M.E., W.J. Matthews, and A.J. Stewart. 1985. Grazing minnows, piscivorous bass and stream algae: dynamics of a strong interaction. Ecology 66: 1448–1456.

Prein, M. 1995. Exploitation of seahorses and pipefishes. Naga, ICLARM Newsletter, January 1995.

Price, I.A., and F.J. Scott. 1992. The turf algal flora of the Great Barrier Reef. Part 1. Rhodophyta, James Cook University Press, Townsville, Australia, 266 p.

Provenzano, A.J., Jr. (ed.) 1983. The Biology of crustacea, vol. 6, Pathobiology. Academic Press, New York.

Prudhoe, S. 1985. A monograph of the polyclad Turbellaria. British Museum (Natural History), London.

Purdy, E.G. 1974. Reef configurations: cause and effect. *In* L.F. LaPorte (ed.), Reefs in time and space, pages 9–76. SEPM Spec. Pub. 18. Society of Economic Paleontologists and Mineralogists, Tulsa, Oklahoma.

Quoy, J.R., and J.P. Gaimard. 1825. Memoire sur l'accroissment des polyps lithophytes consider e geologiquement. Annls. Sci. Nat. 6: 373–390.

Ramus, J., and M. Venable. 1987. Temporal ammonium patchiness and growth rate in *Codium* and *Ulva* (Ulvophyceae). J. Phycol. 23: 518–523.

Randall, J.E. 1958. A review of the labrid fish genus *Labroides,* with descriptions of two new species and notes on ecology. Pac. Sci. 12: 327–347.

Randall, J.E. 1961. Overgrazing of algae by herbivorous marine fishes. Ecology 42:812.

Randall, J.E. 1963. An analysis of the fish populations of artificial and natural reefs in the Virgin Islands. Carib. J. Sci. 3: 31–47.

Randall, J.E. 1965. Grazing effects on seagrasses by herbivorous reef fishes in the West Indies. Ecology 46: 255–260.

Randall, J.E. 1967. Food habits of reef fishes of the West Indies. Stud. Trop. Oceanogr. 5: 665–847.

Randall, J.E. 1974. The effects of fishes on coral reefs. Proc. 2nd. Internat. Coral Reef Symp., Brisbane 1: 159–166.

Randall, J.E. 1976. The endemic shore fishes of the Hawaiian Islands, Lord Howe Island and Easter Island. Travauxet Documents del' ORSTOM 47:49–73.

Randall, J.E. 1987. Introductions of marine fishes to the Hawaiian Islands. Bull. Mar. Sci. 41: 490–502.

Randall, J.E. 1995. Zoogeographic analysis of the inshore Hawaiian fish fauna. *In* J.E. Maragos, M.N.A. Peterson, L.G. Eldredge, J.E. Bardach, and H.F. Takeuchi (eds.), Marine and coastal biodiversity in the tropical island Pacific region, vol. 1, Species systematics and information management priorities, pages 193–203. East-West Center, University of Hawaii, Honolulu.

Randall, J.E., G.R. Allen, and R.C. Steene. 1990. Fishes of the Great Barrier Reef and Coral Sea. Crawford House Press, Bathurst NSW; University of Hawaii, Honolulu.

Randall, J.E., R.E. Schroeder, and W.A. Starck. 1964. Notes on the biology of the echinoid *Diadema antillarum.* Carib. J. Sci. 4: 421–433.

Randall, R.H. 1972. Chemical pollution in the sea and the crown-of-thorns starfish *(Acanthaster planci).* Biotropica 4: 132–144.

Randall, R.H., and R.F. Myers. 1983. Guide to the coastal resources of Guam, vol. 2, The corals. University of Guam Press, Guam.

Rapaport, E.H. 1994. Remarks on marine and continental biogeography: an a geographical viewpoint. Philosophical Trans. Roy. Soc. London B 343: 71–78.

Rasmussen, K.A., and E.W. Frankenberg. 1990. Intertidal bioerosion by the chiton *Acanthopleura granulata;* San Salvador, Bahamas. Bull. Mar. Sci. 47: 680–695.

Raven, J.A. 1974. Carbon dioxide fixation. *In* W.D.P. Stewart (ed.), Algal physiology and biochemistry, pages 434–455. Blackwell Scientific, Oxford.

Reaka, M.L., and R.B. Manning. 1987. The significance of body size, dispersal potential, and habitat for rates of morphological evolution in stomatopod Crustacea. Smithson. Contr. Zool. 448: 46 p.

Reaka-Kudla, M.L. 1994. Biodiversity of coral reefs. *In* C. Heasley (ed.), Science and a changing world, page 22. AAAS '94 Program and Abstracts.

Reed, S.A. 1971. Techinques for raising the planula larvae and newly settled polyps of *Pocillopora damicornis. In* H.M. Lenhoff, L. Muscatine, and L.V. Davis (eds.), Experimental coelenterate biology, University of Hawaii Press, Honolulu.

Reese, E.S. 1975. A comparative field study of the social behavior and related ecology of reef fishes of the family Chaetodontidae. Z. Tierpsychol. 37: 37–61.

Reese, E.S. 1977. Coevolution of coral and coral feeding fishes of the family Chaetodontidae. Proc. 3rd Internat. Coral Reef Symp., Miami 1: 267–274.

Reese, E.S. 1981. Predation of corals by fishes of the family Chaetodontidae: Implications for conservation and management of coral reef ecosystems. Bull. Mar. Sci. 31: 594–604.

Rehder, H.A. 1980. The marine mollusks of Easter Island (Islade Pascua) and Salay Gomez. Smithson. Contr. Zool. 289: 1–167.

Reichelt, R.E. 1988. Space and structure effects on coral reefs. Proc. 6th Internat. Coral Reef Symp., Townsville 1: 235–242.

Reichelt, R.E., R.H. Bradbury, and P.J. Moran. 1990a. Distribution of *Acanthaster planci* outbreaks on the Great Barrier Reef between 1966 and 1989. Coral Reefs 9: 97–103.

Reichelt, R.E., W. Greve, R.H. Bradbury, and P.J. Moran. 1990b. *Acanthaster planci* outbreak initiation: a starfish-coral model. Ecol. Modelling 49: 153–177.

Reid, A. 1992. Indonesian fishermen detained in Broome: a report on the social and economic background. *In* Illegal entry! Occasional Paper No. 1/1992. Center for Southeast Asian Studies, Northern Territory University, Canberra.

Reid, R.P., and K.M. Browne. 1991. Intertidal stromatolites in a fringing Holocene reef complex, Bahamas. Geology 19: 15–18.

Reinthal, P.N., and S.M. Lewis. 1986. Social behaviour, foraging efficiency and habitat utilization in a group of tropical herbivorous fish. Anim. Behav. 34: 1687–1693.

Rex, M.A., C.T. Stuart, R.R. Hessler, J.A. Allen, H.L. Sanders, and G.D.F. Wilson. 1993. Global-scale patterns of species diversity in the deep-sea benthos. Nature 365: 636–639.

Reyes-Bonilla, H. 1992. New records for hermatypic corals (Anthozoa: Scleractinia) in the Gulf of California, Mexico, with an historical and biogeographical discussion. J. Nat. Hist. 26: 1163–1175.

Rice, M.E., and I.G. Macintyre. 1972. A preliminary study of sipunculan burrows in rock thin sections. Carib. J. Sci. 12: 41–44.

Rice, M.E., and I.G. Macintyre. 1982. Distribution of Sipuncula in the coral reef community, Carrie Bow Cay, Belize. *In* K. Rützler and I.G. Macintyre (eds.), The Atlantic barrier reef ecosystem at Carrie Bow Cay, Belize I: structure and communities, pages 311–320. Smithson. Contrib. Mar. Sci. 12. 539 p.

Richard, G. (ed.). 1985. Fauna and flora. A first compilation of French Polynesian sea-dwellers. Proc. 5th Internat. Coral Reef Congr., Tahiti 1: 379–518.

Richmond, R.H. 1985. Reversible metamorphosis in coral planula larvae. Mar. Ecol. Progr. Ser. 22: 181–185.

Richmond, R.H. 1987. Energetics, competency, and long-distance dispersal of planula larvae of the coral *Pocillopora damicornis.* Mar. Biol. 93: 527–533.

Richmond, R.H. 1988. Competency and dispersal potential of planula larvae of a spawning versus a brooding coral. Proc. 6th Internat. Coral Reef Symp., Townsville 2: 827–831.

Richmond, R.H. 1990. Relationships among reproductive mode, biogeographic distribution patterns and evolution in scleractinian corals. *In* M. Hoshi and O. Yamashita (eds.), Advances in invertebrate reproduction, vol. 5, pages 317–322. Amsterdam, Elsevier.

Richmond, R.H. 1993a. Fertilization in corals: puzzles and problems. Proc. 7th Internat. Coral Reef Symp., Guam 1: 502 (Abstr.).

Richmond, R.H. 1993b. Coral reefs: present problems and future concerns resulting from anthropogenic disturbance. Am. Zool. 33: 524–536.

Richmond, R.H. 1993c. Effects of coastal runoff on coral reproduction *In* Ginsburg, R.N.

(compiler), Proceedings of the Colloquium and Forum on Global Aspects of Coral Reefs: Health, Hazards, and History, pages 360–364. Rosenstiel School of Marine and Atmospheric Sciences, University of Miami, 420 p.

Richmond, R.H. 1995. Hybridization as an evolutionary force in mass-spawning scleractinian corals. J. Cell. Biochem. 19B: 341.

Richmond, R.H., and C.L. Hunter. 1990. Reproduction and recruitment of corals: comparisons among the Caribbean, the tropical Pacific, and the Red Sea. Mar. Ecol. Prog. Ser. 60: 185–203.

Richmond, R.H., and P.L. Jokiel. 1984. Lunar periodicity in larva release in the reef coral *Pocillopora damicornis* at Enewetak and Hawaii. Bull. Mar. Sci. 34: 280–287.

Ricklefs, R.E., and D. Schluter (eds.). 1993. Species diversity in ecological communities: historical and geographical perspectives. University of Chicago Press, Chicago.

Riding, R. (ed.). 1991. Calcareous algae and stromatolites. Springer-Verlag, New York.

Riebesell, U., D.A. Wolf-Gladrow, and V. Smetacek. 1993. Carbon dioxide limitation of marine phytoplankton growth rates. Nature 361: 249–251.

Rinkevich, B. 1989. The contribution of photosynthetic products to coral reproduction. Mar. Biol. 101: 259–263.

Rinkevich, B., and Y. Loya. 1979a. The reproduction of the Red Sea coral *Stylophora pistillata.* I. Gonads and planulae. Mar. Ecol. Progr. Ser. 1: 133–144.

Rinkevich, B., and Y. Loya. 1979b. The reproduction of the Red Sea coral *Stylophora pistillata.* II. Synchronization in breeding and seasonality of planula shedding. Mar. Ecol. Progr. Ser. 1: 145–152.

Risk, M.J., and J.K. MacGeachy. 1978. Aspects of bioerosion of modern Caribbean reefs. Rev. Biol. Trop. 26 (suppl. 1): 85–105.

Risk, M.J., and P.W. Sammarco. 1982. Bioerosion of corals and the influence of damselfish territoriality: a preliminary study. Oecologia 52: 376–380.

Robblee, M.J., T.R. Barber, P.R. Carlson, Jr., M.J. Durako, J.W. Fourqurean, L.K. Muhlstein, D. Porter, L.A. Yarbro, R.T. Zieman, and J.C. Zieman. 1991. Mass mortality of the tropical seagrass *Thalassia* testudinum in Florida Bay (USA). Mar. Ecol. Progr. Ser. 71: 297–299.

Roberts, C.M. 1995. Effects of fishing on the ecosystem structure of coral reefs. Conserv. Biol. 9(5): 988–995.

Roberts, C.M., and R.F.G. Ormond. 1992. Butterflyfish social behaviour, with special reference to the incidence of territoriality: a review. Env. Biol. Fish. 34: 79–93.

Roberts, C.M., R.F.G. Ormond, and A.R.D. Shepherd. 1988. The usefulness of butterfly fishes as environmental indicators on coral reefs. Proc. 6th Internat. Coral Reef Symp., Townsville 2: 331–336.

Roberts, C.M., and N.V.C. Polunin. 1991. Are marine reserves effective in management of reef fisheries? Rev. Fish Biol. Fisheries 1: 65–91.

Roberts, C.M., and N.V.C. Polunin. 1993. Marine reserves: simple solutions to managing complex fisheries? Ambio 22: 363–368.

Roberts, R.J. (ed.). 1989. Fish pathology. Baillière Tindall, London.

Robertson, D.R. 1982. Fish feces as fish food on a Pacific coral reef. Mar. Ecol. Progr. Ser. 7: 253–265.

Robertson, D.R. 1991. Increases in surgeonfish populations after mass mortality of the sea urchin *Diadema antillarum* in Panama indicate food limitation. Mar. Biol. 111: 437–444.

Robertson, D.R., and S.D. Gaines. 1986. Interference competition structures habitat use in a local assemblage of coral reef surgeon fishes. Ecology 67: 1372–1383.

Robertson, D.R., N.V.C. Polunin, and K. Leighton. 1979. The behavioral ecology of three Indian Ocean surgeon fishes (*Acanthurus lineatus, A. leucosternum,* and *Zebrasoma scopas*): their feeding strategies, and social and mating systems. Env. Biol. Fish. 4: 125–170.

Robertson, D.R., H.P.A. Sweatman, E.A. Fletcher, and M.G. Cleland. 1976. Schooling as a mechanism for circumventing the territoriality of competitors. Ecology 57: 1208–1220.

Robertson, R. 1970. Review of the predators and parasites of stony corals, with special reference to symbiotic prosobranch gastropods. Pac. Sci. 24: 43–54.

Robins, C.R. 1991. Regional diversity among Caribbean reef fishes. BioScience 41: 458–459.

Röckel, D., W. Korn, and A.J. Kohn. 1995. Manual of the living Conidae. Vol. 1: Indo-Pacific region. Christa Hemmen Verlag, Wiesbaden, Germany.

Rogers, C.S. 1979. The effect of shading on coral reef structure and function. J. Exp. Mar. Biol. Ecol. 41: 269–288.

Rogers, C.S. 1983. Sublethal and lethal effects of sediments applied to common Caribbean corals in the field. Mar. Pollut. Bull. 14: 378–382.

Rogers, C.S. 1985. Degradation of Caribbean and western Atlantic coral reefs and decline of associated fisheries. Proc. 5th Internat. Coral Reef Congr., Tahiti 6: 491–496.

Rogers, C.S. 1990. Responses of coral reefs and reef organisms to sedimentation. Mar. Ecol. Prog. Ser. 62: 185–202.

Rogers, C.S., L. McLain, and E. Zullo. 1988. Damage to coral reefs in Virgin Islands National Park and Biosphere Reserve from recreational activities. Proc. 6th Internat. Coral Reef Symp., Townsville 2: 405–410.

Rogers, C.S., T.H. Suchanek, and F.A. Pecora. 1982. Effects of hurricanes David and Frederick (1979) on shallow *Acropora palmata* communities, St. Croix, U.S. Virgin Islands. Bull. Mar. Sci. 32: 532–548.

Rohde, K. 1993. Ecology of marine parasites. CAB International, Wallingford, United Kingdom.

Rose, C.S., and M.J. Risk. 1985. Increase in *Cliona deletrix* infestation of *Montastrea cavernosa* heads on an organically polluted portion of the Grand Cayman fringing reef. Mar. Ecol. 6: 345–363.

Rosen, B.R. 1981. The tropical high diversity enigma—the corals' eye view. *In* P.L. Forey (ed.), The evolving biosphere: chance, change, and challenge, pages 103–129. Cambridge University Press.

Rosen, B.R. 1984. Reef coral biogeography and climate through the Late Cainozoic: just islands in the sun or a critical pattern of islands? *In* P. Brenchley (ed.), Fossils and climate, pages 201–262. Wiley, New York.

Rosen, B.R. 1988. Progress, problems and patterns in the biogeography of reef corals and other tropical marine organisms. Helgoländer Meeresuntersuchungen 42: 269–301.

Rosen, B.R., and A.B. Smith. 1988. Tectonics from fossils? An analysis of reef-coral and sea-urchin distributions from late Cretaceous to Recent, using a new method. *In* M.G. Audley-Charles and A. Hallam (eds.), Gondwana and Tethys, pages 275–306. Geol. Soc. Special Publ. No. 37.

Rosen, B.R., and J.D. Taylor. 1969. Reef coral from Aldabra: a new mode of reproduction. Science 166: 119–121.

Rosenblatt, R.H., J.E. McCosker, and I. Rubinoff. 1972. Indo-West Pacific fishes from the Gulf of Chiriqui, Panama. Contr. Sci., Nat. Hist. Mus., Los Angeles County 234: 1–18.

Rosenblatt, R.H., and R.S. Waples. 1986. A genetic comparison of allopatric populations of shore fish species from the eastern and central Pacific Ocean: dispersal and vicariance. Copeia 1986: 275–284.

Rosenzweig, M.L., and Z. Abramsky. 1993. How are diversity and productivity related? *In* R.E. Ricklefs and D. Schluter, (eds.), Species diversity in ecological communities: historical and geographical perspectives, pages 52–65. University of Chicago Press, Chicago.

Rosewater, J. 1965. The family Tridacnidae in the Indo-Pacific. Indo-Pacific Mollusca 1: 347–396.

Ross, M.A. 1984. A quantitative study of the stony coral fishery in Cebu, Philippines. Mar. Ecol. 5(1): 75–91.

Rougerie, F., and B. Waulthy. 1993. The endo-upwelling concept: from geothermal convection to reef construction. Coral Reefs 12: 19–30.

Rowan, R. 1991. Molecular systematics of symbiotic algae. J. Phycol. 27: 661–666.

Rowan, R., and D.A. Powers. 1991a. A molecular classification of zooxanthellae and the evolution of animal–algal symbioses. Science 251: 1348–1351.

Rowan, R., and D.A. Powers. 1991b. Molecular genetic identification of symbiotic dinoflagellates (zooxanthellae). Mar. Ecol. Progr. Ser. 71: 65–73.

Rowan, R., and D.A. Powers. 1992. Ribosomal RNA sequences and the diversity of symbiotic dinoflagellates (zooxanthellae). Proc. Natl. Acad. Sci. 89: 3639–3643.

Rowe, F.W.E., and J. Gates. 1995. Echinodermata. Zoological Catalogue of Australia, vol. 33. CSIRO Australia, Melbourne.

Roy, K., and S. Smith. 1971. Sedimentation and coral development in turbid water: Fanning Lagoon. Pac. Sci. 25: 234–248.

Ruddle, K. 1990. The continuity of traditional management practises: the case of Japanese coastal fisheries. *In* K. Ruddle and R.E. Johannes (eds.), Traditional marine resource management in the Pacific Basin: an anthology, pages 263–285. UNESCO, Jakarta.

Ruddle, K., and R.E. Johannes. (eds.). 1990. Traditional marine resource management in the Pacific Basin: an anthology. UNESCO, Jakarta.

Rudman, W. 1987. The Chromodoridae (Opisthobranchia: Mollusca) of the Indo-West

Pacific: *Chromodoris epicuria, C. aureopurpurea, C. annulata, C. coi,* and *Risbecia tryoni* color groups. Zool. J. Linn. Soc. 90: 305–407.

Russ, G.R. 1984a. A review of coral reef fisheries. UNESCO Reports in Marine Sciences 27: 74–92.

Russ, G.R. 1984b. Distribution and abundance of herbivorous grazing fishes in the central Great Barrier Reef. I. Levels of variability across the entire continental shelf. Mar. Ecol. Progr. Ser. 20: 23–34.

Russ, G.R. 1985. Effects of protective management on coral reef fishes in the central Philippines. Proc. 5th Internat. Coral Reef Congr., Tahiti 4: 219–224.

Russ, G.R. 1987. Is rate of removal of algae by grazers reduced inside territories of tropical damsel fishes? J. Exp. Mar. Biol. Ecol. 110: 1–17.

Russ, G.R. 1991. Coral reef fisheries: effects and yields. *In* P.F. Sale (ed.), The ecology of fishes on coral reefs, pages 601–635. Academic Press, San Diego, 754 p.

Russ, G.R., A.C. Alcala, and A.S. Cabanban. 1993. Marine reserves and fisheries management on coral reefs with preliminary modelling of the effects on yield per recruit. Proc. 7th Internat. Coral Reef Symposium, Guam 2: 978–985.

Russ, G.R., and J. St. John. 1988. Diets, growth rates and secondary production of herbivorous coral reef fishes. Proc. 6th Internat. Coral Reef Symp., Townsville 2: 37–43.

Russo, A.R. 1980. Bioerosion by two rock boring echinoids (*Echinometra mathaei* and *Echinostrephus aciculatus*) on Enewetak Atoll, Marshall Islands. J. Mar. Res. 38: 99–110.

Rützler, K. 1974. The burrowing sponges of Bermuda. Smithson. Contrib. Zool. 165: 1–32.

Rützler, K. 1975. The role of burrowing sponges in bioerosion. Oecologia 19: 203–216.

Rützler, K. 1988. Mangrove sponge disease induced by bacterial symbionts: failure of a primitive immune system? Diseases Aquat. Org. 5: 143–149.

Rützler, K. 1990. Associations between Caribbean sponges and photosynthetic organisms. pages 455–466. *In* K. Rützler (ed.), New perspectives in sponge biology. Smithsonian, Washington, D.C.

Rützler, K., and I.G. Macintyre (eds.). 1982. The Atlantic barrier reef ecosystem at Carrie Bow Cay, Belize I: structure and communities. Smithson. Contrib. Mar. Sci. 12. 539 p.

Rützler, K., and G. Rieger. 1973. Sponge burrowing: fine structure of *Cliona lampa* penetrating calcareous substrata. Mar. Biol. 21: 144–162.

Rützler, K., D.L. Santavy, and A. Antonius. 1983. The black band disease of Atlantic reef corals. III. Distribution, ecology, and development. P.S.Z.N.I. Mar. Ecol. 4: 329–358.

Ruyter van Steveninck, E.D. 1984. The composition of algal vegetation in and outside damselfish territories on a Florida reef. Aquat. Bot. 20: 11–19.

Ruyter van Steveninck, E.D. de., and R.P.M. Bak. 1986. Changes in abundance of coral-reef bottom components related to mass mortality of the sea urchin *Diadema antillarum.* Mar. Ecol. Progr. Ser. 34: 87–94.

Ryther, J.H. 1963. Geographic variations in productivity. *In* M.N. Hill (ed.), The sea, vol. 2. Wiley, New York.

Ryther, J.H. 1969. Photosynthesis and fish production in the sea. Science 166: 72–76.

Sadd, J.L. 1984. Sediment transport and $CaCO_3$ budget on a fringing reef, Cane Bay, St. Croix, U.S. Virgin Islands. Bull. Mar. Sci. 35: 221–238.

Sadovy, Y. in press. Grouper stocks of the western central Atlantic: the need for management and management needs. Proc. Gulf Carib. Fish. Instit. 43.

Sakai, T. 1976. Crabs of Japan and the adjacent seas. Kodansha, Tokyo.

Sale, P.F. 1991. The ecology of fishes on coral reefs. Academic Press. San Diego, 754 p.

Salm, R.V. 1984. Man's use of coral reefs. *In* R.A. Kenchington and B.E.T. Hudson (eds.), Coral reef management handbook, pages 15–21. UNESCO, Jakarta, Indonesia.

Salm, R.V., and J.R. Clark. 1984. Marine and coastal protected Areas: a guideline for planners and managers. IUCN, Gland, Switzerland.

Salvat, B. (ed.). 1987. Human impacts on coral reefs: facts and recommendations. Antenne Museum E.P.H.E., French Polynesia, 200 p.

Salvat, B. 1992a. Coral reefs—a challenging ecosystem for human societies. Global Environmental Change 2: 12–18.

Salvat, B. 1992b. Blanchissement et mortalite des scleractiniaires surles recifs de Moorea (Archipel de la Societe) en 1991. CR Acad. Sci. Paris 314: 105–111.

Sammarco, P.W. 1977. The effects of grazing by *Diadema antillarum* Philippi on a shallow-water coral reef community. Ph.D. dissertation, SUNY, Stony Brook, New York.

Sammarco, P.W. 1980. *Diadema* and its relationship to coral spat mortality: grazing, competition, and biological disturbances. J. Exp. Mar. Biol. Ecol. 45: 245–272.

Sammarco, P.W. 1982a. Effects of grazing by *Diadema antillarum* Philippi (Echinodermata: Echinoidea) on algal diversity and community structure. J. Exp. Mar. Biol. Ecol. 65: 83–105.

Sammarco, P.W. 1982b. Polyp bail-out: an escape response to environmental stress and a new means of reproduction in corals. Mar. Ecol. Progr. Ser. 10: 57–65.

Sammarco, P.W. 1983. Effects of fish grazing and damselfish territoriality on coral reef algae. I. Algal community structure. Mar. Ecol. Prog. Ser. 13: 1–14.

Sammarco, P.W. 1985. The Great Barrier Reef vs. the Caribbean: comparisons of grazers, coral recruitment patterns, and reef recovery. Proc. 5th Internat. Coral Reef Congr., Tahiti 4: 391–397.

Sammarco, P.W., and J.H. Carleton. 1981. Damselfish territoriality and coral community structure: reduced grazing, coral recruitment, and effects on coral spat. Proc. 4th Internat. Coral Reef Symp., Manila 2: 525–535.

Sammarco, P.W., J.C. Coll, and S. LaBarre. 1985. Competitive strategies of soft corals (Coelenterata: Octocorallia). II. Variable defensive responses and susceptibility to scleractinian corals. J. Exp. Mar. Biol. Ecol. 91: 199–215.

Sammarco, P.W., J.C. Coll, S. LaBarre, and B. Willis. 1983. Competitive strategies of soft corals (Coelenterata: Octocorallia): allelopathic effects on selected scleractinian corals. Coral Reefs 1: 173–178.

Sammarco, P.W., J.S. Levington, and J.C. Ogden. 1974. Grazing and control of coral reef community structure by *Diadema antillarum* Philippi (Echinodermata: Echinoidea): a preliminary study. J. Mar. Res. 32: 47–53.

Sammarco, P.W., and A.H. Williams. 1982. Damselfish territoriality: influence on *Diadema* distribution and implications for coral community structure. Mar. Ecol. Prog. Ser. 8: 53–59.

Sandberg, P. 1983. An oscillating trend in Phanerozoic non-skeletal carbonate mineralogy. Nature 305: 19–22.

Sano, M., M. Shimizu, and Y. Nose. 1987. Long-term effects of destruction of hermatypic coral by *Acanthaster planci* infestation on reef fish communities of Iriomote Island, Japan. Mar. Ecol. Progr. Ser. 37: 191–199.

Sansone, F.J., C.A. Andrews, R.W. Buddemeier, and G.W. Tribble. 1988. Well point sampling of reef interstitial water. Coral Reefs 7: 19–22.

Sargent, M.C., and T.S. Austin. 1949. Organic productivity of an atoll. Trans. Amer. Geophys. Union 30: 245–249.

Schaffer, W.M. 1981. Ecological abstraction: the consequences of reduced dimensionality in ecological models. Ecol. Monogr. 51: 383–401.

Scheltema, R.S. 1971. Larval dispersal as a means of genetic exchange between geographically separated populations of shallow-water benthic marine gastropods. Biol. Bull. 140: 284–322.

Scheltema, R.S. 1988. Initial evidence for the transport of teleplanic larvae of benthic invertebrates across the East Pacific barrier. Biol. Bull. 174: 145–152.

Schlager, W. 1981. The paradox of drowned reefs and carbonate platforms. Geol. Soc. Amer. Bull. 92: 197–211.

Schlanger, S.O., and I. Premoli Silva. 1981. Tectonic, volcanic, and paleogeographic implications of redeposited reef faunas of late Cretaceous and Tertiary age from the Nauru Basin and the Line Islands. *In* Initial Reports of the Deep Sea Drilling Project, pages 817–827. vol. 61. U.S. Govt. Printing Office, Washington D.C.

Schlapak, B.R., and J.B. Herbich. 1978. Characteristics of coral and coral dredging. Center for Dredging Studies Report No. 213. Texas A&M University, College Station, Texas.

Schluter, D., and R.E. Ricklefs. 1993. Species diversity: an introduction to the problem. *In* R.E. Ricklefs and D. Schluter (eds.), Species diversity in ecological communities: historical and geographical perspectives, pages 1–10. University of Chicago Press, Chicago.

Schmale, M.C. 1991. Prevalence and distribution patterns of tumors in bicolor damselfish (*Pomacentrus partitus*) on South Florida reefs. Mar. Biol. 109: 203–212.

Schmale, M.C., G.T. Hensley, and L.R. Udey. 1986. Neurofibromatosis in the bicolor damselfish *Pomacentrus partitus* as a model of von Recklinghausen neurofibromatosis. Annals of the New York Academy of Science 486: 386–402.

Schmitt, R.J. 1987. Indirect interactions between prey: apparent competition, predator aggregation, and habitat segregation. Ecology 68: 1887–1897.

Schoener, T.W. 1983. Field experiments on interspecific competition. Am. Nat. 122: 240–285.

Schoener, T.W. 1989. Foodwebs from the small to the large. Ecology 70: 1559–1589.

Schoener, T.W. 1993. On the relative importance of direct versus indirect effects in ecological communities. *In* H. Kawanabe, J.E. Cohen, and K. Iwasaki (eds.), Mutualism and community organization: behavioral, theoretical and food-web approaches, pages 365–411. Oxford University Press, Oxford, 426 p.

Scholle, P.A., D.G. Bebout, and C.H. Moore (eds.), 1983. Carbonate depositional environments. Memoir 33. American Association of Petroleum Geologists, Tulsa, Oklahoma.

Schopf, J.W., and D.Z. Oehler. 1976. How old are the eukaryotes? Science 193: 47–49.

Schroeder, J.H., and B.H. Purser. 1986. Reef diagenesis. Springer-Verlag, Berlin, 455 p.

Schroeder, J.H., and H. Zankl. 1974. Dynamic reef formation: a sedimentological concept based on studies of Recent Bermuda and Bahamian reefs. Proc. 2nd Internat. Coral Reef Symp., Brisbane 2: 413–428.

Schupp, P.J., and V.J. Paul. 1994. Calcium carbonate and secondary metabolites in tropical seaweeds: variable effects on herbivorous fishes. Ecology 75: 1172–1185.

Scoffin, T.P., C.W. Stearn, D. Boucher, P. Frydl, C.M. Hawkins, I.G. Hunter, and J.K. MacGeachy. 1980. Calcium carbonate budget of a fringing reef on the west coast of Barbados. Part II—Erosion, sediments and internal structure. Bull. Mar. Sci. 30: 475–508.

Scott, B.D., and H.R. Jitts. 1977. Photosynthesis of phytoplankton and zooxanthellae on a coral reef. Mar. Biol. 41: 307–315.

Scott, F.J., and G.R. Russ. 1987. Effects of grazing on species composition of the epilithic algal community on coral reefs of the central Great Barrier Reef. Mar. Ecol. Progr. Ser. 39: 239–304.

Scott, M.E. 1988. The impact of infection and disease on animal populations: implications for conservation biology. Conserv. Biol. 2: 40–56.

Scott, P.J.B. 1984. The corals of Hong Kong. Hong Kong University Press, 111 p.

Scott, P.J.B., and M.J. Risk. 1988. The effect of *Lithophaga* (Bivalvia: Mytilidae) boreholes on the strength of the coral *Porites lobata.* Coral Reefs 7: 145–151.

Scott, P.J.B., M.J. Risk, and J.D. Carriquiry. 1988. El Niño, bioerosion and the survival of east Pacific reefs. Proc. 6th Internat. Coral Reef Symp., Townsville 2: 517–520.

Shackleton, N.J. 1984. Oxygen isotope evidence for Cenozoic climatic change. *In* P. Benchley (ed.), Fossils and climate, pages 27–34. Wiley, Chichester, United Kingdom.

Shelley, C.C., J.S. Glazebrook, E. Turak, L. Winsor, and G.R.W. Denton. 1988. Trematode (Digenea: Bucephalidae) infection in the burrowing clam *Tridacna crocea* from the Great Barrier Reef. Diseases Aquat. Org. 4: 143–147.

Sheppard, C.R.C. 1988. Similar trends, different causes: responses of corals to stressed environments in Arabian Seas. Proc. 6th Internat. Coral Reef Symp., Townsville 3: 297–302.

Sheppard, C.R.C., A. Price, and C. Roberts. 1992. Marine ecology of the Arabian region. Academic Press, San Diego, 359 p.

Sheppard, C.R.C., and A.L.S. Sheppard. 1991. Corals and coral communities of Arabia. Fauna of Saudi Arabia 12: 3–10.

Sherman, K. 1994. Sustainability, biomass yields, and health of coastal ecosystems: an ecological perspective. Mar. Ecol. Progr. Ser. 112: 277–301.

Sherr, B.F., and E.B. Sherr. 1984. Role of the terozoic protozoa in carbon and energy flows in aquatic ecosystems. *In* M.J. Klugg and C.A. Reddy (eds.), Current perspectives in microbial ecology, pages 412–423. American Society of Microbiologists, Washington, D.C.

Shinn, E.A. 1988. The geology of the Florida Keys. Oceanus 31: 47–53.

Shlesinger, Y., and Y. Loya. 1985. Coral community reproductive patterns: Red Sea versus the Great Barrier Reef. Science 228: 1333–1335.

Shulman, M.J. 1985. Recruitment of coral reef fishes: effects of distribution of predators and shelter. Ecology 66: 1056–1066.

Shulman, M.J., and J.C. Ogden. 1987. What controls tropical reef fish populations: recruitment or benthic mortality? An example in the Caribbean reef fish *Haemulon flavolineatum.* Mar. Ecol. Prog. Ser. 39: 233–242.

Shulman, M.J., J.C. Ogden, J.P. Ebersole, W.N. McFarland, S.L. Miller, and N.G. Wolf. 1983. Priority effects in the recruitment of juvenile coral reef fishes. Ecology 64: 1508–1513.

Siboga-Expeditie Monographie. 1901–1961. Volumes 1–39. Brill, Leiden.

Sih, A., P. Crowley, M. McPeek, J. Petranka, and K. Strohmeier. 1985. Predation, competition, and prey communities: a review of field experiments. Ann. Rev. Ecol. Syst. 16: 269–311.

Sih, A., and L.B. Kats. 1991. Effects of refuge availability on the responses of salamander larvae to chemical cues from predatory green sunfish. Anim. Behav. 42: 330–332.

Sih, A., J.W. Petranka, and L.B. Kats. 1988. The dynamics of prey refuge use: a model and tests with sunfish and salamander larvae. Am. Nat. 132: 463–483.

Silva, P.C. 1992. Geographic patterns in diversity of benthic algae. Pac. Sci. 46: 429–437.

Simkiss, K. 1964. Phosphates as crystal poisons of calcification. Biol. Rev. 39: 487–505.

Sims, R.W. (ed.). 1980. Animal identification, a reference guide, vol. 1, Marine and brackish water animals. British Museum (Natural History). Wiley, New York.

Sindermann, C.J. 1990. Principal diseases of marine fish and shellfish, vol. 1 and 2. Academic Press, San Diego.

Skelton, P.W. 1976. Functional morphology of the Hippuritidae. Lethaia 9: 83–100.

Skelton, P.W., E. Gili, and J.-P. Masse. 1992. Rudists as successful sediment-dwellers, not reef-builders, on Cretaceous carbonate platforms. *In* Fifth North American Paleontological Convention Abstracts and Program. Program, Paleontological Society Spec. Publ. 6: 271.

Skinner, R.H., and W. Kandrashoff. 1988. Abnormalities and diseases observed in commercial fish catches from Biscayne Bay, Florida. Water Resources Bull. 24(5): 961–966.

Slobodkin, L.B., and L. Fishelson. 1974. The effect of the cleaner fish *Labroides dimidiatus* on the point diversity of fishes on the reef front at Eilat. Am. Nat. 108: 369–376.

Smith, C., and J. Tyler. 1972. Space resource sharing in a coral reef fish community. Bull. Nat. Hist. Mus. Los Angeles County. 14: 125–170.

Smith, F.J.W. 1971. Atlantic reef corals. University of Miami Press, Coral Gables, Florida.

Smith, M.M., and P.C. Heemstra (eds.). 1986. Smith's seafishes. MacMillan South Africa, Johannesburg.

Smith, S.V. 1978. Coral-reef area and the contributions of reefs to processes and resources in the world's oceans. Nature 273: 225–226.

Smith, S.V. 1984. Phosphorous versus nitrogen limitation in the marine environment. Limnol. Oceanogr. 29: 1149–1160.

Smith, S.V. 1988. Mass balance in coral reef-dominated areas. *In* B.-O. Jansson (ed.), Lecture notes on coastal and estuarine studies 22, pages 209–226. Springer-Verlag, Berlin.

Smith, S.V., and R.W. Buddemeier. 1992. Global change and coral reef ecosystems. Ann. Rev. Ecol. Syst. 23: 89–118.

Smith, S.V., and J.T. Harrison. 1977. Calcium carbonate production of the Mare Incognitum, the upper windward reef slope, at Eniwetak Atoll. Science 197: 556–559.

Smith, S.V., and P.L. Jokiel. 1978. Water composition and biogeochemical gradients in the Canton Atoll lagoon. Atoll Res. Bull. 221: 15–53.

Smith, S.V., W.J. Kimmerer, E.A. Laws, R.E. Brock, and T.W. Walsh. 1981. Kaneohe Bay sewage diversion experiment: perspectives on ecosystem response to nutritional perturbation. Pac. Sci. 35: 279–402.

Smith, S.V., W.J. Kimmerer, and T.W. Walsh. 1986. Vertical flux and biogeochemical turnover regulate nutrient limitation of net organic production in the North Pacific Gyre. Limnol. Oceanogr. 31: 161–167.

Smith, S.V., and D.W. Kinsey. 1976. Calcium carbonate production, coral reef growth and sea level change. Science 194: 937–939.

Smith, S.V., and D.W. Kinsey. 1988. Why don't budgets of energy, nutrients and carbonates always balance at the level of organisms, reefs and tropical oceans? Proc. 6th Internat. Coral Reef Symp., Townsville 1: 115–122.

Smith, T.J., III, J.H. Hudson, M.B. Robblee, G.V.N. Powell, and P.J. Isdale. 1989. Fresh water flow from the Everglades to Florida Bay: a historical reconstruction based on fluorescent banding in the coral *Solenastrea bouroni*. Bull. Mar. Sci. 44: 274–282.

Soest, R.W.M., van. 1990. Shallow-water reef sponges of Eastern Indonesia. *In* K. Rützler (ed.), New perspectives in sponge biology, pages 302–308. Smithsonian, Washington, D.C.

Sorensen, J.C., S.T. McCreary, and M.J. Hershman. 1984. Institutional arrangements for management of coastal resources. Renewable Resources Information Series Coastal Management Publication No. 1. U.S. Agency for International Development.

Sorokin, Yu.I. 1990. Plankton in reef ecosystems. *In* Z. Dubinsky (ed.), Ecosystems of the world, vol. 25: Coral reefs, pages 291–327. Elsevier Science Publishing, New York, 550 p.

Sorokin, Yu.I. 1993. Coral reef ecology. Ecological Studies 102. Springer-Verlag, Berlin, 465 p.

Soule, D.F., and J.D. Soule. 1976. The status of faunistic information on tropical reef bryozoans. Micronesica 12: 157–164.

Sousa, W.P. 1991. Can models of soft-sediment community structure be complete without parasites? Am. Zool. 31: 821–830.

South, G.R. 1993. Seaweeds. *In* A. Wright and L. Hill (eds.), Nearshore marine resources of the South Pacific, pages 683–710. Forum Fisheries Agency, Honiara, 710 p.

Sparks, A.K. 1985. Synopsis of invertebrate pathology exclusive of insects. Elsevier Science Publishing, Amsterdam.

Springer, V.G., and J.T. Williams. 1990. Widely distributed Pacific plate endemics and lowered sea-level. Bull. Mar. Sci. 47: 631–640.

Spurgeon, J.P.G. 1992. The economic valuation of coral reefs. Mar. Pollut. Bull. 24: 529–536.

Squires, D.F. 1958. The Cretaceous and Tertiary corals of New Zealand. Palaeontol. Bull., New Zealand Geological Survey 29: 1–107.

Stallard, M.O., and D.J. Faulkner. 1974a. Chemical constituents of the digestive gland of the seahare *Aplysia californica*—I. Importance of diet. Compar. Biochem. Physiol. 49B: 25–35.

Stallard, M.O., and D.J. Faulkner. 1974b. Chemical constituents of the digestive gland of the seahare *Aplysia californica*—II. Chemical transformations. Compar. Biochem. Physiol. 49B: 37–41.

Stanley, G.D., Jr. 1981. Early history of scleractinian corals and its geological consequences. J. Geol. 9: 507–511.

Stanley, G.D., Jr. 1992. Tropical reef ecosystems and their evolution. Encyclopedia of Earth System Science 4: 375–388.

Stanley, G.D., Jr., and J.A. Fagerstrom (eds.). 1988. Ancient reef ecosystems theme issue. Palaios 3: 256 p.

Stanley, S.M. 1979. Macroevolution. W.H. Freeman, San Francisco.

Stanley, S.M. 1986. Anatomy of a regional mass extinction: Plio-Pleistocene decimation of the Western Atlantic bivalve fauna. Palaios 1: 17–36.

Stanley, S.M., and L.D. Campbell. 1981. Neogene mass extinction of western Atlantic molluscs. Nature 293: 457–459.

Stearn, C.W., and T.P. Scoffin. 1977. Carbonate budget of a fringing reef, Barbados. Proc. 3rd Internat. Coral Reef Symp., Miami 2: 471–477.

Stebbing, A.R.D. 1981. Stress, health and home ostasis. Mar. Pollut. Bull. 12(10): 326–329.

Steele, J.H. 1991. Marine functional diversity. BioScience 41: 470–474.

Stehli, F.G., A.L. McAlester, and C.E. Helsley. 1967. Taxonomic diversity of Recent bivalves and some implications for geology. Geol. Soc. Amer. Bull. 78: 455–466.

Stehli, F.G., and J.W. Wells. 1971. Diversity and age patterns in hermatypic corals. Syst. Zool. 20: 115–126.

Steneck, R.S. 1983a. Escalating herbivory and resulting adaptive trends in calcareous algal crusts. Paleobiology 9: 44–61.

Steneck, R.S. 1983b. Quantifying herbivory on coral reefs: just scratching the surface and still biting off more than we can chew. *In* M.L. Reaka (ed.), The ecology of deep and shallow coral reefs, pages 103–111. NOAA Symp. Ser. Undersea Res. 1(1). 149 p.

Steneck, R.S. 1988. Herbivory on coral reefs: a synthesis. Proc. 6th Internat. Coral Reef Symp., Townsville 1: 37–49.

Steneck, R.S., and L. Watling. 1982. Feeding capabilities and limitations of herbivorous molluscs: a functional group approach. Mar. Biol. 68: 299–319.

Stephen, A.C., and S.J. Edmonds. 1972. The phyla Sipuncula and Echiura. British Museum (Natural History), London.

Stephenson, W., and R.B. Searles. 1960. Experimental studies on the ecology of intertidal environments of Heron Island. I. Exclusion of fish from beachrock. Austr. J. Mar. Freshw. Res. 2: 241–267.

Stepien, C.A., J.E. Randall, and R.H. Rosenblatt. 1994. Genetic and morphological divergence of a circumtropical complex of goatfishes: *Mulloidichthys vanicolensis, M. dentatus,* and *M. martinicus.* Pac. Sci. 48: 44–56.

Sterrer, W. (ed.). 1986. Marine fauna and flora of Bermuda: a systematic guide to the identification of marine organisms. Wiley, New York.

Stevens, G.C. 1989. The latitudinal gradienting eographic range: how so many species coexist in the tropics. Am. Nat. 133: 240–256.

Stimson, J.S. 1978. Mode and timing of reproduction in some common hermatypic corals of Hawaii and Enewetak. Mar. Biol. 48: 173–184.

Stimson, J.S. 1984. Possible uses of lipids stored in coral tissues. Am. Zool. 24: 78A.

Stimson, J.S. 1990. Stimulation of fat-body production in the polyps of the coral *Pocillopora damicornis* by the presence of mutualistic crabs of the genus *Trapezia.* Mar. Biol. 106: 211–218.

Stoddart, J.A. 1983. A sexual production of planulae in the coral *Pocillopora damicornis.* Mar. Biol. 76: 279–284.

Stone, L., and A. Roberts. 1991. Conditions for a species to gain advantage from the presence of competitors. Ecology 72: 1964–1972.

Stoskopf, M.K. 1993. Fish medicine. W.B. Saunders Co., Philadelphia.

Strauss, S.Y. 1991. Indirect effects in community ecology: their definition, study and importance. Trends Ecol. Evol. 6: 206–210.

Strong, D.R. 1992. Are trophic cascades all wet? Differentiation and donor-control in speciose ecosystems. Ecology 73: 747–754.

Suchanek, T.H., S.L. Williams, J.C. Ogden, D.K. Hubbard, and I.P. Gill. 1985. Utilization of shallow-water seagrass detritus by Caribbean deep-sea macrofauna: ^{13}C evidence. Deep-Sea Res. 32: 201–214.

Sudekum, A.E., J.D. Parrish, R.L. Radtke, and S. Ralston. 1991. Life history and ecology of large jacks in undisturbed, shallow, oceanic communities. Fish. Bull. (U.S.) 89: 493–513.

Suharsono, R.K. Pipe, and B.E. Brown. 1993. Cellular and ultrastructural changes in the endoderm of the temperate sea anemone, *Anemonia viridis,* Forskal as a result of increased temperature. Mar. Biol. 116: 311–318.

Sukarno, N. Naamin, and M. Hutomo. 1986. *In* Proceedings of MAB-COMAR Regional Workshop on Coral Reef Ecosystems: Their Management Practices and Research/Training Needs, pages 24–33.

Sutcliffe, W.H., Jr. 1972. Some relations of land drainage, nutrients, particulate material, and fish catch in two eastern Canadian bays. J. Fish. Res. Bd. Canada 29: 357–362.

Sutcliffe, W.H., Jr. 1973. Correlations between seasonal river discharge and local landings of American lobster (*Homarus americanus*) and Atlantic halibut *(Hippoglossus hippoglossus)* in the Gulf of St. Lawrence. J. Fish. Res. Bd. Canada 30: 856–859.

Szmant-Froelich, A.M. 1984. Reef coral reproduction: diversity and community patterns. *In* Advances in reef science, pages 122–123. Joint Meeting Intl. Soc. Reef Studies and Atlantic Reef Committee. University of Miami, Miami.

Szmant-Froelich, A.M. 1985. The effect of colony size on the productive ability of the Caribbean coral *Montastrea annularis* (Ellis and Solander). Proc. 5th Internat. Coral Reef Congr., Tahiti 4: 295–300.

Szmant, A.M. 1986. Reproductive ecology of Caribbean reef corals. Coral Reefs 5: 43–54.

Szmant, A.M., L.M. Ferrer, and L.M. FitzGerald. 1990. Nitrogen excretion and O:N ratios in reef corals: evidence for conservation of nitrogen. Mar. Biol. 104: 119–127.

Szmant-Froelich, A., M. Reutter, and L. Riggs. 1985. Sexual reproduction of *Favia fragum* (Esper): lunar patterns of gametogenesis, embryo genesis and planulation in Puerto Rico. Bull. Mar. Sci. 37: 880–892.

Tadem, E., J. Reyes, and L.S. Magno. 1984. Showcases of underdevelopment in Mindanao: fishes, forest, fruits. Alternate Resource Center, Davao City, Philippines.

Taylor, D.L. 1971. Photosynthesis of symbiotic chloroplasts in *Tridachia crispata* (Bergh). Comp. Biochem. Physiol. 38A: 233–236.

Taylor, J.D. 1978. Faunal response to the instability of reef habitats: Pleistocene molluscan assemblages of Aldabra Atoll. Palaeontology 21: 1–30.

Taylor, J.D., and R. Saloman. 1978. Some effects of hydraulic dredging and coastal development in Boca Clega Bay, FL. Fish. Bull. 67: 213–241.

Taylor, J.D., and C.N. Taylor. 1977. Latitudinal distribution of predatory gastropods on the eastern Atlantic shelf. J. Biogeogr. 4: 73–81.

Taylor, W.R. 1960. Marine algae of the eastern tropical and subtropical coasts of the Americas. Univ. Michigan Press, Ann Arbor.

Teichert, C. 1958. Cold- and deep-water coral banks. AAPG Bull. 42: 1064–1082.

Terborgh, J. 1973. On the notion of favorableness in plant ecology. Am. Nat. 107: 481–501.

Thayer, G.W., P.L. Murphy, and M.W. Lacroix. 1994. Responses of plant communities in western Florida Bay to the die-off of seagrasses. Bull. Mar. Sci. 54: 718–726.

Thomassin, B.A. 1976. Feeding behavior of the fetlt-, sponge-, and coral-feeder seastars, mainly *Culcita schmideliana.* Helgoländerwissenschaften Meeresuntersuchungen 28: 51–65.

Thompson, D.B., J.H. Brown, and W.D. Spencer. 1991. Indirect facilitation of granivorous birds by desert rodents: experimental evidence from foraging patterns. Ecology 72: 852–863.

Thompson, J.E., R.P. Walker, S.J. Wratten, and D.J. Faulkner. 1982. A chemical defense mechanism for the nudibranch *Cadlinaluteo marginata.* Tetrahedron 38: 1865–1873.

Thomson, D.A., L.T. Findley, and A.N. Kerstitch. 1979. Reef fishes of the Sea of Cortez. Wiley, New York.

Thresher, R.E. 1991. Geographic variability in the ecology of coral reef fishes: evidence, evolution, and possible implications. *In* P.F. Sale (ed.), The ecology of fishes on coral reefs, pages 401–436. Academic Press, San Diego, 754 p.

Thresher, R.E., and E.B. Brothers. 1985. Reproductive ecology and biogeography of Indo-West Pacific angelfishes (Pisces: Pomacanthidae). Evolution 39: 878–887.

Tomascik, T. 1991. Settlement patterns of Caribbean scleractinian corals on artificial substrata along a eutrophication gradient, Barbados, West Indies. Mar. Ecol. Prog. Ser. 77: 261–269.

Tomascik, T., and F. Sander. 1985. Effects of eutrophication on reef-building corals 1. Growth rate of the reef building coral *Montastrea annularis.* Mar. Biol. 87: 143–155.

Tomascik, T., and F. Sander. 1987a. Effects of eutrophication on reef-building corals II. Structure of scleractinian coral communities on fringing reefs, Barbados, West Indies. Mar. Biol. 94: 53–75.

Tomascik, T., and F. Sander. 1987b. Effects of eutrophication on reef-building corals III. Reproduction of the reef-building coral *Porites porites.* Mar. Biol. 94: 77–94.

Tomascik, T., A.J. Suharsono, and A. Mah. 1993. Case histories: a historical perspective of the natural and anthropogenic impacts in the Indonesian Archipelago with a focus on the Kepulauan Seribu, Java Sea. *In* Ginsburg, R.N. (compiler), Proceedings of the Colloquium and Forum on Global Aspects of Coral Reefs: Health, Hazards, and History, pages J26–J32. Rosenstiel School of Marine and Atmospheric Sciences, University of Miami, 420 p.

Tomlinson, J.T. 1969. The burrowing barnacles (Cirripedia: Order Acrothoracica). U.S. Natl. Mus. Bull. 296: 1–162.

Toomey, D.F. (ed.) 1981. European fossil reef models. Spec. Publ. No. 30. Society of Economic Paleontologists and Mineralogists, Tulsa, Oklahoma.

Totton, A.K., and H.E. Bargmann. 1965. A synopsis of the Siphonophora. British Museum (Natural History), London.

Trench, R.K. 1987. Dinoflagellates in non-parasitic symbioses. *In* F.J.R. Taylor (ed.), The biology of dinoflagellates, pages 531–570. Blackwell, Oxford.

Trench, R.K. 1993. Microalgal-invertebrate symbioses: a review. Endocytobiosis and Cell Res. 9: 135–175.

Tribble, G.W., M.J. Atkinson, F.J. Sansone, and S.V. Smith. 1994. Reef metabolism and endo-upwelling in perspective. Coral Reefs 13: 199–201.

Tribble, G.W., F.J. Sansone, Y.-H. Li, S.V. Smith, and R.W. Buddemeier. 1988. Material fluxes from a reef framework. Proc. 6th Internat. Coral Reef Symp., Townsville 2: 577–582.

Tricas, T.C. 1986. Life history, foraging ecology, and territorial behavior of the Hawaiian butterfly fish *Chaetodon multicinctus.* Ph.D. dissertation, University of Hawaii, Honolulu.

Tricas, T.C. 1989. Determinants of feeding territory size in the corallivorous butterfly fish, *Chaetodon multicinctus.* Anim. Behav. 37: 830–841.

Trono, G.C.J., and E.T. Ganzon-Fortes. 1988. Philippines seaweeds. National Book Store Publications, Manila.

Trudgill, S.T. 1976. The marine erosion of limestones on Aldabra Atoll, Indian Ocean. Z. Geomorph. N.F., Suppl. -Bd. 26: 164–200.

Trudgill, S.T. 1983. Measurements of rates of erosion of reefs and reef limestones. *In* D.J. Barnes (ed.), Perspectives on coral reefs, pages 256–262. Brian Clouston Publ., Manuka, Australia, 277 p.

Tudhope, A.W., and M.J. Risk. 1985. Rate of dissolution of carbonate sediments by micro boring organisms, Davies Reef, Australia. J. Sed. Petrology 55: 440–447.

Tunnicliffe, V. 1979. The role of boring sponges in coral fracture. *In* C. Levi and N. Boury-Esnault (eds.), Biologie des spongiaires, pages 309–315. Centre National de la Recherche Scientifique, Paris. No. 291.

Tunnicliffe, V. 1992. Biodiversity: the marine biota of British Columbia. *In* M.A. Fenger, E.H. Miller, J.A. Johnson, and E.J.R. Williams (eds.), Our living legacy: proceedings of a symposium on biological diversity, pages 191–200. Royal British Columbia Museum, Victoria, Canada.

Turner, S.J. 1994a. The biology and population outbreaks of the corallivorous gastropod *Drupella* on Indo-Pacific reefs. Oceanogr. and Mar. Biol. Ann. Rev. 32: 461–530.

Turner, S.J. 1994b. Spatial variability in the abundance of the corallivorous gastropod *Drupella cornus.* Coral Reefs 13: 41–48.

Twilley, R.R., W.M. Kemp, K.W. Staver, J.C. Stevenson, and W.R. Boynton. 1985. Nutrient enrichment of estuarine submersed vascular plant communities. 1. Algal growth and effects on production of plants and associated communities. Mar. Ecol. Progr. Ser. 23: 179–191.

Udey, L.R., E. Young, and B. Sallman. 1977. Isolation and characterization of an anaerobic bacterium, *Eubacterium tarantellu* spp. nov., associated with striped mullet (*Mugilcephalus*) mortality in Biscayne Bay, Florida. J. Fish. Res. Bd. Canada 34: 402–409.

Uehara, T., H. Hidenori, and Y. Arakaki. 1990. Fertilization block age and hybridization among species of sea urchins. *In* M. Hoshi and O. Yamashita (eds.), Advances in invertebrate reproduction, vol. 5, pages 305–310. Elsevier Science Publishing, Amsterdam.

UNEP/IUCN. 1988a. Coral reefs of the world, vol. 1: Atlantic and Eastern Pacific. UNEP/IUCN, Gland, Switzerland; Cambridge, UK; Nariobi, Kenya.

UNEP/IUCN. 1988b. Coral reefs of the world, vol. 2: Indian Ocean, Red Sea and Gulf. UNEP/IUCN, Gland, Switzerland; Cambridge, UK; Nairobi, Kenya.

UNEP/IUCN. 1988c. Coral reefs of the world, vol. 3: Central and Western Pacific. UNEP/ IUCN, Gland, Switzerland; Cambridge, UK; Nairobi, Kenya.

U.S. Naval Weather Service Command. 1979. Summary of synoptic meteorological observations (SSMO)—Caribbean and nearby island coastal marine areas.

UOGML (University of Guam Marine Laboratory). 1981. A working list of marine organisms from Guam, first edition. Technical Report 70. University of Guam Marine Laboratory, Guam, 88 p.

Vacelet, J. 1985. Coralline sponges and the evolution of Porifera. *In* S. Conway Morris, J.D. George, R. Gibson, and H.M. Platt (eds.), The origins and relationships of lower invertebrates, pages 1–13. Spec. No. 28. Systematic Association Clarendon, Oxford.

Vail, P.R., R.M. Mitchum, Jr., S. Thompson III, R.G. Todd, J.B. Sangree, J.M. Widmier, N.N. Bubb, and W.G. Natelid. 1977. Seismic stratigraphy and global sea-level changes. AAPG Mem. 26: 49–212.

Valencia, M.J. 1985. Oil and gas potential, overlapping claims and potential relations. *In* G. Kent and M. Valencia (eds.), Marine policy in Southeast Asia, pages 155–187. University of California Press, Berkeley.

Valentine, J.W. (ed.). 1985. Phanerozoic diversity patterns. Princeton University Press, Princeton, New Jersey.

Valentine, J.W., and D. Jablonski. 1982. Major determinants of the biogeographic pattern of the shallow-sea fauna. Bull. Soc. Geol. France (7)24: 893–899.

VanAlstyne, K.L. 1988. Herbivore grazing increases polyphenolic defenses in the intertidal brown alga *Fucus distichus.* Ecology 69: 655–663.

VanAlstyne, K.L. 1989. Adventitious branching as a herbivore-induced defense in the intertidal brown alga Fucus distichus. Mar. Ecol. Progr. Ser. 56: 169–176.

Van den Hoek, C., A.M. Breeman, R.P.M. Bak, and G. Van Buurt. 1978. The distribution of algae, corals and gorgonians in relation to depth, light attenuation, water movement and grazing pressure in the fringing coral reef of Curaçao, Netherlands Antilles. Aquat. Bot. 5: 1–46.

Van den Hoek, C., A.M. Cortel-Breeman, and J.B.W. Wanders. 1975. Algal zonation in the fringing coral reef of Curaçao, Netherlands Antillies, in relation to zonation of corals and gorgonians. Aquat. Bot. 1: 269–308.

Vandermeer, J. 1980. Indirect mutualism: variations on a theme by Stephen Levine. Am. Nat. 116: 441–448.

Vandermuelen, J.H. 1974. Studies on reef corals II. Fine structure of planktonic planula larva of *Pocillopora damicornis,* with emphasis on the aboral epidermis. Mar. Biol. 27: 239–249.

van Eepol, R., and D. Grigg. 1970. Effects of dredging at Great Bay, St. John. Water Pollution Rept. 5. Caribbean Research Institute, University of the Virgin Islands.

Van Moorsel, G.W.N.M. 1983. Reproductive strategies in two closely related stony corals (*Agaricia,* Scleractinia). Mar. Ecol. Prog. Ser. 13: 273–283.

van Moorsel, G.W.N.M. 1988. Early maximum growth of stony corals (Scleractinia) after settlement on artificial substrata on a Caribbean reef. Mar. Ecol. Progr. Ser. 50: 127–135.

VanName, W.G. 1945. The North and South American ascidians. Bull. Amer. Mus. Nat. Hist. 84: 1–476.

Vanni, M.J., and D.L. Findlay. 1990. Trophic cascades and phytoplankton community structure. Ecology 71: 921–937.

Vaughan, T.W. 1910. The recent Madreporaria of southern Florida. Yb. Carnegie Inst. Wash. 7: 135–144.

Vermeij, G.J. 1978. Biogeography and adaptation, patterns of marine life. Harvard University Press, Cambridge, 332 p.

Vermeij, G.J. 1982. Phenotypic evolution in a poorly dispersing snail after arrival of a predator. Nature 299: 349–350.

Vermeij, G.J. 1987a. Evolution and escalation, an ecological history of life. Princeton University Press, Princeton, New Jersey.

Vermeij, G.J. 1987b. The dispersal barrier in the tropical Pacific: implications for molluscan speciation and extinction. Evolution 41: 1046–1058.

Vermeij, G.J. 1989. Saving the sea: what we know and what we need to know. Conserv. Biol. 3: 240–241.

Vermeij, G.J., and E.J. Petuch. 1986. Differential extinction in tropical American molluscs: endemism, architecture, and the Panama land bridge. Malacologia 27: 29–41.

Vermeij, G.J., and G. Rosenberg. 1993. Giving and receiving: the tropical Atlantic and donor and recipient region for invading species. Amer. Malacol. Bull. 10: 181–194.

Veron, J.E.N. 1974. Southern geographic limits to the distribution of Great Barrier Reef hermatypic corals. Proc. 2nd Internat. Coral Reef Symp., Brisbane 1: 465–473.

Veron, J.E.N. 1982. The species concept in "Scleractinia of Eastern Australia". Proc. 4th Internat. Coral Reef Symp., Manila 2: 183–186.

Veron, J.E.N. 1986. Corals of Australia and the Indo-Pacific. University of Hawaii, Honolulu.

Veron, J.E.N. 1992a. Conservation of biodiversity: a critical time for the hermatypic corals of Japan. Coral Reefs 11: 13–21.

Veron, J.E.N. 1992b. Environmental control of Holocene changes to the world's most northern hermatypic coral outcrop. Pac. Sci. 46: 405–425.

Veron, J.E.N. 1993. A biogeographic database of hermatypic corals. Species of the central Indo-Pacific, genera of the world. Austr. Institute Mar. Sci. Monogr. Ser. 10: 1–433.

Veron, J.E.N. 1995. Corals in space and time. The biogeography and evolution of the Scleractinia. University of New South Wales Press, Sydney, Australia.

Veron, J.E.N., and R. Kelley. 1988. Species stability in reef corals of Papua New Guinea and the Indo-Pacific. Australasian Assoc. Paleontol. Mem. 6: 1–69.

Veron, J.E.N., and P.R. Minchin. 1992. Correlations between sea surface temperature, circulation patterns and the distribution of hermatypic corals of Japan. Continental Shelf Research 12: 835–857.

Veron, J.E.N., and M. Pichon. 1976. Scleractinia of Eastern Australia. Part I. Families

Thamnasteriidae, Astrocoeniidae, Pocilloporidae. Austr. Institute Mar. Sci. Monogr. Ser. 1: 1–86.

Veron, J.E.N., and M. Pichon. 1980. Scleractinia of Eastern Australia. Part III. Families Agariciidae, Siderastreidae, Fungiidae, Oculinidae, Merulinidae, Mussidae, Pectiniidae, Caryophylliidae, Dendrophylliidae. Austr. Institute Mar. Sci. Monogr. Ser. 4: 1–459.

Veron, J.E.N., and M. Pichon, 1982. Scleractinia of Eastern Australia. Part IV. Family Poritidae. Austr. Institute Mar. Sci. Monogr. Ser. 5: 1–159.

Veron, J.E.N., M. Pichon, and M. Wijsman-Best. 1977. Scleractinia of Eastern Australia. Part II. Families Faviidae, Trachyphylliidae. Austr. Institute Mar. Sci. Monogr. Ser. 2: 1–233.

Veron, J.E.N., and C.C. Wallace. 1984. Scleractinia of Eastern Australia. Part V. Family Acroporidae. Austr. Institute Mar. Sci. Monogr. Ser. 6: 1–485.

Verseveldt, J. 1980. A revision of the genus *Sinularia* May (Octocorallia, Alcyonacea). Zoologische Verhandelingen 179: 1–128.

Verseveldt, J. 1982. A revision of the genus *Sarcophyton* Lesson (Octocorallia, Alcyonacea). Zoologische Verhandelingen 192: 1–91.

Verseveldt, J. 1983. A revision of the genus *Lobophytum* von Marenzeller (Octocorallia, Alcyonacea). Zoologische Verhandelingen 200: 1–103.

Verseveldt, J., and F.M. Bayer. 1988. Revision of the genera *Bellonella, Eleutherobia, Nidalia* and *Nidalopsis* (Octocorallia, Alcyoniidae and Nidaliidae), with descriptions of two new genera. Zoologische Verhandelingen 245: 1–131.

Vicente, V.P. 1989. Regional commercial sponge extinctions in the West Indies: are recent climatic changes responsible? Mar. Ecol. Progr. Ser. 10: 179–191.

Vine, P.J. 1974. Effects of algal grazing and aggressive behaviour of the fishes *Pomacentrus lividus* and *Acanthurus sohal* on coral-reef ecology. Mar. Biol. 24: 131–136.

Vine, P.J. 1986. Red Sea invertebrates. IMMWL Publishing, London.

Vogel, K. 1993. Bioeroders in fossil reefs. Facies 28: 109–114.

Voight, J.R. 1988. Trans-Panamanian geminate octopods (Mollusca: Octopoda). Malacologia 29: 289–294.

Volk, T. 1989. Rise of angiosperms as a factor in long-term climatic cooling. Geology 17: 107–110.

von Cosel, R. 1991. Biogeographic patterns of West African bivalves. *In* American Malacological Union, Program and Abstracts, p. 30.

Walbran, P.D., R.A. Henderson, A.J.T. Jull, and M.J. Head. 1989. Evidence from sediments of long-term Acanthaster planci predation on corals of the Great Barrier Reef. Science 245: 847–850.

Walker, N.D., H.H. Roberts, L.J. Rouse, and O.K. Huh. 1982. Thermal history of reef-associated environments during record cold-air outbreak event. Coral Reefs 1: 83–88.

Wallace, C.C. 1985a. Reproduction, recruitment and fragmentation in nine sympatric species of the coral genus *Acropora*. Mar. Biol. 88: 217–233.

Wallace, C.C. 1985b. Seasonal peaks and annual fluctuation in recruitment of juvenile scleractinian corals. Mar. Ecol. Progr. Ser. 21: 289–298.

Wallace, C.C., and M.J. Dallwitz. 1982. Writing coral identification keys that work. Proc. 4th Internat. Coral Reef Symp., Manila 2: 187–190.

Wallace, C.C., J.M. Pandolfi, A. Young, and J. Wolstenholme. 1991. Indo-Pacific coral biogeography: a case study from the *Acropora selago* group. Austr. J. Bot. 4: 199–210.

Wallace, C.C., and B.L. Willis. 1994. Systematics of the coral genus *Acropora:* implications of new biological findings for species concepts. Ann. Rev. Ecol. Syst. 25: 237–262.

Waller, T.R. 1993. The evolution of *"Chlamys"* (Mollusca: Bivalvia: Pectinidae) in the tropical Western Atlantic and Eastern Pacific. Amer. Malacol. Bull. 10: 195–249.

Walter, C., L. Talaue, and J.N. Pasamonte. 1982. A preliminary quantitative study on emergence of reef associated zooplankton from a Philippine coral reef. Proc. 4th Internat. Coral Reef Symp., Manila 1: 443–451.

Wanders, J.B.W. 1977. The role of benthic algae in the shallow reef of Curaçao (Netherlands Antilles) III: the significance of grazing. Aquat. Bot. 3: 357–390.

Ward, S. 1992. Evidence for broadcast spawning as well as brooding in the scleractinian coral *Pocillopora damicornis.* Mar. Biol. 112: 641–646.

Warford, J.J. 1986. Natural resource management and economic development. Projects Policy Department, World Bank, Washington, D.C.

Warme, J.E. 1975. Borings as trace fossils, and the processes of marine bioerosion. *In* R.W. Frey (ed.), The study of trace fossils, pages 181–227. Springer-Verlag, New York.

Warme, J.E. 1977. Carbonate borers—their role in reef ecology and preservation. *In* S.H. Frost, M.P. Weiss, and J.B. Saunders (eds.), Reefs and related carbonates—ecology and sedimentology, page 261–279. AAPG Stud. Geol. 4. American Association of Petroleum Geologists, Tulsa, Oklahoma.

Warmke, G.L., and R.T. Abbott. 1961. Caribbean sea shells. Livingston Publ. Co., Narberth, Pennsylvania.

Warner, R.R., and T.P. Hughes. 1988. The population dynamics of reef fishes. Proc. 6th Internat. Coral Reef Symp., Townsville 1: 149–155.

Warren, J.W. 1991. Diseases of hatchery fish, 6th edn., U.S. Fish and Wildlife Service, Pacific Region.

Wass, R.C. 1982. The shoreline fishery of American Samoa—past and present. *In* J.L. Munro (ed.), Ecological aspects of coastal zone management, pages 51–83. Proc. Seminar on Marine and Coastal Processes in the Pacific, Motupore Island Research Center, July 1980. UNESCO.

Wass, R.C. 1987. Influence of *Acanthaster*-induced coral kills on fish communities at Fagatele Bay and Cape Larsen. *In* C. Birkeland, R.H. Randall, R.C. Wass, B.D. Smith, and S. Wilkins, Biological resource assesement of the Fagatele Bay National Marine Sanctuary, pages 193–209. Prepared for the Government of American Samoa and the Sanctuary Program Division, NOAA Technical Memoranda NOSMWMD3. U.S. Department of Commerce.

Weaver, C.S., and J.E. duPont. 1970. The living volutes. Delaware Museum of Natural History, Greenville, Delaware.

Weber, J.N., and E.W. White. 1974. Activation energy for skeletal aragonite deposited by the hermatypic coral *Platygyra* spp. Mar. Biol. 26: 353–359.

Weber, J.N., and P.M.J. Woodhead. 1970. Ecological studies of the coral predator *Acanthaster planci* in the South Pacific. Mar. Biol. 6: 12–17.

Weber, P. 1993. Reviving coral reefs. *In* L.R. Brown (ed.), State of the world, pages 42–60. W.H. Norton, New York, 268 p.

Wellington, G.M. 1982. Depth zonation of corals in the Gulf of Panama: control and facilitation by resident reef fishes. Ecol. Monogr. 52: 223–241.

Wellington, G.M., P.W. Glynn, and J.E.N. Veron. 1995. Clipperton Island: a unique atoll in the eastern Pacific. Coral Reefs 14: 162.

Wellington, G.M., and B.C. Victor. 1985. El Niño mass coral mortality: a test of resource limitation in a coral reef damselfish population. Oecologia 68: 15–19.

Wellington, G.M., and B.C. Victor. 1988. Variation in components of reproductive success in an undersaturated population of coral-reef damselfish: a field perspective. Am. Nat. 131: 588–601.

Wells, F.E., and C.W. Bryce. 1993. Seaslugs of Western Australia. Western Australian Museum, Perth.

Wells, J.W. 1955. A survey of the distribution of coral genera in the Great Barrier Reef region. Reports of the Great Barrier Reef Committee 6: 21–29.

Wells, J.W. 1956. Scleractinia. *In* R.C. Moore (ed.), Treatise on invertebrate paleontology, Part F, Coelenterata, pages F328–F444. University of Kansas Press, Kansas City.

Wells, J.W. 1966. Evolutionary development in the scleractinian family Fungiidae. Symp. Zool. Soc. London 16: 223–246.

Wells, J.W. 1983. Annotated checklist of the scleractinian corals of the Galápagos. *In* P.W. Glynn and G.M. Wellington, Corals and coral reefs of the Galápagos Islands, pages 213–291. University of California Press, Berkeley, 330 p.

Wells, S.M. 1982. International trade in ornamental corals and shells. Proc. 4th Internat. Coral Reef Symp., Manila 1: 323–330.

Wells, S., and N. Hanna. 1992. The Greenpeace book of coral reefs. Sterling Publ., New York, 160 p.

Werner, E.E. 1992. Individual behavior and higher-order species interactions. Am. Nat. 140: S5–S32.

Werner, E.E., and D.J. Hall. 1974. Optimal foraging and the size selection of prey by the bluegill sunfish *(Lepomis macrochirus)*. Ecology 55: 1042–1052.

Werner, E.E., and D.J. Hall. 1988. Ontogenetic habitat shifts in bluegill: the foraging rate-predation risk trade-off. Ecology 69: 1352–1366.

Werner, E.E., J.F. Gilliam, D.J. Hall, and G.G. Mittelbach. 1983. An experimental test of the effects of predation risk on habitat use in fish. Ecology 64: 1540–1548.

Westneat, M.W., and J.M. Resing. 1988. Predation on coral spawn by planktivorous fish. Coral Reefs 7: 89–92.

Wexler, M. 1994. The art of growing giants. National Wildlife: 20–26.

White, A.T. 1987. Effects of construction activity on coral reef and lagoon systems. Pages 185–193 *In* B. Salvat (ed.), Human impacts on coral reefs: facts and recommendations. Antenne Museum E.P.H.E., French Polynesia.

White, A.T. 1988a. Chaetodon occurrence relative to coral reef habitats in the Philippines with implications for reef assessment. Proc. 6th Internat. Coral Reef Symp., Townsville 2: 427–431.

White, A.T. 1988b. Marine parks and reserves: management for coastal environments in Southeast Asia. ICLARM Education Series 2, Manila, Philippines.

White, M.J.D. 1978. Modes of speciation. W.H. Freeman and Co., San Francisco.

Whittaker, R.H. 1970. Communities and ecosystems. Collier-Macmillan Ltd., London, 83 p.

Whittaker, R.H. 1972. Evolution and measurement of species diversity. Taxon 21: 213–251.

Wicksten, M.K., and M.E. Hendrickx. 1992. Checklist of penaeoid and caridean shrimps (Decapoda: Penaeoidea, Caridea) from the eastern tropical Pacific. Proc. San Diego Soc. Nat. Hist. 9: 1–11.

Wiebe, W.J. 1987. Nutrient pool dynamics in tropical, marine, coastal environments, with special reference to the Caribbean and Indo-West Pacific regions. *In* C. Birkeland (ed.), Comparison between Atlantic and Pacific tropical marine coastal ecosystems: community structure, ecological processes, and productivity, pages 19–42. Reports in Mar. Sci. 46. UNESCO.

Wiedenmeyer, F. 1977. Shallow-water sponges of the western Bahamas. Experientia Supplementum 28. Birkhäuser Verlag, Basel and Stuttgart, Germany.

Wiegert, R.G., and J. Kozlowski. 1984. Indirect causality in ecosystems. Am. Nat. 124: 293–298.

Wilkinson, C.R. 1983. Role of sponges in coral reef structural processes. *In* D.J. Barnes (ed.), Perspectives on coral reefs, pages 263–274. Brian Clouston Publ., Manuka, Australia, 277 p.

Wilkinson, C.R. 1986. The nutritional spectrum in coral reef benthos—or sponging off one another for dinner. Oceanus 29: 68–75.

Wilkinson, C.R. 1987a. Interocean differences in size and nutrition of coral reef sponge populations. Science 236: 1654–1657.

Wilkinson, C.R. 1987b. Sponge biomass as an indication of reef productivity in two oceans. *In* C. Birkeland (ed.), Comparison between Atlantic and Pacific tropical marine coastal ecosystems: community structure, ecological processes, and productivity, pages 99–103. Reports in Marine Science 46. UNESCO.

Wilkinson, C.R. 1987c. Significance of microbial symbionts in sponge evolution and ecology. Symbiosis 4: 135–146.

Wilkinson, C.R. 1987d. Microbial ecology on a coral reef. Search 18: 31–33.

Wilkinson, C.R. 1993. Coral reefs are facing widespread extinctions: can we prevent these through sustainable management practices? Proc. 7th Internat. Coral Reef Symp., Guam 1: 11–21.

Wilkinson, C.R., and A.C. Cheshire. 1989. Patterns in the distribution of sponge populations across the central Great Barrier Reef. Coral Reefs 8: 127–134.

Wilkinson, C.R., L.M. Chou, A.R. Ridzwan, S. Soekarno and S. Sudara. 1993. Status of coral reefs in Southeast Asia: threats and responses. *In* Ginsburg, R.N. (compiler), Proceedings of the Colloquium and Forum on Global Aspects of Coral Reefs: Health, Hazards, and History, pages J33–J39. Rosenstiel School of Marine and Atmospheric Sciences, University of Miami, 420 p.

Wilkinson, C.R., and P.W. Sammarco. 1983. Effects on fish grazing and damselfish territoriality on coral algae. II. Nitrogen fixation. Mar. Ecol. Progr. Ser. 13: 15–19.

Willan, R.C., and N. Coleman. 1984. Nudibranchs of Australasia. Australasian Marine Photographic Index, Sydney.

Williams, A.B. 1984. Shrimps, lobsters, and crabs of the Atlantic coast of the Eastern United States, Maine to Florida. Smithsonian Institution Press, Washington D.C.

Williams, A.H. 1980. The threespot damselfish: a noncarnivorous keystone species. Am. Nat. 116: 138–142.

Williams, A.H. 1981. An analysis of competitive interactions in a patchy back-reef environment. Ecology 62: 1107–1120.

Williams, D. McB. 1986. Temporal variation in the structure of reef slope fish communities (central Great Barrier Reef): short-term effects of *Acanthaster planci* infestation. Mar. Ecol. Progr. Ser. 28: 157–164.

Williams, D. McB., and A.I. Hatcher. 1983. Structure of fish communities on outer slopes of inshore, mid-shelf and outer shelf reefs of the Great Barrier Reef. Mar. Ecol. Progr. Ser. 10: 239–250

Williams, D. McB., E. Wolanski, and J.C. Andrews. 1984. Transport mechanisms and the potential movement of planktonic larvae in the central region of the Great Barrier Reef. Coral Reefs 3: 229–236.

Williams, E.H., Jr., and L. Bunkley-Williams. 1987. Caribbean mass mortalities: a problem with a solution. Oceanus 30(4): 69–75.

Williams, E.H., Jr., and L. Bunkley-Williams. 1990. The world-wide coral reef bleaching cycle and related sources of coral mortality. Atoll Res. Bull. 335: 1–71.

Williams, E.H., Jr., L. Bunkley-Williams, E.C. Peters, B. Pinto-Rodriguez, R. Matos-Morales, A.A. Mignucci-Giannoni, K.V. Hall, J.V. Rueda-Almonacid, J. Sybesma, I. Bonnelly de Calventi, and R.H. Boulon. 1994. An epizootic of cutaneous fibropapillomas in green turtles *Chelonia mydas* of the Caribbean: part of a panzootic? J. Aquat. Animal Health 6: 70–78.

Williams, E.H., Jr., C. Goenaga, and V. Vicente. 1987. Mass bleaching on Atlantic coral reefs. Science 238: 877–878.

Williams, E.H., Jr., and C.J. Sindermann. 1992. Effects of disease interactions with exotic organisms on the health of the marine environment. *In* M.R. DeVoe (ed.), Introductions and transfers of marine species: achieving a balance between economic development and resource protection, Proceedings of the Conference and Workshop, October 30–November 2, Hilton Head, South Carolina. South Carolina Sea Grant Consortium, Charleston.

Williams, E.H., Jr., and T.J. Wolfe-Waters. 1990. An abnormal incidence of the commensal copepod, *Doridocola astrophyticus* Humes, associated with injury of its host, the basket-star, *Astrophyton muricatum* (Lamarck). Crustaceana 59(3): 302.

Williams, G.C. 1993. Coral reef octocorals. An illustrated guide to the soft corals, sea fans, and sea pens inhabiting the coral reefs of northern Natal. Durban Natural Science Museum, Durban, S. Africa.

Williams, J., and L.S. Hamilton. 1982. Watershed forest influences in the tropics and subtropics. A selected, annotated bibliography. East-West Environment and Policy Institute, East-West Center, Honolulu, Hawaii, 217 p.

Williams, M. 1991. Nutrient inputs to an Amazon floodplain stream following deforestation. Thesis, University of Maryland Center for Environmental and Estuarine Research.

Williams, S.L. 1987. Competition between the seagrasses *Thalassia testudinum* and *Syringodium filiforme* in a Caribbean lagoon. Mar. Ecol. Progr. Ser. 35: 91–98.

Williams, S.L., and R.C. Carpenter. 1988. Nitrogen-limited primary productivity of coral reef algal turfs: the potential contribution of ammonium excreted by *Diadema antillarum* Philippi. Mar. Ecol. Progr. Ser. 47: 145–152.

Willis, B.L., R.C. Babcock, P.L. Harrison, and T.K. Oliver. 1985. Patterns in the mass spawning of corals on the Great Barrier Reef from 1981 to 1984. Proc. 5th Internat. Coral Reef Congr., Tahiti 4: 343–348.

Willis, B.L., R.C. Babcock, P.L. Harrison, and C.C. Wallace. 1993. Experimental evidence of hybridization in reef corals involved in mass spawning events. Proc. 7th Internat. Coral Reef Symp., Guam 1: 504 (Abstr.).

Wilson, B.R. 1993. Australian marine shells. Prosobranch gastropods, vols. 1 and 2. Odyssey Publishing, Kallaroo, W. Australia.

Wilson, B.R., and J. Stoddart. 1987. A thorny problem. Crown-of-thorns starfish in W.A. Landscape 3: 35–39.

Wilson, J.L. 1975. Carbonate facies in geologic history. Springer-Verlag, New York.

Winterbottom, R., and A.R. Emery. 1981. A new genus and two new species of gobiid fishes (Perciformes) from the Chagos Archipelago, Central Indian Ocean. Env. Biol. Fish. 6: 139–149.

Witman, J.D. 1988. Effects of predation by the fireworm *Hermodice carunculata* on milleporid hydrocorals. Bull. Mar. Sci. 42: 446–458.

Wolf, N.G. 1987. Schooling tendency and foraging benefit in the ocean surgeonfish. Behav. Ecol. Sociobiol. 21: 59–63.

Wolff, W.J., A. Gueye, A. Meijboom, T. Piersma, and M.A. Sall. 1987. Distribution, biomass, recruitment and productivity of *Anadara senilis* (L.) (Mollusca: Bivalvia) on the Banc d'Arguin, Mauritania. Netherlands J. Sea Res. 21: 243–253.

Wood, E.A. 1983. Reef corals of the world. T.F.H. Publications, Neptune City, New Jersey, 256 p.

Wood, R. 1990. Reef-building sponges. Amer. Sci. 78: 224–235.

Wood, R., A.Yu. Zhuravlev, and F. Debrenne. 1992. Functional biology and ecology of Archaeocyatha. Palaios 7: 131–156.

Woodley, S.J. 1985. The Great Barrier Marine Park: the management challenge. Proc. 5th Internat. Coral Reef Congr., Tahiti 4: 259–264.

Woodroffe, C.D., and J. Grindrod. 1991. Mangrove biogeography: the role of Quaternary environmental and sea-level change. J. Biogeogr. 18: 479–492.

Wootton, J.T. 1992. Indirect effects, prey susceptibility, and habitat selection: impacts of birds on limpet sandalgae. Ecology 73: 981–991.

Wootton, J.T. 1993. Indirect effect sand habitat use in an intertidal community: interaction chains and interaction modifications. Am. Nat. 141: 71–89.

Worsley, T.R., R.D. Nance, and J.B. Moody. 1986. Tectonic cycles and the history of the earth's biogeochemical and paleoceanographic record. Paleoceanography 3: 233–263.

Worthen, W.B., and J.L. Moore. 1991. Higher-order interactions and indirect effects: a resolution using laboratory *Drosophila* communities. Am. Nat. 138: 1092–1104.

Wulff, J.L. 1984. Sponge-mediated coral reef growth and rejuvenation. Coral Reefs 3: 157–163.

Wyers, S.C. 1985. Sexual reproduction of the coral *Diploria strigosa* (Scleractinia, Favidae) in Bermuda: research in progress. Proc. 5th Internat. Coral Reef Cong., Tahiti 4: 301–306.

Yaldwyn, J.C. 1973. Decapod Crustacea from South Pacific reefs and islands. *In* R. Fraser (ed.), Oceanography of the South Pacific 1972, pages 503–511. New Zealand National Commission for UNESCO, Wellington.

Yamaguchi, M. 1973. Early life histories of coral reef asteroids, with special reference to *Acanthaster planci* (L.). *In* O.A. Jones and R. Endean (eds.), Biology and geology of coral reefs II, pages 369–387. Biology 1. Academic Press, New York.

Yamaguchi, M. 1975. Coral-reef asteroids of Guam. Biotropica 7: 12–23.

Yamaguchi, M. 1986. *Acanthaster planci* infestations of reefs and coral assemblages in Japan: a retrospective analysis of control efforts. Coral Reefs 5: 23–30.

Yamamura, S., and Y. Hirata. 1963. Structures of aplysin and aplysinol, naturally occurring bromo-compounds. Tetrahedron 19: 1485–1496.

Yamazato, K., M. Oshiro, and E. Oshiro. 1975. Reproductive biology of a scleractinian coral, Goniopora queenslandiaedecima. Proc. 13th Pac. Sci. Congr. 1: 135 (Abstr.).

Yap, H.T., and E.D. Gomez. 1985. Coral reef degradation and pollution in the East Asian Seas region. *In* A.L. Dahl and J. Carew-Reid (eds.), Environment and resources in the Pacific, pages 185–207. UNEP Regional Seas Reports and Studies No. 69.

Yodzis, P. 1988. The indeterminacy of ecological interactions as perceived through perturbation experiments. Ecology 69: 508–515.

Yonge, C.M., and A.G. Nicholls. 1931. Studies on the physiology of corals. IV. The structure, distribution and physiology of zooxanthellae. Sci. Rep. Great Barrier Reef Exped. 1: 135–176.

Young, P.C., and H. Kirkman. 1975. The seagrass communities of Moret on Bay, Queensland. Aquat. Bot. 1: 191–202.

Youngbluth, M.J. 1968. Aspects of the ecology and ethology of the cleaning fish, *Labroides phthirophagus* Randall. Z. Tierpsychol. 25: 915–932.

Zann, L.P. 1980. Living together in the sea. Tropical Fish Hobbyist Publications, Neptune, New Jersey.

Zann, L.P., and L. Bolton. 1985. The distribution, abundance and ecology of the blue coral *Heliopora coerulea* (Pallas) in the Pacific. Coral Reefs 4: 125–134.

Zann, L., J. Brodie, and V. Veikila. 1990. History and dynamics of the crown-of-thorns starfish *Acanthaster planci* (L.) in the Suva area, Fiji. Coral Reefs 9: 135–144.

Zann, L.P., R.J. Cuffey, and C. Kropach. 1975. Fouling organisms and parasite associated with the skin of seasnakes. *In* W.A. Dunson (ed.), The biology of sea snakes, pages 151–265. University Park Press, Baltimore.

Zea, S. 1987. Epongas del Caribe Colombiano. Editorial Catálogo Científico., Colombia.

Zeller, D.C. 1988. Short-term effects of territoriality of a tropical damselfish and experimental exclusion of large fishes on invertebrates in algal turfs. Mar. Ecol. Progr. Ser. 44: 85–93.

Zibrowius, H. 1974. *Oculina patagonica,* scléractiniaire hermatypique introduit en Méditerranée. Helgoländerwissenschaften Meeresuntersuchungen 26: 153–173.

Zibrowius, H. 1976. Les scléractiniaires de la Méditerranée et de l'Atlantique Nord-Oriental. D.Sc. thesis, Université

Ziegler, A.M., C.R. Scotese, and S.F. Barrett. 1982. Mesozoic and Cenozoic paleogeographic maps. *In* P. Brosche and J. Sündermann (eds.), Tidal friction and the Earth's rotation II, pages 240–252. Elsevier, Amsterdam.

Zieman, J.C. 1982. The ecology of the sea grasses of South Florida: A community profile. U.S. Fish and Wildlife Service, FWS/OBS-82/25. Office of Biological Services, Washington, D.C., 158 p.

Zieman, J.C., J.W. Fourqurean, and R.T. Zieman. 1989. The Florida Bay sea grass die off: process changes, potential causes, and a conceptual model. Presented at the Tenth International Estuarine Research Conference, 8–12 October 1989, Baltimore, Maryland, p. 92 (Abstr.).

Zieman, J.C., S.A. Macko, and A.L. Mills. 1984. Role of sea grasses and mangroves in estuarine food webs: temporal and spatial changes in stable carbon isotope composition and amino acid content during decomposition. Bull. Mar. Sci. 35: 380–392.

Zieman, J.C., G.W. Thayer, R.B. Robblee, and R.T. Zieman. 1979. Production and export of sea grasses from a tropical bay. *In* R.J. Livingston (ed.), Ecological processes in coastal and marine systems, pages 21–34. Plenum Press, New York.

Zorn, J.G. 1991. Making law in Papua New Guinea: the influence of customary law on the common law. Pac. Stud. 14(4): 1–34.

Index

Abudefduf, 242
Acanthaster, 7, 43
Acanthastrea, 181
Acanthochromis, 242
Acanthophora, 260
Acanthopleura, 82
Acanthosquilla, 334
Acanthuridae, 84, 233
Acanthurids, 225
Acclimatization, 356
Accretion, 10
Acmaea, 82, 205
Acrhelia, 180
Acropora, 8
Actinians, 309
Acute, 7, 373
Adaptation, 356
Advection, 159
Aeromonas, 134
African, 320
Agar, 3
Agaricia, 180
Aggregations, 381
Agriculture, 6, 293
AID, 425
Albula, 382
Alcyonaceans, 118
Alcyoniidae, 305
Alcyoniina, 324
Alga, 5
Algae, 8
Algal lawns, 261
Algal ridges, 54
Allelopathy, 261
Allopatric speciation, 335
Alpheus, 78, 262
Alternate stable states, 157
Amazon, 8
Amazon River, 278
Ammonites, 33
Ammonium, 291
Amphidinium, 98
Amphipods, 205
Ampithoe, 205
Anaerobic conditions, 432
Andaman Sea, 278, 365
Aniculus, 79
Annelida, 198
Anthopleura, 319
Anthosigmella, 75
Aphrodisiacs, 423
Aquaculture, 429
Aquarium trade, 4
Arabian Gulf, 358
Aragonite, 16
Archaeocyathids, 31
Arothron, 85
Arthropoda, 198
Ascidian, 5
Asia, 6
Assimilation, 141
Astreopora, 181
Astropyga, 128, 204
Aswan Dam, 433
Atlantic, 2
Atmosphere, 21
Atolls, 1
Australia, 57
Australs, 285
Autotrophs, 149

Bacteria, 7, 71
Bacterial infection, 9
Bafflers, 15
Balistidae, 84, 208, 232
Barbados, 370
Barnacle, 265
Barnacles, 15, 77
Barracuda, 230
Batrachoidid, 222
Bay of Bengal, 278
Bermuda, 8, 317
Bikini, 46
Binders, 15
Bioassays, 192
Bioeroders, 16
Bioerosion, 7, 61
Biogenic reefs, 1, 13
Biogenic sediments, 14
Bioherm, 13, 60
Biomass, 148
Biomineralization, 32
Bivalve, 305
Black band disease, 87
Black corals, 3
Bleaching, 8, 365
Blennies, 233
Blenniidae, 233
Blooms, 7
Body size, 419
Bolbometopon, 242
Boring bivalves, 9
Bottom-up, 171
Boundary flux, 147
Bounty system, 422
Brachiopods, 31
Brachyuran, 331
Brazil, 341
Breakwaters, 3
British Virgin Islands, 8
Brooding, 178
Broodstock, 344
Browsing, 233
Bryozoans, 4
Budding, 176
Buffering, 295
Bugula, 432
Building materials, 3
Butterflyfish, 241
Byrozoan, 265

$CaCO_3$, 5
Caesionid, 323
Caesionids, 419
Calcification, 111
Calcite, 16
Calcium, 1
Callianassa, 126
Cambrian, 25
Carangids, 419
Carbohydrate, 145
Carbon, 1
Carbon fixation, 145
Carbonate, 13
Carboniferous, 21, 32
Cardiid, 309
Caretta, 131
Caribachlamys, 340
Caribbean, 2
Carnivorous, 84
Carolines, 1
Carrageenan, 3
Cascade, 200
Catch per unit effort (CPUE), 418
Causeways, 382
Cellana, 82
Cement, 3, 391
Cenozoic, 36
Center-of-origin, 334
Centroceras, 235
Cephalopods, 31, 313
Ceramium, 235
Chaetetid tabulate "corals", 305
Chaetitids, 32
Chaetodon, 241
Chamid bivalves, 309
Charonia, 213
Cheilinus, 423
Chelonia, 131
China, 6, 409
Chiton, 82
Chitons, 16
Chloralgal, 15
Chlorophyll, 361
Chlorophyta, 73
Chloroplasts, 309
Chlorozoan, 15
Chromodoris, 332
Chronic, 7, 373
Chuuk (Truk), 196
Ciliates, 309
CITES, 408
Cittarium, 82
Clams, 3
Clay, 69

Cleaning Symbioses, 258
Cliona, 57
Clipperton, 313
CLOD, 118
Coccolithophorids, 21, 34
Cocos, 322
Codiolum, 75
Codium, 280
Colobocentrotus, 206
Colombia, 109, 365
Colonization, 333
Colpophyllia, 118
Commercial fisheries, 380
Compensation depth, 276
Competition, 8, 105, 253
Competitors, 252
Concrete blocks, 375
Conidae, 305
Connected, 249
Constraints, 281
Construction material, 374
Constructors, 16
Consumption, 11
Control volumes, 147
Conus, 315
Copepod, 305
Coral bleaching, 356
Coral mining, 374
Coral rubble, 3
Coral sand, 375
Coralliomorpharians, 309
Corallivores, 200
Corallophilia, 200
Coral reef, 13
Coral-reef fisheries, 2
Coral-reef management, 175
Coriolis effect, 277
Cosmopolitan, 344
Cost assessments, 5
Costa Rica, 109, 365
Cost-benefit analysis, 392
CO_2, 1
Crabs, 77, 200
Cretaceous, 2
Crinoids, 31, 313
Crustacea, 77
Cryptochiridae, 305
Culcita, 204
Curaçao, 369
Curio trade, 3
Curios, 386
Currents, 317
Customs, 382
Cuthona, 332
Cuttlefish, 3
Cyanide, 412
Cyanobacterial, 19
Cyphastrea, 180
Cypraea, 315
Cypraeidae, 305

Damselfish, 130
Damselfishes, 230, 233
Darwin, 45
Decomposition, 140
Dendrogyra, 180
Dendrophyllia, 180
Depth zonation, 244
Destroyers, 16
Detergents, 373
Devonian, 32
Diadema, 7
Diatoms, 312
Dictyosphaeria, 92, 371
Dictyota, 265
Dinoflagellates, 7
Diodontidae, 245
Diploria, 118, 180
Direct effects, 249
Discount rates, 400
Disease, 11
Dispersal, 333
Disturbance, 354
Divergence, 336
Dominance, 291
Dredged, 386
Drilling, 373
Drowned reefs, 291
Drupella, 86, 203
Dynamite, 409

Earth, 1
East Pacific barrier, 322
Easter Island, 319
Echinoderms, 3
Echinometra, 7, 207
Echinophyllia, 181
Echinostrephus, 84
Echinothrix, 204
Ecological economics, 392
Economic yield, 381
Economy, 2
Ecosystem overfishing, 9
Ecotourism, 430

Ectocarpus, 235
Ectodermis, 97
Ectoparasites, 258
Ecuador, 109
Edge, 288
Eels, 261
Egg-sperm packets, 187
Egypt, 5
El Niño, 9
El Niño–Southern Ocean, 109
Embryos, 176
Encrusting coralline algae, 16
Endemicity, 299
Endemism, 323
Endodermis, 97
Endolithic, 156
Endoliths, 71
Enewetak, 46
Enewetak Atoll, 1
ENSO, 109
Entermorpha, 235
Entocladia, 75
Entophysalis, 71
Eocene, 36
Epilithic, 156
Epizootics, 116
Eretmochelys, 131
Eucheuma, 3
Eucidaris, 7
Eugomontia, 75
Eulerian, 147
Euphyllia, 181
Eustatic, 47
Eutrophication, 86, 194, 370
Everglades, 293
Evolutionary arms race, 334
Excess production, 152
Exchange, 149
Export, 167
Export fishery, 382
Externalities, 392
Extinction, 1

Falsilyria, 341
Fanning Island, 58
Favia, 180
Favites, 185
Feces, 231
Fecundity, 193, 417
Fertilization, 187, 291
Fiji, 213
Filamentous algae, 10
Filefishes, 84
Filter, 333
Fishermen, 5
Fishing grounds, 383
Fissurella, 205
Fixation, 1
Flatworms, 309
Florida, 5
Flower Gardens Bank, 8
Fluxes, 140
Food availability, 254
Food web, 275
Food webs, 155
Foraging, 261
Foraminifera, 14
Foramol, 14
Forests, 6
Founder event, 336
Fragmentation, 176, 299, 369
Framework, 60
French Frigate Shoals, 3, 372
French Polynesia, 285
Fringing reefs, 46
Fungi, 71
Fungia, 180
Fungiacava, 82
Fusulinids, 32

Galápagos, 7
Galaxea, 177
Gambiers, 285
Gametes, 176
Gastrochaena, 82
Gastropod, 3
Geminate species, 331
Gene flow, 333
Generalists, 198
Generic age, 334
Geodia, 123
Geographic isolation, 335
Geothermal endo-upwelling, 169
Gilbert Islands, 382
Gillnet fishing, 382
Glaciation, 21, 29
Glacio-eustatic cycles, 338
Gleaning, 380
Global warming, 376
Goatfish, 419
Gobiidae, 232
Goby, 230
Golf courses, 389
Goniastrea, 178, 181

Gonochoric, 177
Gonodactylidae, 305
Gonodactylus, 126
Gorgonaceans, 3
Gorgonia, 260
Grazer, 11
Grazers, 198
Great Barrier Reef, 1
Greenhouse, 50
Gross productivity, 167
Groupers, 230
Growth, 176, 275
Grunts, 289
Guam, 1
Guilds, 312
Gulf of Chiriqui, 330
Gulf Stream, 29

Habitat, 157
Haemulidae, 289
Haemulon, 289
Halimeda, 15
Halo, 263
Harengula, 131
Hawaii, 57
Hawaiian, 3
Heat, 18
Heavy rain, 363
Helen's Reef, 413
Heliopora, 319
Helioporidae, 305
Hemiglyphidodon, 237
Herbivores, 71
Hermaphroditic, 177
Hermodice, 200
Herrings, 131
Heterocentrotus, 206
Heterotrophs, 149
Heterotrophy, 98
Hippopus, 344
Hippospongia, 122
Holacanthus, 283
Holocene, 317
Holothuria, 196
Holothuroids, 155
Human population growth, 9
Hurricane, 8
Hybridization, 189
Hydrocoral, 73
Hydrolithon, 235
Hyella, 71
Hymenocera, 286
Hypnea, 206

Ichthyophonus, 129
Immigration, 427
Import, 167
Indirect effects, 200, 249
Indo-Malayan area, 342
Indo-Malayan triangle, 334
Indonesia, 277
Industrial discharges, 389
Interglacial, 39
Intermediate disturbance, 354
Interstices, 156
Introduced species, 344
Iriomote Island, 398
Isophyllia, 180
Isopods, 205
Isthmus of Panama, 327

Jamaican, 3
Japan, 27
Jenneria, 203
Jewelry, 3
Jurassic, 21
Juveniles, 289

Kaneohe Bay, 370
Kenya, 7
Keystone, 237
Kingdoms, 4
Kiribati, 1
Kosrae, 425
Kuwait, 7

Labridae, 208
Labroides, 259
Labyrinthula, 117
Lagoons, 153
Lagrangian, 147
Larval competency, 178
Laticauda, 131
Latitude, 162
Law, 392
Legislation, 413
Leptoria, 1, 190
Life-history, 282
Light, 18, 56
Light spectrum, 56
Lime, 386
Limestone, 1
Limitation, 152

Limpets, 82, 205
Line Islands, 320
Lipids, 146
Liquidate, 424
Lithophaga, 82
Lithotrya, 78
Livelihoods, 411
Livestock sewage, 389
Lobophora, 224
Lobophyllia, 193
Lobsters, 3
Logging, 58, 392
Longitudinal, 276
Lord Howe Island, 316
Lottia, 341
Lucina, 126
Lutjanidae, 289
Lutjanus, 419
Lysiosquilla, 334
Lytechinus, 207

Magdalena, 8
Maldives, 1, 367, 375
Malé Atoll, 374
Management, 256
Management solutions, 386
Management strategy, 386
Mangroves, 288
Manicina, 180
Manipulative field experiments, 371
Marianas Islands, 27
Marine preserves, 400
Marine Sanctuary, 4
Marine tenure, 382
Market, 431
Marquesas, 285
Marshall Islands, 345
Marshalls, 1
Mastigocoleus, 71
Matrix, 159
Maturity, 193
Meandrina, 180, 341
Medicines, 386
Mediterranean Sea, 278
Megabalanus, 209
Meiofauna, 155
Melanesia, 9
Merulina, 182
Mesozoic, 21
Metabolize, 140
Metamorphosis, 175
Microalgae, 14
Microatolls, 372
Microbial loop, 153
Micronesia, 9, 367
Migrations, 381
Milankovitch cycles, 335
Millepora, 200
Miocene, 37
Mississippi River, 8
Mithrax, 206
Mitridae, 305
Moats, 153
Models, 376
Mollusca, 79
Molluscs, 14
Monacanthidae, 84
Monsoon, 53
Montastrea, 1
Montipora, 38, 190
Morphology, 361
Mortalities, 7
Mother-of-pearl, 3
Mucus, 156, 372
Mulloides, 419
Mulloidichthys, 334
Multiple stresses, 376
Multiple-use, 409
Muricid, 326
Muro-ami, 426
Mussismilia, 341
Mutualism, 246
Mycedium, 183
Mycetophyllia, 180

Natural disturbances, 7
Neighboring habitats, 11
Nematocysts, 261
Neogene, 38
Nephtheidae, 305
Nerita, 82
Net primary productivity, 167
Netherlands Antilles, 83, 204
New Caledonia, 302
Niche, 308
Nile, 278
Nitrates, 8
Nitrogen, 6
Nodilittorina, 82
Norfolk Islands, 319
Nudibranchs, 332
Nurseries, 289
Nutrients, 6

Oceania, 9
Octocorals, 309
Octopus, 3
Oil pollution, 370
Okinawa, 7, 179, 424
Omnivorous, 204
Opisthobranchs, 199
Ordovician, 21, 31
Organic matter, 7
Orinoco, 8
Oscillation, 109
Ostreobium, 75
Oulophyllia, 181
Outbreaks, 7, 213
Outwelling, 291
Overexploitation, 2
Oysters, 52, 309
Ozone, 378

Pachyseris, 183
Pacific, 2
Palatability, 234
Palau, 5
Palawan Island, 392
Paleocene, 36
Paleogene, 22
Paleozoic, 1
Panama, 7, 365
Pangaea, 27
Panulirus, 126
Papua New Guinea, 3
Paradigm, 433
Parasite, 115
Parasites, 114, 199
Parrotfishes, 84, 233
Parthenogenesis, 177
Partial mortality, 378
Patelloida, 82
Pathogen, 43, 114, 224
Pathways, 381
(P/B) ratio, 149
Pearl culture, 3
Pectinia, 181
Pelagic fishes, 3
Pelamis, 131
Periclimenes, 126
Perkinsus, 124
Permian, 32
Persian Gulf, 320
Petroleum, 33
Phaeophila, 75
Phanerozoic, 317
Pharmaceuticals, 4
Phase shift, 368
Philippines, 2, 303
Phormidium, 118
Phoronids, 312
Phosphate, 8
Phosphorus, 6
Photic zone, 273
Photoautotrophy, 98
Photosynthesis, 19
Phyla, 4
Phytoplankton, 7
Pigmentation, 365
Pinctada, 3, 431
Plagioporus, 121
Planktotrophic larvae, 7, 222
Planula, 176
Plasticity, 222
Platygyra, 181
Plectonema, 71
Pleistocene, 38
Plexaura, 5
Pliocene, 37, 327
Poachers, 424
Pocillopora, 179
Poisoning, 7
Pollution, 11
Polychaete, 77
Polynesia, 9
Polyp bailout, 176
Polyps, 96
Pomacanthus, 283
Pomacentridae, 233
Pomacentrus, 130
Population turnover, 417
Porcupinefishes, 245
Porites, 56
Portieria, 5
Postlarvae, 289
Potamocorbula, 344
P/R ratio, 144
Prawns, 3
Precautionary principle, 390
Precipitation of calcium carbonate, 19
Predation, 86
Predators, 198
Primary producers, 149
Prochloron, 309
Production, 11, 140
Prokaryotae, 71
Prokaryote, 309
Propagules, 332

Prosobranch, 332
Prostaglandins, 5
Protandry, 178
Proteins, 146
Proterozoic, 24
Protoctista, 71
Protogyny, 178
Pseudamphithoides, 265
Pseudomonas, 134
Pseudopopulations, 279
Puerto Rico, 57
Puffers, 84
Pyramidellid, 345

Quarrying, 375
Queensland, 6

Rabbitfishes, 233
Radiation, 18
Rare species, 299
Ratchet effect, 425
Recovery, 356
Recruitment, 11, 175
Recruits, 8
Recycling, 144, 278, 415
Red Sea, 320
Red tides, 362
Reef, 13
Reef accretion, 377
Reef fragility, 355
Reef frameworks, 7
Reef mounds, 13
Reef-gleaning, 6
Reef-monitoring, 356
Refuge, 384
Refugia, 199
Replenish, 419
Reproduction, 11, 175
Respiration, 140
Revenue, 3
Revillagigedo, 322
Rhizophora, 291
Rhodophyta, 73
River, 6
Riverine discharge, 293
Rivers, 278
Road building, 393
Robust, 354
Rubble, 68, 363
Rudistids, 33
Runoff, 8
Ryukyu Islands, 203

Saipan, 431
Salinity, 52
Samoa, 3, 367
Sanctuary, 297
Sand, 3, 68
Sargassum, 216
Scale, 7, 356
Scales, 152
Scaridae, 84, 233
Scarids, 225
Scarus, 242
Sclerosponges, 305
Scombrids, 419
Scyllarides, 126
Scyphozoan medusae, 309
Sea anemones, 105
Sea cucumbers, 3
Sea Level, 46
Sea turtles, 287
Seafloor spreading, 26
Seagrass, 15
Seahorses, 423
Sea-level rise, 377
Seascape, 288
Seaweeds, 3
Secondary production, 149
Sediment, 57
Sediments, 6
Seriatopora, 181
Serpulids, 52
Settlement, 175
Sewage, 86
Shading, 58, 261
Sharkfin, 423
Sharks, 230
Shelf, 50
Shelter, 234
Ship grounding, 375
Shrimp, 77
Sibling species, 302
Siderastrea, 180, 328
Siganidae, 233
Silica, 18
Silt, 68
Sinks, 291
Sinularia, 315
Siphonodictyon, 75
Sipuncula, 79
Site selection, 175
Small-scale fishing enterprises, 278
Snail, 3
Snakes, 131

Snappers, 289
Society Islands, 285, 365
Socioeconomic conflicts, 386
Soil erosion, 6
Solenopora, 31
Somalia, 277
South America, 8
Souvenirs, 3
Space, 8
Sparid, 222
Sparisoma, 242
Spawning, 178
Specialists, 198
Speciation, 298
Species, 237
Species richness, 279
Spheciospongia, 75
Sphinctozoans, 305
Spondylus, 3
Sponges, 4
Spongia, 122
Squid, 3
Sri Lanka, 3
St. Croix, 369
Standing stocks, 2, 411
Starfish, 7
Stegastes, 237
Stenopus, 126
Stenorhynchus, 126
Stochastic factor, 227
Stoichiometry, 141
Stomatopods, 304
Storms, 53
Stratigraphic Range, 347
Streptococcus, 129
Stress, 193
Stromatolites, 15
Strombus, 3
Stylasterine, 309
Stylophora, 178, 361
Stypopodium, 260
Sublethal responses, 378
Subsidies, 290, 422
Subsistence, 426
Subsistence fishing, 278, 428
Succession, 283
Successional stage, 234
Sugarcane, 293
Sun, 17
Surgeonfishes, 84, 233
Survivorship, 226
Suspension feeders, 276
Suspension-feeding, 8
Sustainable, 381
Sustainable development, 390
Sustainable yield, 387
Sustainable yields, 174
Symbiodinium, 98, 309
Symbioses, 1
Sympatric species, 338
Symphyllia, 184
Synchronization, 193
Synergistic, 376
Synergistic interaction, 379
Syringodium, 291

Tahiti, 213
Tanaids, 205
Tanzania, 374
Tarawa Atoll, 382
Tectonic, 19
Tectonics, 44
Temperature, 50
Territories, 233
Tertiary, 327
Tethys, 37
Tetradontidae, 208
Tetralia, 262
Tetraodontidae, 84, 232
Thailand, 359
Thalassia, 117, 291
Thermal effluents, 370
Threshold, 358
Timing, 185
Tolypiocladia, 235
Tool, 376
Top-down, 171
Topographic complexity, 284
Topographically complex, 284
Tourism, 4
Tourism industry, 374
Trade winds, 277
Traditional management practices, 384
Tragedy of the commons, 426
Transplants, 241
Trapezia, 200, 262
Trapeziidae, 305
Triassic, 21
Tridacna, 124, 344
Tridacnidae, 305
Trididemnum, 5
Triggerfishes, 84
Tripneustes, 207
Trizopagurus, 79

Trochus, 3, 424
Trophic level, 252
Trophic levels, 285
Tuamotus, 1, 285
Tubastrea, 180
Tubiporidae, 305
Turbidity, 111, 363
Turbinaria ornata, 216
Turbo, 3
Turbulence, 234
Turbulent, 53
Turf algae, 156
Turnover, 147
Turtles, 131

Ultraviolet light, 357
Ulva, 265
Uniformitarianism, 43
Uptake, 167
Uptake kinetics, 147
Upwelling, 9, 273
Urbanization, 9, 427
Urchin, 7

Value for live coral, 5
Vectors, 289
Vegetation, 6
Venezuela, 278
Vermetid gastropods, 309
Vermetids, 52
Vibrio, 129
Vicariance, 332
Virgin Islands, 8
Volcanic eruptions, 362
Volcanic rock, 1
Volutid, 338

Wave action, 5, 363
Wave Energy, 52
Waves, 52
Weathering, 18
White band disease, 43
Worm, 15

Xeniidae, 305

Year-class strength, 415
Yemen, 277
Yield, 275

Zoanthids, 85, 309
Zonation, 53
Zoochlorellae, 309
Zooplankton, 98
Zooxanthellae, 8, 98